Extremophiles

Microbiology and Biotechnology

Edited by

Roberto Paul Anitori

Division of Environmental and Biomolecular Systems
Oregon Health and Science University
Beaverton, OR
USA

Caister Academic Press

Caister Academic Press
Norfolk, UK

www.caister.com

British Library Cataloguing-in-Publication Data
A catalogue record for this book is available from the British Library

ISBN: 978-1-904455-98-1

Cover design adapted from Figure 4.1

Printed and bound in Great Britain

Contents

Contributors

Corien Bakermans
Division of Mathematics and Natural Sciences
Altoona College
Pennsylvania State University
Altoona, PA
USA

cub21@psu.edu

John R. Battista
Department of Biological Sciences
Louisiana State University and A&M College
Baton Rouge, LA
USA

jbattis@lsu.edu

Elizaveta Bonch-Osmolovskaya
Winogradsky Institute of Microbiology
Russian Academy of Sciences
Moscow
Russia

elizaveta.bo@gmail.com

Ron Bradner
Department of Chemistry and Biomolecular Sciences
Macquarie University
Sydney, NSW
Australia

ron.bradner@mq.edu.au

Thomas D. Brock
E.B. Fred Professor of Natural Sciences Emeritus
University of Wisconsin–Madison
Madison, WI
USA

tdbrock@wisc.edu

Don Cowan
Department of Biotechnology
Institute for Microbial Biotechnology and Metagenomics
University of the Western Cape
Cape Town
South Africa

dcowan@uwc.ac.za

Mark Dopson
School of Life Sciences
Linnaeus University
Kalmar
Sweden

mark.dopson@lnu.se

James G. Elkins
BioEnergy Science Center (BESC) and Biosciences Division
Oak Ridge National Laboratory
Oak Ridge, TN
USA

elkinsjg@ornl.gov

Kelley R. Gwin
Department of Biological Sciences
Louisiana State University and A&M College
Baton Rouge, LA
USA

kgwin@tigers.lsu.edu

Ruth Henneberger
Institute for Biogeochemistry and Pollutant Dynamics
Swiss Federal Institute of Technology (ETH)
Zurich
Switzerland

ruth.henneburger@env.ethz.ch

Robert Huber
Kommunale Berufsfachschule fuer biologisch-technische Assistenten
Straubing
Germany

robert.huber@bta-straubing.de

Kazem Kashefi
Microbiology and Molecular Genetics
Biomedical and Physical Sciences
Michigan State University
East Lansing, MI
USA

kashefi@msu.edu

Chiaki Kato
Institute of Biogeosciences
Japan Agency for Marine-Earth Science and Technology
Yokosuka
Japan

kato_chi@jamstec.go.jp

Hans Kristian Kotlar
Senior Specialist Biotechnology
Statoil Norge AS
Trondheim
Norway

hkko@statoil.com

Christopher McRae
Department of Chemistry and Biomolecular Sciences
Macquarie University
Sydney, NSW
Australia

christopher.mcrae@mq.edu.au

Suja Mohammed
Department of Chemistry and Biomolecular Sciences;
Biomolecular Frontiers Research Centre
Macquarie University
Sydney, NSW
Australia

suja.mohammed@mq.edu.au

Christine Moissl-Eichinger
Institute for Microbiology and Archaea center
Regensburg University
Regensburg
Germany

christine.moissl-eichinger@biologie.uni.regensburg.de

Helena Nevalainen
Department of Chemistry and Biomolecular Sciences;
Biomolecular Frontiers Research Centre;
Applimex Systems Pty Ltd
Macquarie University
Sydney, NSW
Australia

helena.nevalainen@mq.edu.au

Tommy J. Phelps
BioEnergy Science Center (BESC) and Biosciences Division
Oak Ridge National Laboratory
Oak Ridge, TN
USA

phelpstj@ornl.gov

Mircea Podar
BioEnergy Science Center (BESC) and Biosciences Division
Oak Ridge National Laboratory
Oak Ridge, TN
USA

podarm@ornl.gov

Babu Raman
Bioprocess R&D
Dow AgroSciences
Indianapolis, IN
USA

braman@dow.com

Mark Paul Taylor
Department of Biotechnology
Institute for Microbial Biotechnology and Metagenomics
University of the Western Cape
Cape Town
South Africa

mtaylor@uwc.ac.za

Junior Te'o
Department of Chemistry and Biomolecular Sciences;
Biomolecular Frontiers Research Centre;
Applimex Systems Pty Ltd
Macquarie University
Sydney, NSW
Australia

junior.teo@mq.edu.au

Marla Tuffin
Department of Biotechnology
Institute for Microbial Biotechnology and Metagenomics
University of the Western Cape
Cape Town
South Africa

marlatuffin@gmail.com

Tatiana A. Vishnivetskaya
The University of Tennessee
Center for Environmental Biotechnology
Knoxville, TN
USA

tvishniv@utk.edu

Sania Wadud
Centre for Marine BioInnovation (CMB)
School of Biotechnology and Biomolecular Sciences (BABS)
University of New South Wales
Sydney, NSW
Australia

sanja.wadud@unsw.edu.au

Lonnie van Zyl
Department of Biotechnology
Institute for Microbial Biotechnology and Metagenomics
University of the Western Cape
Cape Town
South Africa

vanzyllj@gmail.com

This book is dedicated to my late father, Vittorio Anitori,
who was not a scientist by profession,
but certainly possessed the mind of one.

Foreword

How do we define an extreme environment and its inhabitants, the extremophiles? I first made an attempt to do this over 40 years ago, in the pre-genome days (Brock, 1969). However, I think this definition still is valid today:

> It is not appropriate to define [an extreme environment] anthropocentrically, as we should be the first to admit that human life is not everywhere possible. More appropriate is its definition as a condition under which some kinds of organisms can grow, whereas others cannot. If we accept this definition it means that an environmental extreme must be defined *taxonomically*. Instead of looking at single species, or groups of related species, we must examine the whole assemblage of species, microbial and multicellular, living in various environments. When we do this we find that there are environments with high species diversity and others with low species diversity. In some environments with low species diversity we find that whole taxonomic groups are missing. For instance, in saline and thermal lakes there are no vertebrates and no vascular plants, although they may be rich in microorganisms, and very high in the numbers of the species that do live there. In many extreme environments we find conditions approaching pure cultures, with only a single species present.

When I first began to study the biology of Yellowstone hot springs, I was struck by the very visible evidence that there was an upper temperature for photosynthetic life that was *lower* than the upper temperature for microbial life in general. Detailed observations of a large number of hot springs, and an extensive review of the literature, showed that this was a general phenomenon. My work eventually led to a summary of the relationships between taxonomy and the upper temperature for different groups that is shown in Table 1. As far as I know, the relationships developed in this table are still valid.

This table raises some interesting questions. Why, for instance, is there an upper temperature for eukaryotic life at about 60°C, whereas prokaryotes (even phototrophic ones) can function well at considerably higher temperatures? Why are microorganisms able to live at considerably higher temperatures than multicellular ones?

Interestingly, for another environmental factor, low pH, a completely different set of relationships exist (Table 2). Some animals and plants can live well at fairly low pH values, whereas the prokaryotic phototrophs (cyanobacteria) have a distinct lower pH limit of around 4. Indeed, for low pH, many eukaryotic phototrophs thrive at pH values well below those of the cyanobacteria. (Brock, 1973). Even certain multicellular animals and plants will grow at lower pH values than cyanobacteria.

Why are eukaryotes able to thrive at very low pH values, but not at high temperature? Why can heterotrophic and lithotrophic bacteria and archaea grow well at temperatures of 100°C and higher, whereas phototrophic life does not exceed 70–73°C? These are evolutionary questions that derive from a careful study of the ecology of extreme environments and the extremophiles that inhabit them.

As this book shows, there are other environmental factors that can be considered extreme,

Table 1 Upper temperature limits for growth of various taxonomic groups

Major group	Group	Approximate upper temperature limit (°C)
Animals	Fish and other aquatic vertebrates	38
	Insects	45–50
	Ostracods (crustaceans)	49–50
Plants	Vascular plants	45
	Mosses	50
Eukaryotic microorganisms	Protozoa	56
	Algae	55–60
	Fungi	60–62
Prokaryotic microorganisms		
Bacteria	*Cyanobacteria* (Oxygenic)	70–73
	Phototrophic bacteria (anoxygenic)	70–73
	Chemolithotrophic bacteria	>90
	Heterotrophic bacteria	>90
Archaea	Chemolithotrophs	121
	Heterotrophs	110

Modified from Brock (1978, p. 40).

Table 2 Lower pH limits for various taxonomic groups

Major group	Group	Lower pH limit[1]	Examples
Animals	Fish	4	Carp
	Insects	2	Ephydrid flies
Plants	Cyanobacteria	4	*Mastigocladus*, *Synechococcus*
	Vascular plants	2.5–3	*Eleocharis*, *Sellowiana*, *Carex*, Ericacean plants
	Mosses	3	*Sphagnum*
Eukaryotic microorganisms	Protozoa	2	Amoebae, Heliozoans
	Eukaryotic algae	1–2	*Euglena mutabilis*, *Chlamydomonas acidophila*, *Chlorella*
		0	*Cyanidium caldarium*
	Fungi	0	*Acontium velatum*
Prokaryotic microorganisms			
Bacteria		0.8	*Thiobacillus thiooxidans*
			Sulfolobus acidocaldarius
		2–3	*Bacillus*, *Streptomyces*

[1]Lower pH limits are only approximate.

Table based on Brock (1978, p. 392).

and the evolution of organisms capable of thriving (or at least surviving) raises further questions. As Bakermans indicates in Chapter 3, adaptation to low temperatures (psychrophily) is not an uncommon trait, and cold-adapted organisms are found throughout all three domains of life. In fact, several distinct mechanisms for adaptation to low temperatures have evolved.

On the other hand, high ionizing radiation is not a common natural environmental factor. Only since the rise of nuclear physics has this factor existed. As Gwin and Battista discuss in Chapter 2, there is no obvious selective advantage to being resistant to ionizing radiation. Yet a significant number of microorganisms (only microbes; no higher organisms) exhibit this interesting characteristic. For this environmental factor, resistance is probably a chance consequence of another evolutionary pathway.

Finally, high hydrostatic pressure is a very common environmental factor, but difficult to study because of the remote regions where piezophiles (barophiles) live, and the complicated equipment needed to maintain this extreme factor (Kato, Chapter 10). Because the deep oceans are also cold, the piezophiles are also psychrophiles. Piezophiles have been identified in many bottom regions of the world's oceans, and significant advances have been made in understanding the mechanisms of piezophily.

I think that the interesting evolutionary questions raised by extremophily are valid, and I would hope that in this genome age that research on these topics will be carried out. Extreme environments and extremophiles are of enormous biological interest, initially for ecological and evolutionary relationships, and now for biotechnological reasons as well.

Many extremophiles have important practical uses, and the biotechnological aspects discussed in this book have not been neglected by the scientific community and private industry. Although the best example is *Taq* polymerase from *Thermus aquaticus*, this enzyme is only one of a large variety of important economic uses for which extremophiles have been harnessed. It has become a watchword that unique microbes are found in unique environments. It is probable that biotechnology has only scratched the surface in its search for new micro-organisms of practical use.

Thomas D. Brock

Selected bibliography

Bott, T.L., and Brock, T.D. (1969). Bacterial growth rates above 90°C in Yellowstone hot springs. Science *164*, 1411–1412.

Brock, T.D. (1967). Life at high temperatures. Science *158*, 1012–1019.

Brock, T.D. (1973). Lower pH limit for the existence of blue-green algae: evolutionary and ecological implications. Science *179*, 480–483.

Brock, T.D. (1978). Thermophilic microorganisms and life at high temperatures. Springer-Verlag, New York 465 pp. [Out of print but full text available on line at: http://digicoll.library.wisc.edu/cgi-bin/Science/Science-idx?id=Science.BrockTher]

Brock, T.D. (1985). Life at high temperatures. Science *230*, 132–138.

Brock, T.D. (1988). La vie à haute température. La Recherche *198*, 476–485.

Brock, T.D. (1969). Microbial growth under extreme conditions. Symp. Soc. Gen. Microbiol. *19*, 15–41.

Brock, T.D., and Darland, G.K. (1970). Limits of microbial existence: temperature and pH. Science *169*, 1316–1318.

Brock, T.D., and Freeze, H. (1969). *Thermus aquaticus* gen. *n.* and sp. *n.*, a non-sporulating extreme thermophile. J. Bacteriol. *98*, 289–297.

Brock, T.D., Brock, M.L., Bott, T.L., and Edwards, M.R. (1971). Microbial life at 90°C: the sulfur bacteria of Boulder Spring. J. Bacteriol. *107*, 303–314.

Brock, T.D., Brock, K.M., Belly, R.T., and Weiss, R.L. (1972). *Sulfolobus*: a new genus of sulfur-oxidizing bacteria living at low pH and high temperature. Archiv. fur Mikrobiol. *84*, 54–68.

Doemel, W.N., and Brock, T.D. (1970). The upper temperature limit of Cyanidium caldarium. Archiv. fur Mikrobiol. *72*, 326–332.

Tansey, M.R., and Brock, T.D. (1972). The upper temperature limit for eukaryotic organisms. Proc. Nat. Acad. Sci. U.S.A. *69*, 2426–2428.

Preface

Living cells are truly astounding. That is what I like to tell my undergraduate microbiology students. Not just the processes and reactions that occur, but the microscopic scale at which they occur. I therefore think that the term mind boggling would not be inappropriate to describe extremophiles, those microorganisms that conduct these processes and reactions under chemical and physical extremes that are usually lethal to cellular molecules. How can extremophiles possibly cope with, and even thrive, under these conditions? A major part of the 'microbiology' element of this book is devoted to reviewing the latest insights into the mechanisms used for survival by these fascinating organisms, from the ability of acidophiles to maintain a neutral intracellular pH (see Chapter 11 by Dopson), to the way that psychrophiles 'loosen up' their proteins at low temperatures (Bakermans, Chapter 3), to other equally ingenious adaptations in other classes of extremophiles (see Chapter 2 by Gwin and Battista, and Chapter 10 by Kato). Living an extreme lifestyle also imposes metabolic constraints on microbes, and has led to an astounding array of metabolic strategies, as exemplified by those lovers of high temperatures, the (hyper)thermophiles (see Chapter 6 by Bonch-Osmolovskaya, and Chapter 9 by Kashefi).

Tough microbes produce tough molecules. Since their discovery, the practical, biotechnological promise of extremophiles and their molecules has therefore been front and centre for both science and industry. The 'biotechnology' component of this book covers both established and recent, novel applications. Can extremozymes improve on their thus far relatively minor penetration of the enzyme market (Taylor *et al.*, Chapter 1)? Will extremophiles play a significant role in the production of sustainable energy in our current 'green' era (Vishnivetskaya *et al.*, Chapter 7)? How can the oil industry contribute (Chapter 8, Kotlar)? It will be fascinating to follow these and other biotechnology-related issues in the coming years.

The chapters in this book are self-contained, and hence need not be read in the order in which they appear. Most chapters are general review articles, whilst a few (e.g. those by Moissl-Eichinger *et al.*, Chapter 4, and Nevalainen *et al.*, Chapter 5) provide a more focused discourse on specific examples of extremophilic microbes. It is my hope that, taken as a whole, or as individual chapters, they will serve as helpful, up-to-date reference guides to their subject matter. The 'Future trends' and 'Web resources' sections located at the end of each chapter will help the reader keep up-to-date with new developments.

This editing adventure, my first, has been at times challenging and daunting, yet ultimately rewarding. The challenge has not been mine alone, so I would like to thank Julie and Lilyanne for dealing with my numerous excursions into the world of 'the book'.

I am confident that the audience to which this book is targeted, graduate students and researchers, are not immune to the sense of wonder elicited by extremophiles. My hope is that the information in the chapters herein will not only inform and educate, but also astound.

Roberto Paul Anitori
Vancouver, Washington, USA

Current Books of Interest

Bacterial Spores: Current Research and Applications	2012
Small DNA Tumour Viruses	2012
Extremophiles: Microbiology and Biotechnology	2012
Bacillus: Cellular and Molecular Biology	2012
Microbial Biofilms: Current Research and Applications	2012
Bacterial Glycomics: Current Research, Technology and Applications	2012
Non-coding RNAs and Epigenetic Regulation of Gene Expression	2012
Brucella: Molecular Microbiology and Genomics	2012
Molecular Virology and Control of Flaviviruses	2012
Bacterial Pathogenesis: Molecular and Cellular Mechanisms	2012
Bunyaviridae: Molecular and Cellular Biology	2011
Emerging Trends in Antibacterial Discovery: Answering the Call to Arms	2011
Epigenetics: A Reference Manual	2011
Metagenomics: Current Innovations and Future Trends	2011
Nitrogen Cycling in Bacteria: Molecular Analysis	2011
The Biology of Paramyxoviruses	2011
Helicobacter pylori	2011
Microbial Bioremediation of Non-metals: Current Research	2011
Lactic Acid Bacteria and Bifidobacteria: Current Progress in Advanced Research	2011
Viruses and Interferon: Current Research	2011
Essentials of Veterinary Parasitology	2011
Hepatitis C: Antiviral Drug Discovery and Development	2011
Streptomyces: Molecular Biology and Biotechnology	2011
Alphaherpesviruses: Molecular Virology	2011
Recent Advances in Plant Virology	2011
Vaccine Design: Innovative Approaches and Novel Strategies	2011
Salmonella: From Genome to Function	2011
PCR Troubleshooting and Optimization: The Essential Guide	2011
Insect Virology	2010
Environmental Microbiology: Current Technology and Water Applications	2011
Sensory Mechanisms in Bacteria: Molecular Aspects of Signal Recognition	2010
Bifidobacteria: Genomics and Molecular Aspects	2010
Molecular Phylogeny of Microorganisms	2010
Nanotechnology in Water Treatment Applications	2010
Iron Uptake and Homeostasis in Microorganisms	2010
Caliciviruses: Molecular and Cellular Virology	2010
Epstein-Barr Virus: Latency and Transformation	2010
Anaerobic Parasitic Protozoa: Genomics and Molecular Biology	2010
Lentiviruses and Macrophages: Molecular and Cellular Interactions	2010
Microbial Population Genetics	2010
Borrelia: Molecular and Cellular Biology	2010

Extremophiles and Biotechnology: How Far have we Come?

1

Mark Paul Taylor, Lonnie van Zyl, Marla Tuffin and Don Cowan

Abstract

In principle, extremophiles have much to offer the biotechnology industry, from robust, process hardy enzymes to metabolically and physiologically diverse whole cell biocatalysts. However, the penetration of extremophilic organisms and their products into biotechnology markets has been modest at best, with preference given to engineered, cost effective enzyme variants and organisms for which established genetic tools are widely available.

Interest in 'extreme' products has often been dissuaded owing to the unattractive need for the sometimes costly and complicated cultivation equipment and the complexities of culture maintenance. The lack of suitable genetic tools by which to improve, adapt or engineer a process involving an extremophilic host further complicates the issue. Legislative controls over national biological resources and allegations of biopiracy have also retarded commercialization and industry–academia collaborations. However, commercial success stories have been described and form part of this review. Future prospects are optimistic, as several new biotechnology companies involved in the production of biomolecules from renewable resources have based their platform technology on extremophiles.

Introduction

Humankind has for its entire history sought to develop tools to improve its understanding of the natural world, and to use a combination of these advancements to fuel technological development and progress. Modern molecular science, in all its many interdisciplinary facets, has given us unprecedented insights into the development of processes that can improve agriculture, medicine and health, energy supply, and structural and material development. Only relatively recently have concerns been voiced with respect to the environmental and social impact of some chemical processes, focusing particularly on issues such as climate change, global warming, fuel dependency and security, fresh water sustainability and global food security. Scientists, both industrial and academic, have sought to circumvent some of the problems associated with conventional industrial processes (especially those that require environmentally persistent chemical reagents and produce toxic or damaging by-products) through the use of biological agents in the manufacture of conventional and novel products, commonly referred to as biotechnology.

Conventional process technologies for the production of polymers, pharmaceuticals, speciality chemicals and agrochemicals often suffer from high cost, low selectivity and by-product formation (Demirjian *et al.*, 2001). In addition, the application of high reaction temperatures and pressures, the need for expensive chemical catalysts and high input energy requirements make some of these processes unattractive for long-term industrial implementation. Science has turned to biotechnology as a potentially beneficial alternative to mainstream industry and industrial processes. The evolution, transformation and integration of biotechnology into mainstream technological processes and novel bioprocesses has resulted in the distinction of several subsectors such as white biotechnology (Frazzetto, 2003), green biotechnology (food/non-food agriculture; Chappell and Grotewold, 2008) and red biotechnology (medical biotechnology; Bauer, 2002).

Although by no means a new discipline, as conventional microbial fermentations have been used for brewing and baking for many centuries, the biotechnological revolution in the latter half of the twentieth century has introduced fresh thinking and a novel approach to the problems and development of industrial, medical and agricultural processing. Since the discovery of 'extremophilic' organisms in the 1960s, it has been a consistent theme to use these organisms and/or their biomolecules in biotechnology processes. Successful applications typically involve one or more of the following: a reduction in the cost of production, an increase in productivity at the same cost or the ability to achieve a process which has previously not been possible or viable. The latter, considered the most probable pathway for implementation of extremophilic catalysts, is usually achieved by harnessing their diverse array of physiological traits, and in particular, their stability under 'extreme' reaction conditions. For whole cell applications, advantages accrue when the organism has a unique metabolism and/or has evolved to live under specialized conditions (thermophilic, psychrophilic, halophilic, acidophilic, alkaliphilic, piezophilic). Extremophiles can also deliver novel metabolites which could be used as therapeutic agents or as additives and preservatives in a range of industries, from pharmaceutical, research, food and beverage, to cosmetics (Schiraldi and De Rosa, 2002; Hendry, 2006).

Although many potential uses of extremophiles and their products have been suggested and investigated over the past 30 years, relatively few have found their way into industrial processes. The disparity between the large sums of money spent on research and the numbers of products or processes realized is vast, and although this is a general trend in biotechnology, it is particularly substantial when considering the commercialization of extremophile-derived technologies (Bull *et al.*, 2000; Peyton *et al.*, 2001). However, with a few exceptions such as *Taq* DNA polymerase, it is usually difficult to judge how quickly or effectively the study of these organisms and their biomolecules translates to real world success in terms of adoption by industry. As stated by Podar and Reysenbach in a recent publication:

> Owing to the highly competitive nature of industrial R&D, in most cases the path from extremophile to successful commercial application is not documented in peer reviewed scientific publications and can be partially followed through patents, biotechnology meetings and company websites.
>
> (Podar and Reysenbach, 2006)

In order to build on recent reviews of this subject (Margesin and Schinner, 2001; Schiraldi and De Rosa, 2002; Gomes and Steiner, 2004; Antranikian *et al.*, 2005; Egorova and Antranikian, 2005; de Miguel Bouzas *et al.*, 2006; Antranikian and Egorova, 2007; Huston, 2008) we aim to briefly review the general approaches used for cultivation of and gene discovery in extremophiles, as well as discuss a number of historical success stories of extremophile commercialization. We will principally focus on recent biotechnological developments, with particular emphasis on novel whole cell biocatalytic transformations and enzymes with biotechnological application. This review will seek to critically evaluate the penetration into biotechnology markets of commercial products of extremophile origin and the most probable areas to which these organisms will contribute in the future.

Mining the extreme environment

It has become apparent that where biotransformation processes form part of complex synthetic pathways (such as in some pharmaceutical compound syntheses), the use of low stability mesophilic enzymes in their native state is generally uneconomic. The consensus has been to engineer such enzymes for increased stability or improved kinetic properties (Demirjian *et al.*, 2001). With the discovery of extremophilic life and a degree of physiological insight, it was soon realized that enzymes from extremophiles could be a viable alternative to enzyme engineering, since it could be presumed that these 'extremozymes' would naturally be robust and tolerant to various external stresses dependent on their origin, for example, enzymes derived from thermophiles that function well at high temperatures. We propose now to build on this concept by discussing the various extremophilic groups and the unique features that characterize each classification.

Extremophiles are defined as organisms that have adapted to thrive in habitats that are considered to be hostile from a human perspective (Horikoshi, 2007). Examples include the extreme pressures of the deep sea (Abe and Horikoshi, 2001), the cold habitats of the Antarctic (van den Broeke, 2000), the heated waters of hydrothermal vents (Driessen and Albers, 2007), the extremes of pH in alkaline lakes (Hendry, 2006) and acid mines (Doshi, 2006), and the saturated salt solution of salterns (Margesin and Schinner, 2001). The major individual extremophilic groups are defined in greater detail in Table 1.1, although it is important to note that many organisms found in extreme environments demonstrate poly extremism (Rothschild and Mancinelli, 2001; Horikoshi, 2007). Although not forming a part of this review, other less well-known extremophilic groups include oligotrophs, osmophiles and xerophiles, and radiotolerant organisms, none of which have yet found widespread biotechnological application.

Scientific interest in extremophiles has naturally developed from conventional academic interest and a desire to understand the scope of life and its adaptive mechanisms, particularly at the 'outer margins of survival'. 'Extremozymes' have attracted attention owing to their biocatalytic diversity and their functionality under extreme conditions, such as at high temperature or in the presence of solvents (Demirjian *et al.*, 2001; Li *et al.*, 2005; Siddiqui and Cavicchioli, 2006; Huston, 2008). As whole cell biocatalysts, interest has focused on specific metabolic products from extremophiles, often unique and of potential commercial value (Rothschild and Mancinelli, 2001). However, few bioprocesses have been built around extremophiles and those that have often exploit the organisms' robust physiology and process hardiness. For example, the production of ethanol at temperatures close to its boiling point (78°C), facilitating its downstream extraction, and the use of thermophilic lipases at temperatures where the substrate is in the liquid state (Bergquist and Morgan, 1992), are both cited examples of the advantages of thermophilic processes.

Conventional cultivation versus new metagenomic technologies

Conventional methods for studying microbial species and biocatalysts from hostile environments have relied on culture-dependent methods and laborious screening programmes for identification of a particular product, biotransformation or phenotype (Singh and Pelaez, 2008). It is widely acknowledged that this approach accesses only a fraction, generally accepted as <1%, of the viable organisms within a particular niche (Cowan *et al.*, 2004). Those extremophilic organisms that can be readily cultivated in the laboratory may require specialized equipment (e.g. high temperature, high-pressure corrosion resistant bioreactors) which is costly at the laboratory scale and may be prohibitive at production level (Schiraldi and De Rosa, 2002). Such limitations notwithstanding, a number of specialist companies have been established (e.g. Matís, Reykjavik, Iceland; Synthetic Genomics, La Jolla, CA, USA; Actilor, Leuna, Saalekreis, Germany; ExtremoZyme Inc., Huntsville, AL, USA; Verenium Corporation, San Diego, CA, USA) with the aim of developing culturing strategies for these organisms to identify biomolecules which could have application in industrial, pharmaceutical or research processes.

In order to overcome current limitations in the production of native and recombinant extremophilic enzymes, a range of fermentation and genetic technologies have been developed. Three general approaches have been used in the development of these technologies: (i) continuous fermentation bioprocess and culture medium design; (ii) implementation of innovative bioreactors and processing equipment; and (iii) enzyme overproduction in alternative mesophilic hosts. Collectively, these have led to substantial progress being made in biomass production of organisms such as *Pyrococcus furiosus*, *Sulfolobus shibatae* and *Sulfolobus solfataricus* Gθ and in the overproduction of interesting enzymes from

Table 1.1 A brief description of the major extremophile groups and their potential commercial benefits.

Extremophile group	Basis of classification	Enzymes of commercial interest	Current commercial applications
Hyperthermophiles	Optimal growth >70°C	Starch processing enzymes	Lignocellulose deconstruction (value metabolites/biofuels), sugar production (sweeteners)
Thermophiles	Optimal growth between 50 and 70°C	Proteases	Laundry and detergent industry, amino acid production, food industry
Thermotolerant	Optimal growth at mesophilic temperatures but capable of growth up to 50°C	Complex polymeric deconstruction, e.g. cellulases, hemicellulases, chitinases, etc.	Lignocellulose deconstruction (value metabolites/biofuels), detergent additives
		Glycolytic enzymes	Clinical diagnostics
		DNA-associated enzymes	Molecular biology enzymes and scientific applications (e.g. DNA sequencing)
Psychrophiles (cryophiles)	Optimal growth below 15°C	Proteases	Cheese and dairy industry (maturation), laundry and detergent industry
		Lipases	Biosensors, laundry and detergent industry
		Fatty acid dehydrogenases	Laundry and detergent industry
		Complex polymeric deconstruction e.g. cellulases, hemicellulases, chitinases etc.	Lignocellulose deconstruction (value metabolites/biofuels)
Alkaliphiles	Optimal growth in conditions >pH 9	Proteases	Detergent additives
		Lipases	Detergent additives
		Starch processing enzymes	Cyclodextrin production
		Complex polymeric deconstruction e.g. cellulases, hemicellulases, chitinases, etc.	Pharmaceuticals and food industry Detergent additives
Acidophiles	Optimal growth in conditions <pH 3	Sulfur oxidases	Desulfurication of coal in mining industry, biomining
Halophiles	Optimal growth in 3–20% salt	β-Carotene biosynthetic genes	Carotene production for food industry
		Compatible solutes	Compatible solute production (pharmaceuticals)
Piezophiles	Optimal growth at >38 MPa	Lipases	Possible application in the reduction in viscosity of crude oil under high pressures

these species (Schiraldi and De Rosa, 2002; Schiraldi *et al.*, 2002). A recent example of high-level expression of a gene of interest in a heterologous host is that of the thermostable maltodextrin phosphorylase from *P. furiosus* PF1535 for α-d-glucose-1-phosphate production from starch (Nahalka, 2008).

With the development of metagenomic techniques the reliance on culture-dependent methods as a source of novel genes and enzymes has been substantially reduced (Kowalchuk *et al.*, 2007; Kotik, 2009). Metagenomics effectively circumvents cultivation and instead allows direct extraction of DNA from a particular environmental sample, and consequently, access to the majority of genes present within a bacterial community (Blow, 2008). The coverage obtained using the metagenomic approach is a distinct advantage when discussing an extremophilic microbial community, where culture efficiencies may be particularly low. In addition, these methods are well suited to the identification of novel sequence-function and structure–function relationships if coupled to suitable screening methods, such as the use of tributyrin hydrolysis as an indicator of lipase/esterase activity (Heath *et al.*, 2009) or even when using sequence-based PCR methods that have been reported to probe unexpectedly large regions of sequence space (Cowan *et al.*, 2004).

Several reliable methods for the isolation and cloning of genes from environmental DNA and RNA have been developed. One technique involves chloroform extraction of nucleic acid directly from environmental samples and subsequent size-based recovery of DNA/RNA fragments from low-melt agarose gels. Appropriately sized insert clone libraries are then constructed, from which phylogenetic and diversity estimates can be made, as well as high-throughput screens for specific genes/enzymes (Rondon *et al.*, 2000; Feller and Gerday, 2003). These and similar approaches have induced a fundamental paradigm shift in the approach to selecting biocatalysts for industrial biotransformations (Burton *et al.*, 2002), where adapting the reaction conditions to suit the enzyme characteristics is replaced by selection or engineering of a biocatalyst ideally suited to the optimum process conditions. This approach has been facilitated in particular by the development and successful implementation of high-throughput robotic screening technologies, enabling very large ($>10^9$ recombinants) metagenomic and mutant libraries to be processed in an acceptable timeframe. Furthermore, advances in high-throughput sequencing technologies and bioinformatics capabilities have brought with them the ability to rapidly identify multiple genetic elements (Bull *et al.*, 2000; Valdes *et al.*, 2008). The culture-independent recovery of two near-complete microbial genomes (representatives of *Leptospirillum* group II and group III; Tyson *et al.*, 2004) and the partial recovery of three genomes from an environmental sample (Venter *et al.*, 2004) were precursors to a large number of current metagenomic (i.e. community diversity) sequencing projects. Community metagenomic data have been used to identify individual genes (e.g. Heath *et al.*, 2009), to characterize metabolic capabilities (e.g. Leitao *et al.*, 2009) and to identify avenues for further investigation. An excellent example was the *in silico* prediction and *in vitro* confirmation of a long-chain alkane degrading pathway in *Geobacillus thermodenitrificans* NG80-2 from genome sequence data. This capacity is potentially exploitable for oil recovery from wells by reducing the viscosity of crude oil, and in bioremediation of oil-contaminated sites (Feng *et al.*, 2007).

Metagenomic phylogenetic studies have demonstrated a surprisingly high diversity of uncultured microorganisms in some extreme environments (often representing the majority of prokaryotic diversity found in any single environmental sample) and have shed light on many fundamental aspects of bacterial and archaeal phylogeny (Amann *et al.*, 1995; Bull *et al.*, 2000; Lauro and Bartlett, 2008). However, it is widely acknowledged that cultivation of individual species is still essential in order to understand organism physiology and microbial community dynamics, and for providing physical entities that could form the basis of a particular bioprocess or research study. The importance of a cultivation approach still forms significant routes of commercial development, such as the propriety High-throughput Culturing™ technology developed by Diversa (San Diego, CA, USA, which has been shown to access a range of novel species and genera which were previously classed as 'unculturable' (Huston, 2008). In short, this technology recreates an *in vitro* environment that mimics that of the natural source material and therefore establishes a suitable environment in which previously unculturable organisms can grow. Microbial cells are separated by differential centrifugation and encapsulated into agarose microcapsules, in which individual colonies grow. Individual microcapsules are sorted by flow cytometry and aliquoted onto aluminium plates for subsequent study (Zengler *et al.*, 2002, 2005).

Whilst metagenomic and *in silico* approaches allow for the rapid identification of biomolecules and metabolic pathways, it is still necessary to use 'traditional' laboratory methods to confirm functionality and to understand the fundamental processes involved (Simonato *et al.*, 2006). Thus, for the rapid identification, study and commercialization of extremophile derived products or processes, an integrated approach using genome sequence information, bioinformatics, culture-dependent and culture-independent techniques is essential. Some of these steps are usually slow owing to the nature of the work, and will therefore always present bottlenecks in the path to commercialization.

We note that the gene mining of extreme environments has been inhibited, to some extent, by an increasing sensitivity of many countries to biopiracy of their national genetic resources. Several examples of this nature have been highly publicized and have had significant impact on collaborative research, with companies often insisting on 'clean' legal credentials with respect to any organism or environmental sample. An example of such a case occurred in 2004, whereby the Kenyan Wildlife Society and the international centre for insect physiology and ecology (icipe) announced their intention to bring a legal case against Genencor. The accusation was made with respect to violation of intellectual property rights and concerning the commercialization of an enzyme from the alkaline Lake Bogoria, Kenya, and used in the manufacture of stonewash denim. This extremozyme had been discovered by Genencor and commercialized by their business partner, Procter and Gamble. The legal case never materialized, but the incident remains unresolved and a point of debate. Similar controversies also exist concerning the ownership of intellectual property derived from other extreme habitats such as Antarctica and Yellowstone National Park.

Extremophile classification

For the remainder of this review we propose to examine in greater detail the major individual extremophilic classes and highlight the relevant commercial successes that have been reported for particular organisms, enzymes or bioproducts therein. Brief descriptors of each extremophilic classification are included for the readers' convenience.

Hyperthermophiles, thermophiles and thermotolerant organisms

Life in the heat

Perhaps the best understood extremophilic group are the microbial thermophiles. The generic term *thermophiles* is subclassified into more distinct and descriptive groups, namely the hyperthermophiles, the thermophiles and those organisms considered to be thermotolerant but not thermophilic (Table 1.1). The hyperthermophilic group is almost exclusively populated by archaeal species, examples of which include the genera *Sulfolobus*, *Pyrodictium*, *Pyrolobus*, *Pyrobaculum*, *Thermoproteus*, *Thermococcus* and *Pyrococcus* (Stetter, 2006a,b; see also Chapter 9). The thermophilic group comprises a mixture of prokaryotic and archaeal species, and includes amongst others, members of the genera *Geobacillus*, *Thermus*, *Thermoanaerobacter* and *Thermotoga* (Rothschild and Mancinelli, 2001; Wagner and Wiegel, 2008; see also Chapter 6). Mechanisms of molecular and physiological adaptation to high temperatures have recently been reviewed in detail and the reader is directed to the excellent articles by Driessen and Albers (2007), Grosjean and Oshima (2007), and Robb *et al.* (2007). In general, survival at moderate to high temperatures (in biological terms, i.e. 60–100°C) requires the deployment of adaptive mechanisms that stabilize and maintain cellular functions and constituents. This is particularly pertinent to enzyme structural stability, with thermophilic enzymes often possessing a high degree of thermal stability, an attractive physical property in an industrial candidate (Taylor *et al.*, 2004; Li *et al.*, 2005; Turner *et al.*, 2007). In this review we will use the term *thermophile* to encompass all high temperature organisms unless otherwise specified.

The potential technological applications of many thermophilic organisms and their products have been reported in detail elsewhere (Taylor *et al.*, 2004; Antranikian *et al.*, 2005; Egorova and Antranikian, 2005; de Miguel Bouzas *et al.*, 2006; Antranikian and Egorova, 2007; Horikoshi, 2007; Turner *et al.*, 2007; Blumer-Schuette *et al.*, 2008), and are summarized in Table 1.1. It has frequently been stated that the application of extremozymes is favoured because of their often robust and adaptable nature and, particularly in the case of enzymes from thermophiles, their increased thermostability. These characteristics have direct relevance for industrial processing where severe reaction conditions are commonly employed (Demirjian *et al.*, 2001; Gomes and Steiner, 2004). In the following sections we seek to evaluate the commercial penetration of thermophilic enzymes and describe some recent whole cell applications in biofuel production.

Commercial enzymes from thermophiles

The global market for enzymes was estimated to be worth \$2.3 billion in 2007 and is expected to grow each year by a compound average growth rate of 6%, to \$2.7 billion in 2012 (Thakore, 2008). Approximately 50% of this market is composed of commodity and technical enzymes, worth approximately \$1.1 billion in 2007, and is estimated to increase to \$1.4 billion by 2012. Technical enzymes are principally used in the detergent, leather, personal care, pulp and paper, starch, textile, molecular biology, biotransformation, biosynthesis and renewable fuel industries (Thakore, 2008). The remainder of the market is made up by food enzymes and animal feed enzymes worth \$830 and \$280 million, respectively, in 2007.

The market penetration of thermophilic enzymes has historically been low, although notable exceptions include *Taq* DNA polymerase and a handful of *Bacillus* enzymes used in clinical diagnostics and starch liquefaction (Sharp *et al.*, 1992). Currently, a wide range of molecular biology enzymes is available from companies such as New England Biolabs, Ipswich, MA, USA, which supplies 13 restriction endonucleases isolated from thermophilic species of the genera *Geobacillus*, *Thermus*, *Pyrococcus* and *Aeropyrum*. They also supply a number of thermophilic polymerases such as Vent™ (*Thermococcus litoralis*), DeepVent™ (*Pyrococcus* sp. GB-D), Phusion™, Phire™, Therminator™ (*Thermococcus* sp. 9° N-7), DyNAzyme™ (*Thermus brockianus*) and a *Sulfolobus* polymerase as well as a number of ligases, DNA binding proteins and other enzymes of thermophilic origin. Another market leader in the field of molecular biology enzymes, Fermentas, currently supplies over twenty thermophile-sourced restriction enzymes and a range of thermophilic polymerases. Biotechnology companies with a specific interest in extremozymes include: ZyGEM, based in Hamilton, New Zealand; Finnzymes based in Espoo, Finland (emphasis on the development of thermostable polymerases); Swissaustral Biotech SA based in Monthey, Switzerland; KAPA Bioscience, South Africa (new and engineered DNA polymerases); and Biohelix, MA, USA (commercial producers of a thermostable helicase with application in isothermal DNA amplification). Further details of the homepages of these companies are included in the 'Web resources' section of this review.

The rapid growth of technology for biofuel production and the development of products from renewable biomass resources have led to the description and commercial availability of a number of thermostable enzymes that facilitate and mediate various stages of lignocellulose deconstruction (broadly defined as cellulases, hemicellulases and lignin modifying enzymes). The high-volume commercial enzyme market is dominated by the companies Novozyme and Genencor. A number of both native and engineered commercial enzyme products offered by these companies (Table 1.2) have been derived from thermotolerant and thermophilic organisms. Interest in the development of these technologies

Table 1.2 Examples of commercially produced enzymes from thermophiles (note that this table is an elaboration on those enzymes mentioned in the text)

Enzyme	Manufacturer	Organism	Reference
Alcalase 2.4 L (protease)	Novozyme	*Bacillus licheniformis*	Sigma Aldrich enzyme explorer[1]
Lipolase 110 L (lipase)	Novozyme	*Thermomyces lanuginosus*	Sigma Aldrich enzyme explorer
Pentopan Mono BG (xylanase)	Novozyme	*Thermomyces lanuginosus*	Sigma Aldrich enzyme explorer
Termamyl 120 (α-amylase)	Novozyme	*Bacillus licheniformis*	Sharma *et al.* (2009)
Novamyl (maltogenic amylase)	Novozyme	*Geobacillus stearothermophilus*	Sigma Aldrich enzyme explorer
Liquozyme (amylase)	Genencor	*Geobacillus stearothermophilus*	Danisco 'protein processing' publication[2]
Protex 14L (protease)	Genencor	*Bacillus thermoproteolyticus*	Danisco 'protein processing' publication
Multifect AA21L (amylase)	Genencor	*Bacillus licheniformis*	Danisco 'protein processing' publication
Protex 6L (protease)	Genencor	*Bacillus licheniformis*	Danisco 'protein processing' publication
Lipase	Sigma Aldrich	*Thermus flavus*	Sigma Aldrich enzyme explorer
Lipase	Sigma Aldrich	*Thermus thermophilus*	Sigma Aldrich enzyme explorer

[1]http://www.sigmaaldrich.com/life-science/metabolomics/enzyme-explorer/analytical-enzymes.html.

[2]http://www.danisco.com/wps/wcm/connect/aa5f63804fb4e0ffa82cfa45de7f5ba1/Protein+Process+SellSht+v5a+(final).pdf.

has led to major collaborations, including a partnership between Dupont and Genencor (now a division of Danisco) forming the Dupont Danisco cellulosic ethanol LLC (limited liability company), an R&D programme seeking to develop cellulosic ethanol technology.

Whole cell thermophilic biocatalysts

Until recently, reports of the use of thermophilic organisms as whole cell biocatalysts have been limited. Whilst *Pyrococcus furiosus* and *Pyrobaculum islandicum* have been suggested as candidates for the biomining industry (Lloyd *et al.*, 2003), the use of other archaea such as *Acidianus* spp., *Sulfolobus* spp., *Metallosphaera* spp. as well as *Sulfolobales* spp. for the leaching of marginal chalcopyrite ores, whether in heaps (HotHeap™ from GeoBiotics, Lakewood, CO, USA) or tanks, has now become well established. These technologies are routinely operated at temperatures approaching 80°C, as in the case of BHP Billiton's BioCOP® process, Melbourne, VIC, Australia (Mikkelsen *et al.*, 2006). The pilot scale plant (Chuquicamata mine, Chile) used to establish leaching parameters as well as other requirements for industrial scale design, which led to the construction of a 300 m^3 industrial test reactor facility at the Pering mine in South Africa, was estimated to produce 20 kilotons of cathode copper per annum (Schiraldi *et al.*, 2002; Batty and Younger, 2007).

Desulfurication by thermophiles has also been suggested as useful in rubber treatment, coal decontamination and the general bioremediation of contaminated materials (Boogerd *et al.*, 1990). To the authors' knowledge, this technology has not been commercialized, probably for economic reasons. Members of the genera *Pyrococcus*, *Methanococcus*, *Pyrodictium*, *Pyrolobus* and *Thermococcus* have been sought for their ability to accumulate compatible solute compounds such as di-myo-inositol-1,1′-phosphate, which act as cellular protectants and have potential use as biomolecular stabilizers and stress alleviants in industrial processes (Muller *et al.*, 2005; Empadinhas and da Costa, 2006; Rodrigues *et al.*, 2007; Faria *et al.*, 2008).

Interest in the production of value metabolites and high-energy fuels such as alcohols, hydrogen and diesel alternatives, is growing rapidly and has been the focus of several sectors of the biotechnology industry. Hydrogen production by archaea of the genus *Thermococcus* (Smith *et al.*, 2001; Kanai *et al.*, 2005) and bacteria of the genera *Thermoanaerobacter* and *Thermotogales* (Nandi and Sengupta, 1998; Bagi *et al.*, 2007; Hong *et al.*, 2007; Shaw *et al.*, 2009) has been reported. However, particular emphasis has instead been placed on the optimization of processes that utilize pretreated and hydrolysed lignocellulosic material as the fermentation feedstock, principally for alcohol production. We note in particular the recent genetic development of several thermophilic strains from the genera *Thermoanaerobacterium*, *Thermoanaerobacter* and *Geobacillus* for ethanol production. These organisms have been chosen for their thermophilicity, fermentative metabolism and capacity for catabolic promiscuity. Members of all three genera are capable of utilizing hexose sugars, pentose sugars and more complex oligosaccharides that form the carbohydrate quotient of the majority of lignocellulosic hydrolysates (Taylor *et al.*, 2009). The engineering of high temperature genetic systems has led to the development of a stable genetic mutant of *Thermoanaerobacterium saccharolyticum* capable of co-utilizing sugar mixtures (d-glucose, d-xylose, d-mannose and l-arabinose) and producing ethanol at close to theoretical maximum (Shaw *et al.*, 2008, 2009). This strain, its predecessors and the process details have been patented by the biotechnology company Mascoma, based in Boston, Massachusetts (Taylor *et al.*, 2009). A mutant strain of *Thermoanaerobacter mathranii*, together with fermentation process descriptions, has been patented by the Danish company Biogasol (Ballerup, Copenhagen, Denmark). Similar developments in the genetic manipulation of several *Geobacillus* species has formed the basis of TMO Renewables (Guildford, Surrey, UK), a UK-based biotechnology company (Cripps *et al.*, 2009; Taylor *et al.*, 2009). These specific developments and the general field of thermophilic ethanol production have recently been reviewed in greater detail (Taylor *et al.*, 2009) and the reader is directed to the 'web resources' section of this review for relevant company homepages. In a related field, the ability of a number of terrestrial and marine hyperthermophilic archaeal and bacterial species to catabolize glucans, xylans and mannans has been suggested as an important resource for biorefineries, particularly with emphasis on crystalline cellulose and lignocellulosic deconstruction (Turner *et al.*, 2007; Blumer-Schuette *et al.*, 2008).

Psychrophiles and psychrotolerant organisms

Life in the cold

Since approximately 70% of the globe is covered by ocean, much of which is typically 2–5°C, and 20% of the land surface is polar, pyschrophiles represent a significant genetic quotient and potentially valuable biotechnological resource (Cowan *et al.*, 2007). Several recent reviews of cold-active organisms (psychrophiles, psychrotrophs and psychrotolerant organisms) have focused on their isolation and handling (Russell and Cowan, 2006), and biodiversity and adaptive physiologies (Cavicchioli, 2006; Siddiqui and Cavicchioli, 2006; see also Chapter 3). For these microorganisms, their enzymes rather than the organisms themselves have been the major focus of industry, since their optimal activity at low temperatures has led to suggestions that they would be ideal biocatalysts in energy conserving and economical bioprocesses (Demirjian *et al.*, 2001; Siddiqui and Cavicchioli, 2006; Huston, 2008). Commercial application lies in the use of cold-active lipases and proteases in laundry detergents designed for low-temperature wash cycles. Possible applications in molecular biology are largely based on the advantageous fact that many of these enzymes can be easily denatured at elevated temperatures (Huston, 2008). Cold active enzymes have often been proposed for use in the dairy (lactose hydrolysis in milk by β-galactosidase) and food industries (baking, juice extraction, meat tenderization, etc.) (Huston, 2008) but, to the authors' knowledge, have never been extensively commercialized.

Commercial enzymes from psychrophiles and other applications

The commercial penetration of psychrophilic enzymes has historically been low, the well-known examples often having been sourced from the psychrotolerant yeast *Candida antarctica*. One such example is a commercial lipase available from Sigma Aldrich. In recent years, leading enzyme companies have mostly focused on the development of low temperature active enzymes for the detergent and textile industries (Huston, 2008). Examples of commercially significant and engineered cold active enzymes include: two Novozymes products, Kannase™, a protease active between 10 and 20°C and derived from *Bacillus clausii*, and Celluzyme™, a cold active cellulose from *Humicola insolens*; and two cold active proteases from Genencor, Purafect™ from *Bacillus lentus*, and Properase™ from *Bacillus alkaliphilus* (Huston, 2008).

Inroads into the molecular biology and research markets have apparently been minor, but recently New England Biolabs marketed a cold adapted alkaline phosphatase from an Antarctic bacterial strain, specifically produced for the dephosphorylation of nucleic acids, which has the convenience of complete denaturation at 65°C. Key indicators of commercial interest in the further exploitation of psychrophilic enzymes and cellular systems are perhaps the recent genetic advances that have been made in the development and facilitation of low-temperature expression systems. Such systems would be essential for large-scale production of novel pyschrophilic enzymes, and would be expected to stimulate the further application of psychrophilic gene products in the wider molecular biology market (Parrilli *et al.*, 2007).

The pCOLD™ range of vectors from Takara Bio Inc. (Otsu, Shiga, Japan) utilize a promoter from a cold shock protein gene (*csp*A) in the development of a low-temperature expression system, which aids expression of soluble gene products by avoiding the formation of exclusion bodies and limiting insoluble aggregate formation. A similar approach has been taken by Stratagene in the development of the Arctic Express™ range of competent *E. coli* strains. These strains express the chaperonins Cpn10 and Cpn60 from the psychrophilic bacterium, *Oleispira antarctica*, as a means of compensating for the reduction of native chaperonin activity in *E. coli* under low-temperature expression conditions (www.stratagene.com/manuals/230191.pdf).

Psychrophiles have also been the subject of attention in the field of biofuels production. Specific interest has been focused on the development of starch liquefaction, specifically the application of cold active enzymes to mediate the so-called 'cold hydrolysis' process for alcohol production. Such a process, if effective, would substantially reduce energy costs by precluding the use of conventional high temperature, high-pressure liquefaction equipment (Galvez, 2005).

The enzyme company Novozymes has partnered with POET LLC (an American based biorefinery company and the largest dry mill ethanol producer in the USA), in the development of their BPX (Broin project X) technology that specifically seeks the development of such a process. The BPX process is reported to incorporate a range of cold active fungal enzymes that mediate starch liquefaction (Williams, 2006; Huston, 2008). Similar interest in the development of a commercially viable cold hydrolysis process has led to the development of the STARGEN™ line of cold-active fungal enzymes from Genencor, which includes a range of amylases and glucoamylases that hydrolyse starch in a simultaneous saccharification and fermentation process (Williams, 2006; Huston, 2008).

Halophilic organisms

Life in salt

Halophilic microorganisms are broadly defined as those that require at least 0.2 M NaCl for their survival. Extreme halophiles typically require around 2 M salt and can survive in salt concentrations up to saturation (approximately 6 M). The specific adaptive physiologies of these organisms have been reviewed in detail (Madern *et al.*, 2000; Kunte, 2007; Paul *et al.*, 2008). Two adaptive strategies are typically employed: the accumulation of high intracellular concentrations of salt (normally K^+) to achieve osmotic balance with the environment (the extreme halophilic archaeal mechanism), and the accumulation of high concentrations of a variety of organic osmotic solutes (halophilic and halotolerant organisms). Halophilic organisms, which are often by nature of their habitat also alkaliphilic, include members of the genera *Haloferax, Haloarcula, Halococcus* and *Salinibacter*. Indeed, the enzymes from halophiles that have found a commercial niche have done so owing to their alkaliphilic properties rather than their halophilic characteristics. One particularly high-profile industrial endo-cellulase (Puradax™, Genencor, Palto, Alt., CA, USA) was isolated from a haloalkaliphilic *Bacillus* sp., discovered in Lake Bogoria in the Kenya rift valley (Jones, 2004). Since its production in 1998 a host of similar enzymes have been released onto the market including, most recently, cellulase and mannanase enzymes from members of the *Bacillus clarkia-agaradhaerens* subgroup isolated from saline soda lakes (Jones, 2004).

The halophilic membrane-bound retinol pigment, bacteriorhodopsin, has been described as a potential component in optoelectronic material for holographic imaging and electronic displays (Margesin and Schinner, 2001). Another pigment, halorhodopsin from *Halobacterium salinarum*, is currently produced commercially by the German company Actilor NanoBioSecurity for use in photochromic, photoelectric and photochemical applications. Other potential products from halophilic organisms include ether-linked lipids, other pigments, biopolymers (polyhydroxyalkanoates and polysaccharides) and compatible solutes (Margesin and Schinner, 2001).

Commercially available enzymes from true halophiles are rare (haloalkaliphilic examples discussed above are perhaps relevant because of their alkaliphilic properties rather than halophilic). This fact probably reflects several technical issues. Firstly, high salinity in industrial processes deteriorates steel, product purification requires costly salt recovery/removal and saline wastewater streams have a negative environmental impact. Furthermore, most intracellular enzymes from extreme halophiles have an obligate requirement for high salt concentrations in order to maintain molecular stability and function. This requirement precludes the use of well-developed expression hosts, and imposes a significant barrier to industrial large-scale protein production.

Whole cell applications

Halophilic bacteria have long been used in the production of traditionally fermented salted foods, especially fish and seafoods. In the production of salted fish products, halophilic organisms of the genera *Tetragenococcus* and *Halomonas* have been reported to mediate the pickling process through the production of lactate and other organic acids (Hernandez-Herrero *et al.*, 1999; Margesin and Schinner, 2001; Simon-Colin *et al.*, 2008).

An indirect industrial application of halophiles is the commercial production of salt from crystallization ponds (Niehaus *et al.*, 1999; Margesin and Schinner, 2001; Wang *et al.*, 2007). The presence of red-pigmented halophilic archaeal communities in saltern crystallization ponds promotes the rate of evaporation and salt production through the metabolic entrapment of solar energy (and direct solar gain). This effect is principally mediated by *Dunaliella, Salinibacter* and various archaeal species such as *Haloferax* spp. (Margesin and Schinner, 2001; Wang *et al.*, 2007).

The ability to grow in high salinities has obvious applications in the bioremediation field, particularly of marine waters contaminated with various pollutants such as oil and halogenated hydrocarbons (Margesin and Schinner, 2001). Many haloarchaeal species have been reported to degrade aromatic compounds (Garcia *et al.*, 2004), insecticides and DDT (Le Borgne *et al.*, 2008) and a variety of other organic pollutants (Le Borgne *et al.*, 2008). It has been widely reported that industrial wastewaters are often saline, particularly those from processes involved in the production of pesticides, pharmaceuticals and herbicides, or water used in oil and gas recovery, so it is fair to assume that these organisms have a great deal of bioremediation potential (Le Borgne *et al.*, 2008).

The products ectoine and β-carotene represent two halophilic biotechnology success stories (Raja *et al.*, 2007; Graf *et al.*, 2008). Ectoine, an example of the organic compounds that serve as cytoplasmic solutes in organisms capable of growth in high salt environments (Graf *et al.*, 2008), is commercially produced by the German based company Bitop AG. This company, whose commercial focus is the production of 'extremolytes', has a portfolio of patents relating to ectoine and hydroxyectoine production processes, principally from the organisms *Halomonas elongata* and *Marinococcus* M52. The commercial market is predominantly the cosmetic industry, and Ectoine is included in the Merck product RonaCare Ectoine™, claimed to be particularly effective on aged, dried or sun-damaged skin. Bitop AG patents also describe novel applications for ectoine and hydroxyectoine such as inclusion in both nasal sprays for irritated nasal mucosa and new dermatological products (MedaEctoine Syxyl™). Recently, a new product called Glycoin®, a formulation containing 2-*O*-alpha-d-glucopyranosylglycerin, has been marketed as a cosmetic to combat ageing. Ectoine also has possible application as a DNA-stabilizing agent (PCR additive), protein stabilizer (crystallography), and cell stabilizer (storage). The production of β-carotene, marketed as a nutritional supplement, by *Dunaliella salina* is currently commercialized by several companies (Table 1.3), and production scales range from simple but cost effective pond cultivations to more sophisticated photobioreactor-based processes (Raja *et al.*, 2007).

Acidophiles and alkaliphiles

Life in low and high pH

Even though acidophilic and alkaliphilic organisms thrive in low and high pH environments respectively, they usually have a circum-neutral intracellular pH (Baker-Austin and Dopson, 2007; Ferrer *et al.*, 2008). Therefore, it is usually only the proteins displayed on the cell surface or those that are secreted into the extracellular milieu that are acid or alkali stable. Notable exceptions are proteins isolated from *Ferroplasma acidiphilum* and the halophilic bacterium *Halobacterium* sp. NRC-1, which has an average predicted pI of 4.9 for its proteins (DasSarma *et al.*, 2006; Ferrer *et al.*, 2008). The physiological adaptations employed by acidophiles (reviewed recently by Baker-Austin and Dopson, 2007; see also Chapter 11) are principally aimed at the maintenance of a pH gradient across the cellular membrane while concurrently generating ATP by the influx of protons through the F(0)F(1) ATPase. Other notable features include reversed membrane potentials, impermeable cell membranes and prolific use of secondary transporters (Baker-Austin and Dopson, 2007). Alkaliphiles must also maintain pH homeostasis and in alkaliphilic *Bacillus* spp. this has been linked to a sodium transport cycle, where Na^+/H^+ antiporters are used to achieve a net intracellular H^+ accumulation (coupled to Na^+ efflux) and Na^+ then re-enters the cell via Na^+/solute symporters (Krulwich *et al.*, 2001; Yumoto, 2002). A number of enzymes from alkaliphiles have found commercial use in the detergent, textile and laundry industries (Horikoshi, 1999), whereas acidophiles have found major application in the biomining industry (Breirley, 2008).

Acidophilic enzymes and biomolecules

Most of the acid-stable enzymes (≤pH 4) currently used in industry are not derived from extremophilic organisms, but instead come from *Aspergillus* species which usually grow at

Table 1.3 Commercial producers of β-carotene from *Dunaliella salina*

Company	Location	Cultivation mode	Description
Nature beta technologies	Israel	Intensive	Controlled pond based cultivation. Cultures mixed by paddles in raceways up to 3000 m^2
Henkel-Cognis	Australia	Intensive	Little process control of algal farms located at Hutt lagoon and Whyalla (800 ha). Low operational costs but slow product formation
Inner Mongolia Biological Eng. Co	China	Semi intensive	Partial process control in small scale ponds (5 ha)
PBL Ltd	Spain	Highly intensive	Fully controlled closed photobioreactor cultivations
Tianjin Lantai Biotechnology Inc.	China	Intensive	No details available

pH4.0–5.0. These include an alpha-amylase from *A. kawachi* (SPEZYME®, Genencor), and two products from Erbslöh Geisenheim: an acid and heat stable phytase (Phytase 5000) and a proto-pectinase (Fructozym® P 6-L). These are applied in a range of industries including ethanol fuel production, animal feed, Japanese style food production and fruit juice clarification. These applications demonstrate that although the premise targeting enzymes from a particular environment in order to suit the technology is both logical and intuitive, enzymes suitable for 'extreme' reaction conditions are not necessarily always derived from extremophiles.

Several enzymes suitable for starch hydrolysis have been isolated from acidophilic organisms, including glucoamylases from *Thermoplasma acidophilum, Picrophilus torridus* and *Picrophilus oshimae*. These enzymes are optimally active at pH 2.0 and 90°C but have not, to the authors' knowledge, been used commercially (Serour and Antranikian, 2002). The German company Ascenion markets several acidophilic enzymes, in particular esterases and glucosidases, which are proposed for use in the clinical, medical, and analytical chemistry, as well as starch saccharification, textile, food, fermentation, paper, brewing and distilling industries. An α-glucosidase, isolated from *Ferroplasma acidiphilium*, has a demonstrated activity over a pH range of 2.4 to 3.5 (Ferrer *et al.*, 2005).

Whole cell acidophilic biocatalysts

Only a few examples exist of industries making use of acidophiles as whole cell biocatalysts. Biohydrometallurgy is one technology that has very successfully exploited the physiological and extremophilic properties of thermoacidophiles (polyextremophiles adapted to both high temperature and low pH). Over the past two decades, BHP Billiton and other mining companies have successfully developed large-scale processes using a consortium of iron- and sulfur-oxidizing bacteria and archaea for the recovery of copper, gold, cobalt and uranium from metal sulfide ores. These organisms include *Leptospirillum ferriphilum, Leptospirillum ferrooxidans, Acidithiobacillus ferrooxidans, Acidithiobacillus thiooxidans, Acidithiobacillus caldus, Sulfobacillus thermosulfidooxidans, Acidimicrobium* spp., *Ferroplasma* spp. and various *Sulfolobus* spp. This technology, now broadly known as biomining, has a multi-billion dollar global value previously estimated at greater than $10 billion (Johnson, 1998). Whilst the topic has been covered extensively in several excellent reviews and will therefore not be discussed in great detail here (Rawlings *et al.*, 2003; Rawlings, 2005; Rawlings and Johnson, 2006, 2007), we highlight some aspects of the technology to demonstrate the basis of its commercial success. Further details on biomining can be found in Chapter 11 of this book (Dopson).

Several factors have contributed to the remarkable success of biomining, including low capital and operating cost (dependent on the type of ore and location of the mine) when compared with traditional smelting, roasting or pressure oxidation techniques. In the case of heap bioleaching, the process is fairly simple. Although it has been much refined, especially with regards to design and engineering, in essence a heap leaching operation only requires the ore to be stacked in heaps, irrigated with an acidic liquor solution and a method for extraction of the metal of interest from the pregnant leach liquor (solvent extraction/electrowinning plants, where needed, are usually already in place). The organisms involved in the process, at least the mesophiles, are found ubiquitously in the environment and the process merely exploits the organisms' natural metabolism (Rawlings and Johnson, 2006, 2007; Breirley, 2008).

One of the acknowledged benefits of extremophilic fermentations is a reduced incidence of contamination (Rawlings, 2005; Rawlings and Johnson, 2007). In some instances, where sterility is not a critical factor, substantial cost reductions through simpler fermentor technology may accrue. Biomining processes, where no attempt to maintain sterility is made, are such examples. Typically, the extreme conditions of high temperature, low pH, high concentrations of metals and metalloids greatly reduce problems of contamination, and generally require the process organisms to adapt to orebody conditions. Such adaptation has been seen in the increase in arsenic resistance in both *A. caldus* and *L. ferrooxidans* when cultured on ores containing high arsenic concentrations of up to 13 g/l. Both bacteria are thought to have independently acquired arsenic resistance genes carried on transposons from the so-called 'horizontal gene pool' of bacterial genomes randomly introduced to the system (Tuffin *et al.*, 2005, 2006).

In biooxidation tanks the leaching process operates as a continuous-flow system where the crushed and partially processed ore is split and pumped into three equally sized primary oxidation tanks at the same time, after which the total content of these tanks are moved to the first of three secondary oxidation tanks (Rawlings, 2005; Rawlings and Johnson, 2006; Breirley, 2008). The pulp then moves from the first to second and finally on to the third secondary oxidation tanks for leaching. This process continually selects for those organisms capable of optimal growth. Slower growing microorganisms are washed out of the system and the continual rounds of selection ensure that the best adapted community of organisms for the conditions is selected. These natural selection processes have largely overcome the need for genetically engineered strains, particularly since the population used to seed an oreheap or reactor is usually sourced from another bioleaching or biooxidation operation (Rawlings, 2005; Rawlings and Johnson, 2006; Breirley, 2008).

The BIOX™ process developed by Gencor has been in use since 1986 when first employed for the treatment of refractory gold ores at the Fairview mine in Baberton, South Africa (Rawlings, 1997). Commercial BIOX™ plants have been established at eight mines worldwide as of the end of 2005, while several more were commissioned during 2006 and 2007. The technology of biomining undergoes continual refinement, such as the recent addition of the GEOCOAT®/GEOLEACH® technology for the pretreatment of refractory gold-bearing sulfidic ores and for the efficient leaching of recalcitrant chalcopyrite ores at elevated temperatures (above 70°C). In the GEOCOAT® process the ore is coated with the mixture of microorganisms as it leaves the conveyor belt during stacking onto biooxidation heaps. BIOPRO™ is an alternative heap inoculation method developed by Newmont Gold Company (Rawlings and Johnson, 2006; Rawlings and Johnson, 2007). The three abovementioned technologies belong to the company GeoBiotics. The company BacTech, with its proprietary technology BACOX™, reportedly has plans to expand into bioremediation of old mine sites with the recently established company, Cleantech, and has interests in black oil shale bioleaching, potentially a very lucrative new source for the recovery of base metals.

The use of biotechnology is often presented as a means of introducing 'cleaner, greener' products or processes. There are, however, drawbacks to many of these technologies, some of which derive from the unfavourable environmental impacts (e.g. air, ground and water pollution) of the by-products of these 'biological' reactions. For example, 'acid mine drainage' (AMD) is a major concern in the biomining industry (Johnson and Hallberg, 2005; Johnson *et al.*, 2008). In general, the term refers to acidic, highly mineralized waters leaching from mines and mine waste dumps. AMD is principally caused by the dissolution of pyrite (FeS_2, found in appreciable amounts in these waters) by the iron oxidizing bacteria normally associated with the leaching process itself, most often *A. ferrooxidans* and *L. ferrooxidans*, in the presence of moisture and oxygen, resulting in acid formation (eqn 1.1).

$$FeS_2 + 3.75O_2 + 3.5H_2O \rightarrow Fe(OH)_3 + 4H^+ + 2SO_4^{2-} \qquad (1.1)$$

Although many techniques exist for the remediation of a variety of acidic wastewater streams, two approaches that have seen considerable success are constructed wetlands and the use of bioreactors. The use of sulfate-reducing bacteria (SRB; e.g. *Desulfovibrio, Desulfomicrobium, Desulfobulbus, Desulfobacterium, Desulfococcus, Desulfosarcina, Desulfonema, Desulfotomaculum, Thermodesulfobacterium, Thermodesulfovibrio* and *Archaeoglobus*) in wetlands or bioreactors for the removal of sulphate and neutralization of acidic wastewater is well documented (Doshi, 2006), but no such systems using acidophilic SRB are known to be in commercial operation. While these organisms (*Desulfosporosinus* spp.) have been used in fixed-bed bioreactors to assess their efficacy for remediation of acidic, metal-rich effluents (Kolmert and Johnson, 2001; Kimura *et al.*, 2006), sulfate conversion rates were approximately 20-fold lower than those achieved with similar bioreactors operating at near neutral pH (Kolmert *et al.*, 1997). However, the need for incorporating acidophilic bacteria into the process is still seen as a priority: a ten month shutdown of the Wheal Jane pilot AMD passive treatment plant was caused, in part, by the inability of the SRB to cope with the acidic influent waste stream (Kolmert and Johnson, 2001; Johnson and Hallberg, 2005).

An alternative approach is termed 'bioshrouding'. Here the mineral tailings are first colonized by species of heterotrophic acidophiles (*Acidiphilium, Acidocella* and *Acidobacterium* spp.). These bacteria reduce ferric iron (Fe^{3+}) under oxygen-limited conditions, thereby removing the primary oxidant of sulfide minerals with a concomitant reduction in acid production. This process represents another potential large-scale application of extremophiles in biotechnology. Both techniques are relatively low cost operations and have shown great promise in the laboratory (Johnson and Hallberg, 2005; Johnson *et al.*, 2008), yet it remains to be seen if they will eventually be employed at an industrial scale.

Finally, a recently described microbial fuel cell for the production of electricity, using the iron reducing acidophile *Acidiphilium cryptum* as an anode biocatalyst, could be the first step towards a new technology (Borole *et al.*, 2008).

Alkaliphilic biomolecules and whole cell biocatalysts

Historical examples of alkaliphile-based industrial processes include the alkaliphilic cyclomaltodextrin glucanotransferases (CTases) which produce cyclodextrins from starch based feedstocks. These enzymes, produced from the alkaliphilic *Bacillus* strain 38-2 (Horikoshi, 1999), have been widely described and employed in the food and pharmaceutical industries. Alkaliphilic enzymes have also been used in the hide dehairing process (Horikoshi, 1999). Two more recent commercially available enzymes derived from alkaliphilic organisms are the cellulase Puradax HA (from *Bacillus* spp.), used for stain removal and colour preservation in textiles, and INDIAGE™ Neutra (alkaliphilic *Streptomyces* spp.) for producing a stonewashed effect on jeans. Both of these enzymes are produced by Genencor, which also produces several enzymes for the paper and pulp industry which are active over a wide pH range (pH 3–9). We note that the special functional properties of the latter enzymes are the result of engineering rather than a product of an extremophilic source. Proteases, amylases, lipases, cellulases and other enzymes that are resistant to and active at high pH and high chelator concentrations are desirable for inclusion in modern household detergents and the specific applications in this industry have been reviewed elsewhere (Horikoshi, 1999).

The stability of bio-molecules in cosmetic preparations under a variety of physical conditions is important to maintain their efficacy. Haloarchaeal membrane lipids (biosurfactants) that have very low melting temperatures have been described (Kebbouche-Gana *et al.*, 2009). These are thought to have application in the cosmetics industry for the production of novel liposomes, given their resistance to acids, alkalis, heat and their ability to produce good water–oil emulsions. Together these properties should enhance product stability and shelf life.

Another area where enzymes from these organisms may find a use is in wastewater remediation. A thermo-alkaline stable catalase, isolated from *Thermus brockianus*, has been suggested for use in the bioremediation of textile and paper industry wastewater where hydrogen peroxide is used in the bleaching of clothes and paper (Thompson *et al.*, 2003). The US based company, Batelle, currently has the patent rights to this enzyme, and although no currently known remediation process makes use of this biomolecule, it remains a promising prospect. Wastewater bioremediation plants appear to contain a high load of alkaliphilic organisms, a number of which have been isolated and described (Berendes *et al.*, 1996; Tseng *et al.*, 2002; Ulukanli and Digrak, 2002; Lee and Park, 2008). It could therefore be argued that whether deliberately introduced or whether selected for by the conditions found in the remediation plant, these organisms play a vital role in remediation of several industrial wastewaters. This observation does not necessarily suggest that purified enzymes are probably useful in high volume bioremediation. In the authors' opinion, the economics of wastewater and other bioremediation processes are wholly incompatible with the economics of enzyme production and application.

The reduction and decolourization of indigo, used in the textile dyeing industry, is a longstanding application of alkaliphilic bacteria, particularly in Japan where indigo was traditionally reduced by alkaliphilic microbial consortia in pools of sodium carbonate, calcium and potassium hydroxide (Horikoshi, 1999). The microbial decolourization of indigo has since been replaced by chemical processing using hydrosulfite, but the traditional biological reduction technology remains in place in the small scale craft industry in Japan (Horikoshi, 1999; Aino *et al.*, 2008).

Piezophiles

Life under pressure

The potential biotechnological use of piezophilic organisms, usually found in deep-sea environments, has only recently been explored (Abe and Horikoshi, 2001; Yano and Poulos, 2003). This fact has mainly been due to their inaccessibility and difficulty in culturing. Adaptation responses in piezophiles have been reported to include the accumulation, in response to hydrostatic pressure, of a novel class of osmolytes called piezolytes (e.g. beta-hydroxybutyrate) (Martin *et al.*, 2002). Also known as barophiles, the diversity and properties of these pressure resistant organisms have been reviewed in detail elsewhere (Kato and Bartlett, 1997; Bartlett, 1999, 2002; Martin *et al.*, 2002). Detailed information on piezophiles is provided by Kato in Chapter 10.

Piezophilic enzymes and whole cell biocatalysts

The expression levels of several genes from piezophilic organisms including *ompH* (*Photobacterium profundum*), ORF1, 2, *cydD* (*Shewanella benthica* and *Shewanella violacea*) and RNA polymerase σ^{54} (*S. violacea*) have been found to be regulated by changes in hydrostatic pressure (Kato *et al.*, 1996; Ikegami *et al.*, 2000; Abe and Horikoshi, 2001). The promoters from these genes may in future be engineered into expression systems or used in the native host to allow control over gene expression by using changes in hydrostatic pressure (Abe and Horikoshi, 2001; Simonato *et al.*, 2006). Apart from the direct use of piezophiles or their biomolecules, the study of these organisms has shown how the application of high hydrostatic pressure affects reaction rates, types of product formed and enzyme stability (Abe and Horikoshi, 2001). The behaviour of proteins under pressure is governed by Le Chatelier's principle, thus application of pressure to a system in equilibrium forces the system to favour the equilibrium state with a smaller volume. The effect of pressure on the rate of a reaction is described by eqn 1.2, where k_p corresponds to the rate at pressure p, k_0 the rate at atmospheric pressure, and $\Delta V^{\#}$ the activation molar volume for the transition state and represents the difference in the volume between the reactants and the transition state.

$$k_p = k_0 \cdot e^{-p\Delta V\#/RT} \quad (1.2)$$

Depending on the nature of the transition state involved ($\Delta V^{\#} > 0$ or $\Delta V^{\#} < 0$), applying pressure may accelerate or decrease the rate of the reaction. In other words, when enzymatic reactions are accompanied by an increase in the volume of activation ($V^{\#}$), they are inhibited by elevated pressures and vice versa. This also applies to the type of products produced or substrates used and can therefore potentially be controlled through changes in pressure (Fig. 1.1).

The example of α-chymotrypsin catalysis serves to demonstrate how pressure may be used to control a reaction. This enzyme catalyses the hydrolysis of both anilides ($\Delta V^{\#} < 0$) and esters ($\Delta V^{\#} > 0$). Therefore, application of pressure to a reaction mixture containing the enzyme and both substrates would favour the hydrolysis of the

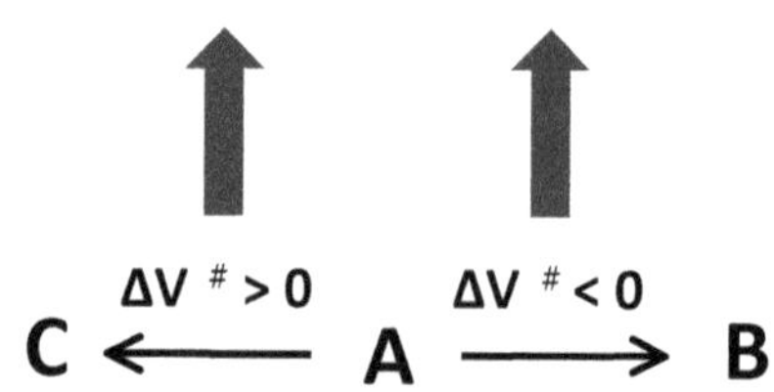

Figure 1.1 Dark arrows indicate either an increase (down arrow) or decrease (up arrow) in pressure. A represents a substrate and B and C are potential products. Modified from Abe and Horikoshi (2001).

anilide and suppress hydrolysis of the ester (Abe and Horikoshi, 2001; Yano and Poulos, 2003).

Several patents describing enhanced recovery of crude oil from oil wells (e.g. StatoilHydro – see Chapter 8) and biological sensors using enzymes isolated from piezo-thermophilic/psychrophilic organisms already exist. In the case of microbially enhanced oil recovery (MEOR) these patents, as it relates to piezophilic organisms, usually describe methods for cultivating obligate piezophilic microbes isolated from oil wells under high pressure. The patents also allude to the use of these organisms for the reduction of the viscosity of the crude oil and hence improve its pumping/handling properties for downstream recovery and processing. The discovery of a wider range of deep sea obligately piezophilic bacteria and archaea probably stimulates the future applications of these organisms and their products in commercial processes.

Future prospects for extremophilic biotechnology

In this review we have sought to critically evaluate the real impact of extremophiles in biotechnology industries and to assess future commercialization trends. We conclude that, with the exception of DNA modifying enzymes, penetration into enzyme markets has been modest in terms of the paucity of products and processes successfully finding a place in a related market. In most, if not all cases, the gene of interest has been cloned from the parent organism and expressed in a heterologous host for production scale-up. This approach works well for compatible hosts, but issues of codon usage and the presence of suitable folding pathways (Goodarzi *et al.*, 2008; Taylor *et al.*, 2008; Thompson *et al.*, 2008) may have limited the commercialization of many potentially valuable proteins, particularly those aimed at the commodity market where decagram/L production is essential. One solution could be to engineer the parent organism to produce the required enzyme or biomolecule at higher levels, but this approach requires suitable genetic systems (vectors, selectable markers and transformation protocols) to have been established. It can be suggested that extremophile genetics is currently an under-developed field. These systems are not necessarily straightforward to develop due to a lack of well-characterized plasmid and phage vectors, difficult transformations and complex fermentations. Nevertheless, such systems are essential for the fuller exploitation of extremophilic products, and a number have been or are in various stages of development. These include *Pseudoalteromonas haloplanktis, Halobacterium* sp. NRC1, *Geobacillus thermoglucosidasius, Thermoanaerobacterium* spp., *A. ferrooxidans, A. thiooxidans, A. caldus* and several *Sulfolobus* spp. (Liu *et al.*, 2000; Mai and Wiegel, 2000; Berquist and DasSarma, 2003; Tian *et al.*, 2003a,b; Aucelli *et al.*, 2006; Cieslinski *et al.*, 2008; Taylor *et al.*, 2008; van Zyl *et al.*, 2008). Those processes involving biofuel production have shown particular promise, with accelerated development of genetic methods reported over the last 5 years.

With the huge current interest in products from renewable resources (e.g. biofuels), there is a significant focus on the generation and characterization of enzymes involved in the complex processes of lignocellulose digestion. Perhaps because of the demand for flexibility and adaptability in process design and evolution, it appears that the major commercial enzyme companies are turning more towards non-conventional resources in order to find novel products that meet the demands of bioprocess engineers and industrialists. However, it is certain that industrialists will not abandon enzyme engineering as a means to achieve the high degree of performance they require and It is probable that the engineering of mesophilic templates to impart specific 'extremophilic' properties, as has been the trend in recent years, will continue.

For extremophile whole cell biocatalysis, it is clear that only a limited number of specific processes, such as thermoacidophiles in biomining, have made the transition from laboratory- to full-scale industrial implementation. The inception of a number of new, smaller-scale biotechnology companies, based almost exclusively on extremophiles and their value products, may change this position in the next few years. The growth of these companies is due in no small part to the development of fundamental genetic systems that have in turn allowed the potential of organism, and product sought, to be realized. There is still significant progress to be made in this area: for example, the successful high-level expression of some extremophilic genes is hampered, as noted above, by the paucity of appropriate genetic systems. We suggest that the appearance of companies such as TMO Renewables and Mascoma, which are strongly dependent on the development of fundamental genetic tools for engineering their particular 'cell factories', represents a trend for the future success of niche biotechnology industries.

While it is reasonable to predict that extremophiles and their products will continue to contribute successful commercial products and processes (Bull *et al.*, 2000; Simonato *et al.*, 2006; Lauro and Bartlett, 2008), we believe that for extremophiles to impact more significantly in industry in the future, important criteria inevitably must be met. They must comply with the economic imperatives of industry, be deliverable at a competitive price, and concerns relating to international bioprospecting require demonstration of a full and legally mandated set of credentials of a product, its developmental history and its origins. It is debatable whether extremozymes will ever impact significantly on some industrial markets, when the engineering of mesophilic enzymes can be effective both in terms of process economics and development. Extremozymes may only impact when the physical parameters of a process move beyond those achievable by current engineering processes.

Successful commercial examples of whole cell biocatalytic applications exist, but essentially rely on close correspondence between the physiological properties of the organism and the physical requirements of the process. Perhaps this is the path forward for extremophiles: in the development of processes that require whole cell bioremediation or biotransformation specifically in an inhospitable environment, whether natural or artificial. With current interest in delivering products from renewable resources, extremophiles have a significant opportunity for impact in a growing, sustainable and potentially highly profitable field. We therefore suggest, with due caution, that we may yet be witness to the beginnings of a new extreme revolution.

Web resources

Many of the homepages of biotechnology companies with the specific remit to commercialize extremophile biotechnology include technological overviews and concepts, and they also highlight relevant publicity and media articles: www.tmo-group.com, www.biogasol.dk, www.mascoma.com, www.bitop.de, www.swissaustral.ch, www.actilor.com, http://www.finnzymes.com, http://www.biohelix.com, http://zygem.com, www.syntheticgenomics.com, www.diversa.com, www.ascenion.de, www.matis.is, http://www.kapabiosystems.com and www.extremozyme.com. The enzyme focused companies Novozymes (www.novozymes.com) and Genencor (www.genencor.com) are interesting resources for information on enzyme development, although specifics of extremophilic origins are generally not provided in consideration of intellectual property rights. However, they remain a resource for general information with respect to developments in the biofuel industry and products from renewable resources. The website of New England Biolabs (www.neb.com) contains many useful and easy to use programmes for manipulating DNA sequences. The website for the company Fermentas, who provide a variety of technical enzymes and molecular biology reagents, is www.fermentas.com. Further information on the exploration of extremophiles and the study of the 'outer margins of survival' can be found at www.astrobiology.com/extreme.html. An excellent overview for both the novice and experienced researcher of extremophiles is located at http://serc.carleton.edu/microbelife/index.html, created and maintained by The Marine Biology Laboratory, Woods Hole, MA and Montana State University Geoscience education web development team. The page of the Thermal Biology Institute at Montana State University is equally informative (http://www.tbi.montana.edu/). Patentstorm and Patentscope are useful browsers for tracking patents filed in the US and internationally (www.patentstorm.us, www.wipo.int/pctdb/en/search-adv.jsp). Battelle is an international science and technology enterprise that explores emerging areas of science, develops and commercializes technology, and manages laboratories for their customers (www.battelle.org).

Other websites of companies mentioned in the text include http://www.ddce.com, http://www.takara-bio.com, http://www.dupont.com, http://www.danisco.com and http://www.ascenion.de.

Acknowledgements

The authors wish to acknowledge the University of the Western Cape and the NRF, South Africa National Research Foundation for research funding. We also wish to acknowledge Brian E. Jones for valuable discussion and industrial insight.

Disclaimer

The views expressed in this article are the authors own and are in no way a reflection of any particular company or industrial collaborator.

References

Abe, F., and Horikoshi, K. (2001). The biotechnological potential of piezophiles. Trends Biotechnol. *19*, 102–108.

Aino, K., Hirota, K., Matsuno, T., Morita, N., Nodasaka, Y., Fujiwara, T., Matsuyama, H., Yoshimune, K., and Yumoto, I. (2008). *Bacillus polygoni* sp. nov., a moderately halophilic, non-motile obligate alkaliphile isolated from indigo balls. Int. J. Syst. Evol. Microbiol. *58*, 120–124.

Amann, R.I., Ludwig, W., and Schleifer, K.H. (1995). Phylogenetic identification and *in situ* detection of individual microbial cells without cultivation. Microbiol. Rev. *59*, 143–169.

Antranikian, G., and Egorova, K. (2007). Extremophiles, a unique resource of biocatalysis for industrial biotechnology, In Physiology and Biochemistry of Extremophiles, Gerday, C., and Glansdorff, N., eds. (Washington, USA, ASM Press), pp. 361–409.

Antranikian, G., Vorgias, C.E., and Bertoldo, C. (2005). Extreme environments as a resource for microorganisms and novel biocatalysts. Adv. Biochem. Eng. Biotechnol. *96*, 219–262.

Aucelli, T., Contursi, P., Girfoglio, M., Rossi, M., and Cannio, R. (2006). A spreadable, non-integrative and high copy number shuttle vector for *Sulfolobus solfataricus* based on the genetic element pSSVx from *Sulfolobus islandicus*. Nucleic Acids Res. *34*, e114.

Bagi, Z., Acs, N., Balint, B., Horvath, L., Dobo, K., Perei, K.R., Rakhely, G., and Kovacs, K.L. (2007). Biotechnological intensification of biogas production. Appl. Microbiol. Biotechnol. *76*, 473–482.

Baker-Austin, C., and Dopson, M. (2007). Life in acid: pH homeostasis in acidophiles. Trends Microbiol. *15*, 165–171.

Bartlett, D.H. (1999). Microbial adaptations to the psychrosphere/piezosphere. J. Mol. Microbiol. Biotechnol. *1*, 93–100.

Bartlett, D.H. (2002). Pressure effects on *in vivo* microbial processes. Biochim. Biophys. Acta *1595*, 367–381.

Batty, L.C., and Younger, P.L. (2007). The effect of pH on plant litter decomposition and metal cycling in wetland mesocosms supplied with mine drainage. Chemosphere *66*, 158–164.

Berendes, F., Gottschalk, G., Heine-Dobbernack, E., Moore, E.R.B., and Tindal, B.J. (1996). *Halomonas desiderata* sp. nov., a new alkaliphilic, halotolerant and denitrifying bacterium isolated from a municipal sewage works. Syst. App. Microbiol. *19*, 158–167.

Bergquist, P.L., and Morgan, H.W. (1992). The molecular genetics and biotechnological application of enzymes from extremely thermophilic eubacteria, In Molecular Biology and Biotechnology of Extremophiles, Herbert, R.A., and Sharp, R.J., eds. (New York, Blackie), pp. 44–75.

Berquist, B.R., and DasSarma, S. (2003). An archaeal chromosomal autonomously replicating sequence element from an extreme halophile, Halobacterium sp. strain NRC-1. J. Bacteriol. *185*, 5959–5966.

Blow, N. (2008). Metagenomics: exploring unseen communities. Nature *453*, 687–690.

Blumer-Schuette, S.E., Kataeva, I., Westpheling, J., Adams, M.W., and Kelly, R.M. (2008). Extremely thermophilic microorganisms for biomass conversion: status and prospects. Curr. Opin. Biotechnol. *19*, 210–217.

Boogerd, F.C., Bos, P., Kuenen, J.G., Heijnen, J.J., and van der Lans, R.G. (1990). Oxygen and carbon dioxide mass transfer and the aerobic, autotrophic cultivation of moderate and extreme thermophiles: A case study related to the microbial desulfurization of coal. Biotechnol. Bioeng. *35*, 1111–1119.

Borole, A.P., O'Neill, H., Tsouris, C., and Cesar, S. (2008). A microbial fuel cell operating at low pH using the acidophile *Acidiphilium cryptum*. Biotechnol. Lett. *30*, 1367–1372.

Breirley, C.L. (2008). How will biomining be applied in the future? Trans. Nonferrous Met. Soc. China *18*, 1302–1310.

van den Broeke, M. (2000). Notes and correspondence on the interpretation of Antarctic temperature trends. J. Climate *13*, 3885–3889.

Bull, A.T., Ward, A.C., and Goodfellow, M. (2000). Search and discovery strategies for biotechnology: the paradigm shift. Microbiol. Mol. Biol. Rev. *64*, 573–606.

Cavicchioli, R. (2006). Cold-adapted archaea. Nat. Rev. Microbiol. *4*, 331–343.

Chappell, J., and Grotewold, E. (2008). Plant biotechnology-predictive, green and quantitative. Curr. Opin. Biotechnol. *19*, 129–130.

Cieslinski, H., Werbowy, K., Kur, J., and Turkiewicz, M. (2008). Molecular characterization of a cryptic plasmid from the psychrotrophic antarctic bacterium *Pseudoalteromonas* sp. 643A. Plasmid *60*, 154–158.

Cowan, D.A., Arslanoglu, A., Burton, S.G., Baker, G.C., Cameron, R.A., Smith, J.J., and Meyer, Q. (2004). Metagenomics, gene discovery and the ideal biocatalyst. Biochem. Soc. Trans. *32*, 298–302.

Cowan, D.A., Casanueva, A., and Stafford, W. (2007). Ecology and biodiversity of cold adapted microorganisms. In Physiology and Biochemistry of Extremophiles, Gerday, C., and Glansdorff, N., eds. (Washington, ASM Press), pp. 119–133.

Cripps, R.E., Eley, K., Leak, D.J., Rudd, B., Taylor, M., Todd, M., Boakes, S., Martin, S., and Atkinson, T. (2009). Metabolic engineering of *Geobacillus thermoglucosidasius* for high yield ethanol production. Metab. Eng. *11*, 398–408.

DasSarma, S., Berquist, B.R., Coker, J.A., DasSarma, P., and Muller, J.A. (2006). Post-genomics of the model haloarchaeon *Halobacterium* sp. NRC-1. Saline Syst. *2*, 3.

Demirjian, D.C., Moris-Varas, F., and Cassidy, C.S. (2001). Enzymes from extremophiles. Curr. Opin. Chem. Biol. *5*, 144–151.

Doshi, S.M. (2006). Bioremediation of acid mine drainage using sulfate-reducing bacteria USA Environmental Protection Agency (http://cluin.info/download/studentpapers/S_Dosshi-SRB.pdf accessed on 3 May 2010).

Driessen, A., and Albers, S.A. (2007). Membrane adaptations of (hyper)thermophiles to high temperatures, In Physiology and Biochemistry of Extremophiles, Gerday, C., and Glansdorff, N., eds. (Washington, ASM Press), pp. 104–119.

Egorova, K., and Antranikian, G. (2005). Industrial relevance of thermophilic Archaea. Curr. Opin. Microbiol. *8*, 649–655.

Empadinhas, N., and da Costa, M.S. (2006). Diversity and biosynthesis of compatible solutes in hyper/thermophiles. Int. Microbiol. *9*, 199–206.

Faria, T.Q., Mingote, A., Siopa, F., Ventura, R., Maycock, C., and Santos, H. (2008). Design of new enzyme stabilizers inspired by glycosides of hyperthermophilic microorganisms. Carbohydr. Res. *343*, 3025–3033.

Feller, G., and Gerday, C. (2003). Psychrophilic enzymes: hot topics in cold adaptation. Nat. Rev. Microbiol. *1*, 200–208.

Feng, L., Wang, W., Cheng, J., Ren, Y., Zhao, G., Gao, C., Tang, Y., Liu, X., Han, W., Peng, X., Liu, R., and Wang, L. (2007). Genome and proteome of long-chain alkane degrading *Geobacillus thermodenitrificans* NG80–2 isolated from a deep-subsurface oil reservoir. Proc. Natl. Acad. Sci. U.S.A. *104*, 5602–5607.

Ferrer, M., Golyshina, O.V., Plou, F.J., Timmis, K.N., and Golyshin, P.N. (2005). A novel alpha-glucosidase from the acidophilic archaeon *Ferroplasma acidiphilum* strain Y with high transglycosylation activity and an unusual catalytic nucleophile. Biochem. J. *391*, 269–276.

Ferrer, M., Golyshina, O.V., Beloqui, A., Bottger, L.H., Andreu, J.M., Polaina, J., De Lacey, A.L., Trautwein, A.X., Timmis, K.N., and Golyshin, P.N. (2008). A purple acidophilic di-ferric DNA ligase from *Ferroplasma*. Proc. Natl. Acad. Sci. U.S.A. *105*, 8878–8883.

Frazzetto, G. (2003). White biotechnology. EMBO Rep. *4*, 835–837.

Galvez, A. (2005). Analyzing cold enzyme starch hydrolysis technology in new ethanol plant design. Ethanol producer magazine (http://www.ethanolproducer.com/article.jsp?article_id=975 accessed on 3 May 2010).

Garcia, M.T., Mellado, E., Ostos, J.C., and Ventosa, A. (2004). *Halomonas organivorans* sp. nov., a moderate halophile able to degrade aromatic compounds. Int J. Syst. Evol. Microbiol. *54*, 1723–1728.

Gomes, J., and Steiner, W. (2004). The biocatalytic potential of Extremophiles and Extremozymes. Food Technol. Biotechnol. *42*, 223–225.

Goodarzi, H., Torabi, N., Najafabadi, H.S., and Archetti, M. (2008). Amino acid and codon usage profiles: adaptive changes in the frequency of amino acids and codons. Gene *407*, 30–41.

Graf, R., Anzali, S., Buenger, J., Pfluecker, F., and Driller, H. (2008). The multifunctional role of ectoine as a natural cell protectant. Clin. Dermatol. *26*, 326–333.

Grosjean, H., and Oshima, T. (2007). How nucleic acids cope with high temperatures, In Physiology and Biochemistry of Extremophiles, Gerday, C., and Glansdorff, N., eds. (Washington, ASM Press), pp. 39–57.

Heath, C., Hu, X.P., Cary, S.C., and Cowan, D. (2009). Identification of a novel alkaliphilic esterase active at low temperatures by screening a metagenomic library from antarctic desert soil. Appl. Environ. Microbiol. *75*, 4657–4659.

Hendry, P. (2006). Extremophiles:Theres more to life. Environ. Chem. *3*, 75–76.

Hernandez-Herrero, M.M., Roig-Sagues, A.X., Rodriguez-Jerez, J.J., and Mora-Ventura, M.T. (1999). Halotolerant and halophilic histamine-forming bacteria isolated during the ripening of salted anchovies (*Engraulis encrasicholus*). J. Food Prot. *62*, 509–514.

Hong, S.Y., Lee, J.S., Cho, K.M., Math, R.K., Kim, Y.H., Hong, S.J., Cho, Y.U., Cho, S.J., Kim, H., and Yun, H.D. (2007). Construction of the bifunctional enzyme cellulase-beta-glucosidase from the hyperthermophilic bacterium *Thermotoga maritima*. Biotechnol. Lett. *29*, 931–936.

Horikoshi, K. (1999). Alkaliphiles: some applications of their products for biotechnology. Microbiol. Mol. Biol. Rev. *63*, 735–750.

Horikoshi, K. (2007). Foreword. In Physiology and Biochemistry of Extremophiles, Gerday, C., and Glansdorff, N., eds. (Washington, ASM Press), pp. 3–13.

Huston, A.L. (2008). Biotechnological aspects of cold adapted enzymes. In Psychrophiles: From Biodiversity to Biotechnology, Margensin, R., Schinner, F., Marx, J.C., and Gerday, C., eds. (Berlin, Germany, Springer Academic Press), pp. 347–365.

Ikegami, A., Nakasone, K., Fujita, M., Fujii, S., Kato, C., Usami, R., and Horikoshi, K. (2000). Cloning and characterization of the gene encoding RNA polymerase sigma factor sigma(54) of deep-sea piezophilic *Shewanella violacea*. Biochim. Biophys. Acta *1491*, 315–320.

Johnson, D.B. (1998). Biodiversity and ecology of acidophilic microorganisms. FEMS Microbiol. Ecol. *27*, 307–317.

Johnson, D.B., and Hallberg, K.B. (2005). Biogeochemistry of the compost bioreactor components of a composite acid mine drainage passive remediation system. Sci. Total Environ. *338*, 81–93.

Johnson, D.B., Yajie, L., and Okibe, N. (2008). 'Bioshrouding': a novel approach for securing reactive mineral tailings. Biotechnol. Lett. *30*, 445–449.

Jones, B.E. (2004). Industrial enzymes: Do halophiles and alkaliphiles have a role to play? In Halophilic Microorganisms, Ventosa, A., ed. (Berlin, Springer-Verlag), pp. 275–285.

Kanai, T., Imanaka, H., Nakajima, A., Uwamori, K., Omori, Y., Fukui, T., Atomi, H., and Imanaka, T. (2005). Continuous hydrogen production by the hyperthermophilic archaeon, *Thermococcus kodakaraensis* KOD1. J. Biotechnol. *116*, 271–282.

Kato, C., and Bartlett, D.H. (1997). The molecular biology of barophilic bacteria. Extremophiles *1*, 111–116.

Kato, C., Tamegai, H., Ikegami, A., Usami, R., and Horikoshi, K. (1996). Open reading frame 3 of the barotolerant bacterium strain DSS12 is complementary with cydD in *Escherichia coli*: cydD functions are required for cell stability at high pressure. J. Biochem. *120*, 301–305.

Kebbouche-Gana, S., Gana, M.L., Khemili, S., Fazouane-Naimi, F., Bouanane, N.A., Penninckx, M., and Hacene, H. (2009). Isolation and characterization of halophilic Archaea able to produce biosurfactants. J. Ind. Microbiol. Biotechnol. *36*, 727–738.

Kimura, S., Hallberg, K.B., and Johnson, D.B. (2006). Sulfidogenesis in low pH (3.8–4.2) media by a mixed population of acidophilic bacteria. Biodegradation *17*.

Kolmert, A., and Johnson, J.B. (2001). Remediation of acidic waste waters using immobilised, acidophilic sulfate-reducing bacteria. J. Chem. Technol. Biotechnol. *76*, 836–843.

Kolmert, A., Henrysson, T., Hallberg, R., and Mattiasson, B. (1997). Optimization of sulphide production in an anaerobic continuous biofilm process with sulphate reducing bacteria. Biotechnol. Lett. *19*.

Kotik, M. (2009). Novel genes retrieved from environmental DNA by polymerase chain reaction: Current genome-walking techniques for future metagenome applications. J. Biotechnol. *144*, 75–82.

Kowalchuk, G.A., Speksnijder, A.G., Zhang, K., Goodman, R.M., and van Veen, J.A. (2007). Finding the needles in the metagenome haystack. Microb. Ecol. *53*, 475–485.

Krulwich, T.A., Ito, M., and Guffanti, A.A. (2001). The Na(+)-dependence of alkaliphily in *Bacillus*. Biochim. Biophys. Acta *1505*, 158–168.

Kunte, H.J. (2007). Osmoregulated solute transport in halophilic bacteria. In Halophilic Microorganisms, Ventosa, A., ed. (Berlin, Germany, Springer Academic Press), pp. 155–165.

Lauro, F.M., and Bartlett, D.H. (2008). Prokaryotic lifestyles in deep sea habitats. Extremophiles *12*, 15–25.

Le Borgne, S., Paniagua, D., and Vazquez-Duhalt, R. (2008). Biodegradation of organic pollutants by halophilic bacteria and archaea. J. Mol. Microbiol. Biotechnol. *15*, 74–92.

Lee, S.H., and Park, D.H. (2008). Isolation and physiological characterization of *Bacillus clausii* SKAL-16 isolated from wastewater. J. Microbiol. Biotechnol. *18*, 1908–1914.

Leitao, E., Moradas-Ferreira, P., and De Marco, P. (2009). Evidence of methanesulfonate utilizers in the Sargasso Sea metagenome. J. Basic. Microbiol. *49 Suppl 1*, S24–30.

Li, W.F., Zhou, X.X., and Lu, P. (2005). Structural features of thermozymes. Biotechnol. Adv. *23*, 271–281.

Liu, Z., Guiliani, N., Appia-Ayme, C., Borne, F., Ratouchniak, J., and Bonnefoy, V. (2000). Construction and characterization of a recA mutant of *Thiobacillus ferrooxidans* by marker exchange mutagenesis. J. Bacteriol. *182*, 2269–2276.

Lloyd, J.R., Lovley, D.R., and Macaskie, L.E. (2003). Biotechnological application of metal-reducing microorganisms. Adv. Appl. Microbiol. *53*, 85–128.

Madern, D., Ebel, C., and Zaccai, G. (2000). Halophilic adaptation of enzymes. Extremophiles *4*, 91–98.

Mai, V., and Wiegel, J. (2000). Advances in development of a genetic system for *Thermoanaerobacterium* spp.: expression of genes encoding hydrolytic enzymes, development of a second shuttle vector, and integration of genes into the chromosome. Appl. Environ. Microbiol. *66*, 4817–4821.

Margesin, R., and Schinner, F. (2001). Potential of halotolerant and halophilic microorganisms for biotechnology. Extremophiles *5*, 73–83.

Martin, D.D., Bartlett, D.H., and Roberts, M.F. (2002). Solute accumulation in the deep-sea bacterium *Photobacterium profundum*. Extremophiles *6*, 507–514.

de Miguel Bouzas, T., Barros-Velazquez, J., and Villa, T.G. (2006). Industrial applications of hyperthermophilic enzymes: a review. Protein Pept. Lett. *13*, 645–651.

Mikkelsen, D., Kappler, U., McEwan, A.G., and Sly, L.I. (2006). Archaeal diversity in two thermophilic chalcopyrite bioleaching reactors. Environ. Microbiol. *8*, 2050–2056.

Muller, V., Spanheimer, R., and Santos, H. (2005). Stress response by solute accumulation in archaea. Curr. Opin. Microbiol. *8*, 729–736.

Nahalka, J. (2008). Physiological aggregation of maltodextrin phosphorylase from *Pyrococcus furiosus* and its application in a process of batch starch degradation to alpha-D-glucose-1-phosphate. J. Ind. Microbiol. Biotechnol. *35*, 219–223.

Nandi, R., and Sengupta, S. (1998). Microbial production of hydrogen: an overview. Crit. Rev. Microbiol. *24*, 61–84.

Niehaus, F., Bertoldo, C., Kahler, M., and Antranikian, G. (1999). Extremophiles as a source of novel enzymes for industrial application. Appl. Microbiol. Biotechnol. *51*, 711–729.

Parrilli, E., Duilio, A., and Tutino, M.L. (2007). Heterologous protein expression in psychrophilic hosts. In Psychrophiles: From Biodiversity to Biotechnology, Margensin, R., Schinner, F., Marx, J.C., and Gerday, C., eds. (Berlin, Germany, Springer Academic Press), pp. 365–381.

Paul, S., Bag, S.K., Das, S., Harvill, E.T., and Dutta, C. (2008). Molecular signature of hypersaline adaptation: insights from genome and proteome composition of halophilic prokaryotes. Genome Bio.l 9, R70.

Peyton, B.M., Mormile, M.R., and Petersen, J.N. (2001). Nitrate reduction with *Halomonas campisalis*. Kinetics of denitrification at pH 9 and 12.5% NaCl. Water Res. *35*, 4237–4242.

Podar, M., and Reysenbach, A.L. (2006). New opportunities revealed by biotechnological explorations of extremophiles. Curr. Opin. Biotechnol. *17*, 250–255.

Raja, R., Hemaiswarya, S., and Rengasamy, R. (2007). Exploitation of *Dunaliella* for beta-carotene production. Appl. Microbiol. Biotechnol. *74*, 517–523.

Rawlings, D.E. (1997). Biomining: Theory, Microbes and Industrial Processes. (Heidelberg, Germany, Springer Academic Press).

Rawlings, D.E. (2005). Characteristics and adaptability of iron- and sulfur-oxidizing microorganisms used for the recovery of metals from minerals and their concentrates. Microb. Cell Fact. *4*, 13.

Rawlings, D.E., and Johnson, D.B. (2006). Biomining (Heidelberg, Germany, Springer Academic Press).

Rawlings, D.E., and Johnson, D.B. (2007). The microbiology of biomining: development and optimization of mineral-oxidizing microbial consortia. Microbiology *153*, 315–324.

Rawlings, D.E., Dew, D., and du Plessis, C. (2003). Biomineralization of metal-containing ores and concentrates. Trends Biotechnol. *21*, 38–44.

Robb, F., Antranikian, G., Grogan, D., and Driessen, A. (2007). Thermophiles:Biology and technology at high temepratures, In Thermophiles:Biology and Technology at High Temepratures, Robb, F., Antranikian, G., Grogan, D., and Driessen, A., eds. (Florida, USA, CRC Press), pp. 3–7

Rodrigues, M.V., Borges, N., Henriques, M., Lamosa, P., Ventura, R., Fernandes, C., Empadinhas, N., Maycock, C., da Costa, M.S., and Santos, H. (2007). Bifunctional CTP:inositol-1-phosphate cytidylyltransferase/ CDP-inositol:inositol-1-phosphate transferase, the key enzyme for di-myo-inositol-phosphate synthesis in several (hyper)thermophiles. J. Bacteriol. *189*, 5405–5412.

Rondon, M.R., August, P.R., Bettermann, A.D., Brady, S.F., Grossman, T.H., Liles, M.R., Loiacono, K.A., Lynch, B.A., MacNeil, I.A., Minor, C., *et al.* (2000). Cloning the soil metagenome: a strategy for accessing the genetic and functional diversity of uncultured microorganisms. Appl. Environ. Microbiol. *66*, 2541–2547.

Rothschild, L.J., and Mancinelli, R.L. (2001). Life in extreme environments. Nature *409*, 1092–1101.

Russell, N.R., and Cowan, D.A. (2006). Methods for the handling of psychrophilic microorganisms, In Extremophiles, volume 35 (Methods in Microbiology), Oren, A., and Rainey, F., eds. (London, UK, Elsevier Academic Press), pp. 371–395.

Schiraldi, C., and De Rosa, M. (2002). The production of biocatalysts and biomolecules from extremophiles. Trends Biotechnol. *20*, 515–521.

Schiraldi, C., Giuliano, M., and De Rosa, M. (2002). Perspectives on biotechnological applications of archaea. Archaea *1*, 75–86.

Serour, E., and Antranikian, G. (2002). Novel thermoactive glucoamylases from the thermoacidophilic Archaea *Thermoplasma acidophilum*, *Picrophilus torridus* and *Picrophilus oshimae*. Antonie Van Leeuwenhoek *81*, 73–83.

Sharp, R.J., Riley, P.W., and White, D. (1992). Hetrotrophic thermophilic *Bacilli*. In Thermophilic Bacteria, Kristjansson, J.K., ed. (Florida, USA, CRC Press), pp.19–51.

Shaw, A.J., Podkaminer, K.K., Desai, S.G., Bardsley, J.S., Rogers, S.R., Thorne, P.G., Hogsett, D.A., and Lynd, L.R. (2008). Metabolic engineering of a thermophilic bacterium to produce ethanol at high yield. Proc. Natl. Acad. Sci. U.S.A. *105*, 13769–13774.

Shaw, A.J., Hogsett, D.A., and Lynd, L.R. (2009). Identification of the [FeFe]-hydrogenase responsible for hydrogen generation in *Thermoanaerobacterium saccharolyticum* and demonstration of increased ethanol yield via hydrogenase knockout. J. Bacteriol. *191*, 6457–6464.

Siddiqui, K.S., and Cavicchioli, R. (2006). Cold-adapted enzymes. Annu. Rev. Biochem. *75*, 403–433.

Simon-Colin, C., Raguenes, G., Cozien, J., and Guezennec, J.G. (2008). *Halomonas profundus* sp. nov., a new PHA-producing bacterium isolated from a deep-sea hydrothermal vent shrimp. J. Appl. Microbiol. *104*, 1425–1432.

Simonato, F., Campanaro, S., Lauro, F.M., Vezzi, A., D'Angelo, M., Vitulo, N., Valle, G., and Bartlett, D.H. (2006). Piezophilic adaptation: a genomic point of view. J. Biotechnol. *126*, 11–25.

Singh, S.B., and Pelaez, F. (2008). Biodiversity, chemical diversity and drug discovery. Prog. Drug Res. *65*, 141, 143–174.

Smith, E.T., Odom, L.D., Awramko, J.A., Chiong, M., and Blamey, J. (2001). Direct electrochemical characterization of hyperthermophilic *Thermococcus celer* metalloenzymes involved in hydrogen production from pyruvate. J. Biol. Inorg. Chem. *6*, 227–231.

Stetter, K.O. (2006a). History of discovery of the first hyperthermophiles. Extremophiles *10*, 357–362.

Stetter, K.O. (2006b). Hyperthermophiles in the history of life. Philos. Trans. R. Soc. Lond. B Biol. Sci. *361*, 1837–1842; discussion 1842–1833.

Taylor, I.N., Brown, R.C., Bycroft, M., King, G., Littlechild, J.A., Lloyd, M.C., Praquin, C., Toogood, H.S., and Taylor, S.J. (2004). Application of thermophilic enzymes in commercial biotransformation processes. Biochem. Soc. Trans. *32*, 290–292.

Taylor, M.P., Esteban, C.D., and Leak, D.J. (2008). Development of a versatile shuttle vector for gene expression in *Geobacillus* spp. Plasmid *60*, 45–52.

Taylor, M.P., Eley, K.L., Martin, S., Tuffin, M.I., Burton, S.G., and Cowan, D.A. (2009). Thermophilic ethanologenesis: future prospects for second-generation bioethanol production. Trends Biotechnol. *27*, 398–405.

Thakore, Y. (2008). Enzymes for industrial applications BIO030E (http://www.bccresearch.com/report/BIO030E.html accessed 3 May 2010).

Thompson, A.H., Studholme, D.J., Green, E.M., and Leak, D.J. (2008). Heterologous expression of pyruvate decarboxylase in *Geobacillus thermoglucosidasius*. Biotechnol. Lett. *30*, 1359–1365.

Thompson, V.S., Schaller, K.D., and Apel, W.A. (2003). Purification and characterization of a novel thermo-alkali-stable catalase from *Thermus brockianus*. Biotechnol. Prog. *19*, 1292–1299.

Tian, K., Lin, J., Liu, X., Liu, Y., and Zhang, C. (2003a). Expression of phosphofructokinase gene from *Escherichia coli* K-12 in obligately autotrophic bacterium *Acidithiobacillus thiooxidans*. Wei Sheng Wu Xue Bao *43*, 592–598.

Tian, K.L., Lin, J.Q., Liu, X.M., Liu, Y., Zhang, C.K., and Yan, W.M. (2003b). Conversion of an obligate autotrophic bacteria to heterotrophic growth: expression of a heterogeneous phosphofructokinase gene in the chemolithotroph *Acidithiobacillus thiooxidans*. Biotechnol. Lett. *25*, 749–754.

Tseng, M., Yap, M., Ratanakhanokchai, K., Kyu, K.L., and Chen, S. (2002). Purification and characterisation of two cellulase free xylanases from an alkaliphilic *Bacillus firmus*. Enzyme Micro. Technol. *30*, 590–595.

Tuffin, I.M., de Groot, P., Deane, S.M., and Rawlings, D.E. (2005). An unusual Tn21-like transposon containing an ars operon is present in highly arsenic-resistant strains of the biomining bacterium *Acidithiobacillus caldus*. Microbiology *151*, 3027–3039.

Tuffin, I.M., Hector, S.B., Deane, S.M., and Rawlings, D.E. (2006). Resistance determinants of a highly arsenic-resistant strain of *Leptospirillum ferriphilum* isolated from a commercial biooxidation tank. Appl. Environ. Microbiol. 72, 2247–2253.

Turner, P., Mamo, G., and Karlsson, E.N. (2007). Potential and utilization of thermophiles and thermostable enzymes in biorefining. Microb. Cell Fact. *6*, 9.

Ulukanli, Z., and Digrak, M. (2002). Alkaliphilic micro-organisms and habitats. Turk. J. Bio. *26*, 181–191.

Valdes, J., Pedroso, I., Quatrini, R., Dodson, R.J., Tettelin, H., Blake, R., second, Eisen, J.A., and Holmes, D.S. (2008). *Acidithiobacillus ferrooxidans* metabolism: from genome sequence to industrial applications. BMC Genomics *9*, 597.

Wagner, I.D., and Wiegel, J. (2008). Diversity of thermophilic anaerobes. Ann. N.Y. Acad. Sci. *1125*, 1–43.

Wang, C.Y., Ng, C.C., Chen, T.W., Wu, S.J., and Shyu, Y.T. (2007). Microbial diversity analysis of former salterns in southern Taiwan by 16S rRNA-based methods. J. Basic. Microbiol. *47*, 525–533.

Williams, J. (2006). Break it down now. Ethanol producer magazine. (http://www.ethanolproducer.com/article.jsp?article_id=319 accessed 3 May 2010)

Yano, J.K., and Poulos, T.L. (2003). New understandings of thermostable and peizostable enzymes. Curr. Opin. Biotechnol. *14*, 360–365.

Yumoto, I. (2002). Bioenergetics of alkaliphilic *Bacillus* spp. J. Biosci. Bioeng. *93*, 342–353.

Zengler, K., Toledo, G., Rappe, M., Elkins, J., Mathur, E.J., Short, J.M., and Keller, M. (2002). Cultivating the uncultured. Proc. Natl. Acad. Sci. U.S.A. *99*, 15681–15686.

Zengler, K., Walcher, M., Clark, G., Haller, I., Toledo, G., Holland, T., Mathur, E.J., Woodnutt, G., Short, J.M., and Keller, M. (2005). High-throughput cultivation of microorganisms using microcapsules. Methods Enzymol. *397*, 124–130.

van Zyl, L.J., van Munster, J.M., and Rawlings, D.E. (2008). Construction of arsB and tetH mutants of the sulfur-oxidizing bacterium *Acidithiobacillus caldus* by marker exchange. Appl. Environ. Microbiol. *74*, 5686–5694.

Ionizing Radiation-resistant Microorganisms

2

Kelley R. Gwin and John R. Battista

Abstract

Of all the phenotypes associated with microorganisms, ionizing radiation resistance – the ability to survive exposure to high-dose γ-radiation – is perhaps the most difficult to rationalize in terms of the natural world. There is no obvious selective advantage to being ionizing radiation resistant on Earth, as average yearly exposures to ionizing radiation from cosmic rays and radioactive decay are extremely low. Yet a significant number of genera exhibit this characteristic, displaying a remarkable capacity to tolerate levels of damage to cellular macromolecules that eradicates other forms of life. We argue that ionizing radiation resistance is an incidental characteristic, an inadvertent consequence of an evolutionary path that permitted these species to survive a selective pressure capable of damaging the cell in a manner similar to that of ionizing radiation. The phylogenetic distribution of ionizing radiation-resistant species argues that these events occurred multiple times during the evolution of the Bacteria and Archaea, suggesting that different mechanisms may mediate ionizing radiation resistance.

Ionizing radiation

High-energy radiation and its effect on cellular macromolecules

Ionizing radiation is a generic term that applies to any form of radiation with sufficient energy to eject one or more orbital electrons from a molecule (von Sonntag, 1987; Spinks and Woods, 1990; Jonah and Rao, 2001). From a biologist's point of view, the localized release of energy resulting from the interaction of ionizing radiation with biological materials is the most significant feature of this form of radiation. Each ionization event dissipates far more energy than is needed to break the bonds that hold macromolecules together.

Ionizing radiation is usually classified as electromagnetic or particulate. Electromagnetic radiation is most simply thought of as photons of light and includes ionizing and non-ionizing forms of radiation. The energy of a photon is inversely related to its wavelength, the shorter the wavelength the higher the energy and the greater the propensity to cause ionization. Historically, studies of the effects of ionizing radiation on biological systems have most often involved X-rays and γ-rays, photons with wavelengths shorter than 10^{-6} cm. X-rays and γ-rays differ in the manner by which they are most frequently generated: X-rays are produced by electrical devices, γ-rays are produced during the decay of radioactive isotopes. From a biologist's perspective, these two types of electromagnetic radiation have identical properties. Particulate radiations include alpha particles, electrons, protons, neutrons, and heavy charged nuclei. Particles produce ionization by depositing energy along relatively short tracks as they pass through matter. Although particulate radiations play an important role in human radiotherapy, their effects on microorganisms are not as well documented as that of high-energy electromagnetic radiation, and will not be discussed in this review.

Electromagnetic radiations do not cause biological damage directly (von Sonntag, 2006). Instead, the energy of the photons is absorbed by the irradiated material and this event results in a transfer of energy to an orbital electron, triggering one of two processes (Spinks and Woods, 1990; Jonah and Rao, 2001). The Compton process predominates when high energy photons such as those generated by the decay of ^{60}Co are involved. In this process, the incident photon transfers part of its energy to a loosely bound electron, which is accelerated as it is ejected from its orbit. The energy of the incident photon is reduced by the amount of energy transferred to the electron during the collision. Depending on the energy loss, this photon may undergo additional Compton processes within the irradiated matter. The photoelectric process results when a tightly bound electron absorbs sufficient energy to be ejected from an inner orbital. The incident photon gives up all its energy during a photoelectric process, accelerating the electron and creating a vacancy in this inner orbital. An electron from an occupied outer orbital will collapse into the vacancy and in the process emit a photon with a wavelength characteristic of the material that has been irradiated. The accelerated electrons (also called fast electrons) generated during the Compton and photoelectric processes are responsible for the ionizations that lead to a biological effect.

Fast electrons have two modes of action in biological systems: direct and indirect. The direct action of fast electrons is easily understood; the fast electron interacts with a biological macromolecule causing an ionization or excitation event that leads to modification and possible inactivation of that macromolecule. If the macromolecule is critical to a cellular function, that function is lost and there is a biological effect. Alternatively, the fast electron can alter the properties of cellular macromolecules indirectly by modifying other molecules in the cell, making them chemically reactive and potentially toxic. In the indirect mode of action, the fast electrons ultimately produce free radicals from the abundant small molecules present in the cell, and the most important of these small molecules is water. Fast electrons easily generate highly reactive hydroxyl radicals from water, and since approximately 80% of a living cell is water much of the energy deposited by fast electrons will result in the formation of hydroxyl radicals. It is estimated that between 75% and 80% of the biological damage caused by high-energy ionizing radiation is mediated indirectly by hydroxyl radical formation. These radicals are electrophilic (a term describing molecules that accept an electron pair, readily forming a covalent bond) and capable of reacting with most biologically important molecules found within a cell at diffusion-controlled rates. Detailed studies of the effect of hydroxyl radicals on carbohydrates, nucleobases and nucleic acids, amino acids and proteins, and lipids have been reported and excellent summaries of this work exist (see von Sonntag, 1987, 2006).

While it may be assumed that all biological macromolecules are subject to modification when exposed to ionizing radiation, there is some controversy as to what constitutes the most critical intracellular target – the macromolecule that when inactivated ultimately results in cell death. Much of the literature suggests that destruction of genomic DNA is responsible for loss of cell viability, and three reasons are most often cited for this assumption: (i) The genome is the largest target in the cell and so is most exposed to the direct and indirect actions of ionizing radiation; (ii) Since the genome is present in relatively low copy number, the number of ionization events needed to eliminate a functional genome is lower than that of other macromolecules that are present in higher numbers; and (iii) The genome is vital to reproductive integrity; all cellular processes rely on a functional genome. There is no other cellular structure that can compensate if the genome is irreparably damaged. While these are powerful arguments, there is a body of evidence (Daly *et al.*, 2004, 2007; Fredrickson *et al.*, 2008; Daly, 2009) developing that suggests focusing on the genome *per se* may be misleading. This work indicates that proteins (possibly proteins necessary for the repair of DNA) are the most critical intracellular targets of ionizing radiation. In essence, the proponents of this work argue that in the presence of these proteins DNA is protected from the actions of ionizing radiation, and that the loss of these proteins permits unchecked DNA damage, leading to cell death.

Ionizing radiation-induced damage to carbohydrates

Carbohydrates, both low molecular weight precursors and polysaccharides, are found in all microorganisms. These compounds are susceptible to modification by hydroxyl radicals, but with the important exception of those modifications that alter the ribose sugars found in nucleic acids there is no evidence that ionizing radiation-induced damage to carbohydrates significantly impacts cell viability (von Sonntag, 1980).

Ionizing radiation-induced damage to DNA

Ionizing radiation results in many types of DNA damage: base damage, single-strand breaks, double-strand breaks, and DNA–DNA or DNA–protein crosslinks. Hydroxyl radicals readily add to C–C and C–N double bonds and can easily abstract protons, modifying a large number of potential sites on the base and sugar moieties of the nucleic acid (Hutchinson, 1985). In DNA, the nucleobase is most often altered, and depending on experimental conditions as many as two dozen products derived from hydroxyl radical-mediated base modifications are detectable by gas chromatography/mass spectrometry. These lesions are subject to base and nucleotide excision repair and so may be reversed, but if unrepaired prior to DNA replication the lesion can result in mutation. Hydroxyl radical-induced changes in the sugar can lead to strand breaks. In the presence of oxygen, abstraction of a proton from position C-4′ or C-5′ of the deoxyribose leads to a sequence of events that breaks the sugar-phosphate backbone, generating strands that may be phosphorylated at their 5′ and 3′ ends (Dizdaroglu *et al.*, 1975a,b; von Sonntag and Schulte-Frohlinde, 1978; von Sonntag, 2006). Fragmentation of the deoxyribose may also occur, leaving a glycolate or phosphoglycolate residue at these breaks.

When these forms of damage are found in one strand of the DNA, most of the lesions are accurately repaired. The random nature of ionizing radiation-induced DNA damage ensures that most damage occurs in one strand, and DNA repair processes utilize the redundant information present in the undamaged strand to template accurate restoration of the genome. However, circumstances can arise that complicate and confound these repair processes. When DNA damage occurs in close proximity on both strands of the DNA, a potentially lethal event may arise (Freifelder and Trumbo, 1969). DNA double-strand breaks can occur when single-strand breaks arise in close proximity on opposite strands, or following excision repair when there is enzymatic digestion of juxtaposed base damage. DNA double-strand breaks can result in significant loss of genetic information and, if not repaired, will prevent replication of the prokaryotic genome (Krasin and Hutchinson, 1977). How an organism deals with double-strand breaks appears to be critical to that organism's capacity to survive exposure to ionizing radiation.

Ionizing radiation-induced damage to proteins

Proteins are highly variable combinations of the 20 naturally occurring amino acids, linked through amide bridges. The varied amino acids react differently with hydroxyl radicals, but the precise reaction mechanisms are poorly understood (Kempner, 1993; Stadtman, 1993). As a consequence, the effect of ionizing radiation on a specific protein is very difficult to predict. Studies of enzyme inactivation by ionizing radiation indicate that multiple processes are occurring in oxygenated solution and it is assumed that these same processes operate *in vivo*:

1. There is evidence of covalently cross-linked peptides post irradiation, presumably arising from radicals forming within the peptide after proton abstraction by hydroxyl radical. For example, aromatic amino acids within peptides have been shown to form biphenyl bridges following hydroxyl radical attack.
2. When irradiated proteins are observed by chromatography or electrophoresis, typically there are a number of discreet bands indicating that hydrolysis of the peptide backbone is occurring.
3. Intact proteins that are not cross-linked can exhibit reduced activity, indicative of an allosteric change brought about by

modification of one or more amino acids that affect the enzyme active site or access to the active site.

Clearly, all of these processes will adversely affect protein function; the relative contribution of each presumably depends on the composition of the protein (Kempner, 1993; Stadtman, 1993).

Ionizing radiation-induced damage to lipids and membranes

The effects of ionizing radiation on bacterial membranes are poorly characterized, but changes to the membrane of a bacterial cell are correlated with increased sensitivity to ionizing radiation (Shenoy *et al.*, 1970). Gamma irradiation induces a number of measurable changes associated with the membrane: there is evidence of lipid peroxidation (Mead, 1976; Pryor, 1978), increased ion permeability, decreased membrane fluidity, and oxidation of sulfhydryl groups found on membrane proteins (von Sonntag, 1987). Changes in fatty acid composition affect susceptibility to lipid peroxidation (polyunsaturated fatty acids being much more easily modified by hydroxyl radicals than saturated fatty acids) and in some studies bacteria with altered fatty acid composition display altered sensitivity to ionizing radiation (Shenoy *et al.*, 1970).

The wealth of information available concerning alterations of the membrane and ionizing radiation-induced lethality is difficult to interpret and the significance of membrane changes in causing lethality remains the subject of a great deal of speculation. Most of the data available concerning how membranes react to ionizing radiation were obtained from model membranes *in vitro*. Living membranes are much more complex structures that function in an environment that contains a number of significant biological defences: natural antioxidants such as α-tocopherol, and protective enzymes, including peroxidases and superoxide dismutase, reduce the potential effect of ionizing radiation-induced radicals (von Sonntag, 1987). Among bacteria that are relatively sensitive to ionizing radiation such as *Escherichia coli*, the dose of ionizing radiation needed to cause lethality is lower than that needed to initiate membrane changes observed in model systems. In such bacteria, the changes in lipid composition reported *in*

Box 2.1 The D_{37} dose

The D_{37} dose is derived from the mathematics associated with simple target theory. If N_0 is the number of organisms initially present and N is the number of organisms surviving after the dose (D), then for every increment of dose (dD) administered, N will be decreased by dN in an amount which is proportional to N. Mathematically this is represented as:

$$-dN/dD = kN$$

an expression that integrates to

$$N/N_0 = e^{-kD}$$

If the fraction of surviving cells is S, then

$$S = e^{-kD}$$

Assuming that the distribution of hits caused by ionizing radiation follows a Poisson distribution, then it can be shown that the proportionality constant k is equal to $1/D_0$, where D_0 is the dose that gives an average of one hit per target. Since e^{-1} is 0.37 when the dose D has been given such that $D/D_0 = 1$, this dose is called the D_{37}.

vitro will occur in living cells, but it is not obvious that these changes affect viability (Shenoy *et al.*, 1970).

The phenomenon of ionizing radiation resistance

The radioresistance of microorganisms is typically compared by determining D_{37} or D_{10} doses, the absorbed doses at which 37% and 10% of the cells in the irradiated population survive, respectively (Spinks and Woods, 1990; Jonah and Rao, 2001). At the D_{37} dose, each cell on average has experienced a lethal event (see Box 2.1); those that survive are balanced by other cells with two or more lethal events. Absorbed dose is usually reported in grays (Gy) or kilo-grays (kGy); 1 gray corresponds to energy absorption of 10,000 erg/g. The D_{37} dose for almost all actively growing, aerobic, vegetative microorganisms lies between 0.5 and 10 kGy (Silverman and Sinskey, 1968). As indicated by this range of inactivating doses, microorganisms vary greatly in their response to ionizing radiation.

Sensitivity to ionizing radiation is strongly influenced by a number of extrinsic and intrinsic factors (Block, 1983). The extrinsic or environmental factors involved are well characterized and their effects on radioresistance are predictable and easily controlled. Extrinsic factors explain differences in ionizing radiation resistance observed in a single species. A given species does not exhibit the same sensitivity to ionizing radiation at all times. In other words, the ionizing radiation resistance of an individual is variable and can be manipulated by altering extrinsic factors. The most significant of the extrinsic factors relate to growth conditions. In general, actively growing cultures (those in exponential phase growth) are more sensitive to ionizing radiation. Anaerobic or anoxic cultures, including cultures in stationary phase, are naturally more resistant than aerobic cultures; the presence of oxygen potentiates ionizing radiation-induced killing. Oxygen is paramagnetic and readily reacts with carbon-centred radicals formed during gamma-irradiation of macromolecules, generating peroxyl radicals that can mediate a number of secondary reactions, including cleavage of DNA (Dizdaroglu *et al.*, 1975a,b; von Sonntag, 2006). In addition, desiccating a normally sensitive cell can increase its D_{37} dose dramatically, making it appear more ionizing radiation resistant than it would be in a hydrated state, provided a species can survive drying. As indicated previously, the homolytic cleavage of water following exposure to ionizing radiation produces hydroxyl radicals, which are responsible for most of the indirect effects of ionizing radiation *in vivo*. Desiccation eliminates most of the water from the cell. Intrinsic factors are heritable differences between species that provide a measure of protection against ionizing radiation that is not easily predicted and may be considered species-specific. Intrinsic factors become obvious when irradiated species are compared under identical growth conditions. For example, spores, which form during a developmental process that extrudes most of the water found in the cell, are usually much more resistant to ionizing radiation than non-spore forming bacteria (Setlow, 1992). The low water content of spores avoids damage that would arise if water were present. Thus, the ability to form a spore is an intrinsic characteristic that influences the radioresistance of a given species.

There are a number of microorganisms that do not form spores that are naturally more resistant to the lethal effects of ionizing radiation when compared with the majority of other vegetative species. This phenotype was originally identified in 1956 with the discovery of *Micrococcus* (now *Deinococcus*) *radiodurans* (Anderson *et al.*, 1956) in tins of canned irradiated meat, but a number of other similar species have been identified in the past 50 years (Cox and Battista, 2005). These vegetative radioresistant species includes the most ionizing radiation-resistant organisms ever described (Ferreira *et al.*, 1999), but unlike the straightforward explanation for the ionizing radiation resistance of spores, the reasons that this group of microorganisms is highly resistant remain enigmatic. Most of these species are poorly characterized, but studies of *D. radiodurans* suggest that they may express novel mechanisms of tolerance to the effects of ionizing radiation. Fig. 2.1 illustrates what is arguably one of the most remarkable phenomena in nature – the recovery of the *D. radiodurans* genome following exposure to 5000 Gy of γ-radiation. This dose reduces the

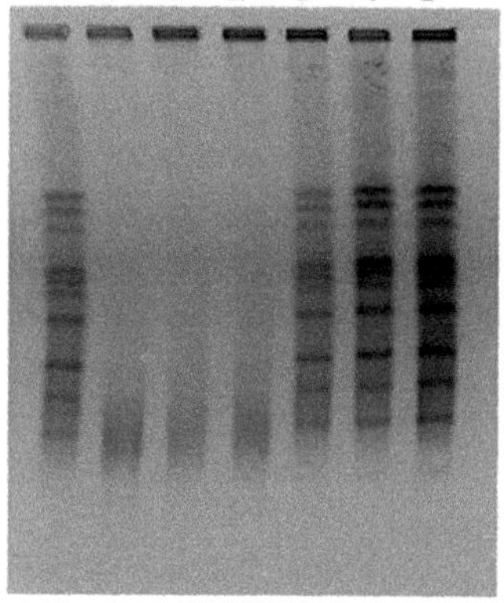

Figure 2.1 *NotI* digestion of *D. radiodurans* R1 genome after exposure to 5000 Gy of γ radiation, separated by pulsed field gel electrophoresis. Lanes correspond to the un-irradiated control (C), immediately after irradiation (0), and 1, 2, 3, 4, and 24 hours post irradiation (lanes 1–5).

genome into fragments of between 10,000 and 50,000 bp immediately post-irradiation (Fig. 2.1, lane marked 0), but within three hours the genome is restored to full size and function in a manner that avoids lethality and mutation. The assumption has been that other ionizing radiation-resistant species utilize equally effective mechanisms to survive ionizing radiation, and the majority of investigations of extreme radioresistance have focused on defining those mechanisms. The remainder of this review focuses on these species, and will explore the questions of how species protect themselves against exposures that sterilize most bacterial cultures and why ionizing radiation resistance evolved on Earth.

Phylogenetic distribution of ionizing radiation resistance

For the purposes of this review, we arbitrarily define ionizing radiation resistance as a characteristic of species that exhibit a D_{10} value of greater than or equal to 1000 Gy during vegetative growth. We recognize that any attempt to define ionizing radiation resistance is problematic in that this characteristic is highly variable and we acknowledge that some species that are radioresistant will be left out of this discussion. We settled on this definition because it delineates the most charismatic and mysterious of the ionizing radiation-resistant species known. Collectively, this group illustrates the most extreme form of ionizing radiation resistance.

Based on our definition of ionizing radiation resistance and a search of the literature, there are nine lineages (nine phyla with fourteen genera; Table 2.1) that exhibit ionizing radiation resistance within the domains *Archaea* and *Bacteria*. After inspection of this list, two observations become obvious.

1. There is a large range of sensitivities to ionizing radiation among these species. In the absence of evidence to the contrary, these differences could suggest that multiple mechanisms are responsible for ionizing radiation resistance or that the effectiveness of a universal mechanism is strongly influenced by one or more species-specific factors.
2. There is no evidence of vertical inheritance among the genera we have classified as ionizing radiation resistant. For example, despite being part of a deeply branching lineage, the genus *Deinococcus* does not appear to have given rise to any of the more recent branches with radioresistant genera (Fig. 2.2). In addition, all of the genera listed in Table 2.1 include species that are not ionizing radiation resistant. This pattern of descent is indicative of either convergent evolution or horizontal transfer of the trait, suggesting that the ionizing radiation-resistant phenotype may have arisen on multiple occasions during microbial evolution.

The habitats of ionizing radiation-resistant species

Ionizing radiation-resistant organisms have been identified in a wide range of environments. Species have been isolated from soil (Murray, 1992), thermal springs (Ferreira *et al.*, 1997, 1999), hydrothermal vents (Jolivet *et al.*, 2003, 2004), animal faeces (Ito *et al.*, 1983), processed meats (Davis *et al.*, 1963; Grant and Patterson, 1989), dried foods (Lewis, 1971), room dust (Christensen and Kristensen, 1981), medical instruments (Christensen and Kristensen, 1981), textiles (Kristensen and Christensen, 1981; Christensen *et al.*, 1991a,b),

Table 2.1 Ionizing radiation resistance[1] among the *Archaea* and *Bacteria*

Phylum	Genus	Species	Estimated D_{10} value (Gy)[2]	Reference
Archaea				
Euryarchaeota	*Thermococcus*	*T. gammatolerans*	6500	Jolivet *et al.* (2003)
		T. stetteri	6500	Jolivet *et al.* (2003)
		T. marinus	4500	Jolivet *et al.* (2004)
		T. radiotolerans	5200	Jolivet *et al.* (2004)
	Pyrococcus	*P. furiosus*	3000	DiRuggiero *et al.* (1997)
		P. abyssi	3000	Jolivet *et al.* (2003)
	Halobacterium	strain NRC-1	5000	Kottemann *et al.* (2005)
Bacteria				
Deinococcus–Thermus	*Deinococcus*	*D. aerius*	4900	Yang *et al.* (2009)
		D. aetherius	8000	Yang *et al.* (2010)
		D. apachensis	–	Rainey *et al.* (2005)
		D. aquaticus	2800	Im *et al.* (2008)
		D. aquiradiocola	3000	Asker *et al.* (2009)
		D. caeni	2800	Im *et al.* (2008)
		D. deserti	8000	de Groot *et al.* (2005)
		D. ficus	–	Lai *et al.* (2006)
		D. geothermalis	6100	Ferreira *et al.* (1997)
		D. gobiensis	8000	Yuan *et al.* (2009)
		D. grandis	3600	Oyaiza *et al.* (1987)
		D. hohokamensis	–	Rainey *et al.* (2005)
		D. hopiensis	–	Rainey *et al.* (2005)
		D. maricopensis	6000	Rainey *et al.* (2005)
		D. misasensis	2000	Asker *et al.* (2008)
		D. murrayi	9100	Ferreira *et al.* (1997)
		D. navajonensis	–	Rainey *et al.* (2005)
		D. papagonensis	–	Rainey *et al.* (2005)
		D. pimensis	–	Rainey *et al.* (2005)
		D. piscis	7400	Shashidhar and Bandekar (2009)
		D. proteolyticus	10,000	Davis *et al.* (1963)
		D. radiodurans	10,000	Anderson *et al.* (1956)
		D. radiophilus	5000	Lewis (1971)
		D. radiopugnans	12,000	Battista and Rainey (2001)
		D. roseus	2000	Asker *et al.* (2008)
		D. sonorensis	–	Rainey *et al.* (2005)
		D. xinjiangensis	<5000	Peng *et al.* (2009)
		D. yavapaiensis	–	Rainey *et al.* (2005)
		D. yunweiensis	–	Zhang *et al.* (2007)
	Truepera	*T. radiovictrix*	–	Albuquerque *et al.* (2005)

Table 2.1 continued

Phylum	Genus	Species	Estimated D_{10} value (Gy)[2]	Reference
Actinobacter	*Kineococcus*	*K. aurantiacus*	2000	Phillips *et al.* (2002)
		K. radiotolerans	2000	Phillips *et al.* (2002)
	Rubrobacter	*R. radiotolerans*	11,000	Ferreira *et al.* (1999)
		R. xylanophilus	5500	Ferreira *et al.* (1999)
	Kocuria	*K. rosea*	2000	Rainey *et al.* (1997)
Alphaproteobacteria	*Methylobacterium*	*M. extorquens*	2100	Romanovskaya and Rokitko (2002)
		M. mesophilicum	2200	Romanovskaya and Rokitko (2002)
		M. radiotolerans	1000	Green and Bousfield (1983); Ito and Iizuka (1971)
Gammaproteobacteria	*Acinetobacter*	*A. radioresistens*	2.000	Nishimura and Uchida (1994)
Firmicutes	*Bacillus*	*B. subtilis*	4400[3]	Romanovskaya and Rokitko (2002)
Cyanobacteria	*Chroococcidiopsis*	Strain 015	4000	Billi *et al.* (2000)
		Strain 057	–	–
		Strain 101	–	–
		Strain 171	–	–
Planctomycetes	*Gemmata*	*G. obscuriglobus*	>1000	Lieber *et al.* (2009)
Cytophaga–Flexibacter–Bacteriodes	*Hymenobacter*	*H. actinosclerus*	3500	Collins *et al.* (2000)

[1]Species are included in Table 2.1 only if evidence of substantial ionizing radiation resistance is offered in the associated reference. This evidence takes different forms. In many situations a survival curve was provided. If the strain was isolated from environmental samples exposed to greater than 15,000 Gy, the species was listed whether or not a survival curve was available. A number of *Deinococcus* spp. were excluded from the table, because no evidence of ionizing radiation resistance is, at present, available.

[2]The D_{10} value defines the dose needed to eradicate 90% of the irradiated population. This value is dependent on growth conditions and can vary between studies. The values listed, if not explicitly calculated in the reference given, are estimated from the data provided in that reference.

[3]It is not obvious that the D_{10} estimate for *B. subtilis* was accomplished in a manner that prevents spores from contributing to the survival of the culture.

sewage (Ito and Iizuka, 1971; Ito *et al.*, 1983) and effluent from paper mills (Vaisanen *et al.*, 1998). For many years, the isolation of ionizing radiation-resistant species was the result of fortuitous accidents, byproducts of attempts to use ionizing radiation to sterilize a variety of materials, but the natural habitat of these species remained obscure. Several systematic searches for ionizing radiation-resistant species during the past two decades have revealed two ecological niches in which these species could be reliably identified: dry soils (Rainey *et al.*, 2005) and high-temperature aqueous environments (Ferreira *et al.*, 1997, 1999; Jolivet *et al.*, 2003, 2004).

A number of environments have been examined for the presence of ionizing radiation-resistant species using techniques that include selective (involving the administration of high-dose ionizing radiation to the sample) (Rainey *et al.*, 2005) and non-selective methods (Callegan *et al.*, 2008). Collectively, these studies reveal that hydrothermal vents, geothermal springs, and

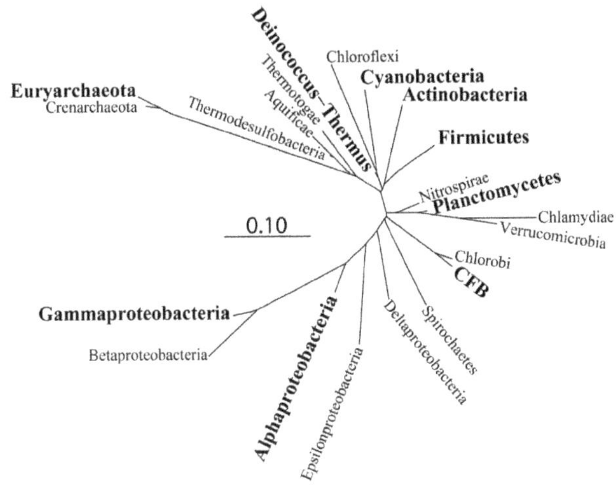

Figure 2.2 A 16S rRNA gene sequence-based phylogeny of the main lineages of the domain *Archaea* and *Bacteria*. The branches with names in bold are those in which ionizing radiation-resistant taxa have been described. The tree was constructed using software available through The ARB Project (http://www.arb-home.de/). CFB is an abbreviation for *Cytophaga–Flexibacter–Bacteroides*. The scale bar is inferred nucleotide substitutions per 100 nucleotides.

arid soils include a higher percentage of ionizing radiation-resistant species than other environments. All of the hyperthermophilic archaea that have been identified as ionizing radiation resistant were isolated from deep-sea hydrothermal vents (DiRuggiero *et al.*, 1997; Jolivet *et al.*, 2003, 2004). Species of *Deinococcus* (Ferreira *et al.*, 1997) and *Trupera* (Albuquerque *et al.*, 2005) have been identified during surveys of hot spring sites in Portugal, Italy, and the Azores. Arid soils are rich in species that survive exposure to doses above 5000 Gy (Rainey *et al.*, 2005). In one large-scale study, ionizing radiation-resistant representatives of eleven distinct genus-level taxonomic groups were found in soil samples from the Sonoran desert in southern Arizona in the United States. The authors compared this environment to forest soil in southern Louisiana, USA, and concluded that ionizing radiation resistance is a phenotype more typically found associated with a dry environment, suggesting that the capacity to tolerate ionizing radiation enhances the species' ability to survive in the arid environment (Rainey *et al.*, 2005).

Assuming that their native habitats are restricted to the environments listed, their widespread distribution argues that they are easily dispersed. It has been suggested that in a desiccated state these microorganisms can be moved on the wind and when deposited in a favourable location, colonize that environment (Battista and Rainey, 2001). Direct evidence supporting this scenario for dispersal has not been provided.

Proposed mechanisms of ionizing radiation resistance

The *Deinococcaceae* (Battista and Rainey, 2001) are a bacterial family that currently consists of 44 validly named species (http://www.bacterio.cict.fr/d/deinococcus.html). The family is very deeply rooted in the domain *Bacteria* and part of the *Deinococcus-Thermus* phylum, which is divided into the orders *Deinococcales* (genera: *Deinococcus* and *Trupera*) and *Thermales* (genera: *Thermus, Marinithermus, Meiothermus, Oceanithermus,* and *Vulcanithermus*). In general, the *Deinococcales* exhibit ionizing radiation resistance, whereas the *Thermales* do not exhibit this characteristic. *Deinococcus radiodurans* was the first ionizing radiation-resistant bacterium identified (Anderson *et al.*, 1956) and at present it is the most extensively studied radioresistant microorganism. Fortuitously, *D. radiodurans* is relatively easy to grow and can be genetically manipulated through natural transformation, properties that make it a popular choice for investigation. Other species, even among the *Deinococcales,* receive very little attention relative to *D. radiodurans.* Unless otherwise stated, discussion of specific mechanisms of ionizing radiation resistance is based on work done with *D. radiodurans.* Experimental evidence supports the idea that the recovery of this species from substantial DNA damage relies on both passive features of deinococcal physiology and repair enzymes. We do not attempt to assess the relative contribution of these different mechanisms to recovery from DNA damage. Readers are referred to the recent review by Blasius *et al.* (2008) for a more detailed treatment of this topic. Instead we summarize the relevant information with the intent of giving the reader a sense of the unique characteristics associated with ionizing radiation resistance in *D. radiodurans.* The studies described are broken into two broad categories: genome sequence and sequence-enabled analysis, and experimental (genetic, biochemical, and physiological) studies.

We note that there is a tendency on the part of those who study ionizing radiation resistance to generalize the results found in *D. radiodurans,* suggesting that once the mechanism responsible for this species' radioresistance is characterized that mechanism will define other radioresistant species as well. While such a scenario may be true, it is certainly not guaranteed. Given the number of ionizing radiation-resistant species that appear to have arisen independently, it is prudent to assume that mechanisms identified as mediating ionizing radiation resistance in one species are not necessarily responsible for radioresistance in a different species until proof of their involvement is established.

Genome sequence analyses of *Deinococcus* species

The genomes of three species of *Deinococcus* have been sequenced to date: *D. radiodurans* (White *et al.*, 1999; Makarova *et al.*, 2001), *D. geothermalis* (Makarova *et al.*, 2007), and *D. deserti* (de Groot *et al.*, 2009). Each species displays comparable ionizing radiation resistance. All three genomes are between 3 and 4 Mbp and all are segmented, consisting of three or four genetic elements (two chromosomes, a megaplasmid and a plasmid). Each organism is predicted to encode between 3000 and 3500 gene products.

The details of the genome sequences of these species, in the absence of other experimental work, have not provided a great deal of insight into the mechanisms of ionizing radiation resistance. However, there are three points that we feel provide context to the discussion of mechanism that follows:

1. Assuming that each species of *Deinococcus* utilizes the same mechanisms for ionizing radiation resistance, investigators have attempted to narrow the list of genes involved in the process by identifying loci that are unique to the genus and found in all three species. This analysis narrowed the candidates to 230 putative gene products (de Groot *et al.*, 2009) that at present are distinctive, having no homologue in greater than 1000 other prokaryotes with sequenced genomes. Most of these unique loci encode proteins of unknown function.
2. There is nothing obvious from inspection and comparison of these deinococcal genomes that suggest mechanisms responsible for ionizing radiation resistance. To the contrary, these species seem to encode most of the conventional complement of identifiable prokaryotic proteins. Comparison of the *Deinococcus* genomes with the genomes of *Thermus* species (Omelchenko *et al.*, 2005), the closest non-radioresistant relative of *Deinococcus* failed to reveal any insights into the phenotype, and no overlap was found among gene products encoded by *Deinococcus* species and other radioresistant species (e.g. *Thermococcus gammatolerans* (Zivanovic *et al.*, 2009). If specific proteins are responsible for the ionizing radiation resistance of *Deinococcus,* the change in protein function is either too subtle to be detected by examining sequence data or the proteins responsible mediate an uncharacterized function. This function could be unique to the Deinococci

(and other radioresistant bacteria), catalysing a previously unknown activity that enhances the survival of these species. Alternatively, the unique proteins of these species might be modifying repair mechanisms that are known to participate in recovery following ionizing radiation-induced damage; the cell would use repair systems common to almost all species, but those repair systems are made more effective in the presence of the unique proteins.

3 The *Deinococcaceae* are very distant relatives of most *Bacteria* and the details of what we know of other species do not necessarily translate to the biochemistry and physiology of this species. In this circumstance, gene prediction based on sequence similarity should always be viewed with healthy scepticism. Such predictions are hypotheses and until they are directly tested, it is prudent to avoid assuming a protein performs the same role in every species that expresses a homologue of that protein. When the genome sequence of *D. radiodurans* became available, it was clear that the species encoded RecA and two LexA-like proteins (White *et al.*, 1999). In *E. coli*, the multi-functional RecA protein plays an integral role in the organism's ability to respond to stress. RecA becomes activated as it recognizes signs of DNA damage (ssDNA and dNTPs) and once activated interacts with and stimulates the proteolytic cleavage of the LexA repressor (Walker, 1984, 1985). When cleaved, LexA cannot interact with its normal binding sites in the genome, de-repressing transcription of a number of loci in what is known as the SOS response. In *E. coli*, the SOS regulon consists of approximately 40 genes involved in DNA repair and replication including *recA* and *lexA*, and is typified by an auto-regulated feedback loop. If you assumed (as some did) that *D. radiodurans* expresses an SOS-like response involving LexA (White *et al.*, 1999), this assumption would be incorrect. LexA plays no role in response to DNA damage in this species (Narumi *et al.*, 2001; Bonacossa de Almeida *et al.*, 2002), and its exact function has not been defined.

Genome-enabled studies of gene expression and protein synthesis

The availability of genome sequence has permitted the study of global gene expression in *D. radiodurans* using DNA microarray technology. A microarray is a series of microscopic spots placed in an ordered pattern on a solid support. Each spot consists of specific DNA sequences, known as probes that are typically short oligonucleotides. In theory, a *D. radiodurans* microarray will contain probes that correspond to each open reading frame in the *D. radiodurans* genome. Probes are used to hybridize to cDNA made from mRNA isolated from the cell. The cDNA is labelled with a fluorophore, which can be detected when bound to the probe, and so microarray studies allow the investigator to follow differential gene expression *in vivo*. Gene expression under normal growth conditions may be compared with cultures recovering from a number of stresses including ionizing radiation. Liu *et al.* (2003) were the first to publish a study that followed global gene expression in *D. radiodurans* in response to ionizing radiation. Stationary phase cultures of *D. radiodurans* were exposed to 15 kGy, and allowed to recover over a 24-hour time course. The group reported that 832 genes were induced and 451 genes were repressed two-fold or greater at one or more of nine time points during the recovery period. These data were very difficult to interpret, as there is not a perfect correlation between genes that are essential for a given phenotype such as ionizing radiation resistance, and those genes that appear to be expressed in response to that stimulus (Winzeler *et al.*, 1999; Badarinarayana *et al.*, 2001). The very large number of genes reported to change post-irradiation made it impossible to point to a particular gene or set of genes and link them directly to radioresistance.

Tanaka *et al.* (2004) characterized the transcriptional response of *D. radiodurans* R1 to 3000 Gy, a sublethal dose of ionizing radiation, during exponential phase growth, and compared this expression profile with R1 cultures recovering from desiccation. Since the process of desiccation and re-hydration introduces DNA damage, it was assumed that some of the proteins needed to repair ionizing radiation-induced damage, including DNA double-strand breaks, would be identical to proteins used to mend DNA damage introduced following desiccation. In contrast to the study by Liu *et al.* (2003), they found that the response of *D. radiodurans* to ionizing radiation and desiccation was not characterized by large changes in the transcriptional profile. Their studies defined a subset of 32 genes that were induced in response to both stresses, but the identity of the genes was unexpected. Only four genes (*recA, ruvB, uvrA,* and *uvrB*) were identified that encode proteins previously associated with DNA repair in other species, indicating that increased transcription of most known DNA repair proteins is not necessary for the radioresistance of *D. radiodurans*. Instead, the five most highly induced transcripts generated in response to each stress were derived from genes encoding proteins whose functions are poorly understood. Each of these five genes, designated *ddrA, ddrB, ddrC, ddrD,* and *pprA*, were replaced with different drug cassettes and all possible combinations of double mutants of these alleles were examined for the ability to survive exposure to ionizing radiation at doses to 13 kGy (Tanaka *et al.*, 2004). Analysis of this collection of mutants established that the five gene products contribute to ionizing radiation resistance in *D. radiodurans*, and that their activities are redundant.

The proteomes of *D. radiodurans* (Lipton *et al.*, 2002, 2006; Zhang *et al.*, 2005) and *D. deserti* (de Groot *et al.*, 2009) have also been studied, but these analyses have been used primarily as a method of verifying the assignment of genes in the genome sequence, thus providing no insight into the mechanism of radioresistance. Zhang *et al.* (2005) did look at protein synthesis in *D. radiodurans* 1 hour after exposure to 1000 Gy. Using mass spectrometry to analyse spots removed from a two-dimensional gel, they reported significant increases in 21 identifiable proteins. There was very little overlap with the gene expression data reported by Tanaka *et al.* (2004) but elevated amounts of PprA were reported. It is not known how these changes in the proteome relate to ionizing radiation resistance.

As indicated by the preceding discussion, analyses of the genome sequence of the deinococci have not identified the mechanisms of radioresistance in these species, but these studies have greatly facilitated experimental studies of this group during the past decade. Genetic manipulation is simplified by knowing the sequence of every open reading frame, allowing researchers to clone these sequences for purposes of protein expression or gene inactivation. In addition, analyses of the global transcriptome and proteome have been useful in identifying proteins that are potentially involved in the cell's response to ionizing radiation and other stresses. These genome analyses have led to specific testable predictions about the biochemistry and molecular biology of ionizing radiation resistance, allowing for more rapid experimental progress.

Experimental studies in *D. radiodurans*: avoiding cellular damage

Genome copy number

D. radiodurans is multi-genomic. Depending on growth phase the species contains from four to ten genome equivalents (Hansen, 1978; Harsojo *et al.*, 1981), and the presence of redundant genetic information has been invoked as a possible explanation for radioresistance (Battista, 1997; Battista *et al.*, 1999). Increasing numbers of genome copies are known to enhance ionizing radiation resistance in prokaryotic and eukaryotic cells (Mortimer, 1958; Krasin and Hutchinson, 1977). Genome redundancy could protect the cell in two ways. First, it is more difficult to inactive critical targets in a species with multiple genomes. The probability of inactivating all copies of a given locus is equal to p^N, where p is the probability of inactivating a gene in a single copy of the genome and N is the number of genome copies present (Cox and Battista, 2005). Clearly, as the number of genome copies increases it is less and less likely that a specific target will be inactivated. Second, the excess genetic information serves as a reservoir of information that is used by DNA repair systems to correctly repair DNA damage, restoring regions lost to deletions caused by double-strand breaks (Krasin and Hutchinson, 1977). Pathways that function to repair deletions in one genome copy rely on duplicating the same region from another genome copy that remained undamaged.

The first possibility seems unlikely to contribute significantly to ionizing radiation resistance.

Many bacteria contain more than one genome copy, particularly during exponential growth. For example, *E. coli* contains between 5 and 18 genome equivalents during exponential-phase growth (Akerlund *et al.*, 1995), and *Azotobacter vinelandii* can accumulate more than 100 genome copies (Maldonado *et al.*, 1994). Neither species is considered ionizing radiation resistant, so genome redundancy alone cannot account for radioresistance. Additionally, in a recent study by Harris *et al.* (2009), increases in genome copy number were not observed when *E. coli* K12 was made 3000-fold more ionizing radiation resistant through natural selection.

Possessing a large number of genome copies cannot, in and of itself, explain ionizing radiation resistance, but it is difficult to envisage how any species can repair DNA double-strand breaks without more than one genome, so it seems probable that genome redundancy is necessary for genome repair. Oddly, this idea has not been directly tested in *D. radiodurans*, and the contribution of genome redundancy to radioresistance in this species (relative to other mechanisms) is unclear.

Nucleoid organization

Stationary phase *D. radiodurans* cells are distinguished by nucleoids that have a distinctive ring-like shape (Levin-Zaidman *et al.*, 2003). The genetic material seems to be condensed, and a hypothesis has been put forward that suggests that this structure contributes to ionizing radiation resistance (Levin-Zaidman *et al.*, 2003; Zimmerman and Battista, 2005). The original proponents of the idea suggested that the structure passively contributed to radioresistance by constraining the release of genome fragments produced following irradiation; holding these fragments in place would avoid loss of genetic information and facilitate rapid repair. The concept was appealing in its simplicity, but subsequent investigations showed (a) that the prevalence of ring-like nucleoids under different growth conditions does not always correlate with ionizing-radiation resistance (Daly *et al.*, 2004); (b) that not all radioresistant species, including at least two species of *Deinococcus*, exhibit ring-like nucleoids (Zimmerman and Battista, 2005); and (c) that cryoelectron microscopy failed to detect an ordered pattern in the genomic DNA of *D. radiodurans* that was consistent with what would need to be present to constraint the movement of fragments (Eltsov and Dubochet, 2005, 2006). Until substantive evidence of the role of nucleoid structure in the ionizing radiation resistance of *D. radiodurans* is found, it should be assumed that this structure does not directly contribute to the phenotype.

Antioxidants: protection of proteins

D. radiodurans R1 contains high intracellular levels of manganese (II) ions [Mn(II)] (Leibowitz *et al.*, 1976; Daly *et al.*, 2004), and the species has been shown to concentrate Mn against a gradient, suggesting that Mn(II) plays an important role in *D. radiodurans* physiology. Daly and colleagues were the first to comprehensively explore the potential involvement of this ion in *D. radiodurans'* radioresistance. In the first of two important papers (Daly *et al.*, 2004), they established (a) that *D. radiodurans'* resistance to ionizing radiation decreases if cultures are starved of Mn(II); and (b) that the numbers of DNA double-strand breaks formed are the same for a specific dose of ionizing radiation both in the presence or absence of Mn(II). Since there are some species with high intracellular Mn(II) concentrations that remain sensitive to ionizing radiation (for example *Neisseria gonorrhoeae;* Mehr and Seifert, 1998), these investigators correctly surmised that high Mn(II) had to be coupled with low intracellular concentrations of iron (II) ions [Fe(II)] to affect radioresistance. They demonstrated that the radioresistant species they examined had high Mn(II)/Fe ratios, and that radiosensitive species did not. They argued that a high Mn(II)/Fe ratio protects the cell from oxidative damage post irradiation by suppressing oxygen radical-mediated reactions, by scavenging oxygen radicals before they have the opportunity to react with cellular macromolecules. Since they had clearly shown that a high Mn(II)/Fe ratio does not prevent DNA damage, they were forced to conclude that the most critical features of radioresistance were independent of DNA damage. This is a remarkable conclusion in that it questions the dogma established by over 50 years of study in radiobiology (Daly, 2009). Damage to DNA has always been considered the reason cells die post irradiation (Ward, 1975; Hutchinson, 1985), and consequently the impact of ionizing radiation on other macromolecules was considered collateral damage. While this study did not rule out the importance of DNA damage, it did indicate that differences in damage to DNA were not the explanation for differences in radiotolerance observed between radiosensitive and radioresistant organisms. Something other than DNA was affected by the alteration in Mn(II), and whatever that something was, it appeared responsible for ionizing radiation resistance.

In their second study (Daly *et al.*, 2007), the Daly group noted that levels of protein oxidation differed significantly between species that were sensitive and resistant to ionizing radiation. Radiosensitive species exhibited higher levels of protein oxidation post-irradiation relative to radioresistant species, and this difference was correlated with the Mn(II)/Fe ratio. The authors hypothesize that the presumed antioxidant effect of a high Mn(II)/Fe ratio shields proteins critical to survival from inactivation following irradiation, making these proteins more effective in carrying out their function. In other words, protein and not DNA was the critical target following ionizing radiation-induced damage. This concept is most easily illustrated by considering the association between DNA and DNA repair proteins. DNA cannot be repaired if these proteins are not functional. Since repairing DNA is universally considered essential to the cell's viability, it follows that anything that enhances that repair is also essential to viability. Desiccation-resistant cells also exhibit the same patterns with respect to Mn(II)/Fe ratio and protein oxidation (Fredrickson *et al.*, 2008) that are associated with radioresistant species, echoing suggestions (see below) that acquisition of desiccation tolerance may lead inadvertently to ionizing radiation resistance.

As might be expected, this hypothesis is not without its critics. First, the entire argument relies on the assumption that reducing Mn(II) sensitizes *D. radiodurans* to ionizing radiation. However, this assumption may be problematic, because there are indications that manganese is essential for the viability of *D. radiodurans* under normal growth conditions (Chou and Tan, 1990). If this is true, then interpretations of studies assessing viability under conditions of reduced Mn(II) post-irradiation are at best equivocal. No attempt has been made to assess the essentiality of Mn in this species. Second, there is the potential for proteins to be continuously replenished through transcription and translation. Why would the cell not simply replace the proteins inactivated through oxidation? To our knowledge this question has not been evaluated experimentally. Third, the proposed model suggests that the altered Mn(II)/Fe ratio reduces protein oxidation through a passive mechanism that generically shields proteins from damage, but this scenario is difficult to reconcile with the observation that DNA is not protected. There is evidence that Mn(II) is sequestered in the nucleoid and along cell septa (Daly *et al.*, 2004), which may imply that that there is differential shielding and that specific compartments and their associated proteins are protected. Presumably, these compartments would not contain DNA. Unfortunately, the identity of the most important proteins affected by ionizing radiation-induced oxidation is unknown. Since all measurements of protein oxidation to date have been made on extracts from whole cells, there is no way of knowing whether specific areas of the cell have been affected relative to other areas. In addition, it is difficult to establish whether the lower levels of protein oxidation observed in radioresistant species is due to passive protection or to removal by enzymes that target and repair or remove oxidized proteins (Blasius *et al.*, 2008). A number of gene products that affect the fate of proteins are induced in response to gamma irradiation in *D. radiodurans*. Among these are enzymes that restore oxidized methionine residues, chaperones that facilitate protein re-folding, and a number of proteases that would degrade oxidized proteins (Liu *et al.*, 2003; Tanaka *et al.*, 2004). There is a formal possibility that these processes acting alone or in concert with the reported anti-oxidant effect of the Mn(II)/Fe ratio could account for the difference in protein oxidation observed.

Experimental studies in *D. radiodurans*: repairing genomic DNA

Recent evidence has shown that DNA double-strand breaks (and presumably other types of

DNA damage) form at the same rate in *E. coli* and *D. radiodurans* if cultures are irradiated under identical conditions. Until the work by Daly and colleagues (Daly *et al.*, 2004, 2007), highly efficient DNA repair was considered the most probable (and possibly only) explanation for the ionizing radiation resistance of *D. radiodurans* (Moseley, 1983; Minton, 1996; Battista, 1997; Battista *et al.*, 1999). The cell's genomic DNA suffers damage and that damage must be repaired if cells are to survive. Treating cultures with actinomycin D or chloramphenicol (compounds that stop mRNA and protein synthesis, respectively) prior to irradiation sensitizes this species to killing (Dean *et al.*, 1970; Driedger *et al.*, 1970). Obviously, *D. radiodurans* requires newly synthesized proteins when recovering from ionizing radiation-induced DNA damage, and by analogy to other bacterial species such as *E. coli*, it has long been assumed that DNA repair proteins are produced post-irradiation.

The survival curves of actively growing cultures of *D. radiodurans* R1 are characterized by a shoulder of resistance in which there is no measurable loss of viability to 5000 Gy (Moseley and Laser, 1965; Moseley, 1983). If cultures are irradiated during stationary phase growth, the shoulder extends to approximately 15,000 Gy (Daly *et al.*, 1994; Daly and Minton, 1996, 1997). The level of intracellular damage inflicted at these doses is massive, and obvious in all macromolecules. While we cannot say with certainty what damage causes cell death, it is apparent that genomic DNA must be repaired. If it is not, the cell is no longer viable. Approximately 275 DNA double-strand breaks form per genome copy per cell at the D_{37} dose (about 6500 Gy) in cells during exponential phase growth, which means that one double-strand break forms on average every 10,000 base pairs in this species (Cox and Battista, 2005). At its D_{37} value, *E. coli* exhibits eight or nine DNA double-strand breaks per genome copy (Krasin and Hutchinson, 1977). DNA double-strand breaks (Burrell and Dean, 1975), DNA single-strand breaks (Dean *et al.*, 1966, 1970; Bonura and Bruce, 1974) and base damage (thymine glycol formation) (Hariharan and Cerutti, 1971, 1972) appear at levels corresponding with the doses of ionizing radiation applied. *D. radiodurans* tolerates this level of DNA damage, whereas radiosensitive species cannot withstand the onslaught. Understanding *D. radiodurans* remarkable ability to survive exposure to ionizing radiation requires that we identify how genomic DNA is restored post-irradiation. This seemingly simple realization has spawned a number of ideas concerning the mechanisms most responsible for the species' radioresistance. While all of these ideas acknowledge the central importance of DNA repair to cell survival, they differ in the details of how DNA repair is augmented in *D. radiodurans*. The four most prevalent explanations for *D. radiodurans*' enhanced DNA repair capacity are summarized below. Each explanation suggests a different strategy for improving the effectiveness of DNA repair, which in turn is believed to increase the species' ionizing radiation resistance.

1 *There may be factors unrelated to DNA repair per se that permit repair proteins in D. radiodurans to function in ways that their homologues in other species cannot function.* In this explanation, the DNA repair proteins in *D. radiodurans* are no different than DNA repair proteins in radiosensitive species. Proponents of this explanation suggest one of two options. The first is the model described above which proposes that limiting protein oxidation and/or restoring oxidized proteins post-irradiation (Daly *et al.*, 2004, 2007; Daly, 2009) is all that is necessary to enhance radioresistance. In this model, cells have all that they need by way of DNA repair capacity, provided the necessary proteins remain functional post-irradiation. The second option posits that the radioresistant cell uses its DNA repair proteins in a novel manner that results in more effective repair. Several recent investigations have indicated that this scenario applies to *D. radiodurans*. Unlike other characterized bacteria, repair of DNA double-strand breaks is a two-step process in *D. radiodurans* (Daly and Minton, 1996). In the first step, there is extensive single-strand annealing followed by DNA synthesis that is not related to genome replication (Zahradka *et al.*, 2006; Slade *et al.*, 2009). As these events occur, the smaller fragments generated by

DNA damage coalesce into much larger fragments. The second step results in assembly of the genome. These steps appear to be carried out by *D. radiodurans* homologues of RecA, DNA polymerases I and III, RadA, and the RecFOR complex (Zahradka *et al.*, 2006; Slade *et al.*, 2009; Bentchikou *et al.*, 2010). Genes encoding these proteins are found in almost all bacterial species, but thus far only *D. radiodurans* is known to use them in this distinctive manner. Of course, these alternative possibilities of protection and novel repair pathways are not mutually exclusive. They could act synergistically to enhance overall DNA repair (Slade *et al.*, 2009).

2 *The DNA repair proteins encoded by D. radiodurans are more adept at performing DNA repair functions.* In its simplest incarnation this idea presumes that the *D. radiodurans* repair proteins function 'better' than their counterparts in more radiosensitive species. Attempts to express DNA repair proteins [DNA polymerase I (Gutman *et al.*, 1994) and UvrA (Agostini *et al.*, 1996)] from *E. coli* in *D. radiodurans* did not affect the ionizing radiation resistance of *D. radiodurans*, suggesting the activities of these proteins are comparable in radiosensitive and radioresistant species. Although there is very little direct evidence supporting this model, the idea persists because of subtle differences detected in the biochemistry of some *D. radiodurans*' repair proteins relative to their homologues found in other species. For example, investigations have revealed that RecA protein of *D. radiodurans* behaves differently relative to its *E. coli* homologue (Kim and Cox, 2002; Kim *et al.*, 2002). *D. radiodurans* RecA binds more efficiently to double-strand DNA relative to single-strand DNA, which is the opposite of what occurs in other characterized species. It is not yet known if this difference relates to radioresistance, but given the critical role of RecA in *D. radiodurans* survival, there is a formal possibility that this RecA is some way superior to other RecA molecules when dealing with extensive DNA damage.

3 *The conventional DNA repair proteins are necessary, but not sufficient for radioresistance, requiring novel accessory functions that facilitate and enhance DNA repair.* This concept was introduced during the discussion of the potential effects of genome copy number and nucleoid organization on radioresistance – each of these features passively increasing the effectiveness of DNA repair. In addition, it has been suggested that *D. radiodurans* may facilitate DNA repair by regulating DNA replication. Immediately after cultures of *D. radiodurans* are irradiated, DNA replication ceases and remains stopped until repairs are completed. This phenomenon appears to be more than what would occur if the DNA polymerase simply encountered a lesion in the DNA that blocked its forward progress. Several studies (Dean *et al.*, 1966; Lett *et al.*, 1967; Moseley and Copland, 1976) have reported that the DNA damage-induced delay in *D. radiodurans*' chromosomal DNA replication is dose dependent, and that the length of the delay always exceeds the time required to repair the DNA damage that caused the inhibition. These observations suggest that the process of DNA replication in *D. radiodurans* is regulated and sensitive to the level of intracellular DNA damage. If such a regulatory system is functioning, it could allow DNA repair to proceed without the potentially lethal complication of initiating genome replication prior to the completion of repair. However, beyond the observation of delays in DNA replication post-irradiation, there is little genetic or biochemical evidence defining such regulation in *D. radiodurans*. An unusually long lag in growth was noted in *D. radiodurans* cells that carry defects in the ClpPX protease (Servant *et al.*, 2007) after exposure to ionizing radiation. These cells are repair proficient, but they do not reinitiate replication following repair, suggesting that ClpPX could be part of a mechanism that senses DNA damage and/or relays that signal to the replication machinery.

4 *There are novel DNA repair processes at work in D. radiodurans that in conjunction with the conventional set of DNA repair proteins mediate this species ionizing radiation resistance.* Most of the open reading frames that are unique

to the three species of *Deinococcus* with characterized genome sequences encode proteins of unknown function, and until those functions are defined there will remain the possibility that the Deinococci express previously uncharacterized DNA repair proteins. This idea is also given credence by the microarray work described earlier, which noted that the most abundant transcripts formed post-irradiation produced novel proteins, designated DdrA, DdrB, DdrC, DdrD, and PprA (Tanaka *et al.*, 2004). These proteins are encoded in all deinococcal species examined and they appear to be coordinately regulated. Each of these proteins participates in ionizing radiation resistance, and subsequent characterization has discovered activities associated with three of these proteins that indicate potential function. DdrA (Harris *et al.*, 2004) and PprA (Narumi *et al.*, 2004) appear to cap free ends of DNA, protecting them from exonuclease digestion, which would be useful in preventing destruction of fragments formed by ionizing radiation. DdrB is a single-strand binding protein of unique structure (Norais *et al.*, 2009). Single strand binding proteins frequently act as mobilization scaffolds, recruiting proteins to sites on the DNA. DdrB probably has a role in organizing genome reconstitution functions, but the proteins it interacts with have not yet been identified.

As the preceding discussion suggests, there are many factors that arguably contribute to *D. radiodurans*' radioresistance. The details of how these different factors interact with each other to produce a radioresistant species remains to be elucidated, but it seems probable that no single process predominates. Given what we know of *D. radiodurans*, it is difficult to envisage how limiting protein oxidation can, in and of itself, confer radioresistance on the species without the novel modifications to this species' DNA repair proficiency. The challenge in future studies will be to assess the relative significance of the different processes identified above on radioresistance, and to determine if all species employ the same mechanisms of ionizing radiation resistance.

Why did ionizing radiation resistance evolve?

Natural selection

Natural selection is the evolutionary process that results in the spread of favourable variations within a population (Barton *et al.*, 2007). Natural selection acts to increase the fitness (reproductive success) of individuals with specific variations in the population, ultimately leading to the predominance of the offspring of those individuals in descendent populations. The specific variations confer a selective advantage on these individuals, allowing them better reproductive success in a given environment or under conditions that constrain the growth of individuals without the specific variations. Rifampicin resistance in *Escherichia coli* provides a trivial example of the process (Garibyan *et al.*, 2003). Spontaneous mutation will modify the *rpoB* gene product, creating a protein that cannot bind rifampicin. Bacteria expressing these modified RpoB proteins are resistant to the lethal effect of the antibiotic. Adding rifampicin to a culture will kill all but those individuals with the appropriately altered *rpoB* genes. These individuals have a selective advantage as the environment for growth changes relative to those individuals without this specific variation.

While the example given above illustrates natural selection, it is too simplistic to assume that changes in a single gene typically results in dramatic increases in fitness. Many of the phenotypes that characterize a species, such as adaptation to an extreme environment, are due to a complex series of changes that incrementally resulted in that phenotype. In asexual microbial populations, genetic change accumulates with time and natural selection sorts through those changes, gradually culling less favourable variants (Bell, 1997).

Clearly, identifying the selective pressure responsible for building ionizing radiation resistance in microorganisms is the key to understanding the evolution of this phenotype. However, as detailed below, evidence argues against a direct role for ionizing radiation in the process.

Natural and manmade sources of ionizing radiation

The existence of ionizing radiation-resistant species is difficult to explain if one assumes that ionizing radiation is the selective pressure needed for these species to evolve. On Earth ionizing radiation results from two natural sources: cosmic rays and the decay of naturally occurring radionuclides (United Nations Scientific Committee on the Effects of Atomic Radiation, 1982). Cosmic rays are classified as either primary cosmic rays, those arising outside the solar system or during solar flares, or secondary cosmic rays, those created when primary cosmic rays react with atoms in our atmosphere. There are two types of radionuclides – cosmogenic and primordial. Cosmogenic radionuclides are a byproduct of the collision of primary cosmic rays with atoms in the atmosphere; they do not significantly contribute to the production of ionizing radiation on Earth. Primordial radionuclides include ^{40}K, ^{87}Rb, ^{238}U, and ^{232}Th. Their distribution ranges from highly localized, as with uranium isotopes, to ubiquitous (^{40}K is found in most environments). The average absorbed doses due to cosmic radiation and radioactive decay worldwide in air at ground level are estimated to be 3×10^{-8} Gy/h and 5×10^{-8} Gy/h, respectively. Of course, these values are averaged over all environments found on the planet. Dose rates change and increases are found particularly in those areas where higher concentrations of primordial radionuclides are found near the Earth's surface, but even the most radioactive environment on Earth (a beach near Guarapari, Brazil) displays an average dose rate of 2×10^{-5} Gy/h. Given the degree of radioresistance observed in some ionizing radiation-resistant species, it does not seem reasonable that naturally occurring ionizing radiation could act as the selective agent in the evolution of these species.

In the 1970s, a group of French scientists provided evidence that during Earth's distant past conditions were such that spontaneous fission reactions occurred and that 15 'fossil' nuclear reactors can be found near present-day Oklo, Gabon (Naudet, 1974). The remains of these reactors exist as uranium deposits, and are distinguished by a distribution of uranium isotopes (particularly ^{235}U) that is characteristic of the spent fuel produced by nuclear power plants. ^{235}Uranium is consumed during a fission reaction and the levels of this isotope are half of what they are elsewhere on Earth. There is general agreement among geologists and physicists that a change this large could only result from a self-sustaining nuclear chain reaction. These reactors formed 1.7 billion years ago and were active for several hundred thousand years (De Laeter *et al.*, 1980; Gauthier-Lafaye, 2002).

The existence of natural nuclear reactors forces a question. Is the ionizing radiation resistance observed in microorganisms today a vestige of mechanisms that arose to protect species that lived near these reactors? As will be discussed later, microorganisms do adapt when accidently or deliberately exposed to ionizing radiation, throwing off strains that are more resistant to this stress. In the absence of other information, it is not difficult to generate a scenario whereby the Oklo reactors provided the selective pressure that led to ionizing radiation resistance. However, at their peak the Oklo reactors are believed to have emitted a dose of approximately 5 Gy per day at their core – a very low value when one considers that the D_{37} dose of the most ionizing radiation-resistant species are more than a 1000-fold higher than this daily exposure (Petrov *et al.*, 2006). In addition, the temperature at the core of the reactors is estimated to have been between 150 and 360°C. The low dose rate and high reactor temperatures argue against involvement of the Oklo reactors in the evolution and natural history of ionizing radiation-resistant species.

Iterative exposure to high-dose ionizing radiation will select for ionizing radiation-resistant species, a fact established in the laboratory (Harris *et al.*, 2009) and in the area surrounding the Chernobyl disaster (Dadachova *et al.*, 2007). However, it is difficult to argue that the modern use of nuclear power and resulting concentration of radioactive materials could have led to the variety and widespread distribution of ionizing radiation-resistant species observed in nature. While individual radioresistant species have been found associated with nuclear waste sites (Fredrickson *et al.*, 2004), such isolations are rare, arguing that the species was already radioresistant and that it was selected from among radiosensitive species. If recently formed radioactive materials are creating resistant species among existing flora, it is expected that they would be present in greater numbers at such sites.

This survey of natural and manmade radioactive sources illustrates the absence of a clearly identifiable, natural, selective pressure capable of encouraging the generation of the multiple independent evolutionary events that appear necessary to explain the natural history of ionizing radiation resistance on Earth. Given the exceedingly low rate of terrestrial exposure to ionizing radiation, it is difficult to rationalize why ionizing radiation resistance ever arose during microbial evolution, at least as a response of exposure to such radiation. Ionizing radiation resistance is perhaps best thought of as a characteristic that is incidental, a fortunate circumstance of another evolutionary process.

Proposed explanations for the existence of ionizing radiation resistance

Panspermia: the exchange of bacteria between planets

A meteoroid is any solid object of extraterrestrial origin that enters Earth's atmosphere, and can be as small as a grain of sand or as large as a boulder. Meteoroids may originate from anywhere in space, and it has been established that if planets suffer a sufficiently large impact, debris can be ejected into space, and travel until it encounters the gravity of another planet where it is pulled down to the surface. There are a number of well-studied meteoroids from Mars that have been found on Earth. [Ron Baalke of NASA's Jet Propulsion Laboratory maintains a website listing all of the known Mars meteoroids (http://www2.jpl.nasa.gov/snc/index.html)] It has been suggested that at least some life on Earth may have originated elsewhere in the Universe and then found its way to this planet, hitchhiking on a meteoroid. The general name for this hypothesis is panspermia, and it has been invoked to explain why ionizing radiation-resistant species exist on Earth (Pavlov *et al.*, 2006). Proponents suggest that it is inappropriate to evaluate ionizing radiation resistance based on parameters that are limited to what occurs on Earth. Instead they argue that there are other planets with much higher fluxes of cosmic radiation, and that on such planets there is appropriate selective pressure to evolve an ionizing radiation-resistant species. Those extraterrestrial microbial species would then be accidentally transferred to Earth, expressing an exotic phenotypic characteristic (ionizing radiation resistance) of limited use to them on this planet. There is no direct evidence supporting or refuting this hypothesis, but it is unlikely that anything approaching proof of this idea will emerge from scientific investigation: the estimated time for the proposed transfers of material between planets are minimally on the order of tens of thousands of years, far longer than can be assessed by direct experimentation.

Co-opted adaptation

As discussed above, there is little advantage to being ionizing radiation resistant in nature, and so we cannot assume that this phenotype is the result of natural selection. Instead, it seems clear that some feature or function in these species allows them to be ionizing radiation resistant. In other words, ionizing radiation resistance is the byproduct of a process that permits the cell to tolerate the damage introduced by ionizing radiation. In the parlance of the evolutionary biologist, the radioresistant cell is adapted to exposure to ionizing radiation because it has co-opted pre-existing functions expressed by this cell (Barton *et al.*, 2007). These functions did not arise in response to ionizing radiation, but they are fortuitously beneficial in dealing with the consequences of exposure to such radiation. Feathers are a classic example of co-opted adaptation. They are believed to have arisen as a mechanism of thermal regulation (insulation), but they later allowed feathered species to glide and then fly, incidentally providing a selective advantage in avoiding predators and searching for food. The ability to fly was not the reason that feathers evolved, but the existence of feathers allowed flight to evolve. It is generally assumed that ionizing radiation is not the reason species are ionizing radiation resistant; instead, it is hypothesized that these species tolerate ionizing radiation-induced cellular damage because their survival requires the ability to tolerate similar damage encountered in their day-to-day existence.

As described above, ionizing radiation-resistant bacteria are found in many environments, but it appears they are found in highest concentrations in arid soils (Rainey *et al.*, 2005) or geothermally heated water (Ferreira *et al.*, 1997, 1999; Jolivet *et al.*, 2003, 2004). This fact has led to the expectation that these species express specific adaptations that allow long-term survival in these ecological niches. Both of these stresses damage macromolecules in a manner similar to that reported when cells are exposed to ionizing radiation (Lindahl, 1993; Potts, 1994). Examination of many ionizing radiation-resistant species has established that they are also resistant to desiccation and/or elevated temperatures, suggesting that ionizing radiation resistance is the consequence of a species capacity to tolerate desiccation or heat-induced cellular damage – the ability to survive this damage significantly contributing to protection against ionizing radiation-induced damage. For example, desiccation will introduce DNA double-strand breaks (considered the most lethal lesion formed in response to ionizing radiation) into the prokaryotic genome (Dose *et al.*, 1991; Mattimore and Battista, 1996; Dose *et al.*, 2001). Species displaying high level resistance to ionizing radiation, including members of the genera *Deinococcus* (Mattimore and Battista, 1996) and *Chroococcidiopsis* (Billi *et al.*, 2000), are also extremely resistant to desiccation. Drying introduces large numbers of DNA double-strand breaks into species within these genera, but the lesions are rapidly repaired by mechanisms that presumably are also used to repair ionizing radiation-induced strand breaks.

Creating ionizing radiation-resistant strains from ionizing radiation sensitive species

A survey of *Bacteria* and *Archaea* identifies nine phyla that contain species with unusually high resistance to the lethal effects of ionizing radiation (Table 2.1; Cox and Battista, 2005). As already pointed out above, these phyla are not closely related to each other, do not share a common lineage, and all include genera that are considered radiosensitive. The existence of so many unrelated and isolated radioresistant species in the phylogenetic tree argues that the molecular mechanisms that protect against ionizing radiation-induced damage evolved independently in these organisms, strongly suggesting that many species have the latent capacity to become radioresistant when subjected to appropriate selective pressure.

Evelyn Witkin was the first to establish that it was possible to increase the resistance of a species to DNA damage (Witkin, 1946). She iteratively exposed cultures of *E. coli* B to higher and higher doses of ultraviolet light, killing most of the population with each exposure. The surviving variants were better able to tolerate UV compared with their parent. Since that report, others have demonstrated that repeated exposure to other DNA damaging agents, including ionizing radiation, will create strains with enhanced resistance to that agent. Increases in ionizing radiation resistance have been reported for *E. coli*, *Salmonella enterica* serovar Typhimurium, and *Bacillus pumulis* (Erdman *et al.*, 1961; Davies and Sinskey, 1973; Parisi and Antoine, 1974).

Harris *et al.* performed a similar study using *E. coli* K12 strain MG1655 (Harris *et al.*, 2009). Starting from the same parent they derived four ionizing radiation-resistant strains through identical, but independent protocols of irradiation. After twenty cycles of irradiation and outgrowth, the populations of cells were between 300–3000-fold more resistant than the parent when exposed to 3000 Gy γ-radiation. The phenotype was stable, but physiological and biochemical characterization of the strain did not reveal what had changed. An isolate from each of the most resistant populations was re-sequenced and compared in an attempt to discern what genetic changes were responsible for the transition to ionizing radiation resistance. The result was unexpected, revealing a relatively small number of mutations (approximately 70) that were predominately base substitutions. With two exceptions, comparison of the genomes that had been modified did not reveal obvious functional overlap among the genes carrying mutations. All of the radioresistant strains examined had a deletion of the e14 prophage [an event induced by any DNA damaging agent that triggers the SOS response (Greener and Hill, 1980)] and a mutation in at least one of the proteins that is part of the replication restart primosome. Alone, these two changes were not sufficient to account for the strain's radioresistance. These results strongly suggest that multiple solutions exist to solve the challenges posed by ionizing radiation, and that *E. coli* and possibly other species can avail themselves of these 'options.'

Concluding remarks

Susan Lovett (Lovett, 2006) perhaps best described *D. radiodurans* radioresistance when she drew an analogy between this organism and a trained athlete. Athletes are superior talents because they do a myriad of little things better than the average person. It appears that *D. radiodurans* uses a combination of more or less subtle changes in prokaryotic physiology to improve its survival, changes that appear to involve passive protection and unconventional use of well known repair proteins. While more exotic explanations for radioresistance are a possibility, we believe that *D. radiodurans* response to ionizing radiation argues that DNA damage is the most important challenge the cell faces post irradiation. Without DNA repair the cell will die and there is little evidence to suggest that the same can be said about other macromolecules. This conclusion is in agreement with a half century of investigation into the reasons cells die when exposed to DNA-damaging agents. How DNA repair is enhanced in the species remains an open question. While studies of *D. radiodurans* offer insight into the mechanisms of ionizing radiation resistance, we must remember that the results may not apply to all radioresistant species. The phylogenetic distribution of ionizing radiation resistance and the apparent ease with which radiosensitive species can be made resistant argue for multiple mechanisms.

The existence of ionizing radiation resistance is counterintuitive. Darwinian evolution argues that phenotypes are built slowly over time, providing the organism expressing that phenotype a reproductive advantage. However, in its history Earth has never experienced a flux of ionizing radiation in the range that these species are capable of tolerating. In fact, yearly terrestrial exposures to ionizing radiation are at least 10^6-fold lower than survivable doses administered to these species in the laboratory. We have argued that ionizing radiation is an example of co-opted adaptation, a process whereby ionizing radiation resistance is an inadvertent consequence of an evolutionary pathway that has nothing to do with exposure to high-dose ionizing radiation. There are other stresses that introduce patterns of cellular damage that are similar to those experienced by cells exposed to ionizing radiation and it is our contention that ionizing radiation-resistant species arose over evolutionary time to survive one or more of these stresses. The increased probability of finding ionizing radiation resistance in species found in arid and geothermally heated aqueous environments supports this idea. These environments are inherently damaging to the cell and to occupy these niches the cell must have means of protecting itself; the ability to survive in these environments also fortuitously protects the cell from ionizing radiation-induced damage. As remarkable a phenotype as it is, ionizing radiation resistance appears to be a subsidiary characteristic that would have remained unknown were it not for the use of manmade radioactive sources.

Future directions

1. Establish whether there is a universal mechanism of ionizing radiation resistance. Given the number of unrelated ionizing radiation-resistant species, this notion seems unlikely. Unequivocally determining how a well-studied species, like *D. radiodurans*, tolerates ionizing radiation is a first and essential step to establishing if other species share that mechanism.
2. If no single mechanism is responsible for ionizing radiation resistance, begin to catalogue all that contributes to the phenotype. Multiple mechanisms potentially contribute to a cell's capacity to survive exposure to high-dose ionizing radiation. Whether one mechanism predominates over the others remains to be determined.
3. Create conditions that mimic the evolution of ionizing radiation resistance in the laboratory. The idea that ionizing radiation resistance is an example of co-opted adaptation related to desiccation tolerance is an inference based on: (a) correlations

between desiccation tolerance and ionizing radiation resistance; (b) the similar patterns of DNA damage generated by both stressors; and (c) the enrichment of ionizing radiation-resistant species found in arid environments. Proving this concept will require creating a desiccation tolerant strain from a desiccation sensitive species, and demonstrating increased ionizing radiation resistance.

Web resources

http://www.usuhs.mil/pat/deinococcus/index_20.htm This web site is operated by the laboratory of Michael J. Daly at the Uniformed Service University of Health Sciences in Bethesda, MD. It is the best maintained web site catering to aficionados of the genus *Deinococcus*.

http://cmr.jcvi.org/tigr-scripts/CMR/GenomePage.cgi?database=gdr The J. Craig Venter Institute maintains the original annotation of the *D. radiodurans* genome.

http://genome.jgi-psf.org/deige/deige.home.html The DOE Joint Genome Institute maintains the original annotation of the *D. geothermalis* genome.

http://www.genoscope.cns.fr/spip/-Deinococcus-deserti-VCD115-.html Genoscope – Centre National de Séquençage in France sequenced *D. deserti*. This site provides background on the species and identifies sources of additional information.

http://genome.jgi-psf.org/rubxy/rubxy.home.html The DOE Joint Genome Institute maintains the original annotation of the *Rubrobacter xylanophilus* genome.

http://genome.jgi-psf.org/kinra/kinra.home.html The DOE Joint Genome Institute maintains the original annotation of the *Kineococcus radiotolerans* genome.

http://www.deinove.com/en.html Deinove is a European company attempting to develop products that exploit *Deinococcus* species industrially.

References

Agostini, H.J., Carroll, J.D., and Minton, K.W. (1996). Identification and characterization of *uvrA*, a DNA repair gene of *Deinococcus radiodurans*. J. Bacteriol. *178*, 6759–6765.

Akerlund, T., Nordstrom, K., and Bernander, R. (1995). Analysis of cell size and DNA content in exponentially growing and stationary-phase batch cultures of *Escherichia coli*. J. Bacteriol. *177*, 6791–6797.

Albuquerque, L., Simoes, C., Nobre, M.F., Pino, N.M., Battista, J.R., Silva, M.T., Rainey, F.A., and da Costa, M.S. (2005). *Truepera radiovictrix* gen. nov., sp. nov., a new radiation resistant species and the proposal of *Trueperaceae* fam. nov. FEMS Microbiol. Lett. *247*, 161–169.

Anderson, A.W., Nordon, H.C., Cain, R.F., Parrish, G., and Duggan, D. (1956). Studies on a radio-resistant micrococcus. I. Isolation, morphology, cultural characteristics, and resistance to gamma radiation. Food Technol. *10*, 575–578.

Asker, D., Awad, T.S., Beppu, T., and Ueda, K. (2008). *Deinococcus misasensis* and *Deinococcus roseus*, novel members of the genus *Deinococcus*, isolated from a radioactive site in Japan. Syst. Appl. Microbiol. *31*, 43–49.

Asker, D., Awad, T.S., Beppu, T., and Ueda, K. (2009). *Deinococcus aquiradiocola* sp. nov., isolated from a radioactive site in Japan. Int. J. Syst. Evol. Microbiol. *59*, 144–149.

Badarinarayana, V., Estep, P.W., third, Shendure, J., Edwards, J., Tavazoie, S., Lam, F., and Church, G.M. (2001). Selection analyses of insertional mutants using subgenic-resolution arrays. Nat. Biotechnol. *19*, 1060–1065.

Barton, N.H., Briggs, D.E.G., Eisen, J.A., Goldstein, D.B., and Patel, N.H. (2007). Evolution (Cold Spring Harbor, Cold Spring Harbor Laboratory Press).

Battista, J.R. (1997). Against all odds: the survival strategies of *Deinococcus radiodurans*. Annu. Rev. Microbiol. *51*, 203–224.

Battista, J.R., and Rainey, F.A. (2001). Phylum BIV. '*Deinococcus-Thermus*' Family 1. *Deinococcaceae* Brooks and Murray 1981, 356,vp emend. Rainey, Nobre, Schumann, Stackebrandt and da Costa 1997, 513. In Bergey's Manual of Systematic Bacteriology, Boone, D.R., and Castenholz, R.W., eds. (New York, Springer), pp. 395–414.

Battista, J.R., Earl, A.M., and Park, M.J. (1999). Why is *Deinococcus radiodurans* so resistant to ionizing radiation? Trends Microbiol. 7, 362–365.

Bell, G. (1997). Selection: The Mechanism of Evolution (New York, Chapman and Hall).

Bentchikou, E., Servant, P., Coste, G., and Sommer, S. (2010). A major role of the RecFOR pathway in DNA double-strand-break repair through ESDSA in *Deinococcus radiodurans*. PLoS Genet. *6*, e1000774.

Billi, D., Friedmann, E.I., Hofer, K.G., Caiola, M.G., and Ocampo-Friedmann, R. (2000). Ionizing-radiation resistance in the desiccation-tolerant cyanobacterium *Chroococcidiopsis*. Appl. Environ. Microbiol. *66*, 1489–1492.

Blasius, M., Sommer, S., and Hubscher, U. (2008). *Deinococcus radiodurans*: what belongs to the survival kit? Crit. Rev. Biochem. Mol. Biol. *43*, 221–238.

Bonacossa de Almeida, C., Coste, G., Sommer, S., and Bailone, A. (2002). Quantification of RecA protein in *Deinococcus radiodurans* reveals involvement of RecA, but not LexA, in its regulation. Mol. Genet. Genomics *268*, 28–41.

Bonura, T., and Bruce, A.K. (1974). The repair of single-strand breaks in a radiosensitive mutant of *Micrococcus radiodurans*. Radiat. Res. *57*, 260–275.

Burrell, A.D., and Dean, C.J. (1975). Repair of double-strand breaks in *Micrococcus radiodurans*. Basic Life Sci. *5B*, 507–512.

Callegan, R.P., Nobre, M.F., McTernan, P.M., Battista, J.R., Navarro-Gonzalez, R., McKay, C.P., da Costa, M.S., and Rainey, F.A. (2008). Description of four novel psychrophilic, ionizing radiation-sensitive *Deinococcus* species from alpine environments. Int. J. Syst. Evol. Microbiol. *58*, 1252–1258.

Chou, F.I., and Tan, S.T. (1990). Manganese(II) induces cell division and increases in superoxide dismutase and catalase activities in an aging deinococcal culture. J. Bacteriol. *172*, 2029–2035.

Christensen, E.A., and Kristensen, H. (1981). Radiation-resistance of micro-organisms from air in clean premises. Acta Pathol. Microbiol. Scand. B *89*, 293–301.

Christensen, E.A., Gerner-Smidt, P., and Kristensen, H. (1991a). Radiation resistance of clinical *Acinetobacter* spp.: a need for concern? J. Hosp. Infect. *18*, 85–92.

Christensen, E.A., Kristensen, H., Hoborn, J., and Miller, A. (1991b). Radiation resistance of microorganisms on unsterilized infusion sets. APMIS *99*, 620–626.

Collins, M.D., Hutson, R.A., Grant, I.R., and Patterson, M.F. (2000). Phylogenetic characterization of a novel radiation-resistant bacterium from irradiated pork: description of *Hymenobacter actinosclerus* sp. nov.. Int. J. Syst. Evol. Microbiol. *50*, 731–734.

Cox, M.M., and Battista, J.R. (2005). *Deinococcus radiodurans* – the consummate survivor. Nat. Rev. Microbiol. *3*, 882–892.

Dadachova, E., Bryan, R.A., Huang, X., Moadel, T., Schweitzer, A.D., Aisen, P., Nosanchuk, J.D., and Casadevall, A. (2007). Ionizing radiation changes the electronic properties of melanin and enhances the growth of melanized fungi. PLoS One *2*, e457.

Daly, M.J. (2009). A new perspective on radiation resistance based on *Deinococcus radiodurans*. Nat. Rev. Microbiol. *7*, 237–245.

Daly, M.J., and Minton, K.W. (1996). An alternative pathway of recombination of chromosomal fragments precedes recA-dependent recombination in the radioresistant bacterium *Deinococcus radiodurans*. J. Bacteriol. *178*, 4461–4471.

Daly, M.J., and Minton, K.W. (1997). Recombination between a resident plasmid and the chromosome following irradiation of the radioresistant bacterium *Deinococcus radiodurans*. Gene *187*, 225–229.

Daly, M.J., Gaidamakova, E.K., Matrosova, V.Y., Vasilenko, A., Zhai, M., Leapman, R.D., Lai, B., Ravel, B., Li, S.M., Kemner, K.M., *et al.* (2007). Protein oxidation implicated as the primary determinant of bacterial radioresistance. PLoS Biol. *5*, e92.

Daly, M.J., Gaidamakova, E.K., Matrosova, V.Y., Vasilenko, A., Zhai, M., Venkateswaran, A., Hess, M., Omelchenko, M.V., Kostandarithes, H.M., Makarova, K.S., *et al.* (2004). Accumulation of Mn(II) in *Deinococcus radiodurans* facilitates gamma-radiation resistance. Science *306*, 1025–1028.

Daly, M.J., Ling, O., and Minton, K.W. (1994). Interplasmidic recombination following irradiation of the radioresistant bacterium *Deinococcus radiodurans*. J. Bacteriol. *176*, 7506–7515.

Davies, R., and Sinskey, A.J. (1973). Radiation-resistant mutants of *Salmonella typhimurium* LT2: development and characterization. J. Bacteriol. *113*, 133–144.

Davis, N.S., Silverman, G.J., and Masurovsky, E.B. (1963). Radiation-Resistant, Pigmented Coccus Isolated from Haddock Tissue. J. Bacteriol. *86*, 294–298.

De Laeter, J.R., Rosman, K.J.R., and Smith, C.L. (1980). The Oklo natural reactor: Cumulative fission yields and retentivity of the symmetric mass region fission products. Earth Plan. Sci. Lett. *50*, 238–246.

Dean, C.J., Feldschreiber, P., and Lett, J.T. (1966). Repair of X-ray damage to the deoxyribonucleic acid in *Micrococcus radiodurans*. Nature *209*, 49–52.

Dean, C.J., Little, J.G., and Serianni, R.W. (1970). The control of post irradiation DNA breakdown in *Micrococcus radiodurans*. Biochem. Biophys. Res. Commun. *39*, 126–134.

DiRuggiero, J., Santangelo, N., Nackerdien, Z., Ravel, J., and Robb, F.T. (1997). Repair of extensive ionizing-radiation DNA damage at 95 degrees C in the hyperthermophilic archaeon *Pyrococcus furiosus*. J. Bacteriol. *179*, 4643–4645.

Dizdaroglu, M., Schulte-Frohlinde, D., and von Sonntag, C. (1975a). Radiation chemistry of DNA, II. Strand breaks and sugar release by gamma-irradiation of DNA in aqueous solution. The effect of oxygen. Z. Naturforsch. C. *30*, 826–828.

Dizdaroglu, M., von Sonntag, C., and Schulte-Frohlinde, D. (1975b). Letter: Strand breaks and sugar release by gamma-irradiation of DNA in aqueous solution. J. Am. Chem. Soc. *97*, 2277–2278.

Dose, K., Bieger-Dose, A., Kerz, O., and Gill, M. (1991). DNA-strand breaks limit survival in extreme dryness. Orig. Life Evol. Biosph. *21*, 177–187.

Dose, K., Bieger-Dose, A., Ernst, B., Feister, U., Gomez-Silva, B., Klein, A., Risi, S., and Stridde, C. (2001). Survival of microorganisms under the extreme conditions of the Atacama Desert. Orig. Life Evol. Biosph. *31*, 287–303.

Driedger, A.A., James, A.P., and Grayston, M.J. (1970). Cell survival and X-ray-induced DNA degradation in *Micrococcus radiodurans*. Radiat. Res. *44*, 835–845.

Eltsov, M., and Dubochet, J. (2005). Fine structure of the *Deinococcus radiodurans* nucleoid revealed by cryoelectron microscopy of vitreous sections. J. Bacteriol. *187*, 8047–8054.

Eltsov, M., and Dubochet, J. (2006). Study of the *Deinococcus radiodurans* nucleoid by cryoelectron microscopy of vitreous sections: Supplementary comments. J. Bacteriol. *188*, 6053–6058; discussion 6059.

Erdman, I.E., Thatcher, F.S., and Macqueen, K.F. (1961). Studies on the irradiation of microorganisms in relation to food preservation. I. The comparative sensitivities of specific bacteria of public health significance. Can. J. Microbiol. *7*, 199–205.

Ferreira, A.C., Nobre, M.F., Rainey, F.A., Silva, M.T., Wait, R., Burghardt, J., Chung, A.P., and da Costa, M.S. (1997). *Deinococcus geothermalis* sp. nov. and *Deinococcus murrayi* sp. nov., two extremely radiation-resistant and slightly thermophilic species from hot springs. Int. J. Syst. Bacteriol. *47*, 939–947.

Ferreira, A.C., Nobre, M.F., Moore, E., Rainey, F.A., Battista, J.R., and da Costa, M.S. (1999). Characterization and radiation resistance of new isolates of *Rubrobacter radiotolerans* and *Rubrobacter xylanophilus*. Extremophiles 3, 235–238.

Fredrickson, J.K., Zachara, J.M., Balkwill, D.L., Kennedy, D., Li, S.M., Kostandarithes, H.M., Daly, M.J., Romine, M.F., and Brockman, F.J. (2004). Geomicrobiology of high-level nuclear waste-contaminated vadose sediments at the Hanford site, Washington State. Appl. Environ. Microbiol. *70*, 4230–4241.

Fredrickson, J.K., Li, S.M., Gaidamakova, E.K., Matrosova, V.Y., Zhai, M., Sulloway, H.M., Scholten, J.C., Brown, M.G., Balkwill, D.L., and Daly, M.J. (2008). Protein oxidation: key to bacterial desiccation resistance? ISME J. *2*, 393–403.

Freifelder, D., and Trumbo, B. (1969). Matching of single-strand breaks to form double-strand breaks in DNA. Biopolymers 7, 681–693.

Garibyan, L., Huang, T., Kim, M., Wolff, E., Nguyen, A., Nguyen, T., Diep, A., Hu, K., Iverson, A., Yang, H., *et al.* (2003). Use of the *rpoB* gene to determine the specificity of base substitution mutations on the *Escherichia coli* chromosome. DNA Repair (Amst) *2*, 593–608.

Gauthier-Lafaye, F. (2002). 2 billion year old natural analogs for nuclear waste disposal: the natural nuclear fission reactors in Gabon (Africa). C. R. Phys. 3, 839–849.

Grant, I.R., and Patterson, M.F. (1989). A novel radiation resistant *Deinobacter* sp. isolated from irradiated pork. Letters Appl. Microbiol. *8*, 21–24.

Green, P.N., and Bousfield, I.J. (1983). Emendation of *Methylobacterium* (Patt, Cole, and Hanson 1976); *Methylobacterium rhodinum* (Heumann 1962) comb. nov.corrig.; *Methylobacterium radiotolerans* (Ito and Iizuka 1971) comb. nov. corrig.; and *Methylobacterium mesophilicum* (Austin and Goodfellow 1979) comb. nov. Int. J. Syst. Bacteriol. *33*, 875–877.

Greener, A., and Hill, C.W. (1980). Identification of a novel genetic element in *Escherichia coli* K-12. J. Bacteriol. *144*, 312–321.

de Groot, A., Chapon, V., Servant, P., Christen, R., Saux, M.F., Sommer, S., and Heulin, T. (2005). *Deinococcus deserti* sp. nov., a gamma-radiation-tolerant bacterium isolated from the Sahara Desert. Int. J. Syst. Evol. Microbiol. *55*, 2441–2446.

de Groot, A., Dulermo, R., Ortet, P., Blanchard, L., Guerin, P., Fernandez, B., Vacherie, B., Dossat, C., Jolivet, E., Siguier, P., *et al.* (2009). Alliance of proteomics and genomics to unravel the specificities of Sahara bacterium *Deinococcus deserti*. PLoS Genet. *5*, e1000434.

Gutman, P.D., Fuchs, P., and Minton, K.W. (1994). Restoration of the DNA damage resistance of *Deinococcus radiodurans* DNA polymerase mutants by *Escherichia coli* DNA polymerase I and Klenow fragment. Mutat. Res. *314*, 87–97.

Hansen, M.T. (1978). Multiplicity of genome equivalents in the radiation-resistant bacterium *Micrococcus radiodurans*. J. Bacteriol. *134*, 71–75.

Hariharan, P.P., and Cerutti, P.A. (1971). Repair of gamma-ray-induced thymine damage in *Micrococcus radiodurans*. Nat. New Biol. *229*, 247–249.

Hariharan, P.V., and Cerutti, P.A. (1972). Formation and repair of gamma-ray induced thymine damage in *Micrococcus radiodurans*. J. Mol. Biol. *66*, 65–81.

Harris, D.R., Tanaka, M., Saveliev, S.V., Jolivet, E., Earl, A.M., Cox, M.M., and Battista, J.R. (2004). Preserving genome integrity: the DdrA protein of *Deinococcus radiodurans* R1. PLoS Biol. *2*, e304.

Harris, D.R., Pollock, S.V., Wood, E.A., Goiffon, R.J., Klingele, A.J., Cabot, E.L., Schackwitz, W., Martin, J., Eggington, J., Durfee, T.J., *et al.* (2009). Directed evolution of ionizing radiation resistance in *Escherichia coli*. J. Bacteriol. *191*, 5240–5252.

Harsojo, Kitayama, S., and Matsuyama, A. (1981). Genome multiplicity and radiation resistance in *Micrococcus radiodurans*. J. Biochem. *90*, 877–880.

Hutchinson, F. (1985). Chemical changes induced in DNA by ionizing radiation. Prog. Nucleic Acid Res. Mol. Biol. *32*, 115–154.

Im, W.T., Jung, H.M., Ten, L.N., Kim, M.K., Bora, N., Goodfellow, M., Lim, S., Jung, J., and Lee, S.T. (2008). *Deinococcus aquaticus* sp. nov., isolated from fresh water, and *Deinococcus caeni* sp. nov., isolated from activated sludge. Int. J. Syst. Evol. Microbiol. *58*, 2348–2353.

Ito, H., and Iizuka, H. (1971). Taxonomic studies on a radio-resistant *Pseudomonas*. XII. Studies on the microorganisms of cereal grain. Agric. Biol. Chem. 35 1566–1571.

Ito, H., Watanabe, H., Takeshia, M., and Iizuka, H. (1983). Isolation and identification of radiation-resistant cocci belonging to the genus *Deinococcus* from sewage sludges and animal feeds. Agric. Biol. Chem. *47*, 1239–1247.

Jolivet, E., L'Haridon, S., Corre, E., Forterre, P., and Prieur, D. (2003). *Thermococcus gammatolerans* sp. nov., a hyperthermophilic archaeon from a deep-sea hydrothermal vent that resists ionizing radiation. Int. J. Syst. Evol. Microbiol. 53, 847–851.

Jolivet, E., Corre, E., L'Haridon, S., Forterre, P., and Prieur, D. (2004). *Thermococcus marinus* sp. nov. and *Thermococcus radiotolerans* sp. nov., two hyperthermophilic archaea from deep-sea hydrothermal vents that resist ionizing radiation. Extremophiles *8*, 219–227.

Jonah, C.D., and Rao, B.S.M. (2001). Radiation Chemistry: Present Status and Future Trends (Amsterdam, Elsevier).

Kempner, E.S. (1993). Damage to proteins due to the direct action of ionizing radiation. Q. Rev. Biophys. *26*, 27–48.

Kim, J.I., and Cox, M.M. (2002). The RecA proteins of *Deinococcus radiodurans* and *Escherichia coli* promote DNA strand exchange via inverse pathways. Proc. Natl. Acad. Sci. U.S.A. *99*, 7917–7921.

Kim, J.I., Sharma, A.K., Abbott, S.N., Wood, E.A., Dwyer, D.W., Jambura, A., Minton, K.W., Inman, R.B., Daly, M.J., and Cox, M.M. (2002). RecA Protein from the extremely radioresistant bacterium *Deinococcus radiodurans*: expression, purification, and characterization. J. Bacteriol. *184*, 1649–1660.

Kottemann, M., Kish, A., Iloanusi, C., Bjork, S., and DiRuggiero, J. (2005). Physiological responses of the halophilic archaeon *Halobacterium* sp. strain NRC1 to desiccation and gamma irradiation. Extremophiles *9*, 219–227.

Krasin, F., and Hutchinson, F. (1977). Repair of DNA double-strand breaks in *Escherichia coli*, which requires recA function and the presence of a duplicate genome. J. Mol. Biol. *116*, 81–98.

Kristensen, H., and Christensen, E.A. (1981). Radiation-resistant micro-organisms isolated from textiles. Acta Pathol. Microbiol. Scand. B *89*, 303–309.

Lai, W.A., Kampfer, P., Arun, A.B., Shen, F.T., Huber, B., Rekha, P.D., and Young, C.C. (2006). *Deinococcus ficus* sp. nov., isolated from the rhizosphere of *Ficus religiosa* L. Int. J. Syst. Evol. Microbiol. *56*, 787–791.

Leibowitz, P.J., Schwartzberg, L.S., and Bruce, A.K. (1976). The *in vivo* association of manganese with the chromosome of *Micrococcus radiodurans*. Photochem. Photobiol. *23*, 45–50.

Lett, J.T., Feldschreiber, P., Little, J.G., Steele, K., and Dean, C.J. (1967). The repair of X-ray damage to the deoxyribonucleic acid in *Micrococcus radiodurans*: a study of the excision process. Proc. R. Soc. Lond. B Biol. Sci. *167*, 184–201.

Levin-Zaidman, S., Englander, J., Shimoni, E., Sharma, A.K., Minton, K.W., and Minsky, A. (2003). Ringlike structure of the *Deinococcus radiodurans* genome: a key to radioresistance? Science *299*, 254–256.

Lewis, N.F. (1971). Studies on a radio-resistant coccus isolated from Bombay Duck (*Harpodon nehereus*). J. Gen. Microbiol. *66*, 29–35.

Lieber, A., Leis, A., Kushmaro, A., Minsky, A., and Medalia, O. (2009). Chromatin organization and radio resistance in the bacterium *Gemmata obscuriglobus*. J. Bacteriol. *191*, 1439–1445.

Lindahl, T. (1993). Instability and decay of the primary structure of DNA. Nature *362*, 709–715.

Lipton, M.S., Pasa-Tolic, L., Anderson, G.A., Anderson, D.J., Auberry, D.L., Battista, J.R., Daly, M.J., Fredrickson, J., Hixson, K.K., Kostandarithes, H., *et al.* (2002). Global analysis of the *Deinococcus radiodurans* proteome by using accurate mass tags. Proc. Natl. Acad. Sci. U.S.A. *99*, 11049–11054.

Lipton, M.S., Romine, M.F., Monroe, M.E., Elias, D.A., Pasa-Tolic, L., Anderson, G.A., Anderson, D.J., Fredrickson, J., Hixson, K.K., Masselon, C., *et al.* (2006). AMT tag approach to proteomic characterization of *Deinococcus radiodurans* and *Shewanella oneidensis*. Methods Biochem. Anal. *49*, 113–134.

Liu, Y., Zhou, J., Omelchenko, M.V., Beliaev, A.S., Venkateswaran, A., Stair, J., Wu, L., Thompson, D.K., Xu, D., Rogozin, I.B., *et al.* (2003). Transcriptome dynamics of *Deinococcus radiodurans* recovering from ionizing radiation. Proc. Natl. Acad. Sci. U.S.A. *100*, 4191–4196.

Lovett, S.T. (2006). Microbiology: resurrecting a broken genome. Nature *443*, 517–519.

Makarova, K.S., Aravind, L., Wolf, Y.I., Tatusov, R.L., Minton, K.W., Koonin, E.V., and Daly, M.J. (2001). Genome of the extremely radiation-resistant bacterium *Deinococcus radiodurans* viewed from the perspective of comparative genomics. Microbiol. Mol. Biol. Rev. *65*, 44–79.

Makarova, K.S., Omelchenko, M.V., Gaidamakova, E.K., Matrosova, V.Y., Vasilenko, A., Zhai, M., Lapidus, A., Copeland, A., Kim, E., Land, M., *et al.* (2007). *Deinococcus geothermalis*: the pool of extreme radiation resistance genes shrinks. PLoS One *2*, e955.

Maldonado, R., Jimenez, J., and Casadesus, J. (1994). Changes of ploidy during the *Azotobacter vinelandii* growth cycle. J. Bacteriol. *176*, 3911–3919.

Mattimore, V., and Battista, J.R. (1996). Radioresistance of *Deinococcus radiodurans*: functions necessary to survive ionizing radiation are also necessary to survive prolonged desiccation. J. Bacteriol. *178*, 633–637.

Mead, J.F. (1976). Free radical mechanisms of lipid damage and conseqeunces for cellular membranes. In Free Radicals in Biology, W.A. Pryor, ed. (New York, Academic press), pp. 51–68.

Mehr, I.J., and Seifert, H.S. (1998). Differential roles of homologous recombination pathways in *Neisseria gonorrhoeae* pilin antigenic variation, DNA transformation and DNA repair. Mol. Microbiol. *30*, 697–710.

Minton, K.W. (1996). Repair of ionizing-radiation damage in the radiation resistant bacterium *Deinococcus radiodurans*. Mutat. Res. *363*, 1–7.

Mortimer, R.K. (1958). Radiobiological and genetic studies on a polyploid series (haploid to hexaploid) of *Sacchromyces cerevisiae*. Radiat. Res. *66*, 158–169.

Moseley, B.E., and Copland, H.J. (1976). The rate of recombination repair and its relationship to the radiation-induced delay in DNA synthesis in *Micrococcus radiodurans*. J. Gen. Microbiol. *93*, 251–258.

Moseley, B.E., and Laser, H. (1965). Repair of X-ray in *Micrococcus radiodurans*. Proc. R. Soc. Lond. B Biol. Sci. *162*, 210–222.

Moseley, B.E.B. (1983). Photobiology and radiobiology of *Micrococcus* (*Deinococcus*) *radiodurans*. Photochem. Photobiol. Rev. *7*, 223–275.

Murray, R.G.E. (1992). The family *Deinococcaceae*. In The Prokaryotes, Ballows, A., Truper, H.G., Dworkin, M., Harder, W., and Scheilefer, K.H., eds. (New York, Springer-Verlag), pp. 3732–3744.

Narumi, I., Satoh, K., Kikuchi, M., Funayama, T., Yanagisawa, T., Kobayashi, Y., Watanabe, H., and Yamamoto, K. (2001). The LexA protein from *Deinococcus radiodurans* is not involved in RecA induction following gamma irradiation. J. Bacteriol. *183*, 6951–6956.

Narumi, I., Satoh, K., Cui, S., Funayama, T., Kitayama, S., and Watanabe, H. (2004). PprA: a novel protein from *Deinococcus radiodurans* that stimulates DNA ligation. Mol. Microbiol. *54*, 278–285.

Naudet, R. (1974). Summary report on the Oklo phenomenon. French CEA Report: Bull. Infor. Scien. Tech. (Eng. Trans.) *193*, 7–85.

Nishimura, Y., and Uchida, K. (1994). Radiation sensitivities of *Acinetobacter* strains isolated from clinical sources. J. Basic Microbiol. *34*, 357–360.

Norais, C.A., Chitteni-Pattu, S., Wood, E.A., Inman, R.B., and Cox, M.M. (2009). DdrB protein, an alternative *Deinococcus radiodurans* SSB induced by ionizing radiation. J. Biol. Chem. *284*, 21402–21411.

Omelchenko, M.V., Wolf, Y.I., Gaidamakova, E.K., Matrosova, V.Y., Vasilenko, A., Zhai, M., Daly, M.J., Koonin, E.V., and Makarova, K.S. (2005). Comparative genomics of *Thermus thermophilus* and *Deinococcus radiodurans*: divergent routes of adaptation to thermophily and radiation resistance. BMC Evol. Biol. *5*, 57.

Oyaiza, H., Stackebrandt, E., Schleifer, K.H., Ludwig, W., Pohla, H., Ito, H., Hirata, A., Oyaizu, Y., and Komagata, K. (1987). A radiation-resistant rod-shaped bacterium, *Deinobacter grandis* gen. nov., sp. nov., with peptidoglycan containing ornithine. Intl. J. Sys. Bact. *7*, 62–67.

Parisi, A., and Antoine, A.D. (1974). Increased radiation resistance of vegetative *Bacillus pumilus*. Appl. Microbiol. *28*, 41–46.

Pavlov, A.K., Kalinin, V.L., Konstantinov, A.N., Shelegedin, V.N., and Pavlov, A.A. (2006). Was Earth ever infected by martian biota? Clues from radioresistant bacteria. Astrobiology *6*, 911–918.

Peng, F., Zhang, L., Luo, X., Dai, J., An, H., Tang, Y., and Fang, C. (2009). *Deinococcus xinjiangensis* sp. nov., isolated from desert soil. Int. J. Syst. Evol. Microbiol. *59*, 709–713.

Petrov, Y.V., Nazarov, A.I., Onegin, M.S., and Sakhnovsky, E.G. (2006). Natural nuclear reactor at Oklo and variation of fundamental constants: Computation of neutronics of a fresh core. Phys. Rev. C. *74*, 64610–64617.

Phillips, R.W., Wiegel, J., Berry, C.J., Fliermans, C., Peacock, A.D., White, D.C., and Shimkets, L.J. (2002). *Kineococcus radiotolerans* sp. nov., a radiation-resistant, Gram-positive bacterium. Int. J. Syst. Evol. Microbiol. *52*, 933–938.

Potts, M. (1994). Desiccation tolerance of prokaryotes. Microbiol. Rev. *58*, 755–805.

Pryor, W.A. (1978). The formation of free radicals and the consequences of their reactions *in vivo*. Photochem. Photobiol. *28*, 787–801.

Rainey, F.A., Nobre, M.F., Schumann, P., Stackebrandt, E., and da Costa, M.S. (1997). Phylogenetic diversity of the deinococci as determined by 16S ribosomal DNA sequence comparison. Int. J. Syst. Bacteriol. *47*, 510–514.

Rainey, F.A., Ray, K., Ferreira, M., Gatz, B.Z., Nobre, M.F., Bagaley, D., Rash, B.A., Park, M.J., Earl, A.M., Shank, N.C., *et al.* (2005). Extensive diversity of ionizing-radiation-resistant bacteria recovered from Sonoran Desert soil and description of nine new species of the genus *Deinococcus* obtained from a single soil sample. Appl. Environ. Microbiol. *71*, 5225–5235.

Romanovskaya, V.A., and Rokitko, P.V. (2002). The Effect of gamma-radiation and desiccation on the viability of the soil bacteria Isolated from the alienated zone around the Chernobyl nuclear power plant. Microbiology *71*, 608–613.

Servant, P., Jolivet, E., Bentchikou, E., Mennecier, S., Bailone, A., and Sommer, S. (2007). The ClpPX protease is required for radioresistance and regulates cell division after gamma-irradiation in *Deinococcus radiodurans*. Mol. Microbiol. *66*, 1231–1239.

Setlow, P. (1992). I will survive: protecting and repairing spore DNA. J. Bacteriol. *174*, 2737–2741.

Shashidhar, R., and Bandekar, J.R. (2009). *Deinococcus piscis* sp. nov., a radiation-resistant bacterium isolated from a marine fish. Int. J. Syst. Evol. Microbiol. *59*, 2714–2717.

Shenoy, M.A., Joshi, D.S., Singh, B.B., and Gopal-Ayengar, A.R. (1970). Role of bacterial membranes in radiosensitization. Adv. Biol. Med. Phys. *13*, 255–271.

Silverman, G.J., and Sinskey, A.J. (1968). The destruction of microorganisms by ionizing radiation. In In Antiseptic Disinfectants, Fungicides and Chemical and Physical Sterilization, Block, S.S., and Lawrence, C.A., eds. (Philadelphia, Lea and Febiger), pp. 741–761.

Slade, D., Lindner, A.B., Paul, G., and Radman, M. (2009). Recombination and replication in DNA repair of heavily irradiated *Deinococcus radiodurans*. Cell *136*, 1044–1055.

von Sonntag, C. (1980). Free radical reactions of carbohydrates as studied by radiation techniques. Adv. Carbohydr. Chem. Biochem. *37*, 7–77.

von Sonntag, C. (1987). The chemical basis of radiation biology (London, Taylor & Francis Ltd.).

von Sonntag, C. (2006). Free radical-induced DNA damage and its repair (Berlin, Springer).

von Sonntag, C., and Schulte-Frohlinde, D. (1978). Radiation-induced degradation of the sugar in model compounds and in DNA. Mol. Biol. Biochem. Biophys. *27*, 204–226.

Spinks, J.W.T., and Woods, R.J. (1990). Introduction to radiation chemistry (New York, Wiley).

Stadtman, E.R. (1993). Oxidation of free amino acids and amino acid residues in proteins by radiolysis and by metal-catalyzed reactions. Annu. Rev. Biochem. *62*, 797–821.

Tanaka, M., Earl, A.M., Howell, H.A., Park, M.J., Eisen, J.A., Peterson, S.N., and Battista, J.R. (2004). Analysis of *Deinococcus radiodurans*'s transcriptional response to ionizing radiation and desiccation reveals novel proteins that contribute to extreme radioresistance. Genetics *168*, 21–33.

United Nations Scientific Committee on the Effects of Atomic Radiation. Effects of Atomic Radiation Report A/RES/37/87. (New York, 1982)

Vaisanen, O.M., Weber, A., Bennasar, A., Rainey, F.A., Busse, H.J., and Salkinoja-Salonen, M.S. (1998). Microbial communities of printing paper machines. J. Appl. Microbiol. *84*, 1069–1084.

Walker, G.C. (1984). Mutagenesis and inducible responses to deoxyribonucleic acid damage in *Escherichia coli*. Microbiol. Rev. *48*, 60–93.

Walker, G.C. (1985). Inducible DNA repair systems. Annu. Rev. Biochem. *54*, 425–457.

Ward, J.F. (1975). Molecular mechanisms of radiation-induced damage to nucleic acids. Adv. Rad. Biol. *5*, 181–239.

White, O., Eisen, J.A., Heidelberg, J.F., Hickey, E.K., Peterson, J.D., Dodson, R.J., Haft, D.H., Gwinn, M.L., Nelson, W.C., Richardson, D.L., *et al.* (1999). Genome sequence of the radioresistant bacterium *Deinococcus radiodurans* R1. Science *286*, 1571–1577.

Winzeler, E.A., Shoemaker, D.D., Astromoff, A., Liang, H., Anderson, K., Andre, B., Bangham, R., Benito, R., Boeke, J.D., Bussey, H., *et al.* (1999). Functional characterization of the *S. cerevisiae* genome by gene deletion and parallel analysis. Science *285*, 901–906.

Witkin, E.M. (1946). Inherited differences in sensitivity to radiation in *Escherichia coli*. Proc. Natl. Acad. Sci. U.S.A. *32*, 59–68.

Yang, Y., Itoh, T., Yokobori, S., Itahashi, S., Shimada, H., Satoh, K., Ohba, H., Narumi, I., and Yamagishi, A. (2009). *Deinococcus aerius* sp. nov., isolated from the high atmosphere. Int. J. Syst. Evol. Microbiol. *59*, 1862–1866.

Yang, Y., Itoh, T., Yokobori, S., Shimada, H., Itahashi, S., Satoh, K., Ohba, H., Narumi, I., and Yamagishi, A. (2010). *Deinococcus aetherius* sp. nov., isolated from the stratosphere. Int. J. Sys.t Evo.l Microbiol. *60*, 776–779.

Yuan, M., Zhang, W., Dai, S., Wu, J., Wang, Y., Tao, T., Chen, M., and Lin, M. (2009). *Deinococcus gobiensis* sp. nov., an extremely radiation-resistant bacterium. Int. J. Syst. Evol. Microbiol. *59*, 1513–1517.

Zahradka, K., Slade, D., Bailone, A., Sommer, S., Averbeck, D., Petranovic, M., Lindner, A.B., and Radman, M. (2006). Reassembly of shattered chromosomes in *Deinococcus radiodurans*. Nature *443*, 569–573.

Zhang, C., Wei, J., Zheng, Z., Ying, N., Sheng, D., and Hua, Y. (2005). Proteomic analysis of *Deinococcus radiodurans* recovering from gamma-irradiation. Proteomics *5*, 138–143.

Zhang, Y.Q., Sun, C.H., Li, W.J., Yu, L.Y., Zhou, J.Q., Xu, L.H., and Jiang, C.L. (2007). *Deinococcus yunweiensis* sp. nov., a gamma- and UV-radiation-resistant bacterium from China. Int. J. Syst. Evol. Microbiol. *57*, 370–375.

Zimmerman, J.M., and Battista, J.R. (2005). A ring-like nucleoid is not necessary for radioresistance in the *Deinococcaceae*. BMC Microbiol. *5*, 17.

Zivanovic, Y., Armengaud, J., Lagorce, A., Leplat, C., Guerin, P., Dutertre, M., Anthouard, V., Forterre, P., Wincker, P., and Confalonieri, F. (2009). Genome analysis and genome-wide proteomics of *Thermococcus gammatolerans*, the most radioresistant organism known amongst the Archaea. Genome Biol. *10*, R70.

Psychrophiles: Life in the Cold

3

Corien Bakermans

Abstract

Psychrophilic, or cold-loving, organisms actively live at low temperatures. Psychrophily is not an uncommon trait; cold-adapted organisms are found throughout the three domains of life and successfully inhabit a wide variety of low-temperature environments. The ongoing investigation of these environments continues to broaden our view of what is possible for life on Earth. Cold-adapted microorganisms have evolved mechanisms to deal with the thermodynamic constraints of low temperatures. To combat the stability and decreased flexibility of macromolecules, psychrophiles generally increase the disorder within macromolecules to maintain fluidity or flexibility and hence function at low temperatures. To contend with reduced water activity and the presence of ice crystals, cryoprotectants are produced. To counteract decreased reaction and diffusion rates, psychrophiles practice efficient growth. Currently, the functional low-temperature limits of psychrophiles are −12°C for reproduction and −20°C for metabolism. The availability of liquid water appears to be the major growth-limiting factor at subzero temperatures. Examination of molecular and physiological adaptations to low temperatures is increasing our comprehension and appreciation of the capabilities of psychrophiles and their contribution to nutrient cycling in low-temperature environments.

Introduction

A large portion of the Earth's surface exists at temperatures below 4°C, including polar oceans, sea ice, glacial ice, the deep sea, snow cover, alpine areas and permafrost (Graumann and Marahiel, 1996; Russell and Hamamoto, 1998). While these environments are inhabited by a variety of cold-adapted organisms such as bacteria, archaea, yeasts, algae, insects, marine and terrestrial invertebrates, fish and plants, microorganisms dominate in terms of species diversity and biomass (Feller and Gerday, 2003). Microorganisms that thrive at temperatures near 0°C are termed psychrophiles from the Greek *psychros* meaning cold and *philos* meaning loving.

By the definition of Morita (1975), psychrophiles have cardinal growth temperatures (minimum, optimum, and maximum) at or below 0, 15 and 20°C, respectively. In contrast, mesophiles have minimum and maximum growth temperatures at 20 and below 45°C, respectively. Psychrophiles that meet the definition of Morita are sometimes called 'true psychrophiles' or stenopsychrophiles due to their narrow range of growth temperatures. Eurypsychrophilic microorganisms, also referred to as psychrotolerant or psychrotrophic, are able to grow at the low temperatures at which psychrophiles grow, including temperatures below 0°C. However, the optimum growth temperature of eurypsychrophiles is above 15°C and they can often grow at temperatures up to 30°C. Some researchers have argued that these definitions of psychrophiles and mesophiles are somewhat arbitrary and not based on biological or ecological considerations (Feller and Gerday, 2003; Cavicchioli, 2006). Here, the term psychrophile will be used to indicate both steno- and eury-psychrophiles.

Microorganisms that actively live at low temperatures must evolve mechanisms to deal with the accompanying thermodynamic constraints. These constraints include the reduction of available liquid water accompanied by the formation of ice crystals, lower rates of catalysis and transport, decreased membrane fluidity, and stabilization of molecular structures (Cavicchioli *et al.*, 2000). Most of these thermodynamic constraints apply to all microorganisms at the lower limits of their growth temperature ranges and have been studied as cold-shock responses (discussion of which is not within the scope of this chapter). Even thermophiles have cold-shock proteins to deal with the 'low temperatures' at the low end of their growth temperature ranges (Weinberg *et al.*, 2005). Of course, the most severe constraints affect those microorganisms (psychrophiles) that grow at the lowest temperatures.

Building on past research and fuelled by new techniques and capabilities, recent investigations of low-temperature environments and their psychrophilic inhabitants are redefining our understanding of physicochemical constraints, cold adaptations, biomolecular function, habitability and the limits of microbial activity (Graumann and Marahiel, 1996; Russell and Hamamoto, 1998; Deming, 2002; Bowman, 2004; D'Amico *et al.*, 2006). Here I summarize and review the current state of knowledge in these areas.

Physicochemical constraints of low temperatures

Microorganisms that live at low temperatures face challenges that arise from the essential nature of thermal energy. Temperature is a measure of the average energy of motion (vibration, rotation, or translation) of the atoms or molecules of a substance. Hence cool molecules move, vibrate and rotate less than hot molecules. Thermal energy affects phase changes from solid to liquid to gas, diffusion of molecules within the gas or liquid phase and conformational changes of enzymes and their substrates, allowing chemical transformations to occur. As temperature decreases the available thermal energy (enthalpy) of a system decreases, resulting in the increased stability and rigidity of molecules, freezing of water, decreased rates of diffusion and decreased chemical reaction rates (Russell, 1990; Cavicchioli, 2006).

The reduced enthalpy of molecules such as lipids, proteins and nucleic acids at low temperatures has biologically relevant effects. As temperature decreases, proteins will become less flexible and eventually, below a threshold point, cold denaturation will occur (D'Amico *et al.*, 2006). Secondary structures of DNA and RNA will become more stable, inhibiting the processes of replication, transcription and translation. As membranes stiffen at low temperatures, embedded proteins cannot function and transport nutrients effectively, causing starvation (Nedwell, 1999; Pomeroy and Wiebe, 2001). When 50–90% of membrane lipids are in the gel (solid) phase, cells will stop functioning (Jackson and Cronan, 1978; Melchior, 1982). Without functional biomolecules, microorganisms cannot perform the biochemical reactions necessary to sustain life.

Temperatures below 0°C can cause water to freeze or solidify. Without liquid water there is no solvent system for enzymes, membranes, etc. to function in or for substrates to diffuse through. Ice crystals can also pierce cell membranes causing extensive damage. Liquid water can be maintained below 0°C by a variety of physicochemical processes such as freezing point depression, ordering effects, supercooling and pressure. In environmental systems, freezing point depression is a dominant process for sustaining liquid water at subzero temperatures, occurs when solutes are present and is proportional to molal concentration. Hence, sea water (3.5% salinity) freezes at −1.9°C, while a saturated (~30%) NaCl solution freezes at −21°C. The mechanism of freezing point depression is as follows: as water freezes, pure water crystallizes first leaving a more concentrated solution that causes the freezing point to drop further. This depression will continue until the solution becomes saturated with solute, at which point further decreases in temperature result in crystallization of both solute and solvent. Freezing point depression allows for the presence of thin films of water in soils, sea ice, and glacial ice down to temperatures of −20°C, −26°C and −56°C, respectively (Price, 2000; Jakosky *et al.*, 2003; Bock and Eicken, 2005). Adsorbed water has a lower freezing point than bulk water due to ordering effects associated with being in close proximity to an ordered surface, and can extend as far as 1 μm from the surface (Drost-Hansen, 2001). Supercooling occurs in the absence of nucleators (a seed crystal or nucleus that usually begin the freezing process) and can cause water to remain liquid to −42°C. Supercooled water is frequently found in clouds. Microorganisms take advantage of this phenomenon by preventing nucleation within cells through the use of antifreeze proteins (Kawahara, 2002; Scotter *et al.*, 2006). Increased pressure also reduces the freezing point, but only slightly, and comes into play primarily beneath ice sheets and glaciers (Hodson *et al.*, 2008). Additionally, water need not enter a crystalline state at low temperatures, but can form a glass through the process of vitrification that occurs through rapid cooling or through the addition of substances, such as organic polymers, that prevent the formation of ice crystals. Cold-adapted microorganisms take advantage of the fact that liquid water can be maintained below 0°C via the aforementioned physicochemical processes. However, psychrophiles must also contend with ice crystals as a physical barrier to the diffusion of molecules (nutrients and wastes) to and from the cell.

Chemical reaction rates are particularly hard hit by the exponential decrease in thermal energy that accompanies decreasing temperatures, as defined by the Arrhenius equation:

$$k = Ae^{\frac{-E}{K_B T}}$$

where k is the reaction rate, A is the pre-exponential term, E is the activation energy, k_B is Boltzmann's constant and T is the absolute temperature in kelvin. Despite the exponential decrease in reaction rates with falling temperatures, microorganisms do not appear to be hampered by low rates of metabolic activity which appear sustainable for long periods of time (10^4 to 10^6 years) in various low-temperature ecosystems (Johnston and Vestal, 1991). For example, cryptoendolithic microbial communities of the Antarctic Dry Valleys (where temperatures rarely exceed 0°C) successfully invade and colonize sandstones over 10^3 to 10^4 years (Sun and Friedmann, 1999). Anomalous gas concentrations in the Greenland Ice Sheet are argued to be the product of extremely low rates of microbial activity at temperatures of −9°C to −11°C over 80,000 years (Sowers, 2001; Tung *et al.*, 2005, 2006; Miteva *et al.*, 2007). Furthermore, very low rates that support significant microbial communities have been reported in other ecosystems independent of low temperatures (D'Hondt *et al.*, 2004; Lin *et al.*, 2006). Sizable microbial communities (10^6 to 10^8 cells cm^{-3}) are maintained by very low net sulfate reduction rates of 1.8×10^{-3} $mol/m^2/year$ within deep subseafloor sediments where temperatures range from 2 to 25°C (Wang *et al.*, 2008). Because rates of reaction, diffusion, and chemical degradation decrease coordinately, microorganisms appear to continue metabolizing at low temperatures, albeit slowly.

Diversity and ecology

Psychrophily is not an uncommon trait, with cold-adapted organisms found throughout the three domains of life. To date, no genus of *Bacteria* or *Archaea* has been identified whose members are exclusively stenopsychrophiles, although at least one genus (*Colwellia*) of *Bacteria* contains only eury- and steno-psychrophiles. Psychrophiles are isolated from predominantly cold environments that are primarily found at high altitudes and latitudes and in the deep ocean, such as: polar oceans, the deep sea, alpine and high-latitude lakes, sea ice, lake ice, glaciers and ice sheets, snow cover and permafrost (Fig. 3.1). Antarctica alone contains a variety of extreme low-temperature environments, including ice-covered lakes, saline lakes, relatively dry soils and rocks, glaciers, subglacial lakes and streams, snow, seawater, sea ice, marine sediments and ice shelves.

A broad diversity of psychrophilic microorganisms inhabit low-temperature ecosystems. An overview of the diversity of several low-temperature environments, focusing on *Bacteria* and *Archaea*, is presented in Table 3.1 (for additional detail refer to Bej *et al.*, 2009). In general, bacteria (particularly *Proteobacteria*) are more dominant and more diverse than archaea in such environments, except at depth in marine systems. Cyanobacteria, such as *Oscillatoria, Phormidium* and *Nostoc commune*, dominate in Antarctica (not

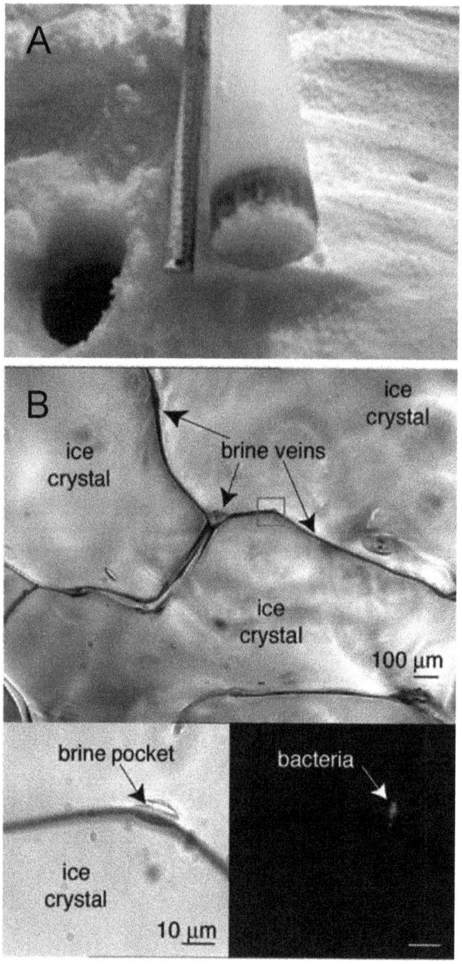

Figure 3.1 (A) Abundant microbial growth evident (dark band) in a core of sea ice. Image from www.arctic.noaa.gov/essay_krembsdeming.html. (B) Light micrograph of sea ice sample demonstrating the veins of liquid water between ice crystals in which microbes persist in frozen environments (Junge *et al.*, 2001). Reprinted from the Annals of Glaciology with permission of the International Glaciological Society and the authors.

detailed in Table 3.1). Psychrophilic fungi can also be found in these environments. Lichens are very common in terrestrial systems, as are yeasts (*Cryptococcus, Leucosporidium, Candida frigida*) in soils (Frisvad, 2008). Not surprisingly, stenopsychrophiles are isolated primarily from very stable low-temperature environments, whereas eurypsychrophiles are regularly isolated from environments with more variable-temperature conditions (Helmke and Weyland, 2004).

Psychrophiles contribute significantly to ecosystem processes within low-temperature environments. As in more temperate environments, psychrophilic algae and cyanobacteria serve as primary producers to a variety of cold-adapted organisms that range from heterotrophic bacteria and fungi to insects, nematodes and copepods (Vincent *et al.*, 2000; Thomas and Dieckmann, 2002). Reaction rates are also comparable to those in temperate environments, an example being carbon turnover and sulfate reduction in Arctic marine sediments (Sageman *et al*, 1998). Methane production in permafrost (which contributes ~25% of the total methane released from natural sources) and the deep sea (which contains massive methane-hydrate deposits) illustrates the magnitude of the impact of psychrophilic methanogens (Kruger *et al.*, 2005; Wagner *et al.*, 2005). Although only limited data about ecosystem processes are available for many low-temperature environments, exceptions are studies of the polar oceans which have provided extensive data on ecosystem dynamics. Despite lower temperatures, polar ocean ecosystem dynamics are comparable to other oceans. Annual productivity, biomass and phytoplankton species succession is similar to that in temperate low-latitude oceans (Smetacek and Nicol, 2005). The photosynthetic efficiencies at *in situ* temperatures of polar and temperate phytoplankton are similar, despite the lower maximum growth rates of the former. As elsewhere, phytoplankton growth is eventually limited by the supply of nutrients like iron (Smetacek and Nicol, 2005). Bacteria and the protozoa that graze on them continue to play an important role in polar ecosystems regardless of the decreased growth rates that result from lower temperatures (Sherr and Sherr, 2003). When bacterial production and biomass is lower in polar ecosystems, this is largely due to lower dissolved organic material inputs rather than lower temperatures (Kirchman *et al.*, 2009). Differences between polar and temperate ecosystems are not due to temperature, but instead due to other variables such as the seasonality of the sea-ice cover or nutrient availability (Smetacek and Nicol, 2005). Food webs of polar oceans are just as extensive and complex as elsewhere.

Table 3.1 Microbial diversity of low-temperature environments

Environment	Dominant features	Microbial inhabitants
Snow (alpine)	Meteoric ice, high porosity, seasonal	*Proteobacteria*, Actinobacteria, and Bacteroides dominate clone libraries. Potentially significant isolates: *Cryobacterium psychrophilum*, *Janthinobacterium lividum*, *Variovorax paradoxus*. Various algae: *Chlamydomonas* and *Chloromonas* (Segawa *et al.*, 2005; Liu *et al.*, 2009)
Glacier ice	Meteoric ice, low nutrient	Very few molecular studies (methodological constraints); however, *Proteobacteria* and *Cytophaga–Flavobacteria–Bacteroides* (CFB) common. *Proteobacteria*, high and low G+C Gram-positive, CFB commonly isolated (Christner *et al.*, 2000, 2001; Miteva *et al.*, 2004; Xiang *et al.*, 2005; Simon *et al.*, 2009)
Permafrost (permanently frozen ground)	High organic content (Arctic), congelation ice	Diverse bacteria (over 30 genera) with *Proteobacteria* dominant. Few studies of archaeal diversity. Methanogens present. Most commonly isolated eukaryotes are yeasts: *Cryptococcus*, *Rhodotorula* and *Saccharomyces* (Shi *et al.*, 1997; Vorobyova *et al.*, 1997; Vishnivetskaya *et al.*, 2000; Steven *et al.*, 2006)
Polar oceans	Moderate salinity (~3.5%), stable temperatures (–2 to 4°C)	Diverse bacteria: *Alphaproteobacteria* (*Roseobacter* and SAR-11), *Gammaproteobacteria* and CFB (*Polaribacter*) dominant. Marine group I Crenarchaeota and marine group II Euryarchaeota (20–25% of total RNA). Ammonia-oxidizers present. Diatoms dominate (DeLong *et al.*, 1994; Murray and Grzymski, 2007)
Sea ice	Moderate to high salinity, congelation ice, seasonal	Diverse bacteria: dominated by *Alpha*- and *Gammaproteobacteria* and CFB in both clone libraries and isolates. Few archaea. Diatoms dominate, *Phaeocystis* sp. also common. (Bowman *et al.*, 1997; Brinkmeyer *et al.*, 2003; Mock and Thomas, 2005; Kaartokallio *et al.*, 2008)
Deep sea	High pressure (up to 110 MPa), saline, stable temperatures (2 to 4°C)	Archaea (especially Crenarchaeota, haloarchaeal group IV) dominate. *Halomonas* common. All cultured piezophiles are *Gammaproteobacteria* (*Shewanella*, *Photobacterium*, *Colwellia*, *Psychromonas*, *Moritella*) (Kaye and Baross, 2000; Karner *et al.*, 2001; Nogi, 2008)
Polar marine sediments	Moderate salinity (~3.5%), stable temperatures (–2 to 0°C)	Diverse bacteria: *Gamma*- and *Deltaproteobacteria* (sulfate reducers) dominate. CFB also common. Lower numbers and diversity of archaea: Euryarchaeota (marine benthic group D) and Crenarchaeota (marine group I) (Ravenschlag *et al.*, 1999; Bowman, 2004; Kendall *et al.*, 2007; Tian *et al.*, 2009)

Adaptations to low temperatures

Psychrophilic microorganisms have evolved a variety of adaptations that allow them to function and thrive at low temperatures: (i) to combat the stability and decreased flexibility of macromolecules, the disorder is generally increased within these molecules to maintain fluidity or flexibility and hence retain function (Feller, 2007); (ii) to contend with reduced water activity and the presence of ice crystals, cryoprotectants are produced; and (iii) to counteract decreased reaction and diffusion rates, cold-adapted organisms practice efficient growth.

Protein adaptations

Because thermal energy from the environment fuels the conformational motions of enzymes that are the basis for their function, low temperatures reduce enzymatic activity. Consider that, 'If the environment is too cold, the enzyme may move so slowly that catalysis no longer occurs at a metabolically useful rate' (Fields, 2001). To facilitate retention of enzymatic activity at low temperatures, proteins from cold-adapted microorganisms have high reaction rates (k_{cat}) that become evident upon comparison with mesophilic and thermophilic homologues. Despite their relatively high temperature optima (30–64°C), psychrophilic enzymes maintain a high percentage (e.g. 60–40%) of their activity at low temperatures when compared with mesophilic and thermophilic enzymes, see Fig. 3.2A (Birolo *et al.*, 2000; Bendt *et al.*, 2001; Collins *et al.*, 2002). Psychrophilic enzymes also have higher k_{cat} than mesophilic enzymes at the same temperature, see Fig. 3.2B (Siddiqui and Cavicchioli, 2006). For example, the k_{cat} of psychrophilic, mesophilic and thermophilic α-amylases at 10°C is 17640, 5820 and 840 min^{-1}, respectively (D'Amico *et al.*, 2003). Increases in k_{cat} (or in the catalytic efficiency, k_{cat}/K_m) of cold-adapted enzymes are often achieved by lowering ΔG^*, the reaction energy (the activation energy barrier between the ground state and the transition state) (Collins, T., *et al.*, 2008). This can be achieved by increasing the disorder of the enzyme-substrate ground state which also results in increased K_m (lower binding affinity) and decreased substrate specificity (Fig. 3.2C).

These characteristics of psychrophilic enzymes, which manifest in the increased disorder of the enzyme-substrate ground state, suggest that cold daptation is realized by increasing the flexibility of the active site. This flexibility is commonly achieved by reducing weak stabilizing interactions, increasing solvent interactions with apolar or interior residues, reducing proline and arginine content or clustering of glycine residues (Feller 1997; Haney, PNAS, 1999; Chakravarty, 2000; Fields, 2001; Tekaia, 2002; Saunders, 2003). The range of currently documented modifications that increase the flexibility of cold-adapted enzymes is summarized in Table 3.2. The increased flexibility of psychrophilic proteins also results in less stable, more thermolabile proteins for which the enthalpy of unfolding (ΔH_{cal}) is reduced. For example, the ΔH_{cal} of psychrophilic, mesophilic and thermophilic homologues of dihydrofolate reductase is 209, 247 and 466 kJ/mol, respectively (Xu *et al.*, 2003). As a result of reduced ΔH_{cal}, psychrophilic enzymes denature or melt at lower temperatures: 44 versus 66°C for psychrophilic

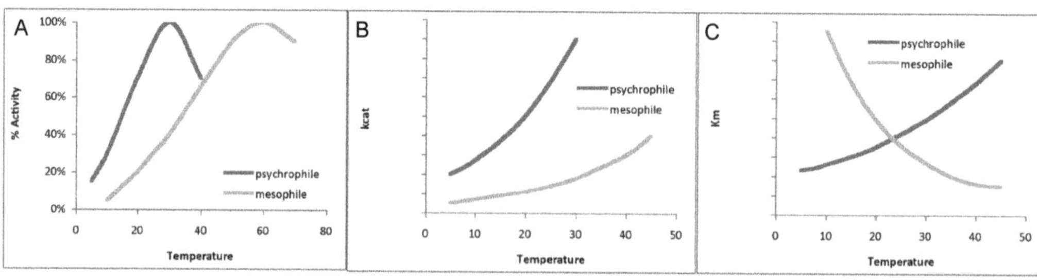

Figure 3.2 Temperature dependence of traits of psychrophilic and mesophilic enzymes. (A) Activity. (B) Specific activity, k_{cat}. (C) Binding affinity, K_m (note that K_m values for mesophiles are plotted on a scale 10 × that of psychrophiles).

Table 3.2 Features that can contribute to protein structural flexibility

	Trend in psychrophiles[1]	Examples
Core hydrophobicity	Buried amino acids (AA) smaller and less hydrophobic	↓ Ile in trypsin, citrate synthase, α-amylase and many more (Aghajari *et al.*, 1998; Russell *et al.*, 1998; Leiros *et al.*, 1999)
Surface hydrophobicity	Higher proportion of non-polar AA on surface	Adenylate kinase (Bae and Phillips, 2004) Xylanase (Van Petegem *et al.*, 2003) Many more
Surface hydrophilicity	Increase in surface charge	β-Lactamase (Feller *et al.*, 1997) Cellulase (Garsoux *et al.*, 2004) Many more
Hydrogen bonds	Fewer	Ala replaced Ser in triosephosphate isomerase (Alvarez *et al.*, 1998)
Arginine-mediated	Reduced Arg/Lys ratio	β-Galactosidase (Hoyoux *et al.*, 2001) Chitobiase (Lonhienne *et al.*, 2001) Many more
Aromatics	Fewer interactions with amino groups or other aromatics	Subtilisin (Davail *et al.*, 1994) β-Lactamase (Feller *et al.*, 1997) α-Amylase (Feller, 2003)
Salt bridges	Fewer	Subtilisin (Davail *et al.*, 1994) α-Amylase (Feller, 2003)
Secondary structure elements	Increased Pro in α-helices	Citrate synthase (Russell *et al.*, 1998)
	Decreased numbers of stabilizing charged AA at termini of α-helices	α-Amylase (Feller, 2003)
Surface loops	Decreased Pro in loops	Many
	More and longer loops connecting α-helices and β-sheets	Triosephosphate isomerase (Alvarez *et al.*, 1998)
Other factors	More Met (halophilic adaptation?)	Many (Thomas and Cavicchioli, 1998)
	Lower affinity for Ca^{2+}	α-Amylase (Feller *et al.*, 1994)
	Fewer disulfide bridges	Equivocal evidence (Feller *et al.*, 1994; Arnorsdottir *et al.*, 2005)

[1]Compared with mesophilic and thermophilic homologues.

versus mesophilic α-amylase (D'Amico *et al.*, 2003). Furthermore, highly flexible proteins may be more prone to cold denaturation, a highly understudied phenomenon.

While a variety of amino acid modifications can result in functional proteins at low temperatures, the actual amino acid modifications made in a particular protein are specific to that protein and cannot be predicted. However, trends in amino acid composition can be seen at the genome and metagenome level of psychrophiles (when compared with mesophiles and thermophiles). These trends are most evident when comparisons are made between buried and solvent-exposed residues of homologous proteins. For example, increases in glutamine and threonine in solvent-accessible regions of proteins were identified in psychrophilic methanogens (Saunders *et al.*, 2003). The trends evident in this study may have been more apparent due to a limited comparison of closely related genera (only methanogens were examined). Comparison of genomes from bacterial psychrophiles, mesophiles and thermophiles also revealed significant trends that included an increase in serine and a decrease in charged surface residues (Methe *et al.*, 2005). While few metagenomic surveys have been conducted in low-temperature ecosystems, one study of Antarctic bacterioplankton observed decreases in salt bridges, proline residues and stabilizing hydrophobic clusters along with increased length of disordered amino acids within identified open reading frames (Grzymski *et al.*, 2006).

Protein structure, and hence function, can also be maintained at low temperatures with the aid of protein chaperones and compatible solutes (see below). Protein chaperones aid in the proper folding of proteins and in the refolding of denatured proteins. The first chaperone encountered by newly synthesized proteins in bacteria is trigger factor, a peptidyl-prolyl *cis–trans* isomerase (PPIase), which counters a rate-limiting step in protein folding (Schmid, 1993). Protein chaperones like PPIases, caseinolytic proteases, DnaJ and DnaK are commonly induced in psychrophiles at low temperatures (Goodchild *et al.*, 2005; Kawamoto *et al.*, 2007). The importance of chaperones to the ability of organisms to grow at low temperatures was highlighted in a study that demonstrated that expression of cold-adapted chaperones (Cpn60/10) in the mesophilic *Escherichia coli* results in its ability to grow at previously non-viable low temperatures such as 4°C (Ferrer *et al.*, 2004). Furthermore, growth at low temperatures apparently proceeds due to Cpn60/10 stabilization of cold-sensitive proteins such as the protein chaperones DnaK and ClpB (Strocchi *et al.*, 2006). These studies highlight the importance of chaperones and correct protein structure to the maintenance of enzyme activity at low temperatures.

Alternatively, increasing the abundance of a particular protein may compensate for reduced levels of activity at low temperatures. Proteomic studies reveal the increased expression of a variety of proteins during low-temperature growth, including isocitrate lyase, elongation factors, ribosomal proteins and transport proteins (Berger *et al.*, 1996; Goodchild *et al.*, 2004; Bakermans *et al.*, 2007). While there are many reasons to increase expression of a particular protein at low temperature (discussed throughout), the abundance of some proteins may be increased to counteract reduced levels of activity. An increase in abundance of transport proteins suggests a means to compensate for reduced substrate binding affinity in addition to low levels of activity. For example, the abundance of six transport-related proteins (FecA, AfuA, Uup, DctP, TolC and LolD) was increased during growth at –4°C in *Psychrobacter cryohalolentis* (Bakermans *et al.*, 2007). The increased K_m of transport proteins presents a special problem for the uptake of nutrients. Organisms in nature are often challenged by low substrate concentrations. The combination of oligotrophic conditions with transport proteins that have a reduced binding affinity for substrates results in severely nutrient-limited organisms (Nedwell, 1999; Pomeroy and Wiebe, 2001). The synergistic effect of low temperatures and low nutrient concentrations may ultimately define the minimum growth temperature of microorganisms, given that there is a physical limit on how many transport proteins can be inserted into the cell membrane.

Nucleic acid adaptations

Low temperatures stabilize nucleic acid secondary structures, thereby inhibiting the processes of transcription, translation and DNA replication. Nucleic acid stability can be affected by the abundance of GC base pairs, which are more stable than AT pairs due to the presence of three hydrogen bonds. Thermophiles typically display an increased G+C content to enhance the stability of duplex nucleic acids, a trend that can be identified at both the whole genome and individual gene level (Nakashima *et al.*, 2003). Unfortunately, the opposite trend (reducing G+C content to maintain conformational flexibility) is not easily discerned in psychrophiles and hence may not be relevant to cold-adaptation. Examination of RNA molecules whose function relies on duplex formation reveals that the G+C content of ribosomal and transfer RNA stems correlates with the optimum growth temperature (OGT) of mesophiles and thermophiles (Galtier and Lobry, 1997). More recent studies have revealed that while G+C content of ribosomal RNA does not correlate well with OGT in psychrophiles, uracil content correlates inversely with OGT of psychrophiles and thermophiles ($r = -0.905$), with notable exceptions within mesophilic microbes (Khachane *et al.*, 2005). In addition, transfer RNAs of psychrophiles show an increase in dihydrouridine content (Dalluge *et al.*, 1997; Noon *et al.*, 2003). Hence, as with proteins, nucleic acids of psychrophiles can show adaptations at the sequence level that contribute to the ability to grow and survive at low temperatures.

Cold-adapted microorganisms can also alleviate stress on transcription, translation and DNA replication with specialized chaperones and helicases (Jiang *et al.*, 1997; Chamot and Owttrim, 2000; Phadtare *et al.*, 2002). For example, CspA is one member of a family of RNA chaperones that bind both RNA and single-stranded DNA and destabilize secondary structure (Jiang *et al.*, 1997; Phadtare *et al.*, 2002). CspA expression is commonly induced during growth at low temperatures in psychrophiles and during cold-shock of mesophiles (Berger *et al.*, 1997; Horton *et al.*, 2000). Cold-inducible DEAD-box helicases unwind duplexes in messenger and ribosomal RNA to ensure successful translation (Lim *et al.*, 2000; Bergholz *et al.*, 2009). Elongation factors (EFs) enable translation by catalysing the translocation of peptidyl-tRNA from the A site to the P site of the ribosome and are induced at low growth temperatures in psychrophiles (Thomas and Cavicchioli, 2000; Bakermans *et al.*, 2007). The increased abundance of EFs at low temperature may compensate for reduced levels of activity. Increased expression of specific ribosomal proteins during growth at low temperatures has been documented, although how this may contribute to ribosome function is unknown (Bayles *et al.*, 2000; Strocchi *et al.*, 2006; Bakermans *et al.*, 2007). While many proteins involved in translation have been studied to identify adaptations (amino acid substitutions) that allow continued function at low temperatures, little is known about potential differences in mechanisms of action between psychrophilic, mesophilic and thermophilic homologues.

Membrane adaptations

At low temperatures membrane lipids can crystallize, severely impairing membrane function. Microorganisms strive to achieve the correct membrane fluidity at all growth temperatures by altering the composition of membrane lipids, a process referred to as homeoviscous adaptation. At low temperatures, membrane fluidity can be retained by reducing membrane packing and hence viscosity (Fig. 3.3).

In most microorganisms, the primary mechanism of maintaining membrane fluidity at low temperatures is by increasing the relative amounts of unsaturated lipids. For example, psychrophilic bacteria commonly increase the proportion of $C_{18:1}$ fatty acids at low temperatures (Russell and Fukunaga, 1990; Russell, 1997). Desaturases, which act directly on membrane lipids to remove hydrogen to a terminal electron acceptor (usually oxygen), are often expressed in response to low temperatures (Murata and Wada, 1995; Bergholz *et al.*, 2009). Desaturation occurs most commonly at the Δ^9 position (the ninth carbon from the carboxyl carbon), with other common positions including Δ^5, Δ^{10} (*Bacteria*) and Δ^6, Δ^{12}, Δ^{15} (yeast, filamentous fungi, algae and some cyanobacteria). The mechanism of desaturation is not reversible and hence is only an advantage

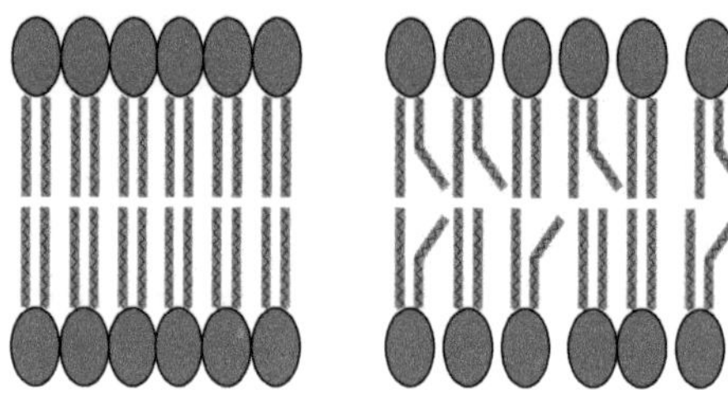

Figure 3.3 Desaturation of lipids reduces membrane packing (chains do not pack as tightly and are less ordered) making it more difficult for the lipids to crystallize or freeze.

for adaptation to a drop in temperature (Russell, 2008).

Desaturation of fatty acids can also be achieved through changes in *de novo* synthesis. The 'anaerobic pathway' of synthesis is a branched pathway containing two β-ketoacyl-ACP-synthases. Synthase II is more efficient at elongating 16:1Δ^9 than 16:0 and is more thermolabile than synthase I (Cronan, Jr., and Rock, 1996). Hence, at low temperatures more 16:1Δ^9 is elongated, while at high temperatures more 16:0 is elongated. Changes in *de novo* synthesis are the only way to decrease acyl chain length and the proportion of branch-chained lipids (Russell, 2008), two additional mechanisms for increasing membrane fluidity. At low temperatures, psychrophilic bacteria commonly increase the proportion of C16 fatty acids (Russell, 1984). The effect of shorter fatty acids (FA) on membrane fluidity is exaggerated when one FA chain of a phospholipid is short and the other is longer. The use of methyl branched-chain FA to increase membrane fluidity is most common in Gram-positive bacteria, with *anteiso*-branched FA being more fluidizing than *iso*-branched FA. In addition, membrane fluidity can be affected by altering polar head groups, although differences tend to be smaller without any obvious pattern (Russell and Fukunaga, 1990). Anaerobic bacteria also reduce membrane packing by altering the alkyl/alkenyl and acyl compositions of ether lipids (Goldfine, 1984). Note that trans-unsaturated and cyclopropane FA are not involved in low-temperature adaptation of cell membranes. Instead these FA serve to strengthen and reduce the fluidity of membranes.

Despite their abundance in cold ocean waters and other low-temperature environments, little is known about how archaea may alter their membrane lipids at low temperatures. The few psychrophilic archaea that have been examined also unsaturate their lipids in response to low temperatures. For example, *Methanococcoides burtonii* appears to control unsaturation of lipids through selected saturation during lipid biosynthesis (Nichols *et al.*, 2004). *Halorubrum lacusprofundi* was also shown to increase the proportion of unsaturated lipids in its membrane at low temperatures (Gibson *et al.*, 2005). Eukaryotic microorganisms maintain membrane fluidity at low temperatures by varying the relative proportion of sterols (which strengthen the membrane and decrease fluidity) and phospholipids in their membranes (Russell, 2008). The different ways that microorganisms alter membrane lipids to achieve homeoviscous adaptation is summarized in Table 3.3.

Cryoprotectants (compatible solutes)

Living at low temperatures brings with it the threat of reduced water availability due to the formation of ice crystals and the concentration of solutes into the remaining liquid water. Psychrophiles can combat the effects of reduced water activity by synthesizing compatible solutes such as sugars, amino acids, alcohols or cryoprotective proteins. Ice crystal formation can be prevented by producing anti-nucleating material (proteins and polysaccharides) and ice binding proteins.

Compatible solutes, also referred to as extrinsic stabilizers or cryoprotectants, depress the freezing point of water within the cell and are excluded from the hydration sphere of proteins, thereby allowing them to retain a layer of free water at their surface and maintain their structure (Welsh, 2000). Compatible solutes act globally and non-specifically, allowing microorganisms to occupy broader thermal niches and to react quickly to changes in temperature (Chin *et al.*, 2010). Common compatible solutes include glycine betaine, glycerol, trehalose, sucrose, proline and

Table 3.3 Changes made to membrane lipids to maintain fluidity

Change	Trend at low temperatures[1]	Mechanism of change	Microorganisms employing change
FA unsaturation	Increases	Desaturases modify existing lipid *or de novo* biosynthesis	Bacteria, archaea, yeasts, filamentous fungi, algae
FA methyl branching	Increases	*De novo* biosynthesis	Bacteria
Anteiso/*iso*-branched ratio	Increases	*De novo* biosynthesis	Bacteria
FA average chain length	Decreases	*De novo* biosynthesis	Bacteria, yeasts, filamentous fungi, algae
Sterol/phospholipid ratio	Decreases	*De novo* biosynthesis	Yeasts, filamentous fungi, algae

[1]As compared with higher temperatures.

FA, fatty acid (acyl).

dimethylsulfoniopropionate (Welsh, 2000). Glycine betaine is known to prevent the aggregation of proteins and aid in maintaining optimal membrane fluidity. Furthermore, the solute glutamate affected the catalytic efficiency, while aspartate increased the GTP binding affinity, of elongation factor 2 at low temperatures in *Methanococcoides burtonii* (Thomas *et al.*, 2001). Genes for the conversion of choline to betaine, the transport of betaine and the symport of proline/Na+ are commonly found in the genomes of psychrophiles (Methe *et al.*, 2005; Bowman, 2008). For example, *Psychromonas ingrahamii* contains the genes for the synthesis of glycine betaine from choline and for the uptake of both choline and glycine betaine (Riley *et al.*, 2008). While genome studies have revealed the potential of psychrophiles to use compatible solutes like glycine betaine, research is needed to confirm which molecules are used as compatible solutes for protection from low temperatures and high salt concentrations.

Cryoprotective proteins surround other proteins to prevent denaturation and have been found in a variety of plants (Lin and Thomashow, 1992), but, to date, have only been found in several ice-nucleating bacteria. For example, the cryoprotective protein Cor26 accumulates in *Pseudomonas fluorescens* KUIN-1 in response to low temperatures, although its manner of action is unclear (Obata *et al.*, 1998).

Anti-nucleating materials, which allow water to supercool to lower temperatures by lowering the activity of ice nucleators, have not been extensively studied in microorganisms. An anti-nucleating protein has been isolated from *Acinetobacter calcoaceticus* which can shift the supercooling point of water to -5°C by inhibiting the formation of ice nuclei (Kawahara *et al.*, 1996). Polyglycerol has also been found to be an effective anti-nucleating material (Wowk and Fahy, 2002), while a polysaccharide from *Bacillus thuringiensis* has been isolated which prevents ice nucleators from acting (Yamashita *et al.*, 2002). Exopolysaccharides (EPS) may be important anti-nucleating materials and are believed to play an important role in the survival of bacteria that inhabit sea ice. EPS can affect ice texture, permeability and brine drainage by preventing recrystallization, impeding anion diffusion and restricting fluid flow which can help organisms resist being flushed from the ice (Krembs and Engel, 2001; Meiners *et al.*, 2004; Collins *et al.*, 2008). EPS may protect organisms by providing a buffer zone between the organism and the bulk water where anion diffusion is impeded and large amounts of bound 'non-freezing' water are contained (Krembs *et al.*, 2000, 2002).

Ice-binding proteins (IBPs) can bind to and inhibit the growth of ice crystals and can impede recrystallization. Anti-freeze proteins (AFPs), one type of IBP, lower the freezing point of water without affecting the melting point. The form of ice crystals is often different in the presence of IBPs and AFPs (in some cases needle-like growth is seen). Recrystallization, a natural process whereby ice crystals grow from small crystals into large crystals, can be inhibited by very small amounts of IBPs or AFPs (0.1 μg/ml). AFPs have not been well studied in bacteria, although they have been isolated from several, including *Pseudomonas putida* GR12-2 and the Antarctic isolate *Marimonas primoryensis* (Xu *et al.*, 1998; Gilbert *et al.*, 2005). In addition, anti-freeze activity has been demonstrated in several genera including *Micrococcus, Rhodococcus, Sphingomonas, Halomonas, Pseudoalteromonas* and *Psychrobacter,* although activity has not been further characterized or identified (Kawahara, 2008).

Efficient growth

Efficient resource utilization may be key to the survival of heterotrophic microbes in low-temperature environments. While little is known about how low temperatures affect resource efficiency in psychrophiles, study of the proteome of *Methanococcoides burtonii* suggests that efficient carbon utilization occurs during growth at low temperatures (Goodchild *et al.*, 2004, 2005). An increase in proteins directing methyl coenzyme M production and concomitant decrease in the oxidative branch of methylotrophic methanogenesis, along with the downregulation of ATP synthase, during growth at 4°C suggests that 'at lower temperatures, *M. burtonii* finely tunes the methanogenesis pathway in order to maintain both a proton motive force and a minimal level of biosynthetic activity' that results in more efficient utilization of carbon (Goodchild *et al.*, 2005). Genome sequence analysis reveals that *Psychrobacter arcticus* 273-4 can conserve resources via the glyoxylate shunt, a bypass of the TCA cycle which allows cells to conserve carbon when growing on two-carbon compounds, as two key enzymes of the glyoxylate shunt (isocitrate lyase and malate dehydrogenase) are present (Ayala-del-Río *et al.*, 2010). In addition, the transcriptome of *P. arcticus* 273-4 indicates that resource conservation occurs during growth at low temperatures, even under nutrient replete conditions, via the upregulation of RNases and peptidases that recycle nucleotides and amino acids (Bergholz *et al.*, 2009).

To date, only psychrophiles have been shown to grow efficiently at temperatures significantly lower than their optimum growth temperature. For example, in two studies of sulfate-reducing bacteria (SRB), a mesophilic SRB was shown to grow most quickly and efficiently at about 31°C (when examined from 5 to 37°C) while a psychrophilic SRB grew equally efficiently at temperatures from −2°C to 7°C, although it grew most quickly at 7°C (Isaksen and Jorgensen, 1996; Knoblauch and Jorgensen, 1999). In addition, many eurypsychrophiles also grow most efficiently at temperatures of 0–8°C (Knoblauch *et al.*, 1999; Sabri *et al.*, 2000; Bakermans and Nealson, 2004). Currently there is a paucity of data on the temperature dependence of growth efficiency. Therefore, it is unknown whether efficient growth at temperatures significantly lower than the optimum is limited to psychrophiles. Growing efficiently at low temperatures might compensate for decreases in available energy and rates, allowing psychrophiles to remain competitive (Harder and Veldkamp, 1968; Veldkamp *et al.*, 1984).

Limits

The functional limits of biomolecules and active metabolism at low temperatures remain ill-defined and understudied. Whether or not microorganisms have already maximized their capacity to alleviate low-temperature stress (which results from the stabilization of secondary structures of nucleic acids) on transcription, translation, and DNA replication remains to be determined and may ultimately depend on the cold adaptation of proteins (see below) like chaperones and helicases. Considering the low melting points of unsaturated fatty acids [oleic acid ($C_{18:1}\Delta^9$) melts at 14°C, while palmitoleic acid ($C_{16:1}\Delta^9$) and linoleic acid ($C_{18:2}\Delta^9\Delta^{12}$) melt at −0.1°C and −5°C, respectively (Williams and Frausto da Silva, 1996)], the functional limits of membrane lipids do not appear to have been reached.

While *in vitro* enzyme activity has been reported at −83°C (Daniel *et al.*, 1998, 2003), very few studies of the functional limits of cold-adapted proteins at low temperatures have been conducted. One study introduced mutations into a psychrophilic α-amylase to restore mesophilic properties like thermostability (D'Amico *et al.*, 2001). However, the mutation (N288V) actually decreased the melting temperature, and at the same time reduced the activity to half that of the wild-type enzyme. Destabilization of the enzyme (an increase in flexibility) resulted in loss of activity at low temperatures. These data suggest that the low-temperature limits have almost been reached for this psychrophilic α-amylase. In contrast, directed evolution of psychrophilic subtilisin S41 created a protein that was more active at low temperatures and less thermolabile than wild-type, but only towards a specific substrate, suggesting that this protein may not have reached its functional low-temperature limit (Miyazaki *et al.*, 2000). In addition, because the specific activity of psychrophilic and mesophilic enzymes is not always the same at their respective optimal temperatures, it could be argued that adaptation to cold in psychrophiles is not complete (assuming that mesophilic enzymes are optimized). Currently there is insufficient data to determine if further evolutionary cold adaptation of proteins is incomplete or not possible.

Have the low-temperature functional limits of proteins, nucleic acids and membranes been reached? Individually, biomolecules appear capable of function at extreme low temperatures; however, the functional limits of biomolecules within complex systems may vary drastically. Cold-adapted microorganisms must coordinate the low-temperature functionality of many biomolecules and their interactions with each other and the environment. Hence, by examining the capabilities of microorganisms inhabiting low-temperature environments on Earth, the functional low-temperature limits of cellular systems can be explored, while recognizing that the low-temperature capabilities that have evolved depend on the particular conditions of the environment and selection pressures experienced.

Currently, the lowest reported temperature for growth and reproduction of microorganisms is –12°C. The sea ice isolate *Psychromonas ingrahamii* has a generation time of 240 h at –12°C as measured by turbidity readings of broth cultures with 50 g/l glycerol added to maintain liquidity of the medium (Breezee *et al.*, 2004). Current studies on reproduction at low temperatures rely on laboratory cultures of isolates and standard measurement techniques such as plate counts or turbidity measurements that may not detect extremely low growth rates. Unfortunately, even under 'ideal' growth conditions in the laboratory, reproduction rates at low temperatures quickly approach long time spans, with generation times of several years predicted for growth at –20°C (Bakermans *et al.*, 2003; Jakosky *et al.*, 2003).

Because reproduction is difficult to measure directly at very low temperatures, it is more common to examine or measure metabolic processes as proxies of microbial activity in both laboratory and environmental studies. Through a variety of techniques, metabolic activity has been demonstrated at temperatures from –9 to –20°C (Table 3.4). These techniques cannot readily distinguish between growth, maintenance, or survival metabolism of microorganisms without knowing if low growth rates correspond to maintenance and survival or just long generation times. Microbial activity at –20°C has been convincingly demonstrated by several laboratories using different techniques that are measures of coordinated metabolism.

Debate continues over whether microbial activity is possible below –20°C. Several intriguing studies have recently been published describing activity of microorganisms at temperatures below –20°C (Christner, 2002; Junge *et al.*, 2006; Panikov *et al.*, 2006; Panikov and Sizova, 2007; Amato and Christner, 2009). Amato and Christner (2009) observed that *Psychrobacter cryohalolentis* (a permafrost isolate) increased the concentration of adenylate in response to temperatures below freezing and could generate ATP while frozen at temperatures from –5 to –80°C.

In Panikov *et al.* (2006), permafrost soils were thawed to +1°C for the addition of ^{14}C-glucose, sequentially incubated at temperatures from 0 to -33°C and the $^{14}CO_2$ released was measured. Permafrost soil at all temperatures produced two orders more $^{14}CO_2$ than autoclaved controls. However, incubation at individual temperatures was brief (two to seven days) and, at temperatures below –20°C, $^{14}CO_2$ release decreased with time. In another study from the same laboratory, permafrost isolates were grown on ethanol mineral medium with microcrystalline cellulose at temperatures from –8 to –35°C and monitored for both CO_2 production and $^{14}CO_2$ uptake for two months (Panikov and Sizova, 2007). At all temperatures cultures produced CO_2 at levels well above the autoclaved controls; however, no

Table 3.4 Activity of microorganisms at subzero temperatures

T (°C)	Environment	Metabolic process analysed	Method
–10	Laboratory culture of permafrost isolate	Cell division	Plate counts, turbidity measurement (Bakermans *et al.*, 2003)
–12	Laboratory culture of sea ice isolate	Cell division	Turbidity measurement (Breezee *et al.*, 2004)
–9	Glacial ice	CO_2/CH_4 production	Correlation of cell counts with CO_2, CH_4 (Tung *et al.*, 2006)
–10	Arctic permafrost	Lipid synthesis	Incorporation of ^{14}C-acetate into lipids (Rivkina *et al.*, 2000)
–11	Glacial ice	CH_4 production	Correlation of cell counts with CH_4 (Tung *et al.*, 2005)
–15	Laboratory cultures of glacial ice isolates	DNA/Protein synthesis	Uptake of ^{3}H-thymidine and ^{3}H-leucine (Christner, 2002)
–15	Arctic permafrost	Respiration	Incorporation of ^{14}C-labelled glucose (Gilichinsky *et al.*, 2003)
–16.5	Arctic permafrost	CH_4 production	CH_4 production from ^{14}C-bicarbonate or ^{14}C-acetate (Rivkina *et al.*, 2002)
–18	Polar lichens	Photosynthesis	CO_2 exchange (Kappen *et al.*, 1996)
–18	Laboratory cultures of permafrost isolates	Respiration	Incorporation of $^{14}CO_2$, release of CO_2 (Panikov and Sizova, 2007)
–20	Sea ice	Respiration	CTC reduction and FISH (Junge *et al.*, 2004)
–20	Laboratory cultures of permafrost isolates	Respiration	Resazurin reduction (Jakosky *et al.*, 2003)
–20	Permafrost soil columns	Respiration	Oxidation of ^{14}C-labelled glucose (Panikov *et al.*, 2006)
–20	Laboratory culture of sea ice isolate	Protein synthesis	Uptake of ^{3}H-leucine (Junge *et al.*, 2006)

sustained activity was evident at temperatures below −18°C after 3 weeks. Interestingly, the 'early' activity at all temperatures occurred at the same rate for temperatures from −16 to −35°C. This temperature-independent early activity may be indicative of some physicochemical process (rather than metabolic activity) or of a burst of metabolic activity that prepared cells for dormancy.

Junge *et al.* (2006) examined the uptake of ^{3}H-leucine into protein at temperatures from 13 to −196°C using laboratory cultures of *Colwellia psychroerythraea*, a sea ice isolate. Uptake of ^{3}H-leucine into the protein fraction of cells was seen at all temperatures and was enhanced by the addition of exopolysaccharides (EPS). No significant activity was seen in TCA- or heat-killed controls, samples with chloramphenicol or sodium azide, *E. coli*, samples with EPS only, and time zero controls (which included the quick freeze and thawing steps). As in Panikov and Sizova (2007), rates were temperature-independent below −15°C. These authors suggest that the temperature-independent activity below −15°C represented the completion of enzyme–substrate interactions that were already in place after the first few seconds of freezing. In contrast, Daniel *et al.* (1998) demonstrated that the activity of a glutamate dehydrogenase did not deviate from Arrhenius behaviour at temperatures from 77°C to −83°C, implying that the enzyme remained functional over a 160°C temperature range. Also, Christner (2002) reported no significant activity at −70°C when examining ^{3}H-thymidine and ^{3}H-leucine incorporation at −15°C and −70°C into Antarctic strains of *Psychrobacter* and *Arthrobacter*.

These studies present tantalizing and sometimes contradictory data suggestive of microbial activity at temperatures below −20°C and await replication and verification by other researchers. Until then, the ability of microorganisms to actively metabolize at temperatures below −20°C will remain uncertain. It has long been recognized that extreme low temperatures are not lethal to organisms—cells are routinely preserved with cryoprotectants at −196°C in liquid nitrogen. More research is needed to understand what is occurring in these systems, what factors enhance low-temperature activity, and how to interpret the temperature-independence of activity below −15°C. Data compiled by Price and Sowers (2004) from a variety of environments suggest (via extrapolation) that metabolic activity capable of sustaining growth is possible down to −40°C. The availability of liquid water appears to be the major growth-limiting factor at subzero temperatures (Beaty *et al.*, 2006).

Considerations for future research

The ongoing investigation of cold-adapted microorganisms and low-temperature environments continues to broaden our view of what is possible for life on Earth. Both the molecular and physiological adaptations of psychrophiles to low temperatures are increasing our comprehension and appreciation of the capabilities of cold-adapted microorganisms and their contribution to nutrient cycling in cold ecosystems. Indeed, environments previously considered to be devoid of life (such as ice, permafrost, and subglacial watersheds) are being re-evaluated for the potential to harbour active microorganisms. While we are beginning to discern how the environment shapes the organism and vice versa, many challenges remain to be overcome in understanding how cold-adaptation is achieved and how microbial activity affects ecosystem dynamics in low-temperature environments.

A thorough understanding of cold adaptation is hampered by the study of a limited number of psychrophiles. Our collection of isolates, genome sequences and environmental metagenomes must be expanded to include a broad diversity of cold-adapted organisms and low-temperature environments. With a larger number of organisms and sequences, observed trends can be confirmed. In addition, while it can be difficult to simulate environmental conditions in the lab due to constraints of time or methodology, more attention should be paid to relevant physiological and environmental factors.

The low-temperature adaptations and capabilities evolved on Earth may not represent the actual functional low-temperature limits of terrestrial life because these adaptations are highly dependent on environmental conditions and selection pressures experienced. For example, cold adaptation may be limited by time due to slow growth rates and a relatively low number of generations for natural selection to act upon. Alternatively, cold adaptation may be limited by the persistence and pervasiveness of low-temperature environments on Earth. Many modern low-temperature environments are young (on a geological or evolutionary time scale) – Siberian permafrost began forming only 3 million years ago (Ma), while the permanent Antarctic ice sheet developed only 15 Ma (Gilichinsky *et al.*, 1992; Barrett, 2003). Just 55 Ma the global climate was warm enough for polar sea surface temperature to reach 23°C (Sluijs *et al.*, 2006). While other low-temperature epochs are evident in the geologic record (34, 150, 300, 450, 600, and 730 Ma), low-temperature adaptations that evolved during glacial periods may or may not be preserved during warm periods. Further studies examining the functional low-temperature limits of psychrophiles are warranted.

Growth at extreme subzero temperatures happens slowly relative to human life times and attention spans. Arctic lichens may be 3700 to 9000 years old, while Antarctic cryptoendoliths and microbes trapped in glacial ice have carbon turnover times of ~10,000 and ~100,000 years, respectively (Denton and Karlen, 1973; Friedmann *et al.*, 1993; Tung *et al.*, 2005). Hence, new measurement techniques must be devised and data carefully interpreted. A geology-based approach, using stable isotope techniques (Sowers, 2001) or concentration gradients of substrates and metabolic products (D'Hondt *et al.*, 2004), may be needed to measure very slow rates and long-term productivity at subzero temperatures. Reliable indicators of historical *in situ* temperature conditions will also be needed. Quantifiable definitions of growth, maintenance, and survival are required that can account for low temperatures, low nutrient concentrations, low diffusion rates, and low reaction rates. Microbial metabolism, whether continuous or sporadic, needs to be active enough to counter chemical degradation (such as amino acid racemization) and other damage (such as that resulting from background radiation). However, reproduction may not be necessary if adaptation and evolution are still possible. After all, why produce a competitor when nutrients and energy are scarce? Continued study of very low rates of metabolism will provide insight on the ecological impact of microbial activity in some of the most untenable low-temperature environments on the planet.

Low-temperature environments are poised to change dramatically in the face of global climate change. The Arctic Council and the International Arctic Science Committee recently reported that (1) 'annual average arctic temperature has increased at almost twice the rate as that of the rest of the world over the past few decades' (p. 10); (2) 'in Alaska and western Canada, winter temperatures have increased as much as 3–4°C in the past 50 years' (p. 22); and (3) 'the boreal forests and arctic tundra contain some of the world's largest land-based stores of carbon' (p. 38) (Hassol, 2004). Recent study of the diminishing sea ice cover has revealed a strong positive feedback between sea ice extent and air temperature that may rapidly cause further increases in Arctic temperatures (Screen and Simmonds, 2010). The effects of rising temperatures on microbial activity in the Arctic are beginning to be documented. Numerous studies have shown significant increases in CO_2 and CH_4 emissions from Arctic tundra and permafrost as temperature increases (Christensen *et al.*, 2003; Schimel *et al.*, 2004; Shaver *et al.*, 2006). For example, CO_2 emissions from thawed permafrost increase with time and are due to respiration of old, previously frozen carbon that eventually results in net carbon release to the atmosphere (Schuur *et al.*, 2009). Presumably this release is facilitated by microorganisms that have become more active as temperatures increase. However, since the microbial component in Arctic tundra, permafrost, sea ice, sea water and ice shelves has yet to be fully described, it is not possible to accurately discern and model the effects of rising temperatures.

Recent discoveries of water and methane on Mars have renewed interest in Astrobiology (the study of the origin, evolution, distribution, and future of life in the universe) and encouraged research in low-temperature terrestrial systems (Boynton *et al.*, 2002; Formisano *et al.*, 2004; Murray *et al.*, 2005; Geminale *et al.*, 2008). Defining the low-temperature adaptations and limits of terrestrial life has a major impact on discerning where in our predominantly cold solar system Earth-like life may exist, and has particular relevance to extraterrestrial sites that may be accessible within the next few decades, such as the cold surface of Mars, the cold-icy surface of Europa and comets.

The continued exploration of low-temperature environments and cold-adaptation will answer many of these questions and substantially increase our understanding of microbial activity and life at low temperatures.

Web resources

None specific to psychrophiles identified.

Long Term Ecological Research Sites with relevance to psychrophiles: Arctic http://www.lternet.edu/sites/arc/

McMurdo Dry Valleys http://www.lternet.edu/sites/mcm/

Palmer Station http://www.lternet.edu/sites/pal/

References

Aghajari, N., Feller, G., Gerday, C., and Haser, R. (1998). Crystal structures of the psychrophilic alpha-amylase from *Alteromonas haloplanctis* in its native form and complexed with an inhibitor. Protein Sci. *7*, 564–572.

Alvarez, M., Zeelen, J.P., Mainfroid, V., Rentier-Delrue, F., Martial, J.A., Wyns, L., Wierenga, R.K., and Maes, D. (1998). Triose-phosphate isomerase (TIM) of the psychrophilic bacterium *Vibrio marinus* – Kinetic and structural properties. J. Biol. Chem. *273*, 2199–2206.

Amato, P., and Christner, B.C. (2009). Energy Metabolism Response to Low-Temperature and Frozen Conditions in *Psychrobacter cryohalolentis*. Appl. Environ. Microbiol. *75*, 711–718.

Arnorsdottir, J., Kristjansson, M.M., and Ficner, R. (2005). Crystal structure of a subtilisin-like serine proteinase from a psychrotrophic *Vibrio* species reveals structural aspects of cold adaptation. FEBS J. *272*, 832–845.

Ayala-del-Río, H.L., Chain, P.S., Grzymski, J.J., Ponder, M.A., Ivanova, N., Bergholtz, P.W., DiBartolo, G., Hauser, L., Land, M., Bakermans, C., *et al.* (2010). The genome sequence of *Psychrobacter arcticus* 273–4, a psychroactive Siberian permafrost bacterium reveals mechanisms for adaptation to low temperature growth. Appl. Environ. Microbiol. *76*, 2304–2312.

Bae, E., and Phillips, G.N. (2004). Structures and analysis of highly homologous psychrophilic, mesophilic, and thermophilic adenylate kinases. J. Biol. Chem. *279*, 28202–28208.

Bakermans, C., and Nealson, K.H. (2004). Relationship of critical temperature to macromolecular synthesis and growth yield in '*Psychrobacter cryopegella*'. J. Bacteriol. *186*, 2340–2345.

Bakermans, C., Tsapin, A.I., Souza-Egipsy, V., Gilichinsky, D.A., and Nealson, K.H. (2003). Reproduction and metabolism at −10°C of bacteria isolated from Siberian permafrost. Environ. Microbiol. *5*, 321–326.

Bakermans, C., Tollaksen, S.L., Giometti, C.S., Wilkerson, C., Tiedje, J.M., and Thomashow, M.F. (2007). Proteomic analysis of *Psychrobacter cryohalolentis* K5 during growth at subzero temperatures. Extremophiles *11*, 345–354.

Barrett, P. (2003). Cooling a continent. Nature *421*, 221–223.

Bayles, D., Tunick, M., Foglia, T., and Miller, A. (2000). Cold shock and its effect on ribosomes and thermal tolerance in *Listeria monocytogenes*. Appl. Environ. Microbiol. *66*, 4351–4355.

Beaty, D., Buxbaum, K., Meyer, M., Barlow, N., Boynton, W., Clark, B., Deming, J., Doran, P.T., Edgett, K., Hancock, S., *et al.* (2006). Findings of the Mars Special Regions Science Analysis Group. Astrobiology *6*, 677–732.

Bej, A.K., Aislabie, J., and Atlas, R.M., eds. (2009). Polar Microbiology: The Ecology, Biodiversity and Bioremediation Potential of Microorganisms in Extremely Cold Environments (CRC Press, Boca Raton, FL).

Bendt, A., Huller, H., Kammel, U., Helmke, E., and Schweder, T. (2001). Cloning, expression, and characterization of a chitinase gene from the Antarctic psychrotolerant bacterium *Vibrio* sp strain Fi: 7. Extremophiles 5, 119–126.

Berger, F., Morellet, N., Menu, F., and Potier, P. (1996). Cold shock and cold acclimation proteins in the psychrotrophic bacterium *Arthrobacter globiformis* SI55. J. Bacteriol. *178*, 2999–3007.

Berger, F., Normand, P., and Potier, P. (1997). CapA, a cspA-like gene that encodes a cold acclimation protein in the psychrotrophic bacterium *Arthrobacter globiformis* SI55. J. Bacteriol. *179*, 5670–5676.

Bergholz, P.W., Bakermans, C., and Tiedje, J.M. (2009). *Psychrobacter arcticus* 273–4 uses resource efficiency and molecular motion adaptations for subzero temperature growth. J. Bacteriol. *191*, 2340–2352.

Birolo, L., Tutino, M.L., Fontanella, B., Gerday, C., Mainolfi, K., Pascarella, S., Sannia, G., Vinci, F., and Marino, G. (2000). Aspartate aminotransferase from the Antarctic bacterium – *Pseudoalteromonas haloplanktis* TAC 125 – Cloning, expression, properties, and molecular modelling. Eur. J. Biochem. *267*, 2790–2802.

Bock, C., and Eicken, H. (2005). A magnetic resonance study of temperature-dependent microstructural evolution and self-diffusion of water in Arctic first-year sea ice. Ann. Glaciol. *40*, 179–184.

Bowman, J.P. (2004). Psychrophilic prokaryote structural–functional relationships, biogeography and evolution within marine sediment. Cell. Mol. Biol. *50*, 503–513.

Bowman, J.P. (2008). Genomic analysis of psychrophilic prokaryotes. In Psychrophiles: from Biodiversity to Biotechnology, Margesin, R., Schinner, F., Marx, J.-C., and Gerday, C., eds. (Berlin Heidelberg, Springer-Verlag), pp. 265–284.

Bowman, J.P., McCammon, S.A., Brown, M.V., Nichols, D.S., and McMeekin, T.A. (1997). Diversity and association of psychrophilic bacteria in Antarctic sea ice. Appl. Environ. Microbiol. *63*, 3068–3078.

Boynton, W.V., Feldman, W.C., Squyres, S.W., Prettyman, T.H., Bruckner, J., Evans, L.G., Reedy, R.C., Starr, R., Arnold, J.R., Drake, D.M., *et al.* (2002). Distribution of hydrogen in the near surface of Mars: Evidence for subsurface ice deposits. Science *297*, 81–85.

Breezee, J., Cady, N., and Staley, J.T. (2004). Subfreezing growth of the sea ice bacterium '*Psychromonas ingrahamii*'. Microbial Ecol. *47*, 300–304.

Brinkmeyer, R., Knittel, K., Jurgens, J., Weyland, H., Amann, R., and Helmke, E. (2003). Diversity and Structure of Bacterial Communities in Arctic versus Antarctic Pack Ice. Appl. Environ. Microbiol. *69*, 6610–6619.

Cavicchioli, R. (2006). Cold-adapted archaea. Nature Reviews Microbiology *4*, 331–343.

Cavicchioli, R., Thomas, T., and Curmi, P.M.G. (2000). Cold stress response in Archaea. Extremophiles *4*, 321–331.

Chamot, D., and Owttrim, G.W. (2000). Regulation of cold shock-induced RNA helicase gene expression in the cyanobacterium *Anabaena* sp. strain PCC 7120. J. Bacteriol. *182*, 1251–1256.

Chin, J.P., Megaw, J., Magill, C.L., Nowotarski, K., Williams, J.P., Bhaganna, P., Linton, M., Patterson, M.F., Underwood, G.J.C., Mswaka, A.Y., *et al.* (2010). Solutes determine the temperature windows for microbial survival and growth. Proc. Natl. Acad. Sci. U.S.A. *107*, 7835–7840.

Christensen, T.R., Ekberg, A., Strom, L., Mastepanov, M., Panikov, N., Mats, O., Svensson, B.H., Nykanen, H., Martikainen, P.J., and Oskarsson, H. (2003). Factors controlling large scale variations in methane emissions from wetlands. Geophys. Res. Lett. *30*, 10–13.

Christner, B., Mosley-Thompson, E., Thompson, L., and Reeve, J. (2001). Isolation of bacteria and 16S rDNAs from Lake Vostok accretion ice. Environ. Microbiol. *3*, 570–577.

Christner, B.C., Mosley-Thompson, E., Thompson, L.G., Zagorodnov, V., Sandman, K., and Reeve, J.N. (2000). Recovery and identification of viable bacteria immured in glacial ice. Icarus *144*, 479–485.

Christner, B.C. (2002). Incorporation of DNA and protein precursors into macromolecules by bacteria at −15°C. Appl. Environ. Microbiol. *68*, 6435–6438.

Collins, R.E., Carpenter, S.D., and Deming, J.W. (2008). Spatial heterogeneity and temporal dynamics of particles, bacteria, and pEPS in Arctic winter sea ice. J. Marine Syst. *74*, 902–917.

Collins, T., Meuwis, M.A., Stals, I., Claeyssens, M., Feller, G., and Gerday, C. (2002). A novel family 8 xylanase, functional and physicochemical characterization. J. Biol. Chem. *277*, 35133–35139.

Collins, T., Roulling, F., Piette, F., Marx, J.-C., Feller, G., Gerday, C., and D'Amico, S. (2008). Fundamentals of Cold-Adapted Enzymes. In Psychrophiles: from Biodiversity to Biotechnology, R. Margesin, F. Schinner, J.-C. Marx, and C. Gerday, eds. (Berlin Heidelberg, Springer-Verlag), pp. 211–227.

Cronan, J., Jr., and Rock, C. (1996). Biosynthesis of membrane lipids. In *Escherichia coli* and *Salmonella typhimurium*. Cellular and Molecular Biology, Neidhardt, F., ed. (Washington DC, American Society for Microbiology Press), pp. 612–636.

D'Amico, S., Gerday, C., and Feller, G. (2001). Structural determinants of cold adaptation and stability in a large protein. J. Biol. Chem. *276*, 25791–25796.

D'Amico, S., Marx, J.-C., Gerday, C., and Feller, G. (2003). Activity-stability relationships in extremophilic enzymes. J. Biol. Chem. *278*, 7891–7896.

D'Amico, S., Collins, T., Marx, J.C., Feller, G., and Gerday, C. (2006). Psychrophilic microorganisms: challenges for life. EMBO Rep. *7*, 385–389.

D'Hondt, S., Jorgensen, B.B., Miller, D.J., Batzke, A., Blake, R., Cragg, B.A., Cypionka, H., Dickens, G.R., Ferdelman, T., Hinrichs, K.U., *et al.* (2004). Distributions of microbial activities in deep subseafloor sediments. Science *306*, 2216–2221.

Dalluge, J.J., Hamamoto, T., Horikoshi, K., Morita, R.Y., Stetter, K.O., and McCloskey, J.A. (1997). Posttransciptional modification of tRNA in psychrophilic bacteria. J. Bacteriol. *179*, 1918–1923.

Daniel, R.M., Smith, J.C., Ferrand, M., Hery, S., Dunn, R., and Finney, J.L. (1998). Enzyme activity below the dynamical transition at 220 K. Biophys. J. *75*, 2504–2507.

Daniel, R.M., Dunn, R.V., Finney, J.L., and Smith, J.C. (2003). The role of dynamics in enzyme activity. Ann. Rev. Biophys. Biomol. Struc. *32*, 69–92.

Davail, S., Feller, G., Narinx, E., and Gerday, C. (1994). Cold adaptation of proteins- purification, characterization, and sequence of the heat-labile subtilisin from the Antarctic psychrophile *Bacillus*TA41. J. Biol. Chem. *269*, 17448–17453.

DeLong, E.F., Wu, K.Y., Prezelin, B.B., and Jovine, R.V.M. (1994). High abundance of Archaea in Antarctic marine picoplankton. Nature *371*, 695–697.

Deming, J.W. (2002). Psychrophiles and polar regions. Curr. Opin. Microbiol. *5*, 301–309.

Denton, G.H., and Karlen, W. (1973). Lichenometry: its application to holocene moraine studies in southern Alaska and Swedish Lapland. Arctic Alpine Res. *5*, 347–372.

Drost-Hansen, W. (2001). Temperature effects on cell-functioning – A critical role for vicinal water. Cell. Mol. Biol. *47*, 865–883.

Feller, G. (2003). Molecular adaptations to cold in psychrophilic enzymes. Cell. Mol. Life Sci. *60*, 648–662.

Feller, G. (2007). Life at low temperatures: is disorder the driving force? Extremophiles *11*, 211–216.

Feller, G., and Gerday, C. (2003). Psychrophilic enzymes: Hot topics in cold adaptation. Nature Rev. Microbiol. *1*, 200–208.

Feller, G., Payan, F., Theys, F., Qian, M.X., Haser, R., and Gerday, C. (1994). Stability and structural analysis of alpha-amylase from the Antarctic psychrophile *Alteromonas haloplanctis* A23. Eur. J. Biochem. *222*, 441–447.

Feller, G., Zekhnini, Z., LamotteBrasseur, J., and Gerday, C. (1997). Enzymes from cold-adapted microorganisms – The class C beta-lactamase from the Antarctic psychrophile *Psychrobacter immobilis* A5. Eur. J. Biochem. *244*, 186–191.

Ferrer, M., Lünsdorf, H., Chernikova, T.N., Yakimov, M., Timmis, K.N., and Golyshin, P.N. (2004). Functional consequences of single:double ring transitions in chaperonins: life in the cold. Mol. Microbiol. *53*, 167–182.

Fields, P.A. (2001). Review: Protein function at thermal extremes: balancing stability and flexibility. Comp. Biochem. Physiol. A. Physiol. *129*, 417–431.

Formisano, V., Atreya, S., Encrenaz, T., Ignatiev, N., and Giuranna, M. (2004). Detection of methane in the atmosphere of Mars. Science *306*, 1758–1761.

Friedmann, E.I., Kappen, L., Meyer, M.A., and Nienow, J.A. (1993). Long-term productivity in the cryptoendolithic microbial community of the Ross Desert, Antarctica. Microb. Ecol. *25*, 51–69.

Frisvad, J.C. (2008). Fungi in cold ecosystems. In Psychrophiles: from Biodiversity to Biotechnology, Margesin, R., Schinner, F., Marx, J.-C., and Gerday, C., eds. (Berlin Heidelberg, Springer-Verlag), pp. 137–156.

Galtier, N., and Lobry, J.R. (1997). Relationships between genomic G+C content, RNA secondary structures, and optimal growth temperature in prokaryotes. J. Mol. Evol. *44*, 632–636.

Garsoux, G., Lamotte, J., Gerday, C., and Feller, G. (2004). Kinetic and structural optimization to catalysis at low temperatures in a psychrophilic cellulase from the Antarctic bacterium *Pseudoalteromonas haloplanktis*. Biochem. J. *384*, 247–253.

Geminale, A., Formisano, V., and Giuranna, M. (2008). Methane in Martian atmosphere: Average spatial, diurnal, and seasonal behaviour. Planet. Space Sci. *56*, 1194–1203.

Gibson, J.A.E., Miller, M.R., Davies, N.W., Neill, G.P., Nichols, D.S., and Volkman, J.K. (2005). Unsaturated diether lipids in the psychrotrophic archaeon *Halorubrum lacusprofundi*. Syst. Appl. Microbiol. *28*, 19–26.

Gilbert, J.A., Davies, P.L., and Laybourn-Parry, J. (2005). A hyperactive, Ca2+-dependent antifreeze protein in an Antarctic bacterium. FEMS Microbiol. Lett. *245*, 67–72.

Gilichinsky, D., Vorobyova, E.A., Erokhina, L.G., Fedorov-Davydov, D.G., and Chaikovskaya, N.R. (1992). Long-term preservation of microbial ecosystems in permafrost. Adv. Space Res. *12*, 255–263.

Gilichinsky, D., Rivkina, E., Shcherbakova, V., Laurinavichuis, K., and Tiedje, J. (2003). Supercooled water brines within permafrost--An unknown ecological niche for microorganisms: a model for astrobiology. Astrobiology 3, 331–341.

Goldfine, H. (1984). The control of membrane fluidity in plasmalogen-containing anaerobic bacteira. In Biomembranes, Vol 12, Membrane Fluidity, Kates, M., and Manson, L., eds. (New York, Plenum Press), pp. 349–377.

Goodchild, A., Saunders, N.F.W., Ertan, H., Raftery, M., Guilhaus, M., Curmi, P.M.G., and Cavicchioli, R. (2004). A proteomic determination of cold adaptation in the Antarctic archaeon, *Methanococcoides burtonii*. Mol. Microbiol. *53*, 309–321.

Goodchild, A., Raftery, M., Saunders, N.F.W., Guilhaus, M., and Cavicchioli, R. (2005). Cold adaptation of the Antarctic archaeon, *Methanococcoides burtonii* assessed by proteomics using ICAT. J. Proteome Res. *4*, 473–480.

Graumann, P., and Marahiel, M.A. (1996). Some like it cold: Response of microorganisms to cold shock. Arch. Microbiol. *166*, 293–300.

Grzymski, J.J., Carter, B.J., DeLong, E.F., Feldman, R.A., Ghadiri, A., and Murray, A.E. (2006). Comparative genomics of DNA fragments from six antarctic marine planktonic bacteria. Appl. Environ. Microbiol. *72*, 1532–1541.

Harder, W., and Veldkamp, H. (1968). Physiology of an obligately psychrophilic marine *Pseudomonas* species. J. Appl. Bacteriol. *31*, 12–23.

Hassol, S.J. (2004). Impacts of a Warming Arctic (Canada, Cambridge University Press).

Helmke, E., and Weyland, H. (2004). Psychrophilic versus psychrotolerant bacteria – Occurrence and significance in polar and temperate marine habitats. Cell. Mol. Biol. *50*, 553–561.

Hodson, A., Anesio, A.M., Tranter, M., Fountain, A., Osborn, M., Priscu, J., Laybourn-Parry, J., and Sattler, B. (2008). Glacial Ecosystems. Ecol. Monograph. *78*, 41–67.

Horton, A.J., Hak, K.M., Steffan, R.J., Foster, J.W., and Bej, A.K. (2000). Adaptive response to cold temperatures and characterization of cspA in *Salmonella typhimurium* LT2. Anton. Leeuw. Int. J. G. *77*, 13–20.

Hoyoux, A., Jennes, I., Dubois, P., Genicot, S., Dubail, F., Francois, J.M., Baise, E., Feller, G., and Gerday, C. (2001). Cold-adapted beta-galactosidase from the Antarctic psychrophile *Pseudoalteromonas haloplanktis*. Appl. Environ. Microbiol. *67*, 1529–1535.

Isaksen, M.F., and Jorgensen, B.B. (1996). Adaptation of psychrophilic and psychrotrophic sulfate-reducing bacteria to permanently cold marine environments. Appl. Environ. Microbiol. *62*, 408–414.

Jackson, M.B., and Cronan, J.E. (1978). Estimate of minimum amount of fluid lipid required for growth of *Escherichia coli*. Biochim. Biophys. Acta *512*, 472–479.

Jakosky, B.M., Nealson, K.H., Bakermans, C., Ley, R.E., and Mellon, M.T. (2003). Subfreezing activity of microorganisms and the potential habitability of Mars' polar regions. Astrobiology 3, 343–350.

Jiang, W., Hou, Y., and Inouye, M. (1997). CspA, the major cold-shock protein of *Escherichia coli*, is an RNA chaperone. J. Biol. Chem. *272*, 196–202.

Johnston, C.G., and Vestal, J.R. (1991). Photosynthetic carbon incorporation and turnover in Antarctic cryptoendolithic microbial communities – are they the slowest-growing communities on Earth? Appl. Environ. Microbiol. *57*, 2308–2311.

Junge, K., Krembs, C., Deming, J., Stierle, A., and Eicken, H. (2001). A microscopic approach to investigate bacteria under *in situ* conditions in sea-ice samples. Ann. Glaciol. *33*, 304–310.

Junge, K., Eicken, H., and Deming, J.W. (2004). Bacterial activity at -2 to -20°C in Arctic wintertime sea ice. Appl. Environ. Microbiol. *70*, 550–557.

Junge, K., Eicken, H., Swanson, B.D., and Deming, J.W. (2006). Bacterial incorporation of leucine into protein down to -20°C with evidence for potential activity in sub-eutectic saline ice formations. Cryobiology *52*, 417–429.

Kaartokallio, H., Tuomainen, J., Kuosa, H., Kuparinen, J., Martikainen, P.J., and Servomaa, K. (2008). Succession of sea-ice bacterial communities in the Baltic Sea fast ice. Polar Biol. *31*, 783–793.

Kappen, L.B., Schroeter, B., Scheidegger, C., Sommerkorn, M., and Hestmark, G. (1996). Cold resistance and metabolic activity of lichens below 0 C. Adv. Space Res. *18*, 119–128.

Karner, M.B., DeLong, E.F., and Karl, D.M. (2001). Archaeal dominance in the mesopelagic zone of the Pacific Ocean. Nature *409*, 507–510.

Kawahara, H. (2002). The structures and functions of ice crystal-controlling proteins from bacteria. J. Biosci. Bioeng. *94*, 492–496.

Kawahara, H. (2008). Cryoprotectants and ice-binding proteins. In Psychrophiles: from Biodiversity to Biotechnology, Margesin, R., Schinner, F., Marx, J.-C., and Gerday, C., eds. (Berlin Heidelberg, Springer-Verlag), pp. 229–246.

Kawahara, H., Nagae, I., and Obata, H. (1996). Purification and characterization of a new anti-nucleating protein isolated from *Acinetobacter calcoaceticus* KINI-1. Biocontrol Sci. *1*, 11–17.

Kawamoto, J., Kurihara, T., Kitagawa, M., Kato, I., and Esaki, N. (2007). Proteomic studies of an Antarctic cold-adapted bacterium, *Shewanella livingstonensis* Ac10, for global identification of cold-inducible proteins. Extremophiles *11*, 819–826.

Kaye, J.Z., and Baross, J.A. (2000). High incidence of halotolerant bacteria in Pacific hydrothermal-vent and pelagic environments. FEMS Microbiol. Ecol. *32*, 249–260.

Kendall, M.M., Wardlaw, G.D., Tang, C.F., Bonin, A.S., Liu, Y.T., and Valentine, D.L. (2007). Diversity of Archaea in marine sediments from Skan Bay, Alaska, including cultivated methanogens, and description of *Methanogenium boonei* sp nov. Appl. Environ. Microbiol. *73*, 407–414.

Khachane, A.N., Timmis, K.N., and dos Santos, V.A.P.M. (2005). Uracil content of 16S rRNA of thermophilic and psychrophilic prokaryotes correlates inversely with their optimal growth temperatures. Nucl. Acids Res. *33*, 4016–4022.

Kirchman, D.L., Moran, X.A.G., and Ducklow, H. (2009). Microbial growth in the polar oceans – role of temperature and potential impact of climate change. Nature Rev. Microbiol. *7*, 451–459.

Knoblauch, C., and Jorgensen, B.B. (1999). Effect of temperature on sulphate reduction, growth rate and growth yield in five psychrophilic sulphate-reducing bacteria from Arctic sediments. Environ. Microbiol. *1*, 457–467.

Knoblauch, C., Sahm, K., and Jorgensen, B. (1999). Psychrophilic sulfate-reducing bacteria isolated from permanently cold Arctic marine sediments: description of *Desulfofrigus oceanense* gen. nov., sp. nov., *Desulfofrigus fragile* sp. nov., *Desulfofaba gelida* gen. nov., sp. nov., *Desulfotalea psychrophila* gen. nov., sp. nov. and *Desulfotalea arctica* sp. nov. Int. J. Syst. Bacteriol. *49*, 1631–1643.

Krembs, C., Gradinger, R., and Spindler, M. (2000). Implications of brine channel geometry and surface area for the interaction of sympagic organisms in Arctic sea ice. J. Exp. Mar. Biol. Ecol. *243*, 55–80.

Krembs, C., and Engel, A. (2001). Abundance and variability of microorganisms and transparent exopolymer particles across the ice–water interface of melting first-year sea ice in the Laptev Sea (Arctic). Mar. Biol. *138*, 173–185.

Krembs, C., Eicken, H., Junge, K., and Deming, J.W. (2002). High concentrations of exopolymeric substances in Arctic winter sea ice: implications for the polar ocean carbon cycle and cryoprotection of diatoms. Deep-Sea Res. Pt. I *49*, 2163–2181.

Kruger, M., Treude, T., Wolters, H., Nauhaus, K., and Boetius, A. (2005). Microbial methane turnover in different marine habitats. Palaeogeography Palaeoclimatology Palaeoecology *227*, 6–17.

Leiros, H.K.S., Willassen, N.P., and Smalas, A.O. (1999). Residue determinants and sequence analysis of cold-adapted trypsins. Extremophiles *3*, 205–219.

Lim, J., Thomas, T., and Cavicchioli, R. (2000). Low temperature regulated DEAD-box RNA helicase from the antarctic archaeon, *Methanococcoides burtonii*. J. Molec. Biology *297*, 553–567.

Lin, C.T., and Thomashow, M.F. (1992). DNA-sequence analysis of a complementary-DNA for cold-regulated *Arabidopsis* gene *cor15* and characterization of the Cor-15 polypeptide. Plant Physiol. *99*, 519–525.

Lin, L.H., Wang, P.L., Rumble, D., Lippmann-Pipke, J., Boice, E., Pratt, L.M., Lollar, B.S., Brodie, E.L., Hazen, T.C., Andersen, G.L., *et al.* (2006). Long-term sustainability of a high-energy, low-diversity crustal biome. Science *314*, 479–482.

Liu, Y.Q., Yao, T.D., Jiao, N.Z., Kang, S.C., Xu, B.Q., Zeng, Y.H., Huang, S.J., and Liu, X.B. (2009). Bacterial diversity in the snow over Tibetan Plateau Glaciers. Extremophiles *13*, 411–423.

Lonhienne, T., Zoidakis, J., Vorgias, C.E., Feller, G., Gerday, C., and Bouriotis, V. (2001). Modular structure, local flexibility and cold-activity of a novel chitobiase from a psychrophilic Antarctic bacterium. J. Mol. Biol. *310*, 291–297.

Meiners, K., Brinkmeyer, R., Granskog, M.A., and Lindfors, A. (2004). Abundance, size distribution and bacterial colonization of exopolymer particles in Antarctic sea ice (Bellingshausen Sea). Aquat. Microb. Ecol. *35*, 283–296.

Melchior, D.L. (1982). Lipid phase-transitions and regulation of membrane fluidity in prokaryotes. Curr. Top. Membr. Transport *17*, 263–316.

Methe, B.A., Nelson, K.E., Deming, J.W., Momen, B., Melamud, E., Zhang, X., Moult, J., Madupu, R., Nelson, W.C., Dodson, R.J., *et al.* (2005). The psychrophilic lifestyle as revealed by the genome sequence of *Colwellia psychrerythraea* 34H through genomic and proteomic analyses. PNAS *102*, 10913–10918.

Miteva, V.I., Sheridan, P.P., and Brenchley, J.E. (2004). Phylogenetic and physiological diversity of microorganisms isolated from a deep Greenland glacier ice core. Appl. Environ. Microbiol. *70*, 202–213.

Miyazaki, K., Wintrode, P.L., Grayling, R.A., Rubingh, D.N., and Arnold, F.H. (2000). Directed evolution study of temperature adaptation in a psychrophilic enzyme. J. Molec. Biol. *297*, 1015–1026.

Mock, T., and Thomas, D.N. (2005). Recent advances in sea-ice microbiology. Environ. Microbiol. *7*, 605–619.

Murata, N., and Wada, H. (1995). Acyl-lipid desaturases and their importance in the tolerance and acclimatization to cold of Cyanobacteria. Biochem. J. *308*, 1–8.

Murray, A.E., and Grzymski, J.J. (2007). Diversity and genomics of Antarctic marine micro-organisms. Philos. Trans. R. Soc. Lond. B. Biol. Sci. *362*, 2259–2271.

Murray, J.B., Muller, J.-P., Neukum, G., Werner, S.C., van Gasselt, S., Hauber, E., Markiewicz, W.J., Head, J.W., III, Foing, B.H., Page, D., *et al.* (2005). Evidence from the Mars Express High Resolution Stereo Camera for a frozen sea close to Mars' equator. Nature *434*, 352–356.

Nakashima, H., Fukuchi, S., and Nishikawa, K. (2003). Compositional changes in RNA, DNA and proteins for bacterial adaptation to higher and lower temperatures. J. Biochem. (Tokyo). *133*, 507–513.

Nedwell, D.B. (1999). Effect of low temperature on microbial growth: lowered affinity for substrates limits growth at low temperature. FEMS Microb. Ecol. *30*, 101–111.

Nichols, D.S., Miller, M.R., Davies, N.W., Goodchild, A., Raftery, M., and Cavicchioli, R. (2004). Cold adaptation in the antarctic archaeon *Methanococcoides burtonii* involves membrane lipid unsaturation. J. Bacteriol. *186*, 8508–8515.

Nogi, Y. (2008). Bacteria in the deep sea: Psychropiezophiles. In Psychrophiles: from Biodiversity to Biotechnology, Margesin, R., Schinner, F., Marx, J.-C., and Gerday, C., eds. (Berlin Heidelberg, Springer-Verlag), pp. 73–82.

Noon, K.R., Guymon, R., Crain, P.F., McCloskey, J.A., Thomm, M., Lim, J., and Cavicchioli, R. (2003). Influence of temperature on tRNA modification in Archaea: *Methanococcoides burtonii* (Optimum Growth Temperature, 23°C) and *Stetteria hydrogenophila* (Topt, 95°C). J. Bacteriol. *185*, 5483–5490.

Obata, H., Ishigaki, H., Kawahara, H., and Yamade, K. (1998). Purification and characterization of a novel cold-regulated protein from an ice-nucleating bacterium, *Pseudomonas fluorescens* KUIN-1. Biosci. Biotech. Bioch. *62*, 2091–2097.

Panikov, N.S., and Sizova, M.V. (2007). Growth kinetics of microorganisms isolated from Alaskan soil and permafrost in solid media frozen down to −35°C. FEMS Microbiol. Ecol. *59*, 500–512.

Panikov, N.S., Flanagan, P.W., Oechel, W.C., Mastepanov, M.A., and Christensen, T.R. (2006). Microbial activity in soils frozen to below −39°C. Soil Biol. Biochem. *38*, 785–794.

Phadtare, S., Inouye, M., and Severinov, K. (2002). The nucleic acid melting activity of Escherichia coli CspE is critical for transcription antitermination and cold acclimation of cells. J. Biol. Chem. *277*, 7239–7245.

Pomeroy, L.R., and Wiebe, W.J. (2001). Temperature and substrates as interactive limiting factors for marine heterotrophic bacteria. Aquat. Microb. Ecol. *23*, 187–204.

Price, P.B. (2000). A habitat for psychrophiles in deep Antarctic ice. PNAS *97*, 1247–1251.

Ravenschlag, K., Sahm, K., Pernthaler, J., and Amann, R. (1999). High bacterial diversity in permanently cold marine sediments. Appl. Environ. Microbiol. *65*, 3982–3989.

Riley, M., Staley, J.T., Danchin, A., Wang, T.Z., Brettin, T.S., Hauser, L.J., Land, M.L., and Thompson, L.S. (2008). Genomics of an extreme psychrophile, *Psychromonas ingrahamii*. BMC Genomics *9*, 210.

Rivkina, E.M., Friedmann, E.I., McKay, C.P., and Gilichinsky, D.A. (2000). Metabolic activity of permafrost bacteria below the freezing point. Appl. Environ. Microbiol. *66*, 3230–3233.

Rivkina, E.M., Laurinavichus, K.S., Gilichinsky, D.A., and Shcherbakova, V.A. (2002). Methane generation in permafrost sediments. Doklady Biol. Sci. *V383*, 179–181.

Russell, N. (1984). The regulation of membrane fluidity in bacteria by acyl chain length changes. In Biomembranes, Vol 12 Membrane Fluidity, Kates, M., and Manson, L., eds. (New York, Plenum Press), pp. 329–347.

Russell, N. (2008). Membrane components and cold sensing. In Psychrophiles: From Biodiversity to Biotechnology, Margesin, R., Schinner, F., Marx, J.-C., and Gerday, C., eds. (Berlin Heidelberg, Springer-Verlag), pp. 177–190.

Russell, N., and Hamamoto, T. (1998). Psychrophiles. In Extremophiles: Microbial Life in Extreme Environments, Horikoshi, K., and Grant, W.D., eds. (New York, Wiley-Liss, Inc.), pp. 25–45.

Russell, N.J. (1990). Cold adaptation of microorganisms. PNAS *326*, 595–608.

Russell, N.J. (1997). Psychrophilic bacteria--Molecular adaptations of membrane lipids. Comp. Biochem. Physiol. Part A: Physiol. *118*, 489–493.

Russell, N.J., and Fukunaga, N. (1990). A comparison of thermal adaptation of membrane-lipids in psychrophilic and thermophilic bacteria. FEMS Microbiol. Rev. *75*, 171–182.

Russell, R.J.M., Gerike, U., Danson, M.J., Hough, D.W., and Taylor, G.L. (1998). Structural adaptations of the cold-active citrate synthase from an Antarctic bacterium. Structure *6*, 351–361.

Sabri, A., Jacques, P., Weekers, F., and Bare, G. (2000). Effect of temperature on growth of psychrophilic and psychrotrophic members of *Rhodotorula aurantiaca*. Appl. Biochem. Biotech. *84–6*, 391–399.

Saunders, N.F.W., Thomas, T., Curmi, P.M.G., Mattick, J.S., Kuczek, E., Slade, R., Davis, J., Franzmann, P.D., Boone, D., Rusterholtz, K., *et al.* (2003). Mechanisms of thermal adaptation revealed from the genomes of the Antarctic archaea *Methanogenium frigidum* and *Methanococcoides burtonii*. Genome Res. *13*, 1580–1588.

Schimel, J.P., Bilbrough, C., and Welker, J.A. (2004). Increased snow depth affects microbial activity and nitrogen mineralization in two Arctic tundra communities. Soil Biol. Biochem. *36*, 217–227.

Schmid, F.X. (1993). Prolyl isomerase – Enzyme catalysis of slow protein-folding reactions. Annu. Rev. Biophys. Biomol. Struct. *22*, 123–143.

Schuur, E.A.G., Vogel, J.G., Crummer, K.G., Lee, H., Sickman, J.O., and Osterkamp, T.E. (2009). The effect of permafrost thaw on old carbon release and net carbon exchange from tundra. Nature *459*, 556–559.

Scotter, A.J., Marshall, C.B., Graham, L.A., Gilbert, J.A., Garnham, C.P., and Davies, P.L. (2006). The basis for hyperactivity of antifreeze proteins. Cryobiology *53*, 229–239.

Screen, J.A., and Simmonds, I. (2010). The central role of diminishing sea ice in recent Arctic temperature amplification. Nature *464*, 1334–1337.

Segawa, T., Miyamoto, K., Ushida, K., Agata, K., Okada, N., and Kohshima, S. (2005). Seasonal change in bacterial flora and biomass in mountain snow from the Tateyama Mountains, Japan, analyzed by 16S rRNA gene sequencing and real-time PCR. Appl. Environ. Microbiol. *71*, 123–130.

Shaver, G.R., Giblin, A.E., Nadelhoffer, K.J., Thieler, K.K., Downs, M.R., Laundre, J.A., and Rastetter, E.B. (2006). Carbon turnover in Alaskan tundra soils: effects of organic matter quality, temperature, moisture and fertilizer. J. Ecol. *94*, 740–753.

Sherr, B.F., and Sherr, E.B. (2003). Community respiration/production and bacterial activity in the upper water column of the central Arctic Ocean. Deep-Sea Res. Pt. I *50*, 529–542.

Shi, T., Reeves, R.H., Gilichinsky, D.A., and Friedmann, E.I. (1997). Characterization of viable bacteria from Siberian permafrost by 16S rDNA sequencing. Microb. Ecol. *33*, 169–179.

Siddiqui, K.S., and Cavicchioli, R. (2006). Cold-adapted enzymes. Annu. Rev. Biochem. *75*, 403–433.

Simon, C., Wiezer, A., Strittmatter, A.W., and Daniel, R. (2009). Phylogenetic diversity and metabolic potential revealed in a glacier ice metagenome. Appl. Environ. Microbiol. *75*, 7519–7526.

Sluijs, A., Schouten, S., Pagani, M., Woltering, M., Brinkhuis, H., Damste, J.S.S., Dickens, G.R., Huber, M., Reichart, G.J., Stein, R., *et al.* (2006). Subtropical arctic ocean temperatures during the Palaeocene/Eocene thermal maximum. Nature *441*, 610–613.

Smetacek, V., and Nicol, S. (2005). Polar ocean ecosystems in a changing world. Nature *437*, 362–368.

Sowers, T. (2001). The N_2O record spanning the penultimate deglaciation from the Vostok ice core. J. Geophys. Res. Atmos. *106*, 31903–31914.

Steven, B., Leveille, R., Pollard, W.H., and Whyte, L.G. (2006). Microbial ecology and biodiversity in permafrost. Extremophiles *10*, 259–267.

Strocchi, M., Ferrer, M., Timmis, Kenneth N., and Golyshin, Peter N. (2006). Low temperature-induced systems failure in *Escherichia coli*: Insights from rescue by cold-adapted chaperones. Proteomics *6*, 193–206.

Sun, H.J., and Friedmann, E.I. (1999). Growth on geological time scales in the Antarctic cryptoendolithic microbial community. Geomicrobiol. J. *16*, 193–202.

Thomas, D.N., and Dieckmann, G.S. (2002). Antarctic sea ice--a habitat for extremophiles. Science *295*, 641–644.

Thomas, T., and Cavicchioli, R. (1998). Archaeal cold-adapted proteins: structural and evolutionary analysis of the elongation factor 2 proteins from psychrophilic, mesophilic and thermophilic methanogens. FEBS Lett. *439*, 281–286.

Thomas, T., and Cavicchioli, R. (2000). Effect of temperature on stability and activity of elongation factor 2 proteins from Antarctic and thermophilic methanogens. J. Bacteriol. *182*, 1328–1332.

Thomas, T., Kumar, N., and Cavicchioli, R. (2001). Effects of ribosomes and intracellular solutes on activities and stabilities of elongation factor 2 proteins from psychrotolerant and thermophilic methanogens. J. Bacteriol. *183*, 1974–1982.

Tian, F., Yu, Y., Chen, B., Li, H.R., Yao, Y.F., and Guo, X.K. (2009). Bacterial, archaeal and eukaryotic diversity in Arctic sediment as revealed by 16S rRNA and 18S rRNA gene clone libraries analysis. Polar Biol. *32*, 93–103.

Tung, H.C., Bramall, N.E., and Price, P.B. (2005). Microbial origin of excess methane in glacial ice and implications for life on Mars. PNAS *102*, 18292–18296.

Tung, H.C., Price, P.B., Bramall, N.E., and Vrdoljak, G. (2006). Microorganisms metabolizing on clay grains in 3-km-deep Greenland basal ice. Astrobiology *6*, 69–86.

Van Petegem, F., Collins, T., Meuwis, M.A., Gerday, C., Feller, G., and Van Beeumen, J. (2003). The structure of a cold-adapted family 8 xylanase at 1.3 angstrom resolution – Structural adaptations to cold and investigation of the active site. J. Biol. Chem. *278*, 7531–7539.

Veldkamp, H., Gemerden, H.v., Harder, W., and Laanbroek, H.J. (1984). Competition among bacteria: An overview. In Current Perspectives in Microbial Ecology, Klug, M.J., and Reddy, C.A., eds. (Washington DC, American Society for Microbiology), pp. 279–290.

Vincent, W.F., Gibson, J.A.E., Pienitz, R., Villeneuve, V., Broady, P.A., Hamilton, P.B., and Howard-Williams, C. (2000). Ice shelf microbial ecosystems in the high Arctic and implications for life on snowball earth. Naturwissenschaften *87*, 137–141.

Vishnivetskaya, T., Kathariou, S., McGrath, J., Gilichinsky, D., and Tiedje, J.M. (2000). Low-temperature recovery strategies for the isolation of bacteria from ancient permafrost sediments. Extremophiles *4*, 165–173.

Vorobyova, E., Soina, V., Gorlenko, M., Minkovskaya, N., Zalinova, N., Mamukelashvili, A., Gilichinsky, D., Rivkina, E., and Vishnivetskaya, T. (1997). The deep cold biosphere: facts and hypothesis. FEMS Microbiol. Rev. *20*, 277–290.

Wagner, D., Lipski, A., Embacher, A., and Gattinger, A. (2005). Methane fluxes in permafrost habitats of the Lena Delta: effects of microbial community structure and organic matter quality. Environ. Microbiol. *7*, 1582–1592.

Wang, G., Spivack, A.J., Rutherford, S., Manor, U., and D'Hondt, S. (2008). Quantification of co-occurring reaction rates in deep subseafloor sediments. Geochem. Cosmochim. Ac. *72*, 3479–3488.

Weinberg, M.V., Schut, G.J., Brehm, S., Datta, S., and Adams, M.W.W. (2005). Cold shock of a hyperthermophilic archaeon: *Pyrococcus furiosus* exhibits multiple responses to a suboptimal growth temperature with a key role for membrane-bound glycoproteins. J. Bacteriol. *187*, 336–348.

Welsh, D.T. (2000). Ecological significance of compatible solute accumulation by micro-organisms: from single cells to global climate. FEMS Microbiol. Rev. *24*, 263–290.

Williams, R., and Frausto da Silva, J. (1996). The Natural Selection of the Chemical Elements: The Environment and Life's Chemistry (New York, Oxford University Press Inc.).

Wowk, B., and Fahy, G.M. (2002). Inhibition of bacterial ice nucleation by polyglycerol polymers. Cryobiology *44*, 14–23.

Xiang, S.R., Yao, T.D., An, L.Z., Wu, G.J., Xu, B.Q., Ma, X.J., Li, Z., Wang, J.X., and Yu, W.S. (2005). Vertical quantitative and dominant population distribution of the bacteria isolated from the Muztagata ice core. Sci. China Ser. D. *48*, 1728–1739.

Xu, H., Griffith, M., Patten, C.L., and Glick, B.R. (1998). Isolation and characterization of an antifreeze protein with ice nucleation activity from the plant growth promoting rhizobacterium *Pseudomonas putida* GR12-2. Can. J. Microbiol. *44*, 64–73.

Xu, Y., Feller, G., Gerday, C., and Glansdorff, N. (2003). Moritella cold-active dihydrofolate reductase: are there natural limits to optimization of catalytic efficiency at low temperature? J. Bacteriol. *185*, 5519–5526.

Yamashita, Y., Kawahara, H., and Obata, H. (2002). Identification of a novel anti-ice-nucleating polysaccharide from *Bacillus thuringiensis* YY529. Biosci. Biotechnol. Biochem. *66*, 948–954.

SM1: A Cold-loving Archaeon with Powerful Nano-grappling Hooks

4

Christine Moissl-Eichinger, Ruth Henneberger and Robert Huber

Abstract

The SM1 euryarchaeon represents an extraordinary microorganism: in the surface waters of cold, sulfidic springs, it lives together with filamentous bacteria, forming the so called string-of-pearls community. In the subsurface however, it can grow partner-independently as a 'monospecies' biofilm. Even though the SM1 euryarchaeon is still uncultivated in the laboratory, it is accessible via an *in situ* cultivation technique using its own biotope as a natural chemostat. This approach allowed the study of its biology and enabled the discovery of unique cell surface appendices with unexpected and unusually high complexity. Each of the archaeal cells is surrounded by approximately 100 protein filaments that are up to 3 μm long and show a high resemblance to barbwire with a tripartite grappling hook at their tip. Based on this structure the appendices were called 'hami' (lat. hamus = (grappling) hook). These hami represent perfectly evolved, natural mechanical nano-tools that could find applications in the growing field of nanobiotechnology.

The Sippenauer Moor: a protected biotope for rare species

Marsh lands and swamps around the globe represent highly specific biotopes and harbour a wide variety of specifically adapted and specialized plants and animals. Many of these marsh areas are nature reserves and therefore provide protected niches for rare and endangered organisms. The Sippenauer Moor is a marsh land that is located about 20 km south of Regensburg in southern Germany and covers an area of approximately 90,000 m^2. Since 1911, the botanical society of Regensburg (Regensburgische Botanische Gesellschaft, founded 1790) has put much effort into the protection of this area, which was announced a nature reserve in 1939. Since then, the marsh has been subjected to profound botanical (and zoological) studies, revealing for example the presence of very rare plants, such as the orchids *Liparis loeselii* and *Dactylorhiza incarnata*.

About 20 sulfide-containing, cold freshwater springs arise within the Sippenauer Moor. During our investigations since 1998, temperature and pH measurements have been performed periodically at the main spring. Throughout the year, the water temperature is nearly constant at 10–11°C and the pH is 7.3. Chemical analyses of the spring water revealed a sulfide concentration of 0.7–1.2 mg/l and a relatively low oxygen concentration of about 1.4 mg/l (Rudolph *et al.*, 2001, 2004). In this spring and many of the smaller springs throughout the marsh area, high microbiological activity is visible by eye, since large colourful mats and microbial assemblages ('streamers') are formed within their streamlets. In initial microbiological ecology studies of selected springs in the late 1990s, samples were taken from biomats and streamers and analysed by fluorescence *in situ* hybridization (FISH), using *Bacteria*- and *Archaea*-directed probes. The microbial assemblages analysed appeared to be unstructured and bulky, and all samples revealed the presence of Bacteria of different sizes and morphologies, with a filamentous morphotype predominating. Besides the presence of rod-shaped, host-dependent Crenarchaeota (Koch *et al.*, 2006), euryarchaeal cocci were found randomly in very low numbers.

Novel archaea that thrive in cold sulfidic springs: the string-of-pearls community

Apart from unstructured biomaterial, the main spring in the Sippenauer Moor also harboured highly unusual, macroscopically visible microbial cell assemblages with a characteristic morphology. These communities resembled a string of pearls with 2–15 tiny, whitish globules, connected by thin, whitish threads that were floating in the sulfidic water, one end connected to solid material like branches, leaves or stones. The pearls had a diameter of about 0.5–3.0 mm (Fig. 4.1A; Rudolph *et al.*, 2001, 2004). Surprisingly, these strings of pearls showed a different microbial composition compared with the other microbial assemblages analysed: the outer part of the pearls and the connecting threads were mainly constituted by filamentous bacteria, while the inner part was dominated by small archaeal coccoid cells. These archaea appeared to be loosely packed, and up to 10^7 archaeal cells were estimated to be present in a single pearl (Fig. 4.1B).

16S rRNA gene analysis revealed that the archaeal component of this unusual microbial community consisted of a single and novel phylotype, the SM1 (Sippenauer Moor) euryarchaeon, which represents a deep and distinct phylogenetic branch within the euryarchaeal phylogenetic tree with no close relationship to cultivated or uncultivated neighbours (Rudolph *et al.*, 2001). Within a pearl, these SM1 cells appeared to be arranged three-dimensionally at defined distances to each other. Furthermore, single cocci seemed to be embedded in fibrous material of unknown chemical composition (Fig. 4.1B; Rudolph *et al.*, 2001).

Phylogenetic analysis of the outer part of the pearls and the connecting threads showed that these parts of the community were mainly composed of a single bacterial phylotype belonging to the genus *Thiothrix*, with a close relationship

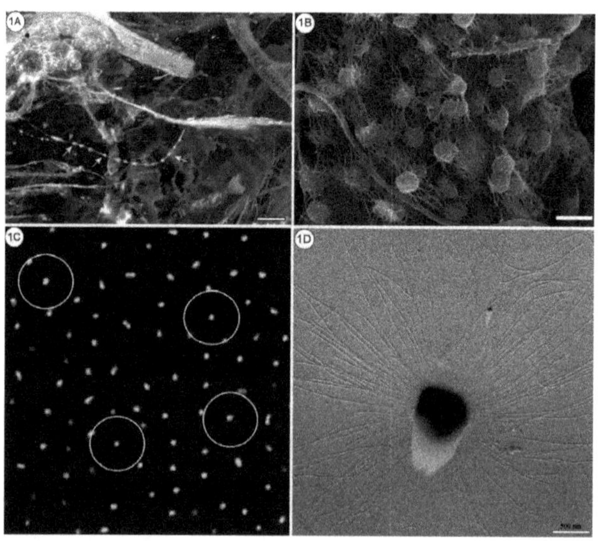

Figure 4.1 (A) Archaeal/bacterial string-of-pearls community in cold, sulfidic spring water. White arrows point to single pearls, greyish arrows point to strings between the pearls. Bar: 10 mm (from Rudolph *et al.*, 2004, with permission). (B) SEM micrograph showing the inner part of a pearl with coccoid SM1 Euryarchaea, embedded in a fibrous matrix. Bar: 1 µm (from Rudolph *et al.*, 2001). (C) Naturally occurring biofilm (Table 4.2), hybridized with an SM1 euryarchaeal fluorescent probe. Cells show an average distance of approximately 4 µm (circles) due to a hami length of 2 µm (from Moissl *et al.*, 2005b). (D) Electron micrograph of an SM1 cell with about 100 pili-like surface appendages (from Moissl *et al.*, 2005a, with permission).

to *Thiothrix unzii* (Moissl *et al.*, 2002; Moissl-Eichinger, unpublished, 2009). Members of the genus *Thiothrix* usually grow as rosettes and possess a complex life cycle, including the formation of gonidial cells and settling on surfaces via holdfast fibres (Larkin and Nelson, 1987). These microbes have gained notoriety in industrial environments because they can cause severe problems concerning activated sludge bulking. Generally, *Thiothrix* is specifically adapted to flowing aquatic environments with higher levels of sulfides and lower oxygen concentrations (Nielsen *et al.*, 2000). This genus is known to grow by aerobic sulfide oxidation, forming amorphous sulfur globules within innercell repositories. In the string-of-pearls microbial community, besides stabilizing the pearls and threads, *Thiothrix* is responsible for the whitish appearance of the community (Moissl *et al*, 2002). Conclusions on growth structures of the SM1 euryarchaeon or its potential role in the appearance of the pearls cannot be drawn due to the very low level of phylogenetic similarity of the SM1 to any cultured or characterized species.

A close and very stable partnership of the SM1 euryarchaeon and *Thiothrix* has been shown by FISH of about 30 different pearls analysed over a period of 3 years. Based on this observation, it seems obvious that the constitution of the string-of-pearls communities is not random. A specific, syntrophic or even symbiotic relationship between these two microorganisms can be envisaged. In many cases, intimately associated microbial communities are based on the exchange of nutrients (Schink, 2002), as has been shown, for example, for the methane oxidizing consortium of sulfate reducing bacteria and methanogenic archaea (Boetius *et al.*, 2000) or phototrophic consortia (Fröstl and Overmann, 2000). A similar mechanism can be proposed for the string-of-pearls community: during metabolism, *Thiothrix* forms different organic compounds (C-sources) and sulfate as a metabolic end product, which could be used as a possible electron acceptor by the putatively anaerobically growing SM1 euryarchaeon. The sulfide produced by the archaea might be oxidized back to sulfate by *Thiothrix*, creating a closed sulfur cycle within a pearl (Fig. 4.2, Moissl *et al*, 2002). In conclusion, the string-of-pearls community represents an extraordinary and stable association of defined groups of microorganisms from different domains of life.

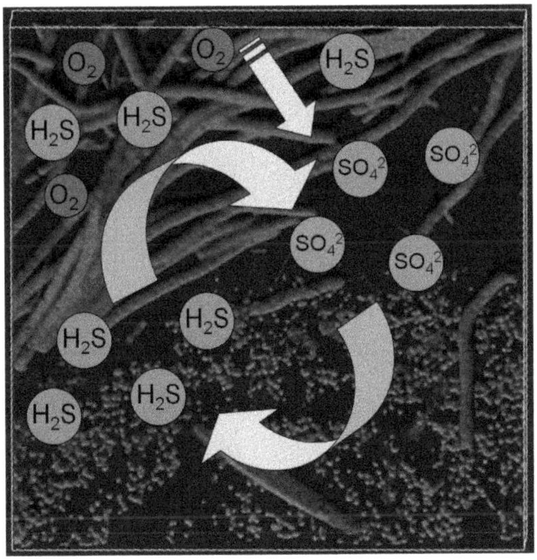

Figure 4.2 Confocal laser scanning micrograph of the edge of a pearl (about 200 µm below surface). Projection (39.8 × 36.9 × 21.4 µm). SM1 euryarchaeal cocci are located within the pearl (small dots), whereas *Thiothrix* sp. forms the outer part (filaments). The proposed sulfur cycle is indicated.

Another biotope of the string-of-pearls community and a new bacterial partner

To obtain insights into the geographical distribution and significance of the string-of-pearls community, microbiological analyses of seven cold (10–13°C), sulfide-containing springs at three distinct geographical locations in Southern Germany were carried out. All springs (close to either Regensburg, Garmisch-Partenkirchen or Passau) showed a neutral pH and sulfide concentrations of about 0.5–4 mg/l. In addition, a warm (38°C), sulfidic, salt-rich spring near Dalyan, Turkey, was included in the studies. From all springs and streamlets, microbial assemblages were sampled that macroscopically resembled the string-of-pearls community from the Sippenauer Moor (Rudolph *et al.*, 2004). These samples were analysed using archaeal 16S rRNA gene sequence analysis. Archaeal sequences were obtained from all locations and the majority of them clustered together with the SM1 sequence within the euryarchaeal 16S rRNA phylogenetic tree (SM1-group). This result indicates a widespread distribution and significant ecological role of members of the so-called SM1-group and other closely related archaea in sulfide-containing terrestrial and marine environments (Rudolph *et al.*, 2004). However, only the sample from one other sulfidic spring, the Islinger Mühlbach located near Regensburg, harboured a sequence that was identical to the SM1 euryarchaeal sequence from the Sippenauer Moor. Furthermore, the SM1 sequence was the sole archaeal sequence detected in this sample.

This result was confirmed by FISH studies of pearl-shaped SM1 euryarchaeal-bacterial communities found in the Islinger Mühlbach. More than 100 pearls from this biotope were analysed by FISH and the presence of SM1 euryarchaeal microcolonies within the pearls was confirmed. Again, the SM1 cells were arranged and grouped as observed for the string-of-pearls communities from the Sippenauer Moor. In addition to identical 16S rRNA gene sequences, the SM1 euryarchaeal intergenic spacer regions (16S/23S rDNA; Achenbach-Richter and Woese, 1988) from both locations displayed a 100% match, suggesting that the two different biotopes harbour the same single SM1 euryarchaeal phylotype.

However, micro- and macroscopical analysis of the bacterial partner identified in the string-of-pearls communities of the Islinger Mühlbach showed morphological differences: the bacterial filaments surrounding the SM1 euryarchaeal microcolonies did not contain intracellular sulfur globules, as seen in *Thiothrix* from the Sippenauer Moor, and the pearls also had a less compact and a more greyish appearance. FISH and 16S rRNA gene sequence analyses revealed that the filamentous bacteria belonged to the *Epsilonproteobacteria*, and had no close relationship to cultivated species (Table 4.1). The closest cultivated neighbour of this so-called IMB1 (Islinger Mühlbach) is *Sulfuricurvum kujiense* (97% sequence identity), an isolate obtained from an underground crude oil storage cavity at Kuji in Iwate, Japan (Kodama and Kazuya, 2003). *S. kujiensi* was described as a facultatively anaerobic, sulfur-oxidizing obligate chemolithotroph, using sulfur compounds and hydrogen as electron donors and nitrate as an electron acceptor. This organism seems to be very sensitive to NaCl and growth is inhibited by salt concentrations above 1% (Kodama and Kazuya, 2003). These characteristics of the closest cultured relative to IMB1 and the metabolic properties of *Thiothrix* support the idea of an internal sulfur cycle within a pearl: like *Thiothrix*, IMB1 also may be a sulfide oxidizing bacterium, probably providing sulfate for the SM1 archaeal metabolism. Furthermore, IMB1 appears to be better adapted to the lower oxygen concentrations of the Islinger Mühlbach water compared with the Sippenauer Moor (0.13 mg/l compared with 1.4 mg/l; sulfide concentration: 0.5 mg/l) and can therefore outcompete *Thiothrix* in the formation of string-of-pearls communities in this biotope (Rudolph *et al.*, 2004).

In situ cultivation of the string-of-pearls community

Thus far, the SM1 euryarchaeon has resisted cultivation on defined media under controlled laboratory conditions. The reasons for this recalcitrance could be manifold and are still unclear. Besides the attempts to grow the SM1

Table 4.1 Comparison of the microbial composition of single pearls from microbial string-of-pearls communities thriving in the Sippenauer Moor and the Islinger Mühlbach (adapted from Rudolph *et al.*, 2004)

	Sippenauer Moor (SM)	Islinger Mühlbach (IM)
Macroscopic appearance	Whitish compact pearls (Ø approx. 0.5–3 mm)	Greyish fluffy pearls (Ø approx. 0.5–3 mm)
Archaeal 16S rRNA gene sequence analysis	SM1 euryarchaeon predominant	Sequence identity to SM1 euryarchaeon
Archaeal 16S-23S rDNA ribosomal intergenic spacer region	SM1 euryarchaeon	Sequence identity to SM1 euryarchaeon
Bacterial 16S rRNA gene sequence analysis	*Thiothrix* sipK4 predominant	IMB1 ε-proteobacterium predominant
Archaeal FISH	SM1 euryarchaeon predominant	SM1 euryarchaeon predominant
Morphology and arrangement of the SM1 euryarchaeon	Microcolonies of small cocci, three-dimensionally arranged in the pearls interior	Microcolonies of small cocci, three-dimensionally arranged in the pearls interior
Bacterial FISH	*Thiothrix* sp. predominant	IMB1 ε-proteobacterium predominant
Phase contrast microscopy of the predominant bacterium	Rosette-forming, straight filaments (Ø 2 μm) with large amounts of refractile globules, consisting of amorphous sulfur	Filamentous rods with a wavy form (Ø 0.5 μm) without visible cell inclusions

FISH, fluorescence *in situ* hybridization.

euryarchaeon in the laboratory, its own biotope has been used as a natural chemostat for *in situ* growth experiments. By supplying polyethylene nets as an artificial growth surface, the string-of-pearls communities were cultivated successfully within their habitat (Moissl *et al.*, 2003). Net material of a total size of up to 10 m^2 was placed in the sulfurous streamlet of the Sippenauer Moor, and within 2–3 days attachment and growth of large quantities of the community was observed. Like the naturally growing string-of-pearls communities, the net communities were also dominated by SM1 euryarchaeal microcolonies. Therein, the cells were again arranged three-dimensionally with defined distances to each other. After about 1 week of incubation, the pearls on the nets increased in size up to 12 mm in diameter, and interestingly, the SM1 euryarchaeal cells within the pearls became randomly distributed and were no longer attached to each other.

At this stage of growth, the microbial cell population was harvested and the SM1 euryarchaeon could subsequently be separated in the laboratory from the other members of the community by density gradient centrifugation (Moissl *et al.*, 2003). After this procedure, bacterial contaminants represented only 2% of the SM1 euryarchaeal fraction, which did not contain any other archaea. About 90% of the purified SM1 euryarchaea exhibited excellent viability signals, when stained with the LIVE/DEAD® BacLight™ dye (Invitrogen, Carlsbad, USA). Therefore, this *in situ* cultivation and purification protocol allowed a fast, reliable and gentle harvest of SM1 euryarchaeal cells directly from the environment for detailed structural, physiological, biochemical and genetic analyses (Moissl *et al.*, 2003). The temperature range for SM1 viability was determined to be approximately between −2°C and +20°C, the cells preferred a neutral pH, and no catalase activity was observed (suggesting an anaerobic life style). Furthermore, no coenzyme F_{420} was detected, clearly distinguishing the SM1 euryarchaeon from methanogenic archaea (DiMarco *et al.*, 1990). The G+C content of the SM1 was determined to be 34.5 mol%. The highly pure SM1 euryarchaeal fraction also allowed for the production of specific antibodies. Cells stained using a new, combined FISH and immunofluorescence procedure (immuno-FISH) exhibited a large fluorescent halo (Moissl *et al.*, 2003).

The SM1 euryarchaeal biofilm: a second, partner-independent lifestyle

After proposing an anaerobic lifestyle of the SM1 euryarchaeon inside the pearls in surface waters, the possible occurrence of the SM1 euryarchaeon (and its specific partners) in subsurface waters was investigated. Polyethylene nets were placed as deep as possible in the wells of different sulfidic springs, and these experiments were also carried out at the source of the Islinger Mühlbach (Henneberger *et al.*, 2006). The sulfidic water of this spring emanates from a metal lined drill hole (diameter 18.5 cm), and forms a small streamlet exhibiting high microbial bioactivity, as evident by the presence of different microbial mats and streamers (Rudolph *et al.*, 2004). The drill hole allowed placement of the nets to a depth of about 35 cm within the emerging water. Within 10 minutes macroscopic droplets were already attached to the nets. The milky material observed was very stringy and attached strongly to the polyethylene fibres. The droplets were collected and analysed by phase-contrast microscopy, revealing that this biomaterial consisted almost exclusively of small coccoid microorganisms, three dimensionally arranged at defined distances to each other. FISH with an SM1 specific probe showed a positive reaction with these coccoid cells. Overall, 95–98% of the total microbial population within the biofilm could be identified as SM1 euryarchaeon. FISH observations were confirmed by 16S rRNA gene and 16S/23S intergenic spacer region sequence analysis, and no other archaea were detected by these molecular methods (Henneberger *et al.*, 2006). Therefore, this community can effectively be considered a 'monospecies' biofilm and represents the first described naturally occurring archaeal biofilm (Fig. 4.1C). A detailed analysis of the SM1 euryarchaeal biofilm is given in Table 4.2.

The archaeal biofilm consisted mainly of SM1 euryarchaeal cells, but contained also up to 5% of morphologically and phylogenetically diverse bacterial microorganisms, as identified by FISH analysis. Since SM1 overwhelmingly dominates the biofilm a competitive relationship between the archaea and bacteria within this community could be envisaged. Molecular diversity analyses showed that the majority of the bacteria were affiliated to sulfate reducing *Deltaproteobacteria*. This fact supports the idea of a sulfate-reducing metabolism of the SM1 euryarchaeon (Henneberger *et al.*, 2006; Probst *et al.*, 2011, unpublished).

The appearance of biofilm droplets on the polyethylene nets after only a few minutes of incubation indicates that the biofilms did not grow on the nets but were washed up from deeper earth layers and attached randomly. This observation suggests that extended SM1 biofilms may occur in the subsurface, where the anaerobic and stable

Table 4.2 Biological properties of the SM1 euryarchaeal biofilm (adapted from Henneberger *et al.*, 2006)

Study	SM1 euryarchaeal biofilm
Visual appearance	Milky, slimy, small drops Ø = 1–5 mm
Harvesting	Harvested using polyethylene nets placed as deep as possible in the sulfidic spring bore hole (Islinger Mühlbach) Attachment of visible, small biofilm drops after only 10 min. Quantitative harvesting after 4 days: 1.5–2 g (wet weight)
Microbial composition	~ 95% SM1 Euryarchaeon, ~5% bacteria (mainly representatives of the *Deltaproteobacteria*)
Structural features	Cells surrounded by filamentous hami and an additional, dense EPS. Hami degradable by proteinase K
EPS composition	Carbohydrates and proteins present in a ratio of 1:1.5
Water content	~ 99.6%
pH	7–7.5
Genome size (SM1 biofilm cells)	~ 3 Mbp
G+C content (SM1 biofilm cells)	35.0 Mol%

milieu of deeper earth layers provides optimal conditions for growth of the SM1 euryarchaeon independently of its specific partners in the string-of-pearls communities, which have been shown to develop and grow within the oxygenated parts of surface waters of sulfidic springs (Henneberger *et al.*, 2006).

Despite these two different lifestyles under aerobic and anaerobic conditions, the spatial arrangement of the SM1 euryarchaeal cells within the string-of-pearls communities and within the SM1 euryarchaeal biofilms was very similar. The biofilm droplets and pearls were remarkably structured and each individual cell showed a constant distance to its neighbouring cells of approximately 4 µm (Fig. 4.1C). Electron microscopy revealed that the surface of each single SM1 euryarchaeal cell was covered by approximately 100 pili-like structures up to 3 µm in length (Fig. 4.1D). As already observed by immuno-FISH, these fibres were arranged like a halo around each individual SM1 cell (Moissl *et al.*, 2005a,b).

Hami: novel and unique prokaryotic cell surface appendices

The presence of fimbriae or pili on cell surfaces is a common phenomenon amongst prokaryotes. These cell appendices vary in their functions and can play significant roles in biofilm formation, attachment of cells to surfaces, motility or the transfer of DNA and proteins (Soto and Hultgren, 1999; Skerker and Shapiro, 2000).

The SM1 cell surface appendices however, are unique and show an unexpectedly high complexity, with a well-defined base-to-top organization that has not been reported previously for other prokaryotes. Two regions can be distinguished along each appendix: the central *prickle region* is a barbwire-like structured filament (7–8 nm in diameter) with three small prickles pointing outwards from the filament, facing the tip at an angle of 45–60° and recurring at regular intervals of 46 nm (Fig. 4.3A). These prickles are thinner than the main filament (approximately 4 nm) and are 30 ± 3 nm in length. Up to 60 of these units make up one filament, counting from the cell surface (Fig. 4.3A; Moissl *et al.*, 2005a), before the filament continues past the final set of prickles into the 152 ± 3 nm-long *hook region* that terminates in a unique tripartite structure. This distal part consists of a plain filament that splits into three separate arms, 4 nm in diameter and 50 nm in length. These arms are bent 180°, facing back to the cell surface, and are slightly thickened at their tips (Fig. 4.3B, Moissl *et al.*, 2005a). This region strongly resembles a man-made fishing hook, barbed hook or anchor. Based on this complex architecture and uniqueness amongst prokaryotic cell appendices we proposed the name 'hamus' (plural 'hami'; Latin for hook, curved process) for this novel class of cell surface structures (Moissl *et al.*, 2005a). Owing to the characteristic appearance of the hami, the SM1 euryarchaeal cells have even been called 'archaeal ninjas' (Schaechtner and Youle, 2007).

The strong resemblance of the hami to hooks, grapples or anchors suggests a function for cell adhesion. However, the prickles in the barbwire-like prickle region are facing the terminal hook, which excludes an adhesive function. It is more probable that this region causes the hami to serve as spacer or connecting elements between individual SM1 cells. This hypothesis is supported by the precise three-dimensional arrangement of SM1 euryarchaeal cells and the regular distance of the cells to each other observed in the SM1 euryarchaeal biofilms and the string-of-pearls communities (Fig. 4.1C; Henneberger *et al.*, 2006). On the other hand, the high resemblance of the terminal hook region to anchors or grappling hooks suggests an adhesion or cell anchoring function for the distal part of the hami. In fact, adhesion studies with an 'optical tweezers' laser (Huber *et al.*, 1999; Huber and Stetter, 2001) revealed strong adhesive forces between SM1 cells and various artificial surfaces as well as between cell-pairs (Moissl *et al.*, 2005a). For these experiments, SM1 euryarchaeal cells were spotted onto glass slides that had been treated with either polylysine, polyglutamate, gelatine, BSA, laminine, fibronectine (each 0.1% w/v) or a bind-silane solution. Any SM1 cells attached to the treated surface could not be removed by the optical tweezers laser (emission wavelength 1064 nm, maximal output power 2.5 W). In additional experiments, single free-floating SM1 cells

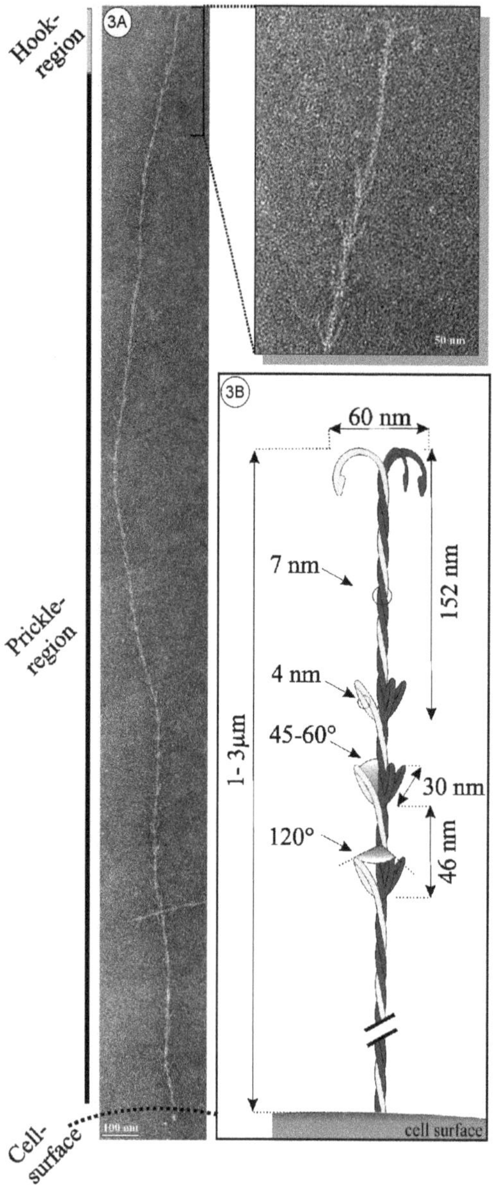

Figure 4.3 (A) Electron micrograph of a full-length hamus (negative contrast). (B) Model of a hamus, showing dimensions (from Moissl *et al.*, 2005a,b, with permission).

in suspension were caught by the laser beam and placed on top of a cell that had been previously fixed to a surface. Subsequent separation of the two cells was not possible, even when maximum laser power was applied. This strong cell-surface and cell–cell attachment clearly shows that the SM1 hami represent perfect adhesion tools, independent of the surface tested.

The hami also showed very high resistance to chemical treatment: isolated hami retained their structure when incubated at temperatures from 0 to 70°C or pH values from 0.5 to 11.5 for more than 1 hour. However, the enzymes proteinase K, protease (subtilisin A), pronase and trypsin digested the hami, indicating a proteinaceous nature thereof (Moissl *et al.*, 2005b).

Hamus structure and formation

Although the hami appear to be highly complex, their basic structure is remarkably simple. Physical treatment of SM1 euryarchaeal cells with glass beads, used to shear the hami for further protein analysis, revealed that the entire hamus is composed of three proto-filaments (Fig. 4.4). The prickles and hooks are formed by bending, curving and curling of the basic filaments (Fig. 4.3B). A major protein subunit of 120 kDa was detected by SDS-PAGE analysis of the purified hami (Moissl *et al.*, 2005a). The corresponding gene sequence encodes a 40 kDa protein, indicating the formation of a stable trimer. Therefore, it seems obvious that the entire hamus structure is built by this single 40 kDa hamus protein, responsible for the formation of the barbwire-like structure and the prickles, as well as for the hook at the distal end. Simply by twisting and untwisting, three unstructured threads create a powerful attachment tool and spacing device. Multiple gene copies have been identified in the SM1 genome by DNA hybridization, but no homologies to previously annotated genes or known proteins were observed. The gene has been cloned and successfully expressed in *E. coli*. However, the formation of inclusion bodies rather than expression on the cell surface of the expression host was observed (Moissl-Eichinger, unpublished, 2010).

It is very probable that the three basic filaments are formed by self-assembly of the protein. Therefore, even cell wall transport and hamus assembly could be a very simple process, not requiring the addition of different proteins at different stages of the construction.

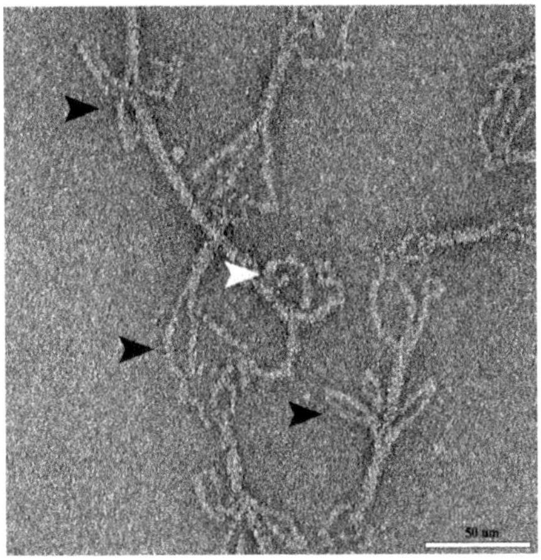

Figure 4.4 Electron micrograph of hami physically treated with glass beads. Three protofilaments are clearly visible (black arrows); the white arrow points to the hook of a hamus.

Barbwire and grappling hooks: perspectives for nanobiotechnology?

Over the past decades, the dimensions of technologically fabricated devices have been decreasing constantly, and technology now allows for the production of, for example, electronic chips as small as 20 nm (Fortina *et al.*, 2005). This so-called nanotechnology is an interdisciplinary field that applies nanoscale materials for developing novel materials, devices and systems using the nano-effects of the original compounds (Reshetilov and Bezborodov, 2008). The molecular machineries of cells can be seen as natural nanotechnological factories, using and producing natural 'nano-devices' with high storage capacity, the capability of self-assembly or replication with extremely low error rates (Lowe, 2000). By combining molecular biology and nanotechnology, the development of a new class of multifunctional devices and materials with extremely high sensibility, precision and flexibility for biological, medical or electronic applications becomes possible (Lowe, 2000; Fortina *et al.*, 2005).

Current nano-biotechnological applications include a range of different technologies in drug discovery, design and delivery, such as the use of fluorescent nano-particles (quantum dots) for the imaging of individual drug receptors in brain cell membranes (Dahan *et al.*, 2003; Jain, 2005). Self-assembled monolayers of organic molecules represent a group of biological nano-structures that can be applied to cover solid surfaces and therefore functionalize them for the development of sensors, molecular electronic devices or protective layers (Lowe, 2000). In particular, microbial cell surface layers (S-layers) are studied extensively in this context (Sleytr *et al.*, 2007), and potential applications of functional S-layer fusion proteins have been reported, e.g. for the coating of microspheres for extracorporeal blood purification (Völlenkle *et al.*, 2004).

Even though nanobiotechnology is a recent phenomenon, it appears that one of the most basic life forms of the microbial world long ago developed a tool that mankind now uses in technology worldwide. The hami of the SM1 euryarchaeon with their barbwire and grappling hooks are a perfectly evolved, natural mechanical nano-tool which are impressively diminutive: in comparison to the hami, a human hair is 15,000 times and a standard fishing hook 250,000 times larger than the nano-grappling hooks (Fig. 4.5). In general, adhesion and immobilization caused by structures at these nano-scales are due to

Figure 4.5 Comparison of the hamus structure with a fish hook and barbwire (left: hamus, model; middle: hamus, cryo-electron tomography, 3D reconstruction; top right: commercial fish hook; bottom right: barbwire). Hamus hook, width: about 60 nm, fish hook: 1 cm. Hami filament, width: 7 nm, barbwire: 0.5 cm (from Moissl *et al.*, 2005b, with permssion).

electrostatic, van-der-Waals interaction or hydrophobic effects. For example, bacterial movement by gliding is mediated by adhesion of type IV pili. Purely mechanical adhesion through molecular hooks has so far been unknown. The amazing physical and chemical stability of the hami opens new possibilities for their use as nano-tools for modern medicine and biotechnology. As a combination of 'high tech' and 'high nature', the unique structure of the hami could find application in the fast-growing field of nanobiotechnology as, for example, a 'nano-Velcro' mechanism.

Acknowledgements

We thank Christian Rudolph, Ariane Briegel, Harald Engelhardt, Reinhard Rachel and Gerhard Wanner for scientific support. We also thank the Deutsche Forschungsgemeinschaft DFG for funding this project (HU 711/2).

References

Achenbach-Richter, L., and Woese, C.R. (1988). The ribosomal gene spacer region in Archaebacteria. System. Appl. Microbiol. *10*, 211–214.

Boetius, A., Ravenschlag, K., Schubert, C.J., Rickert, D., Widdel, F., Gieseke, A., Amann, R., Jørgensen, B.B., Witte, U., and Pfannkuche, O. (2000). A marine microbial consortium apparently mediating anaerobic oxidation of methane. Nature *407*, 623–626.

Dahan, M., Lévi, S., Luccardini, C., Rostaig, P., Riveau, B., and Triller, A. (2003). Diffusion dynamics on glycine receptors revealed by single-quantum dot tracking. Science *302*, 442–445.

DiMarco, A.A., Bobik, T.A., and Wolfe, R.S. (1990). Unusual coenzymes of methanogenesis. Ann. Rev. Biochem. *59*, 355–394.

Fortina, P., Kricka, L.J., Surrey, S., and Grodzinski, P. (2005). Nanobiotechnology: the promise and reality of new approaches to molecular recognition. Trends Biotechnol. *23*, 168–173.

Fröstl, J.M., and Overmann, J. (1998). Physiology and tactic response of the phototrophic consortium '*Chlorochromatium aggregatum*'. Arch. Microbiol. *169*, 129–135.

Henneberger, R., Moissl, C., Amann, T., Rudolph, C., and Huber, R. (2006). New insights in the lifestyle of the cold-loving SM1 euryarchaeon: natural growth as a 'monospecies' biofilm in the subsurface. Appl. Environ. Microbiol. *72*, 192–199.

Huber, R. (1999). Die Laserpinzette als Basis für Einzelzellkultivierungen. BIOspektrum *5*, 289–291.

Huber, R., and Stetter, K.O. (2001). Discovery of hyperthermophilic microorganisms. In: Methods in Enzymology 330: Hyperthermophilic Enzymes, Adam, M.W.W., and Kelly, R.M., eds. (Academic Press Inc., San Diego), pp. 11–24.

Jain, K.K. (2005). The role of nanbiotechnology in drug discovery. Drug Discov. Today *10*, 1435–1442.

Koch, M., Rudolph, C., Moissl, C., and Huber, R. (2006). A cold-loving crenarchaeon is a substantial part of a novel microbial community in cold sulphidic marsh water. FEMS Microbiol Ecol 57, 55–66

Kodama, Y., and Kazuya, W. (2003). Isolation and characterization of a sulfur-oxidizing chemolithotroph growing on crude oil under anaerobic conditions. Appl. Environ. Microbiol. *69*, 107–112.

Larkin, J.M., and Nelson, R. (1987). Mechanism of attachment of swarm cells of *Thiothrix nivea*. J. Bacteriol. *169*, 5877–5879.

Lowe, C.R. (2000). Nanobiotechnology: the fabrication and applications of chemical and biological nanostructures. Curr. Opin. Struct. Biol. *10*, 428–434.

Moissl, C., Rudolph, C., and Huber, R. (2002). Natural communities of novel archaea and bacteria with a string-of-pearls-like morphology: Molecular analysis of the bacterial partners. Appl. Environ. Microbiol. *68*, 933–937.

Moissl, C., Rudolph, C., Rachel, R., Koch, M., and Huber R. (2003). *In situ* growth of the novel SM1 euryarchaeon from a string-of-pearls-like microbial community in its cold biotope, its physical separation and insights into its structure and physiology. Arch. Microbiol. *180*, 211–217.

Moissl, C., Rachel, R., Briegel, A., Engelhardt, H., and Huber, R. (2005a). The unique structure of archaeal „hami', highly complex cell appendages with nano-grappling hooks. Mol. Microbiol. *56*, 361–370.

Moissl, C., Briegel, A., Engelhardt, H., and Huber, R. (2005b). Enterhaken und Stacheldraht: Verblüffende Strukturen aus der archaeellen Nano-Welt. BIOspektrum *6*, 732–733.

Nielsen, P.H., de Muro, M.A., and Nielsen, J.L. (2000). Studies on the *in situ* physiology of *Thiothrix* spp. present in activated sludge. Environ. Microbiol. *2*, 389–98.

Reshetilov, A.N., and Bezborodov, A.M. (2008). Nanobiotechnology and biosensor research. Appl. Biochem. Microbiol. *44*, 1–5.

Rudolph, C., Wanner, G., and Huber, R. (2001). Natural communities of novel Archaea and Bacteria growing in cold sulfurous springs with a string-of-pearls-like morphology. Appl. Environ. Microbiol. *67*, 2336–2344.

Rudolph, C., Moissl, C., Henneberger, R., and Huber R. (2004). Ecology and microbial structures of archaeal/bacterial strings-of-pearls communities and archaeal relatives thriving in cold sulfidic springs. FEMS Microbiol. Ecol. *50*, 1–11.

Schaechter, and Youle. (2007). Archeal ninjas. Available at: http://schaechter.asmblog.org/schaechter/2007/10/barbed-wire-and.html. (Accessed 25 July 2011).

Schink, B. (2002). Synergistic interactions in the microbial world. Antonie Van Leeuwenhoek *81*, 257–261.

Skerker, B., and Shapiro, L. (2000). Identification and cell cycle control of a novel pilus system in *Caulobacter crescentus*. EMBO J. *19*, 3223–3234.

Sleytr, U.B., Huber, C., Ilk, N., Pum, D., Schuster, B., and Egelseer, E.M. (2007). S-layers as a tool kit for nanbiotechnological applications. FEMS Microbiol. Lett. *267*, 131–144.

Soto, G.E., and Hultgren, S.J. (1999). Bacterial adhesins: common themes and variations in architecture and assembly. J Bacteriol *181*, 1059–1071.

Völlenkle C., Weigert, S., Ilk, N., Egelseer, E., Weber, V., Loth, F., Falkenhaben, D., Sleytr, U.B., and Sára, M. (2004). Construction of a functional S-layer fusion protein comprising an immunoglobulin G-binding domain for development of specific adsorbents for extracorporeal blood purification. Appl. Environ. Microbiol. *70*, 1514–1521.

Enzyme Activities and Biotechnological Applications of Cold-active Microfungi

5

Helena Nevalainen, Ron Bradner, Sania Wadud, Suja Mohammed, Christopher McRae and Junior Te'o

Abstract

Fungi are eukaryotic organisms and considered to be less adaptable to extreme environments when compared with bacteria. While there are no thermophilic microfungi in a strict sense, some fungi have adapted to life in the cold. Cold-active microfungi have been isolated from the Antarctic and their enzyme activities explored with a view to finding new candidates for industrial use. On another front, environmental pollution by petroleum products in the Antarctic has led to a search for, and the subsequent discovery of, fungal isolates capable of degrading hydrocarbons. The work has paved the way to developing a bioremedial approach to containing this type of contamination in cold climates. Here we discuss our efforts to map the capability of Antarctic microfungi to degrade oil and also introduce a novel cold-active fungal lipase enzyme.

Introduction

Fungi are achlorophyllous eukaryotic organisms with a unique absorptive mode of nutrition. Modern DNA analysis has demonstrated that the fungal kingdom is represented by one subkingdom, seven phyla and ten subphyla (Hibbett *et al.*, 2007). The fungi contained in the two phyla, the Ascomycota and the Basidiomycota, are commonly the ones recognized as 'fungi'. Within these phyla, fungi whose somatic form is filamentous or hyphal characterize the vast majority of the true fungi, whilst those whose somatic form is unicellular are classified as yeasts.

The subdivision of fungi into microfungi and macrofungi is simply made by reference to the size of their fruiting bodies (variously named fructifications, ascocarps, basidiocarps, sporocarps, sporophores, mushrooms, apothecia, morels, stromata, puffballs, etc.). Macrofungi are species that produce 'large' fructifications, i.e. those visible to the naked eye. The majority of species are from the phyla Basidiomycota, e.g. mushrooms and puffballs, and Ascomycota, e.g. morels and cup fungi. Microfungi, of which moulds comprise the majority, produce mitospores or conidia on relatively simple sporogenous structures. Microfungi can be defined as fungi with either microscopic spore-forming structures or minute fructifications. Soil-dwelling microfungi, for instance, are usually only identifiable after they have been isolated in pure culture on nutrient media and induced to form conidia-bearing structures, conidiophores, which form the basis for morphological identification.

The observed abundance of many species of microfungi identified in the course of ecosystem analyses has led to the view that fungi tend to be universal in their occurrence (Cannon and Sutton, 2004). This view could be somewhat biased since human interaction may at least be partly responsible for the wide distribution of fungi in dung, litter and soil (Gams, 1992), and also because species that sporulate profusely, are easy to isolate into pure culture, and have short generation times will be over represented in the analyses.

Fungi play key roles in ecosystems as decomposers, mutualists, and pathogens, but in most cases the role of individual fungi in nature is unknown. Aside from these roles, fungi have additional economic importance as biocontrol agents and as production organisms for the pharmaceutical and other industries. In a recent study it was estimated that there is a minimum of 712,285 fungal species worldwide (Schmit and Mueller, 2007) and that microfungi associated with terrestrial plants account for over 82% of this number; however, only approximately 13% of these species have been described (Schmit and Mueller, 2007).

Fungi in extreme environments

Fungi have been isolated from many extreme and unusual environments, and new species are being constantly identified. For example, the number of known species that have been isolated from desert environments have roughly doubled in the past 25 years; the isolated species have largely been found to be cosmopolitan (Gams, 2007). A high proportion of the microfungi from these desert soils have also been described as 'thermophilic' and heat-resistant (Mouchacca, 1995). Among the 246 new fungal taxa described in the last 60 years from the Middle East, mainly from desert and salt marsh soils, many are adapted to high temperatures or high salt concentrations (Mouchacca, 2005).

Sampling in remote areas has led to the identification of considerable numbers of novel fungi. For example, in remote locations of Australia, assiduous searches by Summerell *et al.* (1995) and Burgess *et al.* (1996) found several well-known and also a few new species of *Fusarium* (Gams, 2007). Unusual habitats also include the underground seed caches of the North American kangaroo rat, *Dipodomys spectabili,* which have yielded several new *Penicillium* species (Frisvad *et al.*, 1987). Alkaliphilic and alkalitolerant fungi form another ecological group that are adapted to unusual substrata. A range of *Fusarium* and *Acremonium*-like fungal species found in highly alkaline soils in Indonesia grow well on agar media with pH 9.8 (Naga *et al.*, 1995).

There are many natural environments accommodating microfungi with the potential to yield cold-active enzymes of biotechnological significance. These range from the high alpine regions scattered around the globe, the frigid Arctic regions of Eurasia and North America, the lower extreme of South America and last, but by no means least, the Antarctic continent. With the notable exception of the last, each of the others is contiguous with a greater landmass where climates range well into the temperate and are occupied by a vast range of flora and a generally mobile fauna.

The Antarctic environment

The Antarctic continent lies mostly south of the Antarctic Circle and is the coldest, driest, windiest, highest, and most isolated of all the continents. Air temperatures in the maritime Antarctic (West Coast of the Antarctic Peninsula and surrounding islands) are <2°C for 1–4 months of the year, whereas in other coastal regions and in the McMurdo Dry Valleys the average remains below freezing (Chapman and Walsh, 2007). The mean annual soil temperatures in the McMurdo Dry Valleys and other ice-free areas of the Ross Island Region range from −14.6°C to −23.5°C (Seybold *et al.*, 2009), although during December and January the subsurface soil (0–20 cm depth) temperatures in the Dry Valleys are consistently above freezing (Vincent, 1989).[1] Wide thermal fluctuations are a characteristic of the Antarctic climate. Most of the biological and fungal sampling in the Antarctic has occurred in the Peninsula, along other coastal regions and in the Dry Valleys, the closest terrestrial analogue to Mars.

Various climates prevail in Antarctica and microclimates are even more differentiated. It is characteristic of the continent that microclimates and nanoclimates can be very variable and unstable with regards to temperature and water availability, a crucial requirement for microbial life (Walton, 1984). The conditions of temperature and aridity, at least in the ice-free McMurdo Dry Valleys or Ross Desert are comparable to those prevailing on Mars (Marchant and Head, 2004) and therefore place the microorganisms of the region under extreme environmental stress.[2] Antarctica is also one of the most UV-irradiated regions on Earth due to the extended sunlight hours allied with the depletion of the ozone layer which occurs during the austral summer period (Moradas-Ferreira

[1]The following website provides extensive temperature data currently collected from 18 Antarctic research bases http://www.antarctica.ac.uk/met/gjma/ (accessed 3 May 2010)

[2]For a general description of the climate and environmental conditions that prevail in the McMurdo Dry Valleys see the National Aeronautics and Space Administration (NASA) Quest webpage at http://quest.nasa.gov/antarctica/background/NSF/valleys.html (accessed 3 May 2010).

et al., 1996; Abele and Pörtner, 1998; Diaz *et al.*, 2004).[3] This has the potential for adding further to the already high levels of oxidative stress faced by the local microbiota (Joseph-Horne *et al.*, 2001; Kreiner *et al.*, 2002; Lopez-Martinez *et al.*, 2008).

Antarctic microbiota

There is abundant evidence for the presence of microorganisms on the Antarctic continent (Vincent, 1989; Vishniac, 1993; Nichols *et al.*, 1999), and they dominate the biology of Antarctica more than any other continent. Within the Antarctic framework microorganisms fall into a number of categories – invertebrates, bacteria and archaea, algae, cryptogams, yeasts, and filamentous fungi. The origins of the Antarctic microbiota are essentially unknown; without doubt much is indigenous. There are, however, a number of potential avenues by which alien species can be introduced to the region: migratory species provide a ready vehicle to carry foreign material to the continent and so does man, his animals and his material. However, possibly the greatest source of contamination of the Antarctic wilderness is the wind. Various studies have been undertaken to assess the presence of airborne propagules in Antarctic air (Sieburth, 1965; Corte and Daglio, 1994; Marshall and Chalmers, 1997), and a recent detailed study over a 13.5-month period by Marshall (1997b), albeit on Signy Island in the sub-Antarctic South Orkney group, clearly demonstrated the ability of the wind to transport fungal spores.

Filamentous fungi and yeasts

Over 1000 fungal species have been reported from the maritime, continental and sub-Antarctic regions (Bridge *et al.*, 2008). Most of the filamentous fungi and yeasts identified are cosmopolitan species (Vishniac, 1993), but some appear to be new Antarctic species (Onofri *et al.*, 1999). In a recent study by Connell *et al.* (2006) on the fungi in Taylor Valley, ribosomal DNA internal transcribed spacer region (ITS rDNA) sequencing revealed the presence of filamentous fungal species such as *Geomyces* sp., *Chaunopycnis* sp., *Nematoctonus* sp., *Phaseososphaeria* sp. and *Phoma* sp. From these, *Geomyces* sp., a cosmopolitan eurythermal fungus, has been isolated in several other Antarctic studies as well (Sun *et al.*, 1978; Del Frate and Carretta, 1990; Onofri *et al.*, 1994). Amongst the yeasts described were *Rhodotorula* sp., *Diosdzegia* sp., *Cryptococcus yarrowi* and *Cryptococcus saitoi*. Whilst both filamentous and unicellular fungi have been isolated from a diversity of Antarctic soil types, only the unicellular yeasts are known to be endemic in the drier polar desert soils (Vishniac and Klinger, 1986; Vishniac, 1993).

The distribution of fungi in Antarctica is most often associated with strong biotic influences such as birds (penguin rookeries) and vegetation (bryophyte and lichen communities). There are about 130 bryophyte species and around 380 species of lichen in the Antarctic (Alberdi *et al.*, 2002). Lichens, the symbiotic association of certain fungi and (usually) green algae are one of the most conspicuous elements of the terrestrial ecosystem. The bryophytes, although not as widespread as the lichens, have been identified as a rich source of fungal material (Greenfield and Wilson, 1981; Möller and Dreyfuss, 1996; Skotnicki *et al.*, 1996; Azmi and Seppelt, 1998; Tosi *et al.*, 2002). Fungi have also been isolated from soils and even from the carcasses of mummified seals (Azmi and Seppelt, 1998; Aislabie *et al.*, 2001).

In their recent review article, Ruisi *et al.* (2007) noted that in recent decades, Antarctic investigations have mainly concentrated on psychrophilic bacteria and archaea. This trend is now starting to change slowly with the surveys discussed here (e.g. Robinson, 2001; Connell *et al.*, 2006; Bridge *et al.*, 2008). Several Antarctic fungi have also been used as models for exobiology (Onofri *et al.*, 2005). In the main, the Antarctic microfungi have been found to be psychrotrophic (mesophilic psychrotolerant) rather than psychrophilic (Kerry, 1990; Zucconi *et al.*, 1996; Azmi and Seppelt, 1997; Ruisi *et al.*, 2007). Both psychrophilic and psychrotrophic fungi have the ability to grow at 0°C. Psychrophilic fungi have an optimum temperature for growth around 15°C or lower, and a maximum of 20°C or below, whereas psychrotrophic fungi have an optimum growth temperature above 20°C (Robinson, 2001). Table

[3] Comprehensive global ozone and UV data can be obatained from 'World Ozone and Ultraviolet Data Centre (WOUDC)' at http://www.woudc.org/ (accessed 3 May 2010).

Table 5.1 Growth of Antarctic microfungi at 10°C and room temperature on potato dextrose agar (PDA) plates (Bradner, 2003)

Antarctic species	Incubation temperature	Colony diameter (mm) 5 days	10 days	15 days	20 days
Embellisia sp.	10°C	17	40	57	75
	RT	29	63	85[1]	
Cladosporium cladosporioides	10°C	7	15	22	29
	RT	6.5	12	19	22
Helotiales sp.	10°C	6	13	17	25
	RT	5	14	22	na
Phaeosphaeria sp.	10°C	10	16	20.5	28
	RT	10	20	32	38
Geomyces pannorum	10°C	12.5	27	37	52
	RT	9	29	62	73
Podospora minuta	10°C	13	25	31	36
	RT	23	36	47	54
Penicillium corylophilum	10°C	11	29	32	45
	RT	23	na	na	na
Glomerella lagenaria	10°C	5	9	12.5	17
	R/T	7	14	24	28
Penicillium miczynskii	10°C	5	17	26	31
	RT	na	na	na	na

[1]Diameter of plate. na, not available; RT, room temperature.

5.1 shows a comparison of the growth of Antarctic microfungi at room temperature and 10°C.

In summary, there are an increasing number of surveys of Antarctic fungal communities using both culture-based and molecular approaches for the identification of fungi. Both the earlier and more recent surveys have shown that microfungi belonging to the main phyla Ascomycetes and Basidiomycetes have been found in the Antarctic environment, ranging from endolithic to nematophagous species.

Cold adapted enzymes – structure and function

All filamentous fungi acquire their sustenance via the secretion of enzymes into their environment (Johnson and Raven, 2002). It can reasonably be surmised that Antarctic microfungi have acquired adaptations to many of their enzymes sufficient to allow them to remain functional in the face of the intense cold. Although significant progress has been made in elucidating the molecular adaptation mechanism of enzymes produced by extremophiles such as hyperthermophiles (Li *et al.*, 2005; Greaves and Warwicker, 2007; Luke *et al.*, 2007; Robb *et al.*, 2008), the molecular basis of cold adaptation is still relatively poorly understood. However, data now being accumulated on psychrophilic enzymes are starting to shed some light on their functional and structural characteristics (Feller *et al.*, 1997; Gianese *et al.*, 2001; Feller and Gerday, 2003; Bjelic *et al.*, 2008; Aurilia *et al.*, 2009). Enzymes produced by cold adapted microorganisms most commonly display a high catalytic efficiency at low temperatures, measured as k_{cat}/K_M, and a significantly increased thermolability believed to be a consequence of enhanced peptide chain flexibility (Lonhienne *et al.*, 2000; Gianese *et al.*, 2001).

Homology modelling of psychrophilic enzymes has been used to predict their secondary structure and the accessibility of amino acid residues (Gianese *et al.*, 2001). The results obtained showed a clear tendency for the charged residues Arg and Glu in 'hot' enzymes to be replaced at exposed sites on α-helices by Lys and Ala (respectively) in their 'cold' analogues. At buried regions in α-helices, Val is replaced by Ala. Compositional analysis of psychrophilic enzymes shows a significant increase of Ala and Asn and a decrease of Arg at exposed sites. Buried sites in β-strands tend to be depleted of Val. The preferred amino acid exchanges observed between thermo/mesophiles and psychrophiles and compositional analysis both indicate a decreased number of potential side chain H-bonds and salt bridges in cold-adapted enzymes. Rigid secondary structures and disulfide bridges are practically absent, thus accounting for increased thermolability. Cold adaptation strategy seems to apply the reverse principle to that of hot adaptation, namely a decrease in the number of H-bonds/ion pairs when moving from high to low temperatures (Feller *et al.*, 1997; Feller and Gerday, 1997, 2003; Marshall, 1997a). The tertiary and quaternary structures of cold-active enzymes appear to have more open and flexible arrangements, thus providing better access of substrates to the active site at lower temperatures (Nichols *et al.*, 2002).

Just as enzymes from thermophiles often have optimal activity at temperatures higher than found for mesophilic organisms, so cold-active enzymes shift their peak activities to temperature ranges lower than those generally observed for mesophilic enzymes (Brenchley, 1996). Enzymes isolated from cold-adapted microorganisms can be classified into three groups: heat-sensitive but otherwise similar to mesophilic enzymes, heat-sensitive but more active at low temperatures, and the same thermostability as mesophilic enzymes but more active at low temperatures (Ohgiya *et al.*, 1999). The specific activity of a number of cold-adapted enzymes produced by Antarctic and Arctic microorganisms (wild-type and recombinant forms) has been determined (Gerday *et al.*, 2000). Typically, the specific activity of these enzymes is higher than that of their mesophilic counterparts at temperatures of approximately 0–30°C. Denaturation of the cold-active enzyme can be expected to occur at higher temperatures.

Antarctic cold-active enzymes

As indicated above, investigations of Antarctic microorganisms have largely concentrated on bacteria and archaea and to a far lesser degree on fungi. Although a significant catalogue of cosmopolitan and endemic species has been collated over the years from within the Antarctic domain, very little work has come to light in the published arena describing novel enzymes. Not surprisingly, the reported Antarctic cold-active enzymes have been derived from bacterial or archaeal sources, and also some yeasts.

It is not unusual for investigators, upon isolating a new microbial species, to undertake a suite of analytical tests to ascertain its potential enzyme repertoire. However, the subsequent isolation and characterization of enzymes identified, if undertaken, is rarely reported. Table 5.2 details some of the enzymes that have been isolated from Antarctic microorganisms.

Industrial application of cold-adapted enzymes

The value of industrial enzymes was estimated by BCC Research at US$2.3 billion in 2007 (http://bccresearch.com/report/BIO030E.html). Whilst most of the enzymes produced today for industry have been isolated from mesophiles and thermophiles, there is a growing interest in cold adapted enzymes due to their potential industrial applications. Psychrophilic enzymes produced by cold-adapted microorganisms display a high catalytic efficiency that offers considerable potential to the biotechnology industry, making it possible to reduce the amount of enzyme needed in catalytic processes (Demirjian *et al.*, 2001). The low temperature optimum and temperature stability make them preferable for use in reactions performed at low temperatures that require easy inactivation (Yamamura *et al.*, 1999). From the biotechnological point of view, cold-adapted enzymes, which Antarctic microfungi can potentially supply, could in future replace their mesophilic counterparts and help to establish new improved bioprocesses under low-temperature conditions. These enzymes can help to save energy, to save labile or volatile compounds, to prevent the growth of mesophilic contaminants at low temperature,

Table 5.2 Examples of reported cold-active Antarctic enzymes

Enzyme	Source organism	Reference
β-d-galactosidase	*Arthrobacter* sp. 32c	Hildebrandt *et al.* (2009)
	Pseudoalteromonas sp. 22b	Makowski *et al.* (2007)
	Arthrobacter isolate	Coker *et al.* (2003)
Lipase B (CALB)	*Candida antarctica*	Blank *et al.* (2006)
Alkaline phosphatase	*Vibrio* strain G15–21 Antarctic bacterium TAB5	Helland *et al.* (2009) Wang *et al.* (2007)
Catalase	*Vibrio salmonicida*	Riise *et al.* (2006)
β-1,4-d-mannanase	*Cryptopygus antarcticus* (an Antarctic springtail)	Song *et al.* (2008)
Iron superoxide dismutase	*Pseudoalteromonas haloplanktis*	Merlino *et al.* (2008)
	Marinomonas sp. *NJ522*	Zheng *et al.* (2005)
Lip1 lipase	*Pseudoalteromonas haloplanktis* TAC125	de Pascale *et al.* (2008)
Alcohol dehydrogenase	*Flavobacterium frigidimaris* KUC-1	Kazuoka *et al.* (2007)
Serine protease	*Colwellia* sp. NJ341	Wang *et al.* (2007)
DnaK	*Shewanella* sp. Ac10	Yoshimune *et al.* (2005)
Protein-tyrosine phosphatase (CAPTPase)	*Shewanella* sp.	Tsuruta *et al.* (2005)
Cellulase	*Pseudoalteromonas haloplanktis*	Violot *et al.* (2003)
Xylanase	*Pseudoalteromonas haloplanktis*	Collins *et al.* (2002)
Glutamate dehydrogenase	*Psychrobacter* sp. TAD1	Camardella *et al.* (2002), Di Fraia *et al.* (2000)
α-amylase	*Alteromonas haloplanctis*	Feller *et al.* (1999)
Citrate synthase	Bacterial isolate, DS2–3R	Gerike *et al.* (1997)
Protease	*Pseudoalteromonas* sp. NJ276	Wang *et al.* (2008)
Lipase *lip*PA	*Penicillium allii*	Bradner *et al.* (2003)
Polygalacturonase	*Cystofilobasidium capitatum* PPY-1	Nakagawa *et al.* (2005)

and could be easily inactivated by moderate temperatures. The cold-adapted enzymes cover a broad spectrum of biotechnological applications (Feller *et al.*, 1996; Marshall, 1997a; Yamamura *et al.*, 1999; Gerday *et al.*, 2000; Cavicchioli *et al.*, 2002). Some of the potential advantages of using cold-active enzymes in industrial processes are listed in Table 5.3. Antarctic microfungi are probably a valuable source of biotechnologically useful cold-adapted enzymes.

Cold-active lipases

Lipolytic enzymes are ubiquitous in nature, being widely distributed throughout all kingdoms of life (Fojan *et al.*, 2000), although they are more abundant in microorganisms – bacteria, filamentous fungi and yeast (Jaeger *et al.*, 1994). Fungi produce different classes of these enzymes, in particular, carboxyl esterases (EC 3.1.1.1), which hydrolyse short-chain carboxylic acids, and lipases (triacylglycerol acylhydrolases, EC 3.1.1.3), which catalyse the hydrolysis and synthesis of long chain acylglycerides. The distinction between long and short chain fatty acids remains moot, some investigators ascribing the threshold for long chain as C ≥10 (Sunna *et al.*, 2002) whilst others remain more conservative electing C ≥12 (Eggert *et al.*, 2000). However, there is unanimity in classifying lipases as enzymes which are capable of catalysing the hydrolysis of water insoluble (or poorly soluble) triacylglycerides to free fatty acids and glycerol at a water/oil boundary (Brush *et al.*,

Table 5.3 Industrial applications of enzymes at low temperatures

Application	Enzymes[1]	Advantages
Detergent	Protease, lipase, cellulase, etc.	Use in tap or cold water; reduced wear and tear on clothes; energy saving
Various industries		
Modification of constituents	Galactosidase, lipase	Maintain freshness
Improvement of taste and flavour	Protease, lipase, etc.	Maintain freshness
Removal of fish skin	Protease	Product quality
Reduction of viscosity and clarification of fruit juice	Pectinase, cellulase	Maintain fragrance
Preservation	Lysozyme, glucose oxidase	Improved preservative effect
Baking process	Xylanase, amylase	Reduced dough fermentation time
Enzymatic synthesis	Lipase, nitrile hydratase, etc.	For volatile and heat sensitive enzymes, low-temperature acrylamide synthesis
	Alcohol dehydrogenase	Inherent flexibility
Treatment of waste water	Catalase	Lower energy consumption
Biotechnology research		
Protoplast formation	Cell wall digesting enzymes	High protoplast viability
Molecular biology	Phosphatase, uracil DNA glycosylase	Complete heat inactivation

[1]For organisms producing these enzymes, refer to Table 5.2.

1999; Pandey *et al.*, 1999; Schmidt-Dannert, 1999; Brzozowski *et al.*, 2000; Fojan *et al.*, 2000; Sharma *et al.*, 2001; Reetz, 2002; Sunna *et al.*, 2002). The three dimensional structures of a number of lipases and esterases have been elucidated and all are members of the α/β hydrolase superfamily. Lipases display little activity in aqueous solutions containing soluble substrates, whereas esterases show normal Michaelis–Menten kinetics in solution (Sharma *et al.*, 2001).

Cold active lipases have been identified from various Antarctic bacteria including *Acinetobacter* sp., *Pseudomonas* sp., *Psychrophilus* sp. as *Moraxella* sp.; yeasts such as *Candida antarctica* and *C. lipolytica* (reviewed in Joseph *et al.*, 2008), and filamentous fungi including *Penicillium allii* (Bradner *et al.*, 2003), *Aspergillus nidulans* (Mayordomo *et al.*, 2000) and *Geotrichum* sp. SYBC WU-3 (Cai *et al.*, 2009).

Lipases have found applications in numerous industrial areas. The most commercially important field of application for hydrolytic lipases is as additives in household and industrial laundry detergents and in household dishwasher detergents, with sales in 1995 estimated at US$30 million (Jaeger and Reetz, 1998). Lipases also enjoy applications in the food industry – in the production of a variety of products, ranging from fruit juices, bakery foods and vegetable fermentation to the production of desirable flavours in cheese and other foods (Pandey *et al.*, 1999; Sharma *et al.*, 2001). In the pulp and paper industry, lipases are used to remove 'pitch' from the pulp produced for paper making (Jaeger and Reetz, 1998). Additionally, lipases have applications in cosmetics and pharmaceutical processing (Pandey *et al.*, 1999; Sharma *et al.*, 2001) and are the most widely used enzymes in synthetic organic chemistry, catalysing the chemo-, regio- and/or stereoselective hydrolysis of carboxylic acid esters or the reverse reaction, in organic solvents (Reetz, 2002). This use of lipases in fine chemical applications is largely due to the advantages of these catalysts for the production of optically pure compounds (Demirjian *et al.*, 2001).

Development of lipase-based technologies for the synthesis of novel compounds is rapidly expanding the uses of these enzymes (Liese *et al.*, 2000). A limiting factor, however, is the shortage of lipases having the required processing characteristics. Many new and novel lipases remain to be discovered and for this reason prospecting for novel lipase genes is of interest from an industrial and academic standpoint. The extreme environmental conditions to which Antarctic organisms are subjected calls out for an examination to be made of its microfungi in the search for new and different lipolytic enzymes. This is particularly true for cold-adapted enzymes, which have high potential value for the food, detergent and chemico-pharmaceutical industries. One focus of our work, discussed in this book chapter, has been the identification of Antarctic microfungi having the potential to yield economically significant cold-active lipases. Another focus has been the search for fungi capable of mitigating the contaminating effects that can result from pollution by petroleum-based products.

Bioremediation of petroleum-based pollutants

Human exploration of the Antarctic continent began in the early years of the 1ninth century and continues to this day, albeit now in the form of scientific research stations. Allied to the expansion of man into the region was the introduction of his work animals, dogs and occasionally larger animals such as ponies (all now banned under Article 4 of Annex II to the Protocol on Environmental Protection to the Antarctic Treaty). An inevitable consequence of this entry onto the continent by non-indigenous species is their ability to act as a vector for the transmission of alien microorganisms to the Antarctic. Another notable consequence of human intervention is pollution of the previously pristine environment with lead and recalcitrant hydrocarbons originating from petrol, oil and lubricants used in transportation. Pollution by petroleum products has not only been limited to the immediate vicinity of the established research bases but has extended far beyond. In the course of exploration, it has been necessary to establish fuel dumps for both terrestrial and airborne transport and these have inevitably led to more spillage and environmental damage. In the modern era, the presence of tourist vessels along the coastal fringes brings with it the danger of a major maritime mishap with the possibility of a significant oil spill.

It has been known for some time that certain microorganisms display a propensity to degrade hydrocarbon spills and pollution (Baryshnikova *et al.*, 2001; Castro-Ochoa *et al.*, 2005; Coulon *et al.*, 2005). Some of these petroleum-degrading organisms have previously been identified in Antarctic soils in the environs of refuelling sites and bases (Aislabie *et al.*, 2001; Farrell *et al.*, 2003). Unfortunately, much of the microbial activity results in the biodegradation of only the most easily decomposable portions of the oil, such as saturated hydrocarbons and the small molecules of aromatics. In comparison, larger aromatic molecules, resin and asphaltum in the oil are apt to be degraded extremely slowly in the natural degradation process (Leahy and Colwell, 1990).

Filamentous fungi generally offer several potential advantages over other microorganisms in the remediation of petroleum contamination in soils. The movement of fungal hyphae through soil, and access to hydrocarbons, is not restricted by a hydrophobic environment (as it may be with bacteria). Similarly, fungal hyphae can penetrate hydrocarbon-impregnated soil aggregates that may be anoxic internally. The extracellular enzymes which many fungi secrete may aid in the initial metabolism of hydrocarbons. Finally, many fungi are xero- and osmotolerant, which is important to maintain growth in contaminated soils with low water activity (April *et al.*, 2000).

Enzyme activities necessary for crude oil degradation

The initial steps in the catabolism of aliphatic (Singer and Finnerty, 1984), cyclic (Perry, 1984) and aromatic (Cerniglia, 1984) hydrocarbons by bacteria and fungi involve the oxidation of the substrate by oxygenases. For instance, *n*-alkanes are catabolized via mono-terminal oxidation, resulting in the formation of the homologous primary alcohol, aldehyde and carboxylic acid. The fatty acid is then further oxidized via β-oxidation rather than being incorporated directly into cell components (Walker and Cooney, 1973). It has also been shown that catalase activity is a good indicator for evaluating fungal ability to degrade hydrocarbons (Pedrini *et al.*, 2006).

Petroleum hydrocarbons are biodegraded by microorganisms in a quasi-stepwise fashion due to the various susceptibilities of different compound classes to biological action. In general, as the degree of branching and/or the number of rings of an aliphatic or aromatic hydrocarbon increase, the more resistant it becomes to biodegradation. For instance, aliphatic hydrocarbons are degraded in the order *n*-alkanes >alkylcyclohexanes >isoprenoids >bicyclic sesquiterpenes >steranes and hopanes (Leary and Colwell, 1990).

Over the years various investigators have identified a range of fungi that display, to a greater or lesser degree, an ability to degrade the hydrocarbon components of petroleum (Leahy and Colwell, 1990; Oudot *et al.*, 1993; Launen *et al.*, 1995; April *et al.*, 2000). More recent studies have concentrated on species of *Aspergillus* and *Penicillium*, microfungi most commonly associated with oil spills (April *et al*, 2000; Elshafie *et al.*, 2007; George-Okafor *et al.*, 2009). These microfungi have shown an ability to remove the bulk of the aliphatic fraction, specifically in the n-C_{13} to n-C_{26} range. The ability to extensively degrade *n*-alkanes has also been identified in two microfungi, *Fusarium lateritium* and *Drechslera* sp., isolated from the salt marshes of the Kuwaiti desert where they are subjected to both halo-stress and stressful temperature conditions (Obuekwe *et al.*, 2005). Filamentous fungi capable of oxidizing the larger, more complex components of petroleum have been the exception. However, recent work by George-Okafor *et al.* (2009) describes successful isolation of two strains of *Aspergillus* capable of degrading (under experimental conditions) 98% of the polycyclic aromatic hydrocarbon (PAH) moieties found in oil-contaminated Nigerian soil samples.

Oil degrading Antarctic microfungi

Of particular interest is the potential for Antarctic fungi to aid in bioremediation. As noted earlier, an inevitable outcome of man's presence on the continent has been the contamination of his environment with petroleum products in their many guises. Our aim, therefore, has been to identify Antarctic microfungi present with these contaminants which display an ability, not only to decompose the lighter elements of petroleum pollutants but also the larger molecular moieties, and ideally to have the capacity to accomplish this at low temperatures.

Studies with Antarctic microfungi

The focus of our recent studies has been twofold: firstly, to discover Antarctic microfungi able to produce hydrolytic enzymes, in particular, cold-active lipase and secondly, to identify microfungi capable of ameliorating the toxic impact of oil contamination of the environment, particularly in colder climates. Both aspects are discussed in more detail below.

Characterisation of the lipase activity of the Antarctic isolate *Penicillium expansum*

Cold-active lipases are of considerable commercial interest in the applications listed in Table 5.3. Antarctic microfungi provide a new source for such enzymes. We have previously isolated a gene encoding a cold-active lipase produced by *Penicillium allii* (GenBank accession number AY303124) recovered from an Antarctic bryophyte (Bradner, 2003; Bradner *et al.*, 2003). Initial screening of five additional Antarctic isolates [*P. expansum* (AF218786), *P. commune* (AF236103), *Embellisia* sp. (AF212307), *Trichoderma viride* (AF218788) and *Phoma* sp. 34/2a] for lipase activity was carried out using a plate-screening assay. Minimal medium plates at pH 5 and pH 6.5 (Penttilä *et al.*, 1987) were prepared using a lipase substrate (olive oil) as the sole carbon sourcewith the fluorescent dye Rhodamine B included in the medium. Lipase activity was observed as the formation of a fluorescent halo (when viewed under UV light) around the fungal colony, indicating hydrolysis of the olive oil substrate (Fig. 5.1). The plates were incubated at 25°C and examined over 3–4 days for the presence of fluorescent halos.

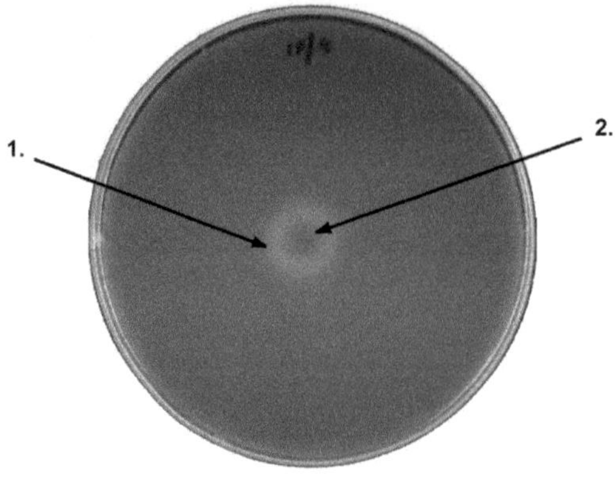

Figure 5.1 A plate containing *P. expansum* showing a fluorescent halo/clearing indicative of lipase activity. The fluorescent halo (1) can be observed surrounding the *P. expansum* colony (2).

From the strains tested, *P. expansum* was selected for detailed examination of the secreted lipase in liquid culture. The results suggested that pH 6.5 was more favourable for lipase production (data not shown).

Lipase production by *P. expansum*

P. expansum was grown in a 250 ml shake flask containing 50 ml of medium (pH 6.5) – Avicel cellulose (2.5% w/v), soy hydrolysate (1.5% w/v), with minimal salts (Penttilä *et al.*, 1987). The medium was inoculated with spores of *P. expansum* and incubated at 25°C with shaking (250 rpm). The supernatant was harvested after seven days and used for colorimetric lipase (lipolytic) activity assays according to the method of Ruiz *et al.* (2004), with some modifications. The assay was carried out using chromogenic synthetic *p*-nitrophenyl ester derivatives of fatty acids with different carbon chain lengths ranging from C_{12} to C_{18} as substrates – *para* (*p*)-nitrophenyl laurate (*p*-NPL, C_{12}), *p*-nitrophenyl myristate (*p*-NPM, C14), *p*-nitrophenyl palmitate (*p*-NPP, C_{16}) and *p*-nitrophenyl stearate (*p*-NPS, C_{18}). The absorbance of the yellow-coloured *p*-nitrophenol released upon substrate hydrolysis was measured at 405 nm.

Zymogram activity analysis of *P. expansum* lipase

Zymogram activity analysis was used to determine more specifically the enzyme activity of proteins following electrophoretic separation. The analysis was carried out using olive oil as the lipase substrate (Castro-Ochoa *et al.*, 2005). Non-reduced protein samples from the culture supernatant were separated by one dimensional SDS-PAGE (Laemmli, 1970), the gel incubated in TritonX-100 (2.5% v/v) to remove SDS and renature the enzymes, and then washed with Milli-Q water. Following incubation of the gel in Tris-HCl buffer, it was overlaid onto a 2% (w/v) agar layer containing olive oil and Rhodamine B (Castro-Ochoa *et al.*, 2005) and left overnight at 25°C. Lipase activity was visualized as a fluorescent band under UV light. The activity band visualized with the olive oil substrate (Fig. 5.2) indicated the presence of a true lipase (~30 kDa) in the culture supernatant of *P. expansum*.

Determination of pH and temperature optimum of the *P. expansum* lipase

As indicated earlier, the distinction between an esterase and a lipase is the ability of the latter to hydrolyse a substrate with a carbon chain length of C ≥ 10 or 12 (Eggert *et al.*, 2000). The results obtained for *P. expansum* (Fig. 5.3A) indicate that

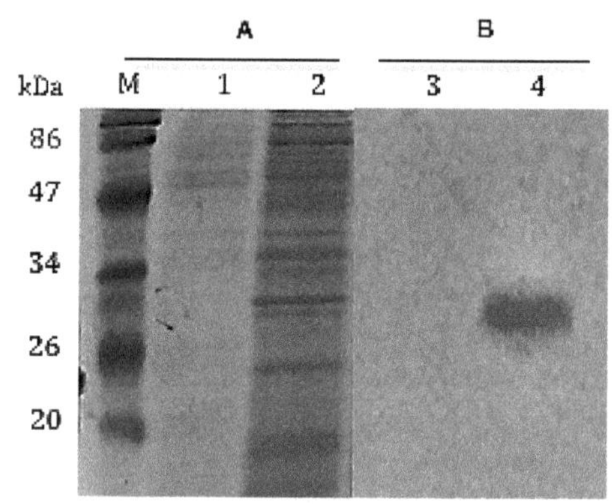

Figure 5.2 Detection of lipolytic activity from the culture supernatant of *P. expansum* using olive oil as substrate. (A) SDS-PAGE analysis of culture supernatant of *P. expansum*. (B) Zymogram showing in-gel lipolytic activity as a dark band against a grey background. Each lane contains 30 µL of culture supernatant. M, molecular weight marker; lanes 1 and 3, negative control (culture medium); lanes 2 and 4, *P. expansum* culture supernatant. Lane 4 contains a lipolytic activity band using olive oil as substrate.

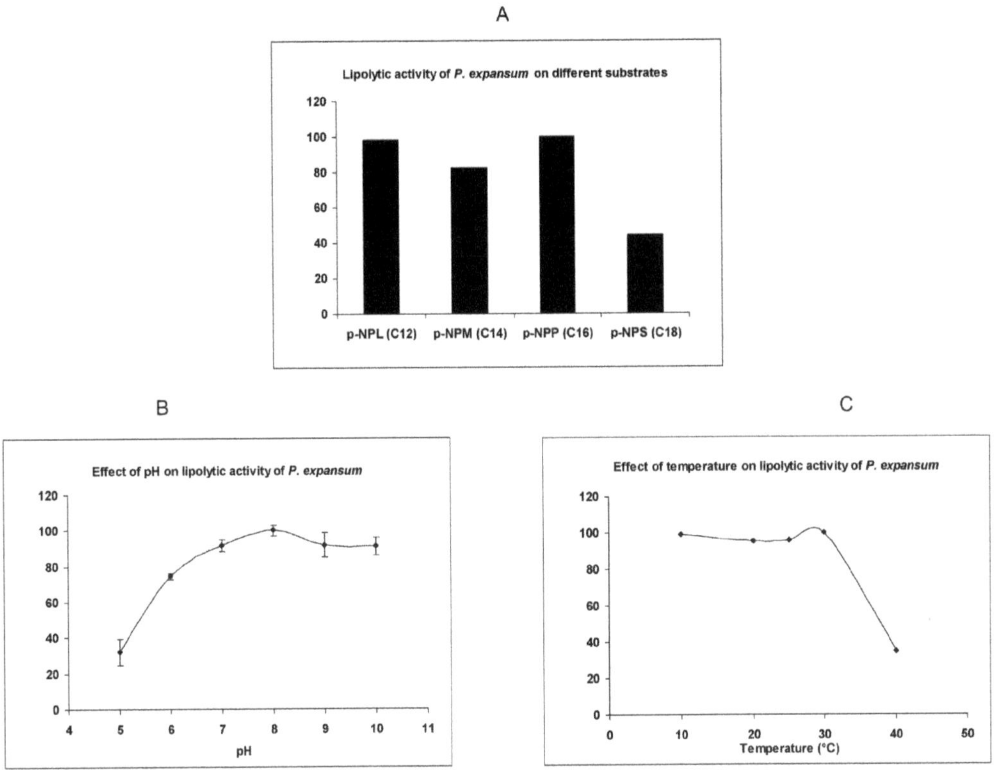

Figure 5.3 Lipolytic activity in the culture supernatant of *P. expansum*, measured using *p*-nitrophenyl substrates. (A) Activity at 25°C and pH 8 using different substrates. (B) Effect of pH using *p*-nitrophenyl laurate (C12) as substrate at 25°C. (C) Effect of temperature using *p*-nitrophenyl laurate (C_{12}) as substrate at pH 8.

the enzyme is a true lipase as it has the ability to hydrolyse C_{12} to C_{18} substrates. The optimum pH and temperature for the lipase activity were determined calorimetrically using *p*-nitrophenyl laurate (C_{12}) as substrate. The results obtained indicated that the lipase from *P. expansum* has optimum activity at pH 8 (Fig. 5.3B). Significantly, the temperature analysis suggests that the *P. expansum* lipase is active in the cold-active temperature range (Fig. 5.3C). The biochemical characteristics indicated above suggest that the *P. expansum* lipase has considerable potential in industrial applications. Its use in laundry detergents, in combination with cold tap water at temperatures of 10–22°C, could result in major energy and cost savings. Another potential application is in the 'clean-up' procedures entailed in the fish or meat processing industries, by hydrolysis of the oily wastes in combination with proteolytic enzymes.

Screening of fungal isolates for crude oil degradation and enzyme production

Another area explored in our work is the ability of Antarctic microfungi to degrade hydrocarbons, especially the more intractable components of crude oils such as resins and asphaltenes. Towards this end, an assortment of Antarctic fungal isolates, extracted from soil removed from various petroleum polluted locations in the vicinity of New Zealand's Scott Base and from refuelling sites in the Dry Valleys region, was screened for their ability to utilize crude oil as a carbon source. The fungi were cultivated at 15°C over a 3-week period on solid medium screening plates containing mineral salts (Penttilä *et al.*, 1987) with crude oil (UN1267, HAZCHEM CODE: 3[Y]E, provided by Caltex Refineries (NSW) Pty Ltd) added at concentrations ranging from 0.5–6% (v/v) as the sole carbon source. Crude oil-degrading ability was assessed as colony growth on each plate and the 11 best growing isolates, tentatively identified based on their morphological characteristics, were chosen for further study.

Growth and analysis of crude oil-degrading fungi

Spore suspensions of the 11 selected fungal isolates were inoculated into 250 ml conical flasks containing 50 ml liquid minimal medium including 6% (v/v) crude oil. Cultivation medium alone, with no inoculum, was used as a control. The cultures were incubated with shaking (170 rpm) at 25°C for 30 days. The addition of 1% (w/v) glucose or the surfactant Tween 80 to the cultures did not improve the growth or crude oil solubilization capacity of the fungi tested. The degree of degradation of the components of the oil that occurred during liquid cultivation (discussed in greater detail below) was initially indicated by the growth pattern exhibited by the fungi and the colour change from dark/brown/black to clear that occurred in the liquid medium (Table 5.4). Following the incubation period the contents of each shake flask were centrifuged (15 min, 10,000× *g*) and the culture supernatants collected for analysis. An internal standard (*o*-dichlorobenzene, 1 ml; 200 mg/l) was added to the supernatants before the crude oil degradation products were quantitatively extracted from the supernatants using dichloromethane. The extracts were analysed by gas chromatography – mass spectrometry (GC-MS) using a Shimadzu GC-17A gas chromatograph fitted with a J&W DB-5 capillary column (30 m × 0.25 mm, 0.25 μm film thickness) interfaced to a Shimadzu QP-5000 mass selective detector. A constant carrier (helium) velocity of 38 cm/s was used, the injector was held at 290°C, and the transfer line held at 310°C. A total of 0.5 μl was injected in splitless mode with a sampling time of 1.4 min. The mass spectrometer was run in full scan mode (40–500 *m/z* at 2000 amu/s). The oven program used was 40°C for 3 min, then ramped at 5°C/min to 310°C and held for 10 min.

Analysis of *n*-alkane oil degradation products

Fig. 5.4 shows the total ion chromatograms (TIC) for the crude oil degradation products and the control. The most notable change to the crude oil is the complete degradation of the *n*-alkanes by all cultures except N5. Straight chain alkanes (*n*-alkanes) are typically the first component of carbonaceous materials to be biodegraded,

Table 5.4 Growth of crude oil-degrading fungal isolates on liquid minimal medium (pH 5.5) supplemented with 6% (v/v) oil

Genus[1]	Strain code	Pattern of growth after 30 days	Capacity for crude oil solubilization[2]
Chrysosporium	A6	MG, no FF	+
Sepedonium	B9	MG, no FF	+
Aureobasidium	C7	MG, FF	++
Chrysosporium	D6	MG, FF	++
Geotrichum	F8	MG, no FF	+
Phoma	H9	MG, FF	++
Sepedonium	I8	MG, no FF	++
Chrysosporium	M3	MG, no FF	+
Geotrichum	M5	MG, FF	++
Aureobasidium	N5	MG, no FF	++
Chrysosporium	P6	MG, FF	++

[1]Identification based on conidiophore morphology. [2]Indicated by culture colour change from black to clear. MG, mycelial growth; FF, floc forming; +, low; ++, good.

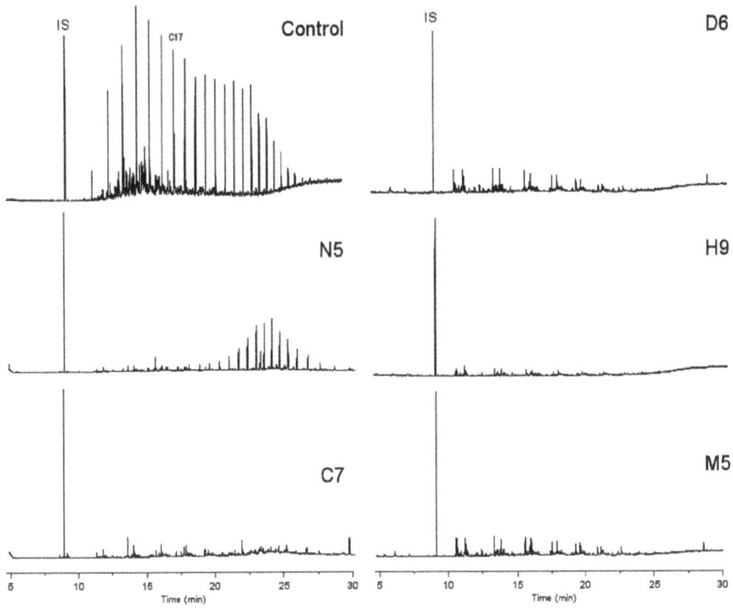

Figure 5.4 Total ion chromatograms (TIC) of the crude oil degradation products of fungal strain codes N5, C7, D6, H9, M5 and the control. Chain length noted on control TIC (IS, internal standard; *o*-dichlorobenzene).

although previous studies have found fungi tend to degrade shorter chain hydrocarbons ($< C_{16}$) in preference to longer chain materials (Bento *et al.*, 2005). Similar behaviour is noted with fungal strain N5, which preferentially degrades *n*-alkanes with chains length less than C_{24}. Each of the remaining strains completely (>99%) degraded all the straight-chain alkyl material, regardless of chain length.

Analysis of other oil degradation products

To examine the ability of the various fungal strains to degrade other classes of compounds, a comparative study was undertaken for each of the degradation extracts. Degradation of various classes of compounds (*n*-alkyl, aromatics, polyaromatics and bicyclic sesquiterpanes) was assessed by evaluating the ratio of representatives of each compound class relative to pristane (an acyclic isoprenoid). To represent *n*-alkyl degradation, the ratio of heptadecane–pristane (C_{17}:Pr) was used; for aromatic degradation, 2,6 + 2,7 dimethylnaphthalenes–pristane (DMN:Pr) and 1,3,5 + 2,3,6 trimethylnaphthalenes–pristane (TMN:Pr); for polyaromatic degradation, 1 + 2 + 3 + 9 methylphenanthrenes–pristane (MP:Pr) and lastly for bicyclic sesquiterpanes degradation the combined areas of two C_{15}-bicyclic-sesquiterpanes–pristane (ST:Pr) was used. Quantitation of each compound class was accomplished using selective ion monitoring (SIM) of characteristic ions for each class, namely m/z 57 for pristane ($I = 1707$) and heptadecane, m/z 156 for dimethylnaphthalenes ($I = 1434$), m/z 170 for trimethylnaphthalenes ($I = 1582$, 1586), m/z 192 for methylphenanthrenes ($I = 1957$, 1965, 1986, 1992) and m/z 193 for bicyclic sesquiterpanes ($I = 1459$, 1498). A shift in these ratios relative to the control is indicative of the degradation of that compound class. For instance, a decrease in a ratio indicates that the compound class represented by that ratio has been degraded preferentially over pristane. As pristane is also biodegradable, the ratios cannot be used directly to determine rate of biodegradation. The ratios can, however, provide a relative indication that the degradation rate of a compound class is either slower (as indicated by a negative ratio difference) or faster (as indicated by a positive ratio difference) than that of pristane.

As indicated in Table 5.5, fungal strains M5, D6 and H9 all produce a significant decrease in *n*-alkyl and aromatic compound class ratios and a minor decrease in the bicyclic sesquiterpane compound class ratios. The minor decrease in the last compound class is to be expected, and is consistent with the general resistance of bicyclic sesquiterpanes to biodegradation. The ability of these fungal strains to strongly degrade both aliphatic and aromatic components, however, is significant in several respects. Firstly, it highlights the resistance of these microorganisms to aromatic compounds. Aromatic compounds typically tend to inhibit the growth of hydrocarbon degrading bacteria (Li *et al.*, 2008). Thus, the potential for these fungal strains to produce enhanced petrochemical degradation via a synergistic effect with native bacteria is high. In contrast, strain N5 shows an appreciably different behaviour to the other isolates. As indicated by the increase in the MP:Pr and ST:Pr ratios, N5 appears to preferentially degrade isoprenoidal material over polyaromatic and bicyclic-sesquiterpanes. This behaviour is, however consistent with other studies of the biodegradation of petrochemicals by fungi which have shown a tendency for fungi to degrade pristane in preference to *n*-heptadecane (Zytner *et al.*, 2006). Fungal strain C7 was particularly notable for its capacity to degrade all compound classes, indicated by the fact that none of the targeted compound classes was detectable in the C7 degradation extract.

Degradation of refined bitumen by crude oil degrading fungi

Crude oil and various oil products are complex multi-component environmental pollutants containing hundreds of individual compounds. Short chain (low-boiling point) components of crude oils and crude oil products are not particularly persistent environmental pollutants as they tend to evaporate from oil spills. Up to 15% of a crude oil spill may be removed by simple evaporation, particularly in summertime. As a consequence, the heavier (higher-boiling point) components of crude oil and its products (asphaltenes and resins) tend to accumulate in the environment. This build-up is further exacerbated by the resistance of resins and asphaltenes to biodegradation (Atlas, 1981; Oudot *et al.*, 1993).

Growth of fungi on refined bitumen

In view of the recalcitrant nature of the asphaltic material, we investigated the synergistic effect of two fungal strain consortia in their ability to degrade refined bitumen. Spores (2.5×10^8) from each of the fungal strains M5, D6 and N5 were pooled with either those of H9 or C7. The consortia were cultured in 250 ml conical flasks containing 50 ml liquid minimal medium (pH 5.5 and 7) and an aliquot (approx. 5 g) of refined bitumen containing 428 mg/g asphaltenes. Controls for each combination contained no spore suspension. As before, the cultures were incubated at 25°C for 30 days and the success of these combinations in the solubilization of the refined

Table 5.5 The change in compound class ratios relative to a control sample. Compound class ratio codes are described in the associated text. (UD, undetected, indicating that the corresponding compound class was not detectable in the GC-MS run)

	Change in compound class ratio (%)				
Strain code	C_{17}:Pr	DMN:Pr	TMN:Pr	MP:Pr	ST:Pr
M5	75	72	42	48	9
D6	59	48	27	60	9
H9	85	44	23	33	2
N5	UD	UD	UD	–89	–15
C7	UD	UD	UD	UD	UD

bitumen was noted. For each of the combinations it was observed that the refined bitumen was partially liquefied at pH 5.5 but completely liquefied at pH 7. The mixed cultures were examined under the light microscope to investigate their growth pattern and this revealed a significant amount of fungal hyphal growth which appeared to have surrounded the bitumen lumps, but also may have resulted from aggregation of the fungal hyphae.

Analysis of refined bitumen degradation

Culture supernatants were centrifuged and the refined bitumen degradation products were quantitatively extracted using dichloromethane. Asphaltene quantification was performed gravimetrically by *n*-pentane precipitation. The dichloromethane from the refined bitumen degradation product extracts was first removed under a stream of dry nitrogen at 30°C. Once the dichloromethane had been removed, the dried extracts were suspended in *n*-pentane by ultrasonication for 30 minutes followed by filtration on preweighed Whatman GF/B glass fibre paper. The degree of asphaltene degradation (Table 5.6) was determined by comparing the asphaltene content of the dichloromethane extract of the consortia with that of the control. Significant degradation of asphaltenes was noted with both consortia. The ready degradation of asphaltic material, which is polyaromatic in nature, is consistent with the previous results obtained from the degradation of crude oil where it was noted that these strains of fungi were resistant to, and could readily degrade, the aromatic material. Further, the relatively larger asphaltene degradation potential of the C7 containing consortium is consistent with the ability of fungal strain C7 to readily breakdown all of the compound classes studied (see earlier).

Production of hydrolases by oil-degrading isolates

As already indicated, the focus of our study has not been limited to identifying Antarctic microfungi capable of degrading petroleum-based pollutants and producing cold-active lipases. We have also sought to examine these microfungi in greater detail with a view to identifying other cold-active hydrolytic enzymes that possess a potential for industrial applications. To this end, the 11 oil-degrading isolates were screened with indicative substrates for a variety of hydrolytic enzyme activities. These were protease (2% w/v glucose, 1% w/v skimmed milk; Mäntylä *et al.*, 1994), β-mannanase (0.5% w/v locust bean gum plus flooding with Congo Red; Teather and Wood, 1982; Rättö and Poutanen, 1988), xylanase (0.5% w/v birchwood xylan plus flooding with Congo Red (Teather and Wood, 1982; Bradner, 2003); lipase –2.5% v/v olive oil, 1% w/v rhodamine B; Kouker and Jaeger, 1987), chitinase (0.3% w/v colloidal chitin; O'Brien and Colwell, 1987), pectinase (0.1% w/v pectin; Hagerman *et al.*, 1985), cellulose (1% w/v Walseth cellulose; Walseth, 1952), β-glucanase (0.1% w/v AZO-CM-Cellulose; Megazyme International Ireland Ltd) and laccase (2% glucose plus flooding with ABTS [2,2′-azinobis-3-ethylbenzthiazoline-6-sulfonate]; Crowe and Olsson, 2001). After incubation, the plates were assessed for respective enzyme activities based on fluorescence (lipase), formation of colour (laccase), halo (protease, chitinase, cellulase) or clearing around the growing colony against a dyed background (β-mannanase, xylanase). The mesophilic microfungus *Trichoderma reesei* Rut-C30 was included for comparison (Montenecourt and Eveleigh, 1979).

All Antarctic oil-degrading isolates produced chitinase, which can be explained by the ample

Table 5.6 Degradation of the asphaltic component of refined bitumen by consortia of Antarctic fungal strains in liquid medium at pH 7

Sample	Asphaltic component degradation (%)
M5 + D6 + N5 + C7	69
M5 + D6 + N5 + H9	50

occurrence of chitin in the Antarctic environment originating from crustaceans such as krill. The majority produced laccase (an enzyme significant in lignin degradation; Eggert *et al.*, 1997) and lipase, but displayed only limited protease activity. None of the crude oil degrading Antarctic isolates produced cellulase, xylanase or β-glucanase activity under the conditions tested, but two isolates showed some ability to degrade β-mannanase. The mesophilic strain *T. reesei* failed to develop colonies on the crude oil plates but, interestingly, was capable of producing and secreting several hydrolases when cultured at 15°C.

Conclusions and future work

The Antarctic environment consists of three main biogeographical zones, the maritime Antarctic, sub-Antarctic and continental Antarctic (Convey, 2007). Common to these areas are low temperatures, wide thermal fluctuations, low water and nutrient availability and high UV irradiance (Onofri *et al.*, 2004). While investigations into Antarctic microorganisms have mainly concentrated on psychrophilic bacteria and archaea, the number of studies addressing Antarctic fungi is on a steady increase (e.g. Robinson, 2001; Connell *et al.*, 2006; Bridge *et al.*, 2008). In addition to studies exploring the physiological and molecular basis of cold-adaptation, identification of cold-active enzymes produced by Antarctic microorganisms including fungi are of interest. Antarctic isolates have been shown to produce a broad range of enzyme activities, summarized in Table 5.3. This exploration is likely to continue.

Current lipase products on the market are of relatively low purity and quality. Cold-active lipases represent a growing market in various industries and cover a broad spectrum of biotechnological applications, for example, additives in food industries and detergents, environmental bioremediations, biotransformation and molecular biology transformations. We have identified a secreted fungal lipase of which the production can be improved further by traditional mutagenesis and screening or by recombinant means. Apart from lipases, we have identified from the Antarctic isolates several other cold-active hydrolytic enzymes that possess potential for industrial application. Our findings indicate that Antarctic microfungi are a rich, yet mostly untapped source of novel cold-active enzymes.

The potential for Antarctic microfungi to be used in environmental bioremediation applications has been demonstrated in this chapter by their strong capacity for degrading hydrocarbonaceous substrates, including asphaltic material. Unlike most fungi, the *Aureobasidium* isolate C7 was capable of degrading all compound classes contained in the crude oil. Additionally, for the refined bitumen, the consortium containing strain C7 was found to be effective in degrading all of the compound classes tested. These findings warrant further studies into the application of Antarctic fungi for *in situ* bioremediation of sites in a cold climate.

Web resources

These are given in the text, footnotes and in the reference section.

References

Abele, D., and Pörtner, H.O. (1998). H_2O_2 accumulation and elevated temperatures cause oxidative stress in Antarctic coastal environments. Paper presented at: International Workshop on Animal Physiology, 4–6 June, 1998 (Borensee, Germany, Humboldt-Universität zu Berlin).

Aislabie, J., Fraser, R., Duncan, S., and Farrell, R.L. (2001). Effects of oil spills on microbial heterotrophs in Antarctic soils. Polar Biol. *24*, 308–313.

Alberdi, M., Bravo, L.A., Gutiérrez, A., Gidekel, M., and Corcuera, L.J. (2002). Ecophysiology of Antarctic vascular plants. Physiol. Plant. *115*, 479–486.

Atlas, R.M. (1981). Microbial degradation of petroleum hydrocarbons: An environmental perspective. Microbiol. Rev. *45*, 180–209.

Aurilia, V., Rioux-Dubé, J.-F., Marabotti, A., Pézolet, M., and Auria, S. (2009). Structure and dynamics of cold-adapted enzymes as investigated by FT-IR spectroscopy and MD. The case of an esterase from *Pseudoalteromonas haloplanktis*. J. Phys. Chem. B *113*, 7753–7761.

Azmi, O.R., and Seppelt, R.D. (1997). Fungi of the Windward Islands, continental Antarctica. Effect of temperature, pH and culture medium on the growth of selected microfungi. Polar Biol. *18*, 128–134.

Azmi, O.R., and Seppelt, R.D. (1998). The broad-scale distribution of microfungi in the Windmill Islands region, continental Antarctica. Polar Biol. *19*, 92–100.

Baryshnikova, L.M., Grishchenkov, V.G., Arinbasarov, M.U., Shkidchenko, A.N., and Boronin, L.M. (2001). Biodegradation of oil products by individual degrading strains and their associations in liquid media. Appl. Biochem. Microbiol. 37, 542–548.

Bento, F.M., Beech, I.B., Gaylarde, C.G., Englert, G.E., and Muller, I.L. (2005). Degradation and corrosive activityies of fungi in a diesel-mild steel-aqueous system. World J. Microbiol. Biotech. *21*, 135–142.

Bjelic, S., Brandsdal, B.O., and Åqvist, J. (2008). Cold adaptation of enzyme reaction rates. Biochemistry *47*, 10049–10057.

Bradner, J.R. (2003). Antarctic Microfungi as a Potential Bioresource. (Macquarie University, Sydney, Australia), p. 185.

Bradner, J.R., Bell, P.J.L., Te'o, V.S.J., and Nevalainen, K.M.H. (2003). The application of PCR for the isolation of a lipase gene from the genomic DNA of an Antarctic microfungus. Curr. Genet. *44* 224–230, 260.

Brenchley, J.E. (1996). Psychrophilic microorganisms and their cold-active enzymes. J. Ind. Microbiol. *17*, 432–437.

Bridge, P., Spooner, B., and Roberts, P. (2008). List of non-lichenized fungi from the Antarctic region (Version 2.3.1; April 2008) http://www.antarctica.ac.uk//bas_research/data/access/fungi/Speciespublic2.html# (Biological Sciences Division at British Antarctic Survey and the Mycology Section, Royal Botanic Gardens, Kew.).

Brush, T.S., Chapman, R., Kurzman, R., and Williams, D. (1999). Purification and characterization of extracellular lipases from *Ophiostoma piliferum*. Bioorg. Med. Chem. *7*, 2131–2138.

Brzozowski, A.M., Savage, H., Verma, C.S., Turkenburg, J.P., Lawson, D.M., Svendsen, A., and Patkar, S. (2000). Structural origins of the interfacial activation in *Thermomyces (Humicola) lanuginosa* lipase. Biochemistry *39*, 15071–15082.

Burgess, L.W., Summerell, A., Backhouse, D., Benyon, F., and Levic, J. (1996). Biodiversity and population studies in *Fusarium*. Sydowia *48*, 1–11.

Camardella, L., Di Fraia, R., Antignani, A., Ciardiello, M.A., di Prisco, G., Coleman, J.K., Buchon, L., Guespin, J., and Russell, N.J. (2002). The Antarctic *Psychrobacter* sp. TAD1 has two cold-active glutamate dehydrogenases with different cofactor specificities. Characterisation of the NAD+-dependent enzyme. Compar. Biochem. Physiol. Part A, Mol. Int. Physiol. *131*, 559–567.

Cannon, P.F., and Sutton, B.C. (2004). Microfungi on wood and plant debris. In Biodiversity of Fungi: Inventory and Monitoring Methods, Mueller, G.M., Bills, G.F., and Foster, M.S., eds. (Elsevier Academic Press, Burlington, MA, USA).

Castro-Ochoa, L.D., Rodriguez-Gomez, C., Valerio-Alfaro, G., and Ros, R.O. (2005). Screening, purification and characterization of the thermoalkalophilic lipase produced by *Bacillus thermoleovorans* CCR11. Enzyme Microbial Tech. *37* 648–654.

Cavicchioli, R., Siddiqui, K.S., Andrews, D., and Sowers, K.R. (2002). Low-temperature extremophiles and their applications. Curr. Opinion Biotech. *13*, 253–261.

Cerniglia, C.E. (1984). Microbial transformation of aromatic hydrocarbons. In Petroleum Microbiology, Atlas, R.M., ed. (Macmillan Publishing Co. , New York), pp. 99–128.

Chapman, W.L., and Walsh, J.E. (2007). A synthesis of Antarctic temperatures. J. Climate *20*, 4096–4117.

Coker, J.A., Sheridan, P.P., Loveland-Curtze, J., Gutshall, K.R., Auman, A.J., and Brenchley, J.E. (2003). Biochemical characterization of a beta-galactosidase with a low temperature optimum obtained from an Antarctic arthrobacter isolate. J. Bacteriol. *185*, 5473–5482.

Connell, L., Redman, R., Craig, S., and Rodriguez, R. (2006). Distribution and abundance of fungi in the soils of Taylor Valley, Antarctica. Soil Biol. Biochem. *38*, 3083–3094.

Corte, A., and Daglio, C.A.N. (1994). A mycological study of Antarctic air. In Biologie Antarctique, Carrick, R., Holdgate, M.W., and Prevost, J., eds. (Hermann, Paris, France), pp. 115–120.

Coulon, F., Pelletier, E., Gourhant, L., and Delille, D. (2005). Effects of nutrient and temperature on degradation of petroleum hydrocarbons in contaminated sub-Antarctic soil. Chemosphere *58*, 1439–1448.

Crowe, J.D., and Olsson, S. (2001). Induction of laccase activity in *Rhizoctonia solani* by antagonistic *Pseudomonas fluorescens* strains and a range of chemical treatments. Appl. Environ. Microbiol. *67*, 2088–2094.

Demirjian, D.C., Morís-Varas, F., and Cassidy, C.S. (2001). Enzymes from extremophiles. Curr. Opin. Chem. Biol. *5*, 144–151.

Di Fraia, R., Wilquet, V., Ciardiello, M.A., Carratore, V., Antignani, A., Camardella, L., Glansdorff, N., and Di Prisco, G. (2000). NADP+-dependent glutamate dehydrogenase in the Antarctic psychrotolerant bacterium *Psychrobacter* sp. TAD1. Characterization, protein and DNA sequence, and relationship to other glutamate dehydrogenases. Eur. J. Biochem. *267*, 121–131.

Diaz, S.B., Deferrari, G., Nelson, D., and Camilion, M.C. (2004). Climatologies of ozone and UV-B irradiances over Antarctica in the last decades. Gayana (Concepción) *68*, 157–160 Available online at: http://www.scielo.cl/scielo.php?pid=S0717-65382004000200029&script=sci_arttext.

Eggert, C., Temp, U., and Eriksson, K.-E.L. (1997). Laccase is essential for lignin degradation by the white-rot fungus *Pycnoporus cinnabarinus*. FEBS Lett. *407*, 89–92.

Eggert, T., Pancreac'h, G., Douchet, I., Verger, R., and Jaeger, K.-E. (2000). A novelle extracellular esterase from *Bacillus subtilis* and its conversion to a monoacylglycerol lipase. Eur. J. Biochem. *267*, 6459–6469.

Farrell, R.L., Rhodes, P.L., and Aislabie, J. (2003). Toluene-degrading Antarctic Pseudomonas strains from fuel-contaminated soil. Biochem. Biophys. Res. Comm. *312*, 235–240.

Feller, G., and Gerday, C. (1997). Psychrophilic enzymes – molecular basis of cold adaptation. Cell. Mol. Life Sci. *53*, 830–841.

Feller, G., and Gerday, C. (2003). Psychrophilic enzymes: Hot topics in cold adaptation. Nat. Rev. Microbiol. *1*, 200–208.

Feller, G., Narinx, E., Arpigny, J.L., Aittaleb, M., Baise, E., Genicot, S., and Gerday, C. (1996). Enzymes from psychrophilic organisms. FEMS Microbiol. Rev. *18*, 189–202.

Feller, G., Arpigny, J.L., Narinx, E., and Gerday, C. (1997). Molecular adaptations of enzymes from psychrophilic organisms. Comp. Biochem. Physiol. *118A*, 495–499.

Feller, G., d'Amico, D., and Gerday, C. (1999). Thermodynamic stability of a cold-active alpha-amylase from the Antarctic bacterium *Alteromonas haloplanctis*. Biochemistry *38*, 4613–4619.

Fojan, P., Jonson, P.H., Petersen, M.T.N., and Petersen, S.B. (2000). What distinguishes an esterase from a lipase: a novel structural approach. Biochimie *82*, 1033–1041.

Frisvad, J.C., Filtenborg, O., and Wicklow, D.T. (1987). Terverticillate penicillia isolated from underground seed caches and cheek pouches of banner-tailed kangaroo rats (*Dipodomys spectabilis*). Can. J. Botany *65*, 765–773.

Gams, W. (1992). The analysis of communities of saprophytic microfungi with special reference to soil fungi. In Fungi in Vegetation Science, Winterhoff, W., ed. (Kluwer, Dordrecht, The Netherlands), pp. 182–223.

Gams, W. (2007). Biodiversity of soil-inhabiting fungi. Biodiv. Conserv. *16*, 69–72.

Gerday, C., Aittaleb, M., Bentahir, M., Chessa, J.-P., Claverie, P., Collins, T., D'Amico, S., Dumont, J., Garsoux, G., Georlette, D., *et al.* (2000). Cold-adapted enzymes: from fundementals to biotechnology. Trends Biotech. *18*, 103–107.

Gerike, U., Danson, M.J., Russell, N.J., and Hough, D.W. (1997). Sequencing and expression of the gene encoding a cold-active citrate synthase from an Antarctic bacterium, strain DS2–3R. Eur. J. Biochem. *248*, 49–57.

Gianese, G., Argos, P., and Pascarella, S. (2001). Structural adaptation of enzymes to low temperatures. Prot. Engin. *14*, 141–148.

Greaves, R.B., and Warwicker, J. (2007). Mechanisms for stabilisation and the maintenance of solubility in proteins from thermophiles. Available at: http://www.biomedcentral.com/1472–6807/7/18.

Greenfield, L., and Wilson, G. (1981). University of Canterbury, Antarctic Expedition no 19, report (University of Canterbury, Christchurch, New Zealand).

Hagerman, A.E., Blau, D.M., and McClure, A.L. (1985). Plate assay for determining the time of production of protease, cellulase and pectinase by germinating fungal spores. Anal. Biochem. *151*, 334–342.

Helland, R., Larsen, R.L., and Asgeirsson, B. (2009). The 1.4 A crystal structure of the large and cold-active *Vibrio* sp. alkaline phosphatase. Biochim. Biophys. Acta *1794*, 297–308.

Hibbett, D.S., Binder, M., Bischoff, J.F., Blackwell, M., Cannon, P.F., Eriksson, O.E., Huhndorf, S., James, T., Kirk, P.M., Locking, R., *et al.* (2007). A higher-level phylogenetic classification of the Fungi. Mycol. Res. *111*, 509–547.

Hildebrandt, P., Wanarska, M., and Kur, J. (2009). A new cold-adapted beta-D-galactosidase from the Antarctic Arthrobacter sp. 32c – gene cloning, overexpression, purification and properties. Available at: http://www.biomedcentral.com/1471–2180/9/151.

Jaeger, K.-E., and Reetz, M. (1998). Microbial lipases form versatile tools for biotechnology. Trends Biotech. *16*, 396–403.

Jaeger, K.-E., Ransac, S., Dijkstra, B.W., Colson, C., van Heuvel, M., and Misset, O. (1994). Bacterial lipases. FEMS Microbiol. Rev. *15*, 29–63.

Johnson, G.B., and Raven, P.H. (2002). Biology, 6th edn (McGraw-Hill Higher Education, USA).

Joseph-Horne, T., Hollomon, D.W., and Wood, P.M. (2001). Fungal respiration: a fusion of standard and alternative components. Biochim. Biophys. Acta *1504*, 179–195.

Kazuoka, T., Oikawa, T., Muraoka, I., Kuroda, S., and Soda, K. (2007). A cold-active and thermostable alcohol dehydrogenase of a psychrotorelant from Antarctic seawater, *Flavobacterium frigidimaris* KUC-1. Extremophiles *11*, 257–267.

Kerry, E. (1990). Effects of temperature on growth rates of fungi from subantarctic Macquarie Island and Casey, Antarctica. Polar Biol. *10*, 423–430.

Kouker, G., and Jaeger, K.E. (1987). Specific and sensitive plate assay for bacterial lipases. Appl. Environ. Microbiol. *53*, 211–213.

Kreiner, M., Harvey, L.M., and McNeil, B. (2002). Oxidative stress response of a recombinant *Aspergillus niger* to exogenous menadione and H_2O_2 addition. Enzy. Microbial Tech. *30*, 346–353.

Laemmli, U.K. (1970). Cleavage of structural proteins during the assembly of the head of the bacteriophage T4. Nature *227*, 680–685.

Leahy, J.G., and Colwell, R.R. (1990). Microbial degradation of hydrocarbons in the environment. Microbiol. Rev. *54*, 305–315.

Li, W.F., Zhou, X.X., and Lu, P. (2005). Structural features of thermozymes. Biotech. Adv. *23*, 271–281.

Li, Y.Q., Liu, H.F., Tian, Z.L., Zhu, L.H., Wu, Y.H., and Tang, H.Q. (2008). Diesel pollution biodegradation: Synergetic effect of mycobacterium and filamentous fungi. Biomedical Environ. Sci. *21*, 181–187.

Liese, A., Seelbach, K., and Wandrey, C. (2000). Industrial Biotransformations: A Collection of Processes (Wiley-VCH, Weinheim, Germany).

Lonhienne, T., Gerday, C., and Feller, G. (2000). Psychrophilic enzymes: revisiting the thermodynamic parameters of activation may explain local flexibility. Biochim. Biophys. Acta *1543*, 1–10.

Lopez-Martinez, G., Elnitsky, M.A., Benoit, J.B., Lee Jr, R.E., and Denlinger, D.L. (2008). High resistance to oxidative damage in the Antarctic midge *Belgica antarctica*, and developmentally linked expression of genes encoding superoxide dismutase, catalase and heat shock proteins. Insect Biochem. Mol. Biol. *38*, 796–804.

Luke, K.A., Higgins, C.L., and Wittung-Stafshedel, P. (2007). Thermodynamic stability and folding of proteins from hyperthermophilic organisms. FEBS J. *274*, 4023–4033.

Makowski, K., Białkowska, A., Szczesna-Antczak, M., Kalinowska, H., Kur, J., Cieśliński, H., and Turkiewicz, M. (2007). Immobilized preparation of cold-adapted and halotolerant Antarctic beta-galactosidase as a highly stable catalyst in lactose hydrolysis. FEMS Microbiol. Ecol. *59*, 535–542.

Mäntylä, A., Saarelainen, R., Fagerström, R., Suominen, P., and Nevalainen, H. (1994). Cloning of the aspartic protease gene of *Trichoderma reesei*. Paper presented at: Second European Conference on Fungal Genetics (Lunteren, The Netherlands).

Marchant, D.R., and Head, J.W. (2004). Microclimate Zones in the Dry Valleys of Antarctica: Implications for Landscape Evolution and Climate Change on Mars. In 35th Lunar and Planetary Science Conference, 15–19 March 2004 (League City, TX, USA).

Marshall, C.J. (1997a). Cold-adapted enzymes. Trends Biotech. *15*, 359–364.

Marshall, W.A. (1997b). Seasonality in Antarctic airborne fungal spores. Appl. Environ. Microbiol. *63*, 2240–2245.

Marshall, W.A., and Chalmers, M.O. (1997). Airborne dispersal of Antarctic terrestrial algae and cyanobacteria. Ecography *20*, 585–594.

Merlino, A., Russo Krauss, I., Castellano, I., De Vendittis, E., Vergara, A., and Sica, F. (2008). Crystallization and preliminary X-ray diffraction studies of a psychrophilic iron superoxide dismutase from *Pseudoalteromonas haloplanktis*. Prot. Pept. Lett. *15*, 415–418.

Möller, C., and Dreyfuss, M.M. (1996). Microfungi from Antarctic lichens, mosses and vascular plants. Mycologia *88*, 922–933.

Montenecourt, B.S., and Eveleigh, D.E. (1979). Selective screening methods for the isolation of high yielding cellulase mutants of *Trichoderma reesei*. Adv. Chem. Series *181*, 289–301.

Moradas-Ferreira, P., Costa, V., Piper, P., and Mager, W. (1996). The molecular defences against reactive oxygen species in yeast. Mol. Microbiol. *19*, 651–658.

Mouchacca, J. (1995). Thermophilic fungi in desert soils: a neglected extreme environment. In Microbial Diversity and Ecosystem Function, Allsopp, D., Colwell, R.R., and Hawksworth, D.L., eds. (CAB International, Wallingford, CT, USA), pp. 265–288

Mouchacca, J. (2005). Mycobiota of the arid Middle East: check-list of novel fungal taxa introduced from 1940 to 2000 and major recent biodiversity titles. J. Arid Environ. *60*, 359–387.

Naga, K., Sakai, T., Rantiatmodjo, R.M., Suzuki, K., Gams, W., and Okada, G. (1995). Studies on the distribution of alkalophilic and alikali-tolerant soil fungi I. Mycoscience *36*, 247–256.

Nichols, D., Bowman, J., Sanderson, K., Nichols, C.M., Lewis, T., McMeekin, T., and Nichols, P.D. (1999). Developments with Antarctic microorganisms: culture collections, bioactivity screening, taxonomy, PUFA production and cold adapted enzymes. Curr. Opin. Biotech. *10*, 240–246.

Nichols, D.S., Sanderson, K., Buia, A., van de Kamp, J., Holloway, P., Bowman, J.P., Smith, M., Nichols, C.M., Nichols, P.D., and McMeekin, T.A. (2002). Bioprospecting and biotechnology in Antarctica. In The Antarctic, Past, Present and Future., Jabour-Green, J., and Haward, M., eds. (Antarctic CRC Research Report #28, Hobart, Tas. Australia), pp. 85–103.

O'Brien, M., and Colwell, R.R. (1987). A rapid test for chitinase activity that uses 4-methylumbelliferyl-*N*-acetyl-ß-d-glucosaminide. Appl. Environ. Microbiol. *53*, 1718–1720.

Ohgiya, S., Hoshino, T., Okuyama, H., Tanaka, S., and Ishizaki, K. (1999). Biotechnology of enzymes from cold-adapted microorganisms. In Biotechnological Applications of Cold-Adapted Organisms, Margesin, R., and Schinner, F., eds. (Springer-Verlag, Berlin, Germany), pp. 17–34.

Onofri, S., Pagano, S., Zucconi, L., Cicalini, A.R., Selbmann, L., and Fenice, M. (1999). Microfungi in Antarctic extreme environments. Extracellular enzyme activities. Paper presented at: IBC's World Congress on Enzyme Technologies, 10–12 March. (San Francisco, CA, USA).

Oudot, J., Dupont, L., Haloui, S., and Roquebert, M.F. (1993). Biodegradation potential of hydrocarbon-assimilating tropical fungi. Soil Biol. Biochem. *25*, 1167–1173.

Pandey, A., Benjamin, S., Soccol, C.R., Nigam, P., Krieger, N., and Soccol, V.T. (1999). The realm of microbial lipases in biotechnology. Biotech. Appl. Biochem. *29*, 119–131.

de Pascale, D., Cusano, A.M., Autore, F., Parrilli, E., di Prisco, G., Marino, G., and Tutino, M.L. (2008). The cold-active Lip1 lipase from the Antarctic bacterium *Pseudoalteromonas haloplanktis* TAC125 is a member of a new bacterial lipolytic enzyme family. Extremophiles *12*, 311–323.

Pedrini, N., Juárez, M.P., Crespo, R., and de Alaniz, M.J.T. (2006). Clues on the role of *Beauveria bassiana* catalases in alkane degradation events. Mycologia *98*, 528–534.

Penttilä, M., Nevalainen, H., Rättö, M., Salminen, E., and Knowles, J. (1987). A versatile transformation system for the cellulolytic filamentous fungus Trichoderma reesei. Gene *61*, 155–164.

Perry, J.J. (1984). Microbial metabolism of cyclic alkanes. In Petroleum Microbiology, Atlas, R.M., ed. (Macmillan Publishing Co., New York), pp. 61–98.

Rättö, M., and Poutanen, K. (1988). Production of mannanase-degrading enzymes. Biotech. Lett. *10*, 661–664.

Reetz, M.T. (2002). Lipases as practical biocatalysts. Curr. Opin. Chem. Biol. *6*, 145–150.

Robb, F., Antranikian, G., Grogan, D., and Driessen, A., eds. (2008). Thermophiles: Biology and Technology at High Temperatures (CRC Press, Boca Raton, FL, USA).

Robinson, C.H. (2001). Cold adaptation of Arctic and Antarctic fungi. New Phytol. *151*, 341–353.

Ruisi, S., Barreca, D., Selbmann, L., Zucconi, L., and Onofri, S. (2007). Fungi in Antarctica. Rev. Environ. Sci. Biotech. *6*, 127–141.

Ruiz, C., Falcocchio, S., Xoxi, E., Javier Pastor, F.I., Diaz, P., and Saso, L. (2004). Activation and inhibition of Candida rugosa and Bacillus-related lipases by saturated fatty acids, evaluated by a new colorimetric microassay. Biochim. Biophy. Acta (BBA) – General Subjects *1672*, 184–191.

Schmidt-Dannert, C. (1999). Recombinant microbial lipases for biotechnological applications. Bioorg. Medic. Chem. *7*, 2123–2130.

Schmit, J.P., and Mueller, G.M. (2007). An estimate of the lower limit of global fungal diversity. Biodiv. Conserv. *16*, 99–111.

Sharma, R., Chisti, Y., and Banerjee, U.C. (2001). Production, purification, characterization, and applications of lipases. Biotech. Adv. *19*, 627–662.

Sieburth, J.M. (1965). Microbiology of Antarctica. In Biogeography and Ecology in Antarctica, van Mieghem, J., and van Oye, P., eds. (Junk, The Hague, The Netherlands), pp. 267–295.

Singer, M.E., and Finnerty, W.R. (1984). Microbial metabolism of straight-chain and branched alkanes. In Petroleum Microbiology, Atlas, R.M., ed. (Macmillan Publishing Co., New York), pp. 1–60.

Skotnicki, M.L., Selkirk, P.M., and Dale, T.M. (1996). RAPD profiling of genetic variation in Antarctic mosses. Paper presented at: Proceedings of an International Workshop on Polar Desert Ecosystems (A. A. Balkema, Rotterdam, Christchurch, New Zealand).

Song, J.M., Nam, K.W., Kang, S.G., Kim, C.G., Kwon, S.T., and Lee, Y.H. (2008). Molecular cloning and characterization of a novel cold-active beta-1,4-D-mannanase from the Antarctic springtail, *Cryptopygus antarcticus*. Compar. Biochem. Physiol. Part B: Biochem. Mol. Biol. *151*, 32–40.

Summerell, B.A., Rugg, C.A., and Burgess, L.W. (1995). Characterization of *Fusarium babinda* sp. nov. Mycol. Res. *99*, 1345–1348.

Sunna, A., Hunter, L., Hutton, C.A., and Bergquist, P.L. (2002). Biochemical characterization of a recombinant thermoalkalophilic lipase and assessment of its substrate enantioselectivity. Enzy. Microbial Tech. *31*, 472–476.

Teather, R.M., and Wood, P.J. (1982). Use of Congo red – polysaccharide interactions in enumeration and characterization of cellulolytic bacteria from bovine rumen. Appl. Environ. Microbiol. *43*, 777–780.

Tosi, S., Casado, B., Gerdol, R., and Caretta, G. (2002). Fungi isolated from Antarctic mosses. Polar Biol. *25*, 262–268.

Vincent, W.F. (1989). Microbial Ecosystems of Antarctica (Cambridge University Press, Cambridge, UK).

Violot, S., Haser, R., Sonan, G., Georlette, D., Feller, G., and Aghajari, N. (2003). Expression, purification, crystallization and preliminary X-ray crystallographic studies of a psychrophilic cellulase from *Pseudoalteromonas haloplanktis*. Acta Crystall. Section D *59*, 1256–1258.

Vishniac, H.S. (1993). The microbiology of Antarctic soils. In Antarctic Microbiology, Friedmann, E.I., ed. (Wiley-Liss, New York), pp. 297–341.

Walker, J.D., and Cooney, J.J. (1973). Pathway of *n*-alkane oxidation in *Cladosporium resinae*. J. Bacteriol. *15*, 635–639.

Walseth, C.S. (1952). Occurrence of cellulases in enzyme preparations from micro-organisms. Tappi J. *35*, 228–232.

Walton, D.W.H. (1984). The terrestrial environment. In Antarctic Ecology, Laws, R.M., ed. (Academic Press, London), pp. 1–60.

Wang, E., Koutsioulis, D., Leiros, H.K., Andersen, O.A., Bouriotis, V., Hough, E., and Heikinheimo, P. (2007). Crystal structure of alkaline phosphatase from the Antarctic bacterium TAB5. J. Mol. Biol. *366*, 1318–1331.

Yamamura, A., Sakaguchi, T., Murakami, Y., Yokoyama, K., and Tamiya, E. (1999). Purification and characterization of cold-active L-glutamate dehydrogenase of NAD(P) and oxygen. J. Biochem. *125*, 760–769.

Yoshimune, K., Galkin, A., Kulakova, L., Yoshimura, T., and Esaki, N. (2005). Cold-active DnaK of an Antarctic psychrotroph *Shewanella* sp. Ac10 supporting the growth of dnaK-null mutant of at cold temperatures. Extremophiles 9, 145–150.

Zheng, Z., Jiang, Y.H., Miao, J.L., Wang, Q.F., Zhang, B.T., and GY., L. (2005). Purification and characterization of a cold-active iron superoxide dismutase from a Psychrophilic Bacterium, Marinomonas sp. NJ522. Biotech. Lett. *28*, 85–88.

Zucconi, L., Pagano, S., Fenice, M., Selbmann, L., Tosi, S., and Onofri, S. (1996). Growth temperature preferences of fungal strains from Victoria Land, Antarctica. Polar Biol. *16*, 53–61.

Zytner, R.G., Salb, A.C., and Stiver, W.H. (2006). Bioremediation of diesel fuel contaminated soil: Comparison of individual compounds to complex mixtures. Soil Sed. Contam. *15*, 277–297.

Metabolic Diversity of Thermophilic Prokaryotes – What's New?

6

Elizaveta Bonch-Osmolovskaya

Abstract

Thermophilic microorganisms, though known since the beginning of the twentieth century, were intensively studied in its last three decades. Natural terrestrial and submarine thermal environments were found to be populated by moderate, extreme and hyperthermophilic microorganisms representing diverse metabolic groups. However, during the past few years this knowledge has been extended, and new metabolic groups of thermophilic prokaryotes described. Among these are ammonia-oxidizing archaea, thermoacidophilic methanotrophs of the phylum *Verrucomicrobia*, microorganisms gaining energy for growth from the disproportionation of sulfur species, and archaea and bacteria metabolizing one carbon (C1) compounds. Other novel metabolic groups, such as thermophilic anammox bacteria, nitrite-oxidizing thermophiles, and microorganisms performing anaerobic methane oxidation in thermal ecosystems, have been detected using molecular or geochemical approaches. These data will, certainly, stimulate further cultivation and isolation efforts.

Introduction

Thermophiles, microorganisms that grow optimally at elevated temperatures, have been known since the beginning of the twentieth century. Up to the mid-1960s, it was believed that they were represented by aerobic and anaerobic organotrophs with optimum and maximum growth temperatures of around 55°C and 70°C, respectively, with thermophilic sulfate reducers constituting the only probable exception. Taxonomically, thermophilic microorganisms represented separate species in the well-known genera *Bacillus, Clostridium,* and *Desulfotomaculum.* However, the following three decades have brought a significant breakthrough in our understanding of the diversity of microbial thermophiles. Not only have the upper temperature limits of optimum and maximum growth temperatures shifted greatly, but the number of metabolic groups that can thrive at high temperatures has expanded significantly. The pioneering works of Thomas D. Brock (Brock, 1978) revealed that the hot springs of Yellowstone National Park were inhabited by microorganisms able to grow at very high temperatures, and possessing peculiar morphologies and special nutrient preferences. Brock and coworkers characterized several novel metabolic groups of thermophilic microorganisms, some of which were later found to represent deep phylogenetic lineages on the tree of life. The characteristics of the new microorganisms included increased optimum (70–75°C) and maximum (80°C) growth temperatures, resulting in a new subdivision of thermophiles into moderate and extreme groups.

The next important step in understanding the diversity of thermophilic microorganisms was made in the 1980s to 1990s and was, in the first place, connected with the names of W. Zillig and K.O. Stetter (Zillig *et al.*, 1981, 1982, 1983a; Stetter *et al.*, 1983, 1988, 1996; Segerer *et al.*, 1986; Huber *et al.*, 2000). Application of anaerobic cultivation to hydrothermal samples resulted in the isolation of hyperthermophiles – microorganisms with an optimal growth temperature exceeding 80°C, and an upper limit as high as 100°C and above. The highest growth temperatures were associated with thermophilic microorganisms inhabiting deep-sea hot vents, where, due to elevated hydrostatic pressure, the temperature of water is far above its boiling point at atmospheric pressure (Jannasch and Mottl, 1985; Stetter, 1996). At present the highest growth temperature reported – 122°C – is for the deep-sea methanogen *Methanopyrus kandleri* incubated at elevated pressure (Takai *et al.*, 2008). Hyperthermophiles are the subject of Chapter 9.

The recognition of much higher temperature limits for life was accompanied by a comprehension of a wider phylogenetic and metabolic diversity of thermophilic microorganisms. Cultured thermophilic archaea and bacteria represent a wide range of metabolic types, including aerobes and anaerobes, phototrophs, lithotrophs, and organotrophs, using different combinations of electron donors and acceptors. The latter capacity is of special importance with respect to microorganisms inhabiting volcanic environments, as many substrates of volcanic origin are available as electron donors or acceptors for microbial growth. The most widely distributed among thermophiles and hyperthermophiles is the capacity to utilize molecular hydrogen of volcanic origin, which can be used in combination with the reduction of diverse electron acceptors, including oxygen, nitrate, elemental sulfur, thiosulfate and sulfate, ferric iron and other metals, and CO_2 (Huber *et al.*, 2000; Shock *et al.*, 2005). Inorganic substrates, such as elemental sulfur, thiosulfate, ferric iron, and CO are also utilized by thermophilic prokaryotes as energy substrates (Brock *et al.*, 1972; Segerer *et al.*, 1986; Svetlichny *et al.*, 1991; Haffenbradl *et al.*, 1996).

By the end of the twentieth century it appeared as though all possible metabolic types of thermophilic prokaryotes had already been discovered and characterized (Table 6.1). However, application of phylogenetic analysis methods to natural DNA samples from thermal sources uncovered the existence of many novel thermophilic prokaryotes, representing high-ranking phylogenetic lineages that were previously unknown from laboratory culturing (Barns *et al.*, 1994; Hugenholz *et al.*, 1998; Takai and Horikoshi, 1999; Takai *et al.*, 2001; and others). Since then, representatives of several novel thermophilic phyla have been successfully cultivated (Reysenbach *et al.*, 2006; Elkins *et al.*, 2008; Mori *et al.*, 2008, and others) while many more remain uncultured and their phenotypic capacities unknown. Some of these new groups are represented by an enormous number of sequences in the databases, suggesting that they are environmentally relevant microorganisms. However, as long as their metabolic characteristics are not elucidated, we will be unable to estimate their environmental impact and the role they may play in, for example, biogeochemical cycles. These roles could involve new thermostable enzymes with bioengineering potential, which, since they are not yet identified, are not available for evaluation. Direct cloning of the respective functional genes from environmental DNA or enrichment cultures ('gene mining') might, at least partly fill this gap. However, the use of PCR primers derived from previously known gene sequences. The latter method would, however, only allow the detection of closely related genes, while truly novel proteins would not be identified.

In this chapter I review recently discovered novel metabolic groups of thermophilic prokaryotes. By 'novel' I mean, first of all, those microbes displaying new environmentally significant processes that have only recently been detected in thermal environments. In some cases, 'novel' will refer to new conditions – for example, hyperthermophilic, for processes previously known to occur only at moderate temperatures, or acidic, for those reported previously only in high-temperature environments with neutral pH. Some new metabolic capacities of thermophilic prokaryotes revealed by complete genome analysis and substantiated by experimental data shall also be described. The phylogenetic diversity of thermophiles, the mechanisms of thermostability of their biomolecules, and possibilities of their application in biotechnology, will not be covered in this chapter as these subjects have been reviewed in numerous publications (for example Stetter, 1996; Jaenicke and Bohm, 1998; Kumar and Nussinov, 2001; Vielle and Zeikus, 2001; Blumer-Schuette *et al.*, 2008).

Table 6.1 Thermophilic microorganisms representing different metabolic groups

	First representatives isolated[1], growing optimally at:		
Metabolic group	50–60°C	70–75°C	>80°C
Aerobic organotrophs	*Bacillus stearothermophilus*, Donk (1920)	*Thermus aquaticus*, Brock and Freeze (1969)	*Aeropyrum pernix*, Sako *et al.* (1996)
Anaerobic organotrophs	*Clostridium thermocellum*, Viljoen *et al.* (1926)	*Thermotoga maritima*, Stetter and Huber (1986)	*Thermoproteus tenax*, Zillig and Stetter (1982)
Aerobic hydrogen oxidizers	*Ralstonia eutropha*, Davis (1969)	*Calderobacterium hydrogenophilum*, Kryukov *et al.* (1984)	*Pyrolobus fumarii*, Blöchl *et al.* (1999)
Sulfur oxidizers	*Sulfobacillus thermosulfidooxidans*, Golovacheva and Karavaiko (1991)	*Sulfolobus acidocaldarius*, Brock *et al.* (1972)	*Acidianus infernus*, Segerer *et al.* (1986)
Methanotrophs	*Methylocaldum szegediense*, Bodrossy *et al.* (1998)	–	–
Nitrifiers	'*Candidatus Nitrososphaera gargensis*', Hatzenpichler *et al.* (2008)	'*Candidatus Nitrosocaldus yellowstonii*', de la Torre *et al.* (2008)	–
Lithotrophic methanogens	*Methanobacterium thermautotrophicum*, Zeikus and Wolfe (1972)	*Methanococcus jannaschii*, Jones *et al.* (1984)	*Methanothermus fervidus*, Stetter (1982)
Acetoclastic methanogens	*Methanosarcina thermophila*, Zinder *et al.* (1985)	–	–
Sulfate reducers	*Desulfotomaculum nigrificans*, Werkman and Weaver (1927)	*Thermodesulfobacterium commune*, Zeikus *et al.* (1983)	*Archaeoglobus fulgidus*, Stetter (1988)
Sulfur reducers	*Desulfurella acetivorans*, Bonch-Osmolovskaya *et al.* (1990)	*Desulfurobacterium thermolithotrophum*, L'Haridon *et al.* (1998)	*Thermoproteus tenax*, Zillig and Stetter (1982)
Denitrifiers	*Bacillus stearothermophilus*, Donk (1920)	*Thermus aquaticus*, Brock and Freeze (1969)	*Pyrobaculum aerophilum*, Völkl *et al.* (1996)
Acetogens	*Moorella thermoacetica*, Fontaine *et al.* (1942)	–	*Archaeoglobus fulgidus*, Stetter (1988)
Iron reducers	*Bacillus infernus*, Boone *et al.* (1995)	*Thermoterrabacterium ferrireducens*, Slobodkin *et al.* (1997)	*Geoglobus ahangari*, Kashefi *et al.* (2002)
Anaerobic CO-oxidizers	*Carboxydocella thermautotrophica*, Sokolova *et al.* (2002)	*Carboxydothermus hydrogenoformans*, Svetlichny *et al.* (1991)	*Archaeoglobus fulgidus*, Stetter (1988)
Anaerobic iron oxidizers	–	–	*Ferroglobus placidus* Hafenbradl *et al.* (1997)

[1]Person(s) responsible for, and date of, first description indicated after taxon name. –, not known.

New metabolic groups of thermophilic lithoautotrophs

Lithoautotrophic thermophilic prokaryotes play an important environmental role in the microbial communities of thermal environments. By assimilating carbon from inorganic substrates of volcanic origin, they support prokaryotic ecosystems based on the utilization of autochthonous organic material. During recent years, several lithotrophic prokaryotes have been described, representing novel taxa of the domains *Archaea* and *Bacteria*. However, the majority fall into previously known metabolic types: lithotrophic reduction of ferric iron and sulfur species, and lithotrophic nitrate reduction. Some of the new metabolic groups I will discuss are not yet represented by validly described isolates, whilst others have not even been obtained in laboratory cultures. However, there is solid evidence of their existence and/or activity in natural thermal environments, based on a combination of different research methods.

Thermophilic nitrification

The process of nitrification, a critical part of the global nitrogen cycle and consisting of the oxidation of ammonia to nitrite and then to nitrate, is carried out by chemolithotrophic prokaryotes. Because ammonia is a common component of hydrothermal fluids (at some sites its concentration can be as high as 50 mM; Reysenbach *et al.*, 2002; Windman *et al.*, 2007), the existence of thermophilic microorganisms able to use it as an energy substrate for growth is probable. Another fact supporting this claim is the presence of nitrate in some hot springs (Zhang *et al.*, 2008) and the existence of many thermophilic bacteria and archaea capable of denitrification. Nevertheless, nitrification was conventionally thought not to proceed at elevated temperatures, with only a few direct indications of the existence of thermophilic ammonia- and nitrite-oxidizing bacteria. An ammonia-oxidizing enrichment culture growing at 55°C was obtained from a Kamchatka hot spring but then lost, so that no pure culture was ever obtained (Golovacheva, 1976). Almost 20 years later, enrichment cultures of ammonia and nitrite oxidizers growing optimally at 50°C were obtained from the hot springs of the Baikal rift zone (Lebedeva *et al.*, 2005).

The search for thermophilic microorganisms with nitrification capacity has gained fresh momentum upon the discovery that the first step of nitrification, i.e. ammonia oxidation to nitrite, is a function not only of *Proteobacteria*, as previously thought, but of a widely distributed mesophilic *Crenarchaeota*, formerly designated as Group I (Delong *et al.*, 1992). Initial evidence came from metagenomic studies in which genes encoding ammonia monooxygenase, the key metabolic enzyme of ammonia oxidation, were found in genomic fragments of *Crenarchaeota* (Schleper *et al.*, 2005; Treusch *et al.*, 2005). The first ammonia-oxidizing archaeal isolate, *Nitrosopumilus maritima* (Könneke *et al.*, 2005), contained ammonia monooxygenase genes highly homologous to those found by metagenomic analysis. Among the three genes encoding subunits of archaeal ammonia monooxygenase, *amoA* became a widely accepted environmental marker of archaeal nitrification, and was also applied in thermal environments. Archaeal *amoA* genes were found in Iceland and Kamchatka hot springs with temperatures ranging from 82 to 97°C and pH from 2.5 to 7.0 (Reigstad *et al.*, 2008), and in thermal sites in the US, China and Russia with temperatures from 41 to 86°C and pH from 3.4 to 9.0 (Zhang *et al.*, 2008). In both cases, the sequences obtained clustered with *amoA* genes of marine or soil groups from non-thermal habitats. Reverse transcription PCR analysis showed that *amoA* genes were transcribed *in situ* and the transcripts were closely related to the *amoA* genes amplified from the same springs (Zhang *et al.*, 2008).

Crenarchaeol, a specific core lipid initially found in mesophilic Group I crenarchaeotes, has been used as another biomarker of archaeal ammonia oxidizers. Its presence in thermal environments of different geographical locations is regarded as evidence of a wide distribution of thermophilic nitrifying crenarchaeota in these habitats (Pearson *et al.*, 2004, 2008; Zhang *et al.*, 2006).

Concurrently with these investigations, the first sustainable cultures of thermophilic ammonia-oxidizing archaea were obtained (de la Torre *et al.*, 2008; Hatzenpichler *et al.*, 2008). As a result of these studies, two species were described based on enrichments of ammonia-oxidizing archaea obtained using ammonia-poor media (1 mM or lower). 16S rRNA analyses showed that the new thermophilic organisms clustered with non-thermophilic crenarchaeotes (Fig. 6.1). Until now, none of these organisms have been isolated in pure culture, and both strains along with the higher taxa they represent have the status of Candidatus. One of the microorganisms, '*Nitrosocaldus yellowstonii*', gen nov., sp. nov. is the first member of '*Nitrosocaldaceae*' fam. nov. and '*Nitrosocaldales*' ord. nov. (de la Torre *et al.*, 2008). As suggested by its species name, the culture of '*N. yellowstonii*' was obtained from sediment of a hot spring in Yellowstone National Park. Phylogenetically, it belongs to the group of mesophilic crenarchaeota. Cells of '*N. yellowstonii*' are small cocci or very short rods that grow between 60 and 74°C and optimally at 65–72°C. Cells of '*N. yellowstonii*' contain crenarchaeol, substantiating its use as a specific marker

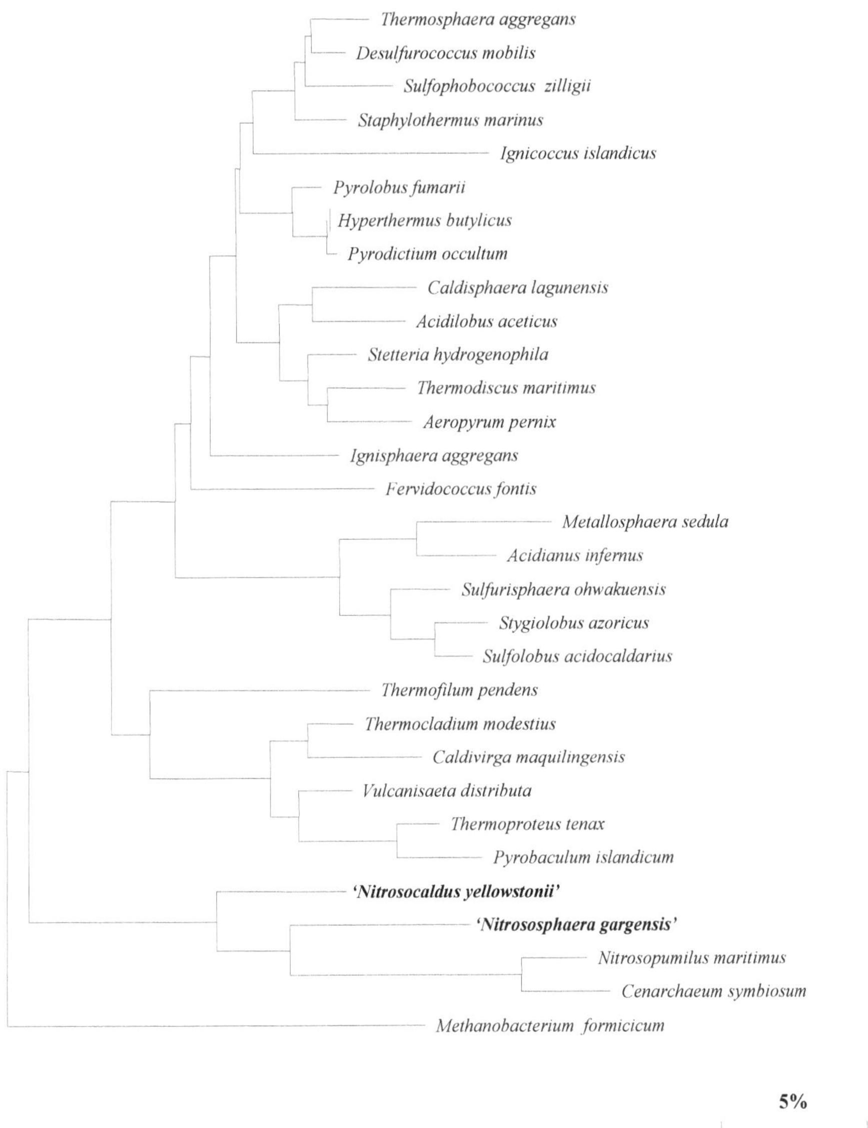

Figure 6.1 Position of new thermophilic ammonia-oxidizing archaea on the phylogenetic tree of *Crenarchaeota*. Courtesy of A.V. Lebedinsky, Winogradsky Institute of Microbiology, RAS.

for ammonia-oxidizing archaea (de la Torre *et al.*, 2008).

The other thermophilic ammonia-oxidizing archaeon, *'Nitrososphaera gargensis'*, was obtained from a sample of cyanobacterial mat in the Garga River (Lake Baikal rift zone, Russia). It was described as a predominant component in a stable ammonia-oxidizing enrichment culture (Hatzenpichler *et al.*, 2008). Its cells are irregular cocci, approximately 1 μm in diameter. The organism can grow lithoautotrophically with ammonia as an energy source and CO_2 as a carbon source, producing nitrite as the end product. However, both its growth and ammonia oxidation were stimulated by the presence of organic carbon (formate, succinate, pyruvate, or acetate). Growth was observed in the temperature range 28–49°C with an optimum at 44–46°C. Core lipids of *'N. gargensis'* consist of crenarchaeol.

Thus, one of the two currently described ammonia-oxidizing archaea (*'N. yellowstonii'*) is a neutrophilic extreme thermophile, growing at temperatures not exceeding 75°C. The other, (*'N. gargensis'*), also a neutrophile, is a moderate thermophile, having a growth temperature optimum at the lower limit of the suggested thermophily range (Reysenbach and Shock, 2002). However, given the range of temperature and pH in the hot springs where *amoA* genes were detected, the diversity of thermophilic archaeal nitrifiers is probably much wider and includes hyperthermophiles, thermoacidophiles, and thermoalkaliphiles (Reigstad *et al.*, 2008; Zhang *et al.* 2008).

As the two archaea mentioned above only carry out the first phase of nitrification (the oxidation of ammonia to nitrite), the presence of nitrate in hot springs cannot be explained by their activity. The rate of biological nitrite oxidation estimated from *in situ* experiments with pool dilution of ^{15}N-nitrate (i.e. the dilution of introduced isotope with the nitrate produced from nitrite in the course of incubation) in two Icelandic hot springs with temperatures of 84°C and 85°C and pH 3.0 was 21.1 and 12.8 μmol nitrate/l/day. These data point to the activity of hyperthermophilic and acidophilic nitrite-oxidizing microorganisms. Interestingly, addition of ammonia was observed to increase the rate of the process by a factor of two, indicating that it was limited by the supply of ammonia (Reigstad *et al.*, 2008).

At present, nitrite-oxidizing thermophiles are represented by bacteria of the phylum *Nitrospira*. 16S rRNA genes related to *Nitrospira* have been detected in hot springs of Thailand and Australia, in the deep-subsurface biosphere (Japan), and hot deep-sea vents at the Mid-Atlantic Ridge (Anitori *et al.*, 2002; Lopez-Garcia *et al.*, 2003; Kanokratana *et al.*, 2004; Hirayama *et al.*, 2005; Weidler *et al.*, 2007). Meanwhile, no *Nitrospira*-specific 16S rRNA genes were detected in the Iceland hot springs mentioned above. Taken together with the *in situ* biological nitrite oxidation data, this finding points strongly to the existence of novel thermoacidophilic nitrite oxidizers that are distinct from *Nitrospira* (Reigstad *et al.*, 2008).

Thermophilic anammox bacteria – evidence of their presence in hot springs

Anammox bacteria, capable of growing by anaerobic ammonia oxidation and simultaneous nitrate reduction (forming molecular nitrogen as the end product) were first detected in an anaerobic mesophilic wastewater treatment plant (Kuenen, 2008). Anammox bacteria turned out to be obligate anaerobes, growing very slowly with a doubling time of around two weeks, and represented a separate group of five genera in the *Planctomycetes* phylum. Later on, these organisms were detected in a large range of fresh-water and marine habitats, where they play an important role in nitrogen turnover. Anammox *Planctomycetes* are lithoautotrophs, assimilating CO_2 via the acetyl CoA pathway. Furthermore, anammox bacteria possess ladderanes, unique lipids believed to protect cells against hydrazine, a toxic intermediate product of anaerobic ammonia oxidation.

The use of biomarkers (16S rRNA genes and lipids) and/or direct activity measurements contributed to the detection of anammox bacteria in diverse environments, including natural thermal habitats. For example, they were found in California and Nevada hot springs (Jaeschke *et al.*, 2009) and in zones neighbouring deep-sea vents at 60–85°C (Byrne *et al.*, 2008). Deeply branching planctomycetes of the annamox cluster have been identified in a hot spring in Oklahoma (Elshahed *et al.*, 2007). Anammox bacteria adapt to changing temperatures by increasing the relative amount of long-chain ladderane fatty acids (Damste *et al.*, 2008). The subsequent finding that ladderane lipids found in hot springs had a higher ratio of long-chain fatty acids suggested the presence of anammox bacteria adapted to high temperatures (Rattray *et al.*, 2010).

Taxonomically, thermophilic anammox bacteria appear to be fairly diverse: deep-sea thermophilic *Planctomycetes* are more closely related to the genus *Kuenenia* (Byrne *et al.*, 2008), while in terrestrial hot springs, their 16S rRNA sequences are closer to those of *Brocadia fulgida*, *Brocadia anammoxidans*, and *Kuenenia stuttgartiensis* (Jaeschke *et al.*, 2009). Both a new cluster positioned between annamox bacteria and other *Planctomycetes*, and uncultured bacteria clustering with *Planctomycetes* were discovered in Nevada hot springs with the highest content of ladderane lipids (Jaeschke *et al.*, 2009). It remains to be clarified whether the new groups of organisms are able to carry out the anammox reaction. Even so, all these quite recent results implicate anammox bacteria as components of thermophilic microbial communities and suggest their participation in the nitrogen cycle in thermal environments.

CO oxidation by thermophilic prokaryotes

Carbon monoxide is a common component of volcanic gases and a readily available energy substrate for thermophilic microorganisms. It can be transformed to methane or acetate by a limited number of thermophilic methanogens or acetogens, respectively, or serve as an energy and carbon source for a large group of thermophilic hydrogenogenic carboxydotrophs (Sokolova *et al.*, 2009). This group of microorganisms is phylogenetically diverse. The first thermophilic organism shown to perform hydrogenogenic carboxydotrophy (reaction 6.1) in the course of growth was *Carboxydothermus hydrogenoformans* (Svetlichny *et al.*, 1991), an extreme thermophile of the phylum *Firmicutes* isolated from a terrestrial hot spring on Kunashir Island, South Kurils.

$$CO + H_2O = H_2 + CO_2 \qquad (6.1)$$

Later on, the ability to grow by means of CO oxidation with concurrent hydrogen production from water was established in many other phylogenetically unrelated members of the *Firmicutes* (Sokolova *et al.*, 2009). All of these organisms were isolated under a 100% CO gas phase.

It was recently discovered that the diversity of thermophilic hydrogenogenic prokaryotes thriving at lower CO concentrations is much higher (Sokolova *et al.*, 2009; Kochetkova *et al.*, 2011). Strain 1512, isolated from a hot spring of the Uzon Caldera, Kamchatka, Russia, can grow at 70°C by hydrogenogenic CO-trophy under a gas phase CO content not exceeding 15% (Kochetkova *et al.*, 2011). However, the presence of 0.2 g/l yeast extract proved to be essential for growth. The optimal growth of strain 1512 was observed at 5% CO in the gas phase. Apart from carboxydotrophic growth, strain 1512 could grow by fermentation of mono- and disaccharides. 16S rRNA gene analysis placed strain 1512 in the genus *Dictyoglomus*, with a 99% similarity to *Dictyoglomus turgidus* (Svetlichny and Svetlichnaya, 1992), also isolated from the Uzon Caldera. Both the type strain and the only other strain of the species *Dictyoglomus thermophilus* (Saiki *et al.*, 1985) are obligately organotrophic hyperthermophilic anaerobes, hydrolysing and fermenting a wide range of organic substrates, but not capable of CO utilization. Thus, hydrogenogenic CO-trophy in the domain *Bacteria* is not an exclusive trait of members of the phylum *Firmicutes*, but can be found in other divisions, including a group that is as deeply branching as the *Dictyoglomi*.

The ability to grow on CO, producing molecular hydrogen from water, was also found in hyperthermophilic archaea having different phylogenetic positions. The first archaeal hydrogenogenic carboxydotroph was *Thermococcus* AM4, isolated from deep-sea vents of the East Pacific Rise (Sokolova *et al.*, 2004). It could grow anaerobically at 85°C under 100% CO through hydrogenogenic oxidation of carbon monoxide. This ability was later identified in several deep-sea hyperthermophilic isolates obtained from diverse hydrothermal habitats, including the East Pacific Rise, the Mid-Atlantic Ridge, and Lau Basin. All new CO-trophic strains were isolated under a 100% CO atmosphere and belonged to the genus

Thermococcus, diverging on the species level. The ability to grow on CO was also shown by *Thermococcus onnurineus* (Lee *et al.*, 2008), but was absent in numerous *Thermococcus* and *Pyrococcus* species (Sokolova *et al.*, 2004).

Another carboxydotrophic hyperthermophilic archaeon isolated from a terrestrial hot spring (located on Moutnovsky Volcano, Kamchatka, Russia) is '*Thermofilum carboxyditrophus*', a member of the phylum *Crenarchaeota* (Kochetkova *et al.*, manuscript in preparation). Isolation of this organism was made possible only because lower CO concentrations were used. Cells of '*T. carboxydotrophus*' are long (15–30 μm) and very thin filaments (0.2 μm). The organism grows chemolithotrophically at 92°C under 45% CO in the gas phase, producing equimolecular amounts of hydrogen and CO_2. Growth of '*T. carboxydotrophus*' is obligately dependent on yeast extract and cysteine (0.2 g/l). This organism is also able to grow by fermenting a range of organic substrates. *Thermofilum pendens*, the only other validly described species of the genus *Thermofilum*, is an organotroph depending for its growth on the presence of cell extracts of hyperthermophilic archaea in the culture medium (Zillig *et al.*, 1983b). It is worth mentioning that the 16S rRNA gene of '*T. carboxyditrophus*' displays 100% similarity with numerous environmental clones obtained from different hot springs in Yellowstone National Park (Barns *et al.*, 1994; Reysenbach *et al.*, 2000). It follows that representatives of the novel species '*T. carboxyditrophus*' are probably widespread in terrestrial hot springs from different geographical locations.

Analysis of complete genomes of diverse thermophilic hydrogenogenic carboxydotrophs and PCR products obtained with specific primers revealed similar organizations of the gene clusters encoding CO-dehydrogenase and energy-conserving hydrogenase, the key enzymes of hydrogenogenic carboxydotrophy (Lebedinsky *et al.*, in preparation). The organization of clusters in terrestrial and marine isolates is not identical, indicating two independent lines of evolution of this metabolic capacity.

In contrast to anaerobic hydrogenogenic carboxydotrophs, aerobic CO-oxidizers, plentiful in non-thermophilic environments, have not until now been isolated from thermal sources. However, a gene encoding aerobic CO-dehydrogenase was recently identified in a recombinant fosmid containing a 16S rRNA gene from an uncultured crenarchaeote inhabiting a subsurface gold mine (Nunoura *et al.*, 2003). Therefore, the phylogenetic and metabolic diversity of thermophilic CO-oxidizing prokaryotes could be much wider than currently recognized.

Lithoautotrophic growth by dismutation of sulfur compounds

The ability to carry out 'inorganic fermentation' or 'sulfur or thiosulfate dismutation or disproportionation' was found in several phylogenetically diverse mesophilic bacteria (Bak and Cypionka, 1987; Thamdrup *et al.*, 1993). These microorganisms were able to gain energy for growth by simultaneous oxidation and reduction of sulfur species according to reaction 6.2 or reaction 6.3:

$$4S^o + 4H_2O \rightarrow 2H^+ + SO_4^{2-} + 3H_2S \qquad (6.2)$$

$$S_2O_3^{2-} + H_2O \rightarrow SO_4^{2-} + H_2S \qquad (6.3)$$

Until recently, *Desulfotomaculum thermobenzoicum* was the only thermophile known to perform thiosulfate disproportionation (Jackson and McInerney, 2000). However, bearing in mind that natural thermal environments typically abound in diverse sulfur species, the existence of distinct thermophilic microorganisms capable of sulfur and thiosulfate disproportionation seems very probable.

A new microorganism (strain S95) that can grow anaerobically on a mineral medium supplemented with elemental sulfur or thiosulfate was recently isolated from a deep-sea hydrothermal vent chimney located at a depth of 1700 m in the East Lau Spreading Centre, Pacific Ocean (Slobodkin *et al.*, manuscript in preparation). The isolate grew in the temperature range 50–92°C and optimally at 75°C on medium containing elemental sulfur and bicarbonate/CO_2 as, respectively, energy and carbon sources. Elemental sulfur was disproportionated to sulfide and sulfate, and optimal growth was observed in the presence of poorly crystalline Fe(III) oxide (ferrihydrite) as a sulfide-scavenging agent. Strain S95 was also able to grow by disproportionation of thiosulfate and sulfite. Analysis of the 16S rRNA gene sequence revealed that the isolate belonged to the phylum *Thermodesulfobacteria* and was equidistantly related to all described genera of this phylum (sequence similarity, 91–92%). Based on its physiological properties and results of phylogenetic analyses, strain S95 is proposed to represent a new species '*Thermosulfurimonas dismutans*' belonging to the novel genus '*Thermosulfurimonas*' (Slobodkin *et al.*, manuscript in preparation). *Thermosulfurimonas dismutans* S95 is the first thermophilic microorganism shown to disproportionate elemental sulfur, a compound common in thermal ecosystems, including deep-sea hydrothermal vents. Microbial sulfur disproportionation in these extreme environments, based on metabolism of a single simple and copious compound, might well constitute a still unrecognized process of primary organic matter production.

New thermophilic microorganisms metabolizing C_1 compounds

C_1 compounds are readily used by aerobic and anaerobic microorganisms, both archaea and bacteria. However, thermophilic prokaryotes able to utilize C_1 compounds have not received much attention until recently. Among the C_1-compounds, methane alone occurs in significant quantities in thermal environments. It is a common component of volcanic gases, usually occurring at a concentration of 0.01–1% (Etiope and Klusman, 2002). Methane originates from buried organic matter and is formed via abiotic reactions. At some sites, methane is present at much higher concentrations, for example, some hot springs in New Zealand contain 1–11% CH_4 (Gigenbach, 1995). Methane in hot springs can also be produced by thermophilic and hyperthermophilic methanogens (Stetter, 1996). Methanol is an energy substrate utilized by many aerobic and anaerobic mesophilic microorganisms and, in natural systems, is most probably generated via pectin degradation accompanied by methanol formation from methyl bridges present in pectin molecules (Bonch-Osmolovskaya and Ilarionov, 1982). The latter could also occur in thermal environments where complex organic substrates are decomposed. The source of formate in deep-sea hydrothermal environments could be its production by *Archaeoglobus fulgidus* in the course of CO uptake (Henstra *et al.*, 2007).

It has been recently found in a study of the Lost City hydrothermal field that methane, methanol, formaldehyde, and formate can be produced abiotically in the course of serpentinization (Martin *et al.*, 2008). In this reaction, the mineral olivine $(Mg.Fe)_2SiO_4$ reacts with water and CO_2 at 300°C beneath the seafloor, forming serpentine $(Mg_{2.85}Fe_{0.15}Si_2O_5(OH)_4)$, magnesium and silicon hydroxides, iron oxide, H_2, CH_4, and hydrocarbons. In addition to the reduced carbon species, formate, methanol, and acetate are produced. Except methane, concentrations of other C_1 products in thermal environments are usually very low or not detectable. Nevertheless, their possible abiotic origin in hydrothermal environments should draw attention to thermophilic organisms potentially able to utilize them as energy substrates.

Thermoacidophilic methanotrophs

Moderately thermophilic, neutrophilic methanotrophs inhabiting terrestrial hot springs are able to grow at temperatures ranging from 37°C to 65°C (optimally at 55–59°C) and belong to the class *Gammaproteobacteria*. Currently, they are represented by several species of the genus *Methylocaldum* (Bodrossy *et al.*, 1998) and by *Methylothermus thermalis* (Tsubota *et al.*, 2005). Despite certain biochemical peculiarities, these organisms possess all typical features of methylotrophic proteobacteria and fit quite well into this relatively homogenous group. However, our understanding of the phylogenetic and metabolic diversity of thermophilic methanotrophs has been significantly changed by recent discoveries.

Three groups of researchers reported isolation of thermoacidophilic methanotrophs from the phylum *Verrucomicrobia*. All isolates grew in the temperature range 40–65°C and pH range 0.8–6.0 and optimally at 55 °C and pH 3.0. The sources of isolation were thermal environments located in geographically separated volcanic areas: Solfatara mud volcano in Italy (Pol *et al.*,

2007), steaming soil in Tikitera, New Zealand, (Dunfield *et al.*, 2007), and an acidic hot spring in Uzon Caldera, Kamchatka (Islam *et al.*, 2008). Before this discovery, the only known acidophilic methanotrophs were members of the genera *Methylocella* and *Methylocapsa*, growing optimally at pH 5.0–5.5 (Dedysh *et al.*, 2000; Dedysh *et al.*, 2002). The latter organisms, isolated from acidic bogs, are psychrotrophic and grow in the temperature range 4–30°C.

Being published independently and almost simultaneously, these three new organisms initially received different genus names: '*Acidimethylosilex*', '*Methylokorus*' and '*Methylacida*' (Pol *et al.*, 2007; Dunfield *et al.*, 2007; Islam *et al.*, 2008). Subsequently, they were reclassified as species of a single genus *Methylacidiphilum* because the level of similarity of their 16S rRNA genes exceeded 98% (Semrau *et al.*, 2008). The genus *Methylacidiphilum* includes at present three species, *M. infernorum*, *M. fumariolicum*, and *M. kamchatkense* and is the only genus of the new family '*Methyloacidiphilaceae*' and the new order '*Methylacidiphilales*' in the phylum *Verrucomicrobia* (Semrau et al., 2008).

Cells of representatives of the genus *Methylacidiphilum* are oval rods, approximately 0.5 × 1 μm in size. Being phylogenetically remote from all known methanotrophs, the new isolates were also found to possess unique phenotypic features, including the absence of intracytoplasmic membrane structures, typical of all methanotrophs, and extreme acidophily. Instead of intracytoplasmic membranes, their cells contain polyhedral membrane structures resembling carboxysomes (Islam *et al.*, 2008). Attempts to amplify the *pmo*A genes, encoding particulate methane monooxygenase, found in all methanotrophs except the genus *Methylocella*, were unsuccessful (Pol *et al.*, 2007). Their pathways of C_1 assimilation are assumed to also be different, representing a combination of serine, tetrahydrofolate and ribulose-bisphosphate pathways (Pol *et al.*, 2007). The fatty acid profiles of methanotrophic thermoacidophilic verrucomicrobia are also different from those of other methanogens, a feature that is useful for their detection in natural microbial communities, including stable isotope probing (SIP) experiments.

16S rRNA sequences of the genus *Methylacidiphilum* cluster with environmental clones from Yellowstone National Park (98% similarity; Islam *et al.*, 2008). Closely related organisms were also detected in hot soil from Uzon Caldera, Kamchatka, by pyrosequencing of the V3 region of 16S rRNA genes (Mardanov *et al.*, 2011). Methanotrophic veruccomicrobia in this habitat represented one of the most numerous groups (37% of all bacteria), sharing this position with representatives of *Acidithiobacillus*. Mesophilic bacteria of the phylum *Verrucomicrobia* are widespread and abundant in different ecological niches, accounting for 1–10% of the total bacterial DNA in soils. However, their metabolism and their place in soil ecosystems remain unclear. Characterization of acidophilic methanotrophic verrucomicrobia from thermal environments might shed some light on the ecological function of the entire group of these bacteria.

Anaerobic methane oxidation in thermal environments

Anaerobic methane oxidation has been the subject of active investigations during the past decade and, despite problems associated with laboratory cultivation of microbial consortia performing this process, significant steps have been made in understanding its nature (Boetius *et al.*, 2000; Orphan *et al.*, 2001). Anaerobic methane oxidation arises from the syntrophic metabolic relationship between CH_4-oxidizing archaea and sulfate reducing bacteria. Most of the sites where anaerobic methane oxidation has been detected are cold sediments (0–15°C) associated with methane hydrates, methane seeps, or mud volcanoes (Strous and Jetten, 2004). The two exceptions are deep-sea hydrothermal areas in Guaymas Basin (Teske *et al.*, 2002; Schouten *et al.*, 2003; Kallmeyer *et al.*, 2004) and the Lost City Hydrothermal Field (Martin *et al.*, 2008). Methane at these sites is of thermogenic origin, and its active utilization by anaerobic archaea is evidenced by an ultralight ^{13}C composition of their lipids, which is diagnostic for methanotrophs (Teske *et al.*, 2002; Schouten *et al.*, 2003). The rate of anaerobic methane oxidation in Guaymas Basin sediments was measured by a radioisotopic method using ^{14}C-labelled methane and found to have a temperature maximum of 60°C (Kallmeyer *et al.*, 2004). The organisms identified by 16S rRNA analysis of environmental samples from this site belonged to anaerobic methane oxidation group ANME1 (subdivided into clusters ANME1a and ANME1b) and to another group of anaerobic methane oxidizers, ANME2c, representing a lineage within *Methanosarcinales* (Teske *et al.*, 2002). The same groups of microorganisms were obtained from a community associated with Lost City (Martin *et al.*, 2008). These findings prove that anaerobic methane oxidation does take place in habitats where thermogenic methane is available, and that thermophilic microorganisms performing this reaction are phylogenetically closely related to their mesophilic counterparts.

Methanol oxidation by thermophilic prokaryotes

Very little is presently known about the utilization of methanol by thermophilic prokaryotes. Thermophilic methanol-utilizing methanogens are scarce and have been found in an anaerobic reactors (Zinder *et al.*, 1985; Jiang *et al.*, 2005) and in oil production water (Cheng *et al.*, 2007). Methane formation from methanol by a thermophilic microbial culture was due to interspecies hydrogen transfer between homoacetogenic *Moorella thermautotrophica* (=*Clostridium thermoautotrophicum*) and *Methanothermobacter* (=*Methanobacterium thermoautotrophicus*) (Ilarionov and Bonch-Osmolovskaya, 1986). In pure culture, *Moorella thermautotrophica* is able to grow on methanol, generating acetate. Recently, a consortium producing methane from methanol at 65°C was obtained from a hot spring in Geyser Valley, Kamchatka (Merkel, unpublished data). It contained two bacterial and one archaeal component. The bacteria were an organism remotely related to *Thermacetogenium phaeum* (16S rRNA similarity of 92%; Hattori *et al.*, 2000) and a member of the order *Planctomycetales* having 99% 16S rRNA similarity with environmental clone OPB17 (unidentified *Planctomycetales*; Hugenholz *et al.*, 1998). The methanogenic archaeal component of the association had a 99% 16S rRNA similarity with *M. thermautotrophicus*. In this consortium, it appears as though the organism related to *T. phaeum* is responsible for hydrogenogenic methanol oxidation and provides *Methanothermobacter* with the growth substrate. The role of planctomycetes in this association is unclear and, thus, even more intriguing. This association was also shown to produce methane from acetate (see the section below entitled *Anaerobic oxidation of non-fermentable substrates by thermophilic prokaryotes*).

Methanol can also be utilized by several phylogenetically distantly related thermophilic microorganisms. Aerobic methanol oxidation is performed by thermophilic methanotrophic verrucomicrobia that can utilize methanol when present in the growth medium at very low concentrations (2–36 μM; Islam *et al.*, 2008). Recently, it was found that strain MDR (isolated from an Uzon Caldera hot spring), having a 99% 16S rRNA similarity with *Desulfurella acetivorans*, could grow anaerobically on methanol (Miroshnichenko, unpublished data). Bacteria of the genus *Desulfurella* (Bonch-Osmolovskaya *et al.*, 1990), which belong to the order *Desulfurellales* in the phylum *Proteobacteria*, are moderately thermophilic anaerobes with sulfur or thiosulfate respiration. The type species of *D. acetivorans* is able to grow lithoautotrophically with molecular hydrogen as the energy substrate and elemental sulfur as the electron acceptor, or by performing complete oxidation of fatty acids from C_2 to C_{16} via sulfur respiration (Bonch-Osmolovskaya *et al.*, 1990). The newly isolated strain MDR demonstrated an even broader range of capacities: it could use nitrate as an electron acceptor and grew on methanol as the sole energy and carbon source, reducing sulfur, thiosulfate or nitrate (Miroshnichenko, unpublished data).

Another new methanol-oxidizing anaerobe is related to a recently described extremely thermophilic bacterium *Caldimicrobium rimae* (Miroshnchenko *et al.*, 2009) representing a novel genus *Caldimicrobium* in the phylum *Thermodesulfobacteria*. The type strain, isolated from an Uzon Caldera hot spring (Fig. 6.2), was found to grow either lithoautotrophically with molecular hydrogen and thiosulfate or elemental sulfur, or by oxidizing several organic acids (malate, succinate, fumarate) and ethanol in the course of thiosulfate respiration (Miroshnichenko *et al.*, 2009). Recently isolated from a hot spring of Pauzhetka

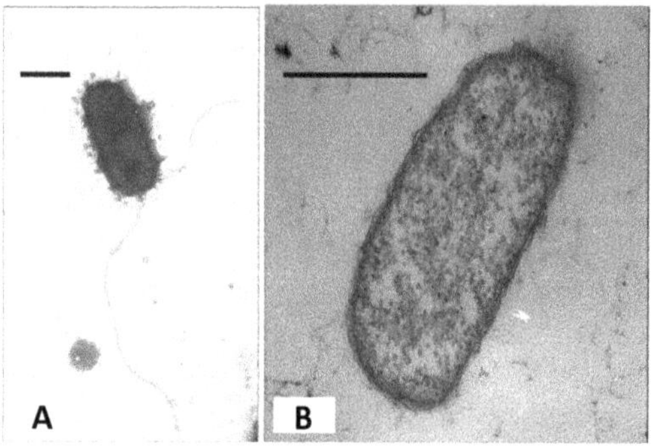

Figure 6.2 Cells of *Caldimicrobium rimae.* (A) Whole cells, (B) thin sections.

(Kamchatka), strain CM was phylogenetically very closely related to *C. rimae* (99% 16S rRNA gene similarity; Miroshnichenko, unpublished data). The novel isolate grows in the temperature range 50–87°C, and optimally at 82–85°C. It is similar to the type strain in its ability to grow lithoautotrophically with molecular hydrogen and thiosulfate or sulfur. However, it also grew on acetate, butyrate, methanol and propanol in the presence of thiosulfate or sulfur. At present, strain CM is the only known methanol-oxidizing hyperthermophilic bacterium.

Both new isolates (strains MDR and CM), utilizing methanol as an energy substrate, represent already known species of thermophilic bacteria. This fact indicates that metabolic versatility could be very wide even at the species level, and that organisms with novel metabolic features do not necessarily represent new taxa. It also suggests that the ability to utilize methanol is, probably, much more common among thermophilic anaerobes than previously thought.

Early work on methane formation in hydrothermal samples from Uzon Caldera demonstrated methane production from methanol at 85°C (Bonch-Osmolovskaya and Karpov, 1987), but no sustainable enrichment culture has ever been obtained. Nevertheless, these findings are tantalizingly suggestive of the existence of hyperthermophilic methanogens (or associations of hyperthermophilic microorganisms) able to produce methane from methanol at high temperatures.

Hydrogenogenic formate oxidation by *Thermococcales*

Possible mechanisms of formate utilization by thermophiles have been recently discussed (Windman *et al.*, 2007). According to this study, formate can either be oxidized to CO_2 in the course of aerobic or anaerobic respiration, or reduced to methane. Meanwhile, a completely new type of metabolism has been quite recently discovered. This process is based on formate oxidation in the absence of electron acceptors, and has never been previously reported, not only for thermophiles, but also for the microbial world at large (Kim *et al.*, 2010). Several strains of the hyperthermophilic archaeal genus *Thermococcus* were able to grow using a reaction (see reaction 6.4 below) previously believed to produce an insufficient amount of energy to support microbial growth ($\Delta G^{o\prime} = +1.3$ kJ/mol; Thauer *et al.*, 1977).

$$HCOOH = H_2 + CO_2 \tag{6.4}$$

In particular, growth of formate-degrading, hydrogen-producing *Desulfovibrio* sp. or *Moorella* sp. in coculture with hydrogen-oxidizing methanogens, *Methanobrevibacter arbophilus* or *Methanothermobacter* sp., used to be described as 'growth at the thermodynamic limit' and explained by interspecies hydrogen transfer, making the process exergonic (Dolfing *et al.*, 2008; Stams and Plugge, 2009). However, under the conditions actually used to culture *Thermococcus* strains (86°C, pH 6.5), $\Delta G'$ varied between −13 and −26 kJ/mol H_2, thus making microbial growth possible without removal of the hydrogen formed (Kim *et al.*, 2010). Formate oxidation to CO_2 was accompanied by generation of equimolar quantities of molecular hydrogen and fairly active growth of microbial cells (Fig. 6.3). This capacity was found in several species of the genus *Thermococcus* that are phylogenetically distantly related, including two with sequenced genomes: *Thermococcus onurinneus* (Lee *et al.*, 2008) and *Thermococcus gammatolerans* (Jolivet *et al.*, 2003). Both genomes contain genes encoding formate dehydrogenase, energy-conserving hydrogenase, and formate transporters, which together make possible growth on exogenous formate. Other formate-oxidizing

Thermococcus species were isolated from deep-sea hot vents of the East Pacific Rise and Lau Basin and these were related to *Thermococcus barophilus* (Kim *et al.*, 2010; Sokolova *et al.*, unpublished).

Members of the genus *Thermococcus* are the most common hyperthermophilic organisms populating continental hot springs, submarine shallow-water and deep-sea hot vents, and deep-subsurface environments (Stetter *et al.*, 1996). They are generally recognized as anaerobic organotrophs, degrading complex organic substrates (see also the following section). Being readily cultured and isolated, they represent one of the best-studied genera of hyperthermophilic archaea, comprising at present more than 20 species. However, recent studies have shown the metabolism of *Thermococcus* species to be more diverse than previously believed. Some species can grow lithotrophically with CO, producing molecular hydrogen from water. Successful isolation of formate-utilizing *Thermococcus* strains from geographically separated deep-sea hydrothermal environments bears evidence to the wide distribution of this metabolic capacity among members of this genus. The ability to utilize formate and CO could account for the large population of *Thermococcales* in high-temperature deep-sea hydrothermal areas, where no allochthonous organic substrates are available.

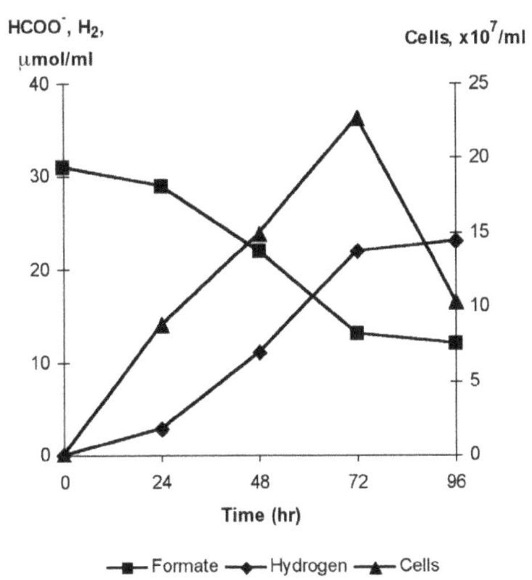

Figure 6.3 Growth of *Thermococcus barophilus* strain Ch5 in marine medium with 2 g/l formate and 0.1 g/l yeast extract.

Degradation of organic substrates by thermophilic prokaryotes

Organic matter of different nature and origin is present in thermal habitats and serves as a source of growth substrates for different organotrophic archaea and bacteria. Organotrophs are often easier to culture than lithoautotrophs, and the number of cultured and described organotrophic thermophilic species is much larger compared with lithotrophs. The most widely used substrate for enrichment and isolation of organotrophic hyperthermophilic archaea is peptone; starch ranks second, whereas amino acids and di- or mono- sugars are used much less frequently and effectively (Schönheit and Schäfer, 1995). In contrast, substrate spectra of organotrophic thermophilic bacteria are much wider and cover a whole variety of organic monomeric and polymeric substrates. Yet, among hyperthermophilic archaea there are also several recently described organisms with broad ranges of organic substrates able to support their growth. For example, the anaerobic, extremely acidophilic archaeon *Acidilobus saccharovorans*, representing a novel order *Acidilobales* (Prokofeva *et al.*, 2009), can grow on a wide spectrum of sugars and polysaccharides. Analysis of complete genomes of thermophilic prokaryotes and, in particular, of hyperthermophilic archaea, revealed the nutrition capacities of these organisms to be much wider than reported in initial phenotypic descriptions based on standard growth experiments (Mardanov *et al.*, 2009, 2010). The substrate specificity of organotrophic hyperthermophiles is reviewed in more detail in Chapter 9. Here, I shall focus on several new findings relating to substrate specificity of hyperthermophilic organotrophs including those derived from genomic data. Another important aspect of the metabolic diversity of organotrophic thermophiles is their environmentally significant capacity to oxidize non-fermentable substrates, making possible the complete decomposition of organic matter in hydrothermal environments.

This issue is discussed in a later section, even though it is also partly covered in Chapter 9.

New hydrolytic activities of hyperthermophilic archaea revealed by complete genome analysis

A hyperthermophilic archaeon, *Thermococcus sibiricus*, was isolated from the Jurassic horizon of a high-temperature oil reservoir in West Siberia (depth 1850 m, temperature 84°C; Miroshnichenko *et al.*, 2001). Its phenotypic description was characteristic of all *Thermococcales*: it utilized peptone, reducing elemental sulfur in the course of growth, and had optima for growth at 85°C and pH 6.5. Analysis of its complete genome sequence (Mardanov *et al.*, 2009) indicated that the catabolic capacities of this organism were much wider. For example, its genome contained many glycosidase genes clustered together and forming a 'saccharolytic island'. These genes are related to the corresponding genes of other archaea and bacteria. Interestingly, the gene encoding thermostable agarase was most closely related to the corresponding gene of the mesophile *Rhodopseudomonas palustris*. Additional growth experiments proved that *T. sibiricus* could grow, although slowly, on several polysaccharides, including microcrystalline cellulose (Avicel) and agarose, and also on olive oil. Table 6.2 shows the consistent difference in cell yields obtained on media containing utilized ($2.21–3.92 \times 10^7$ cells/ml) and non-utilized ($0.23–1.73 \times 10^7$ cells/ml) substrates. Even so, growth on polysaccharides and lipids was much weaker than on peptone, a finding that could be explained by the supposition that peptone contains minor concentrations of essential growth factors that could not be substituted for by yeast extract. The ability of hyperthermophilic archaea to grow on microcrystalline cellulose had already been demonstrated for *Desulfurococcus fermentans* (Perevalova *et al.*, 2005). Genes for β-glucosidase were found in genomes of *Sulfolobus solfataricus* and *Pyrococcus furiosus* and successfully cloned and expressed in *E. coli* (Bauer *et al.*, 1999; Limauro *et al.*, 2001); however, the growth of these hyperthermophiles on cellulosic substrates has never been demonstrated.

Agarose is a polymer of algal origin. Resistance to hydrolysis by the majority of microorganisms has made agarose an essential component of solidified culture media. Nevertheless, there exist organisms, mostly marine mesophilic aerobes, able to readily degrade agarose and agar. Agarose is not a common substrate utilized by thermophilic organisms. The first anaerobic thermophilic agarolytic bacterium to be described was *Caldanaerobacter uzonensis* (Bannikova *et al.*, 2008; Kozina *et al.*, 2010). Agarose is also one of the growth substrates utilized by *Desulfurococcus fermentans* (Perevalova *et al.*, 2005). The ability of *T. sibiricus* to degrade agarose (Table 6.2) suggests that this species, originally isolated from the Jurassic horizon of an oil reservoir, has possessed the ability to degrade algal cell mass since the time it proliferated in the Jurassic ocean.

T. sibiricus can grow anaerobically, degrading lipids in the presence of elemental sulfur (Table 6.2). This result is in agreement with the presence of genes encoding lipases in its genome (Mardanov *et al.*, 2009). Given that no genes coding for enzymes of β-oxidation were found in the *T. sibiricus* genome, it lacked the ability to utilize fatty acids and was therefore assumed to grow at the expense of glycerol oxidation.

The genome of *Acidilobus saccharovorans* (Mardanov *et al.*, 2010) conceals even more surprises. *A. saccharovorans* was described as an obligately anaerobic hyperthermophilic and acidophilic archaeon capable of growth on diverse sugars and polysaccharides in both the absence and presence of elemental sulfur. Complete genome analysis of this organism revealed, unexpectedly, that it possessed numerous extracellular hydrolytic enzymes, including β-glucosidase and lipases. Unlike *T. sibiricus*, this organism possessed all the genes encoding β-oxidation enzymes as well as those involved in the oxidative tricarboxylic acid cycle. However, growth experiments aimed at confirming these metabolic capacities (using a new assortment of substrates) have not yet been successful (Mardanov *et al.*, 2010).

Anaerobic oxidation of non-fermentable substrates by thermophilic prokaryotes

Terminal oxidation of partly reduced fermentation products is an important step in the

Table 6.2 Utilization of substrates by *T. sibiricus*[1]

Substrate	Cell yield[2] (cells/ml)	Utilization of substrate
Control without substrates and with 60 mg/l of yeast extract	1.41×10^7	–
Peptone	4.60×10^8	++
Starch	1.63×10^7	–
Maltose	3.52×10^7	+
Dextran	3.31×10^7	+
Amorphous cellulose	3.25×10^7	+
Carboxymethyl cellulose	2.60×10^7	+
Microcrystalline cellulose	1.73×10^7	–
Cellobiose	2.90×10^7	+
Agarose	3.92×10^7	+
Chitin	1.24×10^7	–
Xylan	1.12×10^7	–
Pectin	1.19×10^7	–
Phenol	0.40×10^7	–
Toluene	0.23×10^7	–
Hexadecane	2.21×10^7	+
Acetone	3.04×10^7	+
Olive oil	3.68×10^7	+
Glycerol	3.39×10^7	+
Sodium linolenate	0.81×10^7	–
Sodium palmitate	0.67×10^7	–
$CO + S^0$	1.48×10^7	–
$CO - S^0$	No cells	–

[1]Adapted from Mardanov *et al.* (2009).

[2]The cell yields represent the amount of the cells in the stationary growth phase after five sequential transfers in the presence of the substrates, 60 mg/l of yeast extract, and elemental sulfur. For substrate concentrations refer to Mardanov *et al.* (2009).

anaerobic degradation process, leading to complete decomposition of organic matter. In this regard, hyperthermophilic anaerobic microbial communities differ significantly from mesophilic ones. In the latter, complete anaerobic decomposition of organic matter requires the presence of several groups of microorganisms, whereas certain hyperthermophiles, for example, *Thermoproteus tenax* (Schönheit and Schäfer, 1995) and *Acidilobus saccharovorans* (Mardanov et al., 2010), are able to completely oxidize complex organic substrates to CO_2 and H_2S via sulfur respiration. At the same time, organic substrates are fermented by many anaerobic thermophiles and hyperthermophiles, generating molecular hydrogen, CO_2, and partly reduced fermentation products, such as acetate and other volatile fatty acids. While several thermophilic microorganisms can anaerobically oxidize acetate at 50–75°C (Bonch-Osmolovskaya *et al.*, 1990; Miroshnichenko *et al.*, 1998, 2003a,b, 2005), no metabolic sink for acetate and other fermentation products at temperatures above 80°C was found until 2002. The first hyperthermophilic microorganism shown to oxidize acetate was *Geoglobus ahangarii*, an iron-reducing hyperthermophile of the phylum *Euryarchaeota*, isolated from deep-sea hydrothermal samples (Kashefi *et al.*, 2002). This organism or other unknown thermophilic and hyperthermophilic prokaryotes might be responsible for the active acetate oxidation at 80–100°C revealed by radioisotopic tracing experiments with ^{14}C2-labelled acetate (using deep-sea hydrothermal samples from the East Pacific Rise; Bonch-Osmolovskaya *et al.*, 2011). ^{14}C-2-labelled acetate was also actively oxidized to CO_2 at 85°C in terrestrial hot springs in Kamchatka (Bonch-Osmolovskaya and Pimenov, unpublished). Strain CM belonging to a recently described species *C. rimae* (mentioned earlier; Miroshnichenko *et al.*, 2009) was shown to oxidize a number of partly oxidized fermentation products – volatile fatty acids and alcohols – at 75–85°C, using elemental sulfur or thiosulfate as electron acceptors. These bacteria might be responsible for active acetate degradation in high-temperature hot springs in the Uzon Caldera.

In the absence of electron acceptors, acetate can be utilized by acetoclastic methanogens. For example, the acetogenic thermophile *T. phaeum* (Hattori *et al.*, 2000) oxidized acetate in coculture with a hydrogenotrophic methanogen. The same process is performed by culture 2012, obtained from a terrestrial hot spring at Kamchatka (Merkel, unpublished data; also see the earlier section *Methanol oxidation by thermophilic prokaryotes*). Another acetate-oxidizing methanogenic enrichment from thermal subsurface environments consisted of *Thermoanaerobacter* sp. and *Methanothermobacter* sp. (Shestakova *et al.*, 2011). This fact is highly surprising because thermophilic bacteria of the genus *Thermoanaerobacter* are widespread in diverse thermal environments (Subbotina *et al.*, 2003), and were previously considered obligate fermentative organotrophs. Their environmental role, therefore, consists not only in fermenting carbohydrates or peptides, but also in contributing to terminal organic matter oxidation. Thus, several independent isolations of thermophilic acetate-utilizing, syntrophic methanogenic associations demonstrate their wide distribution in thermal environments.

The anaerobic thermoacidophilic archaeon *Acidilobus saccharovorans* (Prokofeva *et al.*, 2009) possesses all the genes encoding enzymes implicated in the oxidative tricarboxylic acid cycle and the electron transport chain, which would allow it to metabolize non-fermentable substrates (Mardanov *et al.*, 2010). However, isopropanol was the only substrate among those tested that could support growth of *A. saccharovorans* in the presence of elemental sulfur. Even so, this evidence demonstrates that catabolic abilities of hyperthermophilic archaea could be much wider than initially described.

Electrogenic thermophilic microorganisms

Microbial fuel cells, which convert the energy of degraded organic matter into electricity, have been actively studied over the past decade (Lovley, 2006). Electrogenic microorganisms use the anode as electron acceptor for anaerobic respiration. The ability to reduce the anode usually correlates with an organism's metal reduction capacity. Most mesophilic microorganisms capable of reducing the anode belong to the phylum *Proteobacteria.* A thermophilic microbial community inoculated with sediment from a thermophilic anaerobic digestor and cultured under laboratory conditions was used in a microbial fuel cell, operated at 55°C and fed with acetate (Wrighton *et al.*, 2008). The power density attained was 37 mW/m^2. The community contained members of the phylum *Firmicutes,* and was dominated by bacteria of the genus *Thermincola* (Sokolova *et al.*, 2005). Previously, a species of *Thermincola* had been shown to reduce ferric iron by using acetate as the electron donor (Zavarzina *et al.*, 2007).

A similar experiment was performed directly in a natural hydrothermal environment. Microbial fuel cells were set up in sediments of several hot springs of Uzon Caldera (Gavrilov, unpublished data). The electrochemical potential arising between anoxic sediment zones and surface water was considered to reflect the redox gradient created by geochemical oxidation of sediment reductants and microbial activity in sediment. Accordingly, the measured electrochemical characteristics reflected the maximum energy yield attainable for a microbial community utilizing the anode of a fuel cell as the electron acceptor for growth. In all cases, the initial electrochemical voltage ranged from 200 to 300 mV. Within a week of incubation, the current density increased by 50–70 mA/m^2 and the power density by 10–30 mW/m^2. Analysis of 16S rRNA genes of the anode-associated microbial community revealed a predominance of *Firmicutes* and

Thermaceae, and the presence of representatives of uncultured bacterial phyla.

As demonstrated by the above findings, thermophilic microorganisms can be used efficiently in microbial fuel cells to generate electricity via the degradation of organic matter.

Conclusions

The achievements of the last five years show that, even though the 'golden age' of thermophile exploration (from the 1980s to 1990s) is evidently over, many metabolically unusual thermophilic microorganisms have been detected and, if not isolated in pure culture, successfully cultured as sustainable enrichments. Recent findings have overturned the misconception that persisted for over half a century, relating particularly to methanotrophs and ammonia- and nitrite-oxidizing microorganisms, for which the lack of thermophilic species was conventionally explained by the thermolability of membrane structures essential for their metabolism. The effort to culture ammonia-oxidizing thermophiles was boosted by environmental evidence on the presence in hot springs of the archaeal *amoA* gene. As for thermoacidophilic methanotrophs, they seemed to be waiting patiently for their hour of fame, being discovered simultaneously in three different locations by three different groups of researchers.

New metabolic capacities were found in representatives of new taxa of microorganisms, sometimes of a higher taxonomic level (for example, archaeal nitrifiers). However, maybe equally exciting was the discovery of new metabolic processes performed by well-known and, supposedly, thoroughly studied thermophilic microorganisms, such as representatives of the genera *Thermococcus* (hydrogenogenic formate oxidation) and *Thermoanaerobacter* (syntrophic acetate oxidation).

In some cases, searches for new metabolic groups of thermophiles stemmed from interest in their mesophilic analogues (anaerobic methane oxidation and anammox bacteria). But the reverse is also possible: discoveries made in the thermophilic world might influence research focused on other types of environment. The discovery of thermophilic methanotrophic veruccomicrobia is bound to stimulate interest in the metabolism of abundant and poorly characterized mesophilic veruccomicrobia. The demonstration of *Thermococcales* growth by means of hydrogenogenic formate oxidation is expected to inspire the search for this metabolic type in other groups of organisms. The new findings discussed in this chapter undoubtedly expand our knowledge of the abilities of living microorganisms and of their environmental function in both modern and ancient ecosystems.

Future trends

Analyses of recent achievements in the investigation of the metabolic diversity of thermophiles indicate several major, developing trends for the near future. The combination of molecular and biogeochemical tools applied to natural communities will be used for the detection of novel metabolic groups of thermophilic prokaryotes. Those already detected but not yet cultured (e.g. thermophilic nitrite oxidizers, thermophilic annamox bacteria, thermophilic anaerobic methane oxidizers) will become the subject of intensive cultivation attempts. Another trend that, probably, will become significant in future investigations is a complete characterization of novel taxa without their isolation in pure culture. As with thermophilic ammonia-oxidizing archaea, new organisms will be described as 'Candidatus' on the basis of sustainable enrichment cultures containing only one representative of the group under study. This approach would make it possible to describe microorganisms that are difficult to cultivate. Another important trend in metabolic diversity investigations will be the analysis of complete genomes of thermophilic prokaryotes, and, as a result, the revelation of novel metabolic capacities even in representatives of well-studied microbial taxa.

Web resources

http://alrlab.pdx.edu/mdtables/resources.html
http://serc.carleton.edu/microbelife/extreme/extremeheat/

References

Anitori, R.P., Trott, C., Saul, D.J., Bergquist, P.L., and Walter, M.R. (2002). A culture-independent survey of the bacterial community in a radon hot spring. Astrobiology *2*, 255–270.

Bak, F., and Cypionka, H. (1987) A novel type of energy metabolism involving fermentation of inorganic sulphur compounds. Nature *326*, 891–892.

Bannikova, G.E., Lopatin, S.A., Varlamov, V.P., Kuznetsov, B.B., Kozina, I.V., Miroshnichenko, M.L., Chernyh, N.A., Tourova, T.P., and Bonch-Osmolovskaya, E.A. (2008). Thermophilic bacteria hydrolyzing agar: characterization of thermostable agarase. Appl. Biochem. Microbiol. *44*, 366–371.

Barns, S.M., Fundyga, R.E., Jeffries, M.W., and Pace, N.R. (1994). Remarkable archaeal diversity detected in a Yellowstone National Park hot spring environment. Proc. Natl. Acad. Sci. U.S.A. *91*, 1609–1613.

Bauer, M.W., Driskill, L.E., Callen, W., Snead, M.A., Mathur, E.J., and Kelly, R.M. (1999). An endoglucanase, EglA, from the hyperthermophilic achaeon *Pyrococcus furiosus* hydrolyzes β-1,4 bonds in mixed-linkage (1->3), (1->4)-β-d-glucans and cellulose. J. Bacteriol. *181*, 284–290.

Blöchl, E., Rachel, R., Burggraf, S., Haefenbradl, D., Jannasch, H.W., and Stetter, K.O. (1997). *Pyrolobus fumarii*, gen. and sp. nov., represents a novel group of archaea, extending the upper temperature limit for life to 113°C. Extremophiles *1*, 14–21.

Blumer-Schuette, S.E., Kataeva, I., Westpheling, J., Adams, M.W.W., Kelly, R.M. (2008). Extremely thermophilic microorganisms for biomass conversion. Curr. Opin. Chem. Biol., *19*, 210–217.

Bodrossy, L., Holmes, E.M., Holmes, A.J., Kovacs, K.L., and Murrell, J.C. (1997). Analysis of 16S rRNA and methane monooxygenase gene sequences reveals a novel group of thermotolerant and thermophilic methanotrophs, *Methylocaldum* gen.nov. Arch. Microbiol. *168*, 493–503.

Boetius, A., Ravenschlag, K., Schubert, C.J., Rickert, D., Widdel, F., and Pfannkuche, O. (2000). A marine microbial consortium apparently mediating anaerobic oxidation of methane. Nature *407*, 623–626.

Bonch-Osmolovskaya, E.A., and Ilarionov, S.A. (1982). Methane formation from pectine by a combined culture of *Clostridium pectinofermentans* and *Methanosarcina vacuolata*. Microbiology (Engl. Transl. of Mikrobiologiya) *51*, 413–415.

Bonch-Osmolovskaya, E.A., and Karpov, G.A. (1987). Microbial methane formation in Uzon Caldera hydrotherms. Mikrobiologiya *56*, 516–517. (In Russian).

Bonch-Osmolovskaya, E.A., Sokolova, T.G., Kostrikina, N.A., and Zavarzin, G.A. (1990). *Desulfurella acetivorans* gen. nov. and sp. nov.—a new thermophilic sulfur-reducing eubacterium. Arch. Microbiol. *153*, 151–155.

Bonch-Osmolovskaya, E.A., Perevalova, A.A., Kolganova, T.V., Rusanov, I.I., Jeanthon C., and Pimenov, N.V. (2011). Activity and distribution of thermophilic prokaryotes in hydrothermal fluid, sulfidic structures and sheaths of alvinnelides (East Pacific Rise, 13°N). Appl. Environ. Microbiol. *77*, 2803–2806.

Brock, T.D. (1978). Thermophilic microorganisms and life at high temperatures. (Springer-Verlag, NY).

Brock, T.D., Brock, K.M., Belly R.T., and Weiss, R.L. (1972). *Sulfolobus*: A new genus of sulfur-oxidizing bacteria living at low pH and high temperature. Archiv fur Mikrobiologie *84*, 54–68.

Byrne, N., Strous, M., Crépeau, V., Kartal B., Birrien, J.-L., Schmid, M., Lessohgeur, F., Schouten, S., Jaeschke, A., Jetten, M., Prieur, D., and Godfroy, A. (2009). Presence and activity of anaerobic ammonium oxidizing bacteria at deep-sea hydrothermal vents. ISME J. *3*, 117–123.

Cheng, L., Qiu T.L., Yin X.B., Wu X.L., HuG.Q., Deng Y., and Zhang, H. (2007). *Methermicoccus shengliensis* gen. nov., sp. nov., a thermophilic, methylotrophic methanogen isolated from oil-production water, and proposal of *Methermicoccaceae* fam. nov. Int. J. Syst. Evol. Microbiol. *57*, 2964–2969.

Damste, J.S.S., Rijsta, W.I., Geenevason, J.A.J., Strous, M., and Jetten, M.S.M. (2005). Structural identification of ladderane and other membrane lipids of planctomycetes capable of anaerobic ammonium oxidation. FEBS J. *272*, 4270–4283.

Dedysh, S.N., Liesack W., Khmelenina, V.N., Suzina, N.E., Trotsenko, Y.A., Semrau, J.D., Bares, A.M., Panikov, N.S., and Tiedje, J.M. (2000). *Methylocella palustris* gen. nov., sp. nov., a new methane-oxidizing acidophilic bacterium from peat bogs, representing a novel subtype of serine-pathway methanotrophs. Int. J. Syst. Evol. Microbiol. *50*, 955 – 969.

Dedysh, S.N., Khmelenina, V.N., Suzina, N.E., Trotsenko, Y.A., Semrau, J.D. Liesack, W., and Tiedje, J.M. (2002). *Methylocapsa acidiphila* gen. nov., sp. nov., a novel methane-oxidizing and dinitrogen-fixing acidophilic bacterium from *Sphagnum* bog. Int. J. Syst. Evol. Microbiol. *52*, 251–261.

DeLong, E.F. (1992). Archaea in coastal marine environments. Proc. Natl. Acad. Sci. U.S.A. *89*, 5685–5689.

Dolfing, J., Jiang, B., Henstra, A.M., Stams, A.J.M., and Plugge, C.M. (2008). Syntrophic growth on formate: a new microbial niche in anoxic environments. Appl. Environ. Microbiol. *74*, 6126–613.

Dunfield, P.F., Yuryev, A., Senin, P., Smirnova, A.V., Stot, M.B., Hou, S., Ly, B., Saw, J.H., Zhou, Z., Ren, Y., *et al.* (2007). Methane oxidation by an extremely acidophilic bacterium of the phylum Verrucomicrobia. Nature *450*, 879–882.

Elkins, J.G., Podar, M., Graham, D.E., Makarova, K.S., Wolf, Y., Randau, L., Hedlund, B.P., Brochier-Armanet, C., Kunin, V., Anderson, I., *et al.* (2008). A korarchaeal genome reveals insights into the evolution of the Archaea. Proc. Natl. Acad. Sci. U.S.A. *105*, 8105–8107.

Elshahed, M.S., Youssef, N.H., Luo, Q., Najar, F.Z., Roe, B.A., Sisk, T.M., Buering, S.J., Hinrichs, K.-U., and Krumholz, L.R. (2007). Phylogenetic and metabolic diversity of *Planctomycetes* from anaerobic, sulfide- and sulfur-rich Zodletone Spring, Oklahoma. Appl. Environ. Microbiol. *73*, 4707–4716.

Etiope G., and Klusman, R.W. (2002). Geologic emissions of methane to the atmosphere. Chemosphere *49*, 777–789.

Giggenbach, W. (1995). Variations in the chemical and isotopic composition of fluids discharged from the Taupo Volcanic Zone, New Zealand. J. Volcanol. Geotherm. Res. *68*, 89–116.

Golovacheva, R.S. (1976). Thermophilic nitrifying bacteria from hot springs. Mikrobiologiya *45*, 377–379 [In Russian].

Hatzenpichler, R., Lebedeva, E.V. Spieck, E., Stoecker, K., Richter, A., Daims, H., and Wagner, M. (2008). A moderately thermophilic ammonia-oxidizing crenarchaeote from a hot spring. Proc. Nat. Aca. Sci. U.S.A. *105*, 2134–2139.

Hattori, S., Kamagata,Y., Hanada, S., and Shoun, H. (2000). *Thermacetogenium phaeum* gen. nov., sp. nov., a strictly anaerobic, thermophilic, syntrophic acetate-oxidizing bacterium. Int. J. Syst. Evol. Microbiol. *50*, 1601–1609.

Henstra, A.M., Dijkema, C., and Stams, A.J.M. (2007). *Archaeoglobus fulgidus* couples CO oxidation to sulfate reduction and acetogenesis with formate accumulation. Env. Microbiol. *9*, 1836–1841.

Hirayama, H., Takai, K., Inagaki, F., Yamato, Y., Suzuki, M., Nealson, K.H., and Horikoshi, K. (2005). Bacterial community shift along a subsurface geothermal water stream in a Japanese gold mine. Extremophiles *9*, 169–184.

Huber, R., Huber, H., and Stetter, K.O. (2000). Towards the ecology of hyperthermophiles: biotopes, new isolation strategies and novel metabolic properties. FEMS Microbiol. Rev. *24*, 615–623.

Hugenholz, P., Pitulle, C., Hershberger, K.L., and Pace, N.R. (1998). Novel division level bacterial diversity in a Yellowstone hot spring. J. Bacteriol. *180*, 366–376.

Ilarionov, S.A., and Bonch-Osmolovskaya, E.A. (1986). Methane formation from methanol by associations of microorganisms. Microbiology (Engl. Transl. of Mikrobiologiya) *55*, 224–229.

Islam, T., Jensen, S., Reigstad, L.J., Larsen, O., and Birkeland, N.K. (2008). Methane oxidation at 55°C and pH 2 by a thermoacidophilic bacterium belonging to the *Verrucomicrobia* phylum. Proc. Natl. Acad. Sci. U.S.A. *105*, 300–304.

Jackson, B.E., and McInerney, M. (2000) Thiosulfate disproportionation by *Desulfotomaculum thermobenzoicum*. Appl. Environ. Microbiol. *66*, 3650–3653.

Jaenicke, R., and Bohm, G. (1998). The stability of proteins in extreme environments. Curr. Opin. Struct. Biol. *8*, 738–748.

Jaeschke, A., Op den Camp, H.J.M., Harhangi, H. Klimuk, A., Hopmans, E.C., Jetter, M.S.M., Schouten, S., and Damste, J.S.S. (2009). 16S rRNA gene and lipid biomarker evidence for anaerobic ammonium-oxidizing bacteria (anammox) in California and Nevada hot springs. FEMS Microbiol. Ecol. *67*, 343–350.

Jannasch, H.W., and Mottle, M.J. (1985). Geomicrobiology of deep-sea hydrothermal vents. Science *229*, 717–725.

Jiang, B., Parshina, S.N., Van Doesburg, W., Lomas, B.P., and Stams, A.J.M. (2005). *Methanomethylovorans thermophila* sp. nov., a thermophilic, methylotrophic methanogen from an anaerobic reactor fed with methanol. Int. J. Syst. Evol. Microbiol. *55*, 2465–2470.

Jolivet, E., L'Haridon, S., Corre, E., Forterre, P., and Prieur, D. (2003). *Thermococcus gammatolerans* sp. nov., a hyperthermophilic archaeon from a deep-sea hydrothermal vent that resists ionizing radiation. Int. J. Syst. Evol. Microbiol. *53*, 847–851.

Kallmeyer, J., and Boetius, A,. (2004). Effects of temperature and pressure on sulfate reduction and anaerobic oxidation of methane on hydrothermal sediments of Guaymas Basin. Appl. Environ. Microbiol. *70*, 1231–1233.

Kanokratana, P., Chanapan, S., Pootanakit, K., and Eurwilaichitr, L. (2004). Diversity and abundance of bacteria and archaea in the Bor Khlueng hot spring in Thailand. J. Basic Microbiol. *44*, 430–444.

Kim, Y.J., Lee, H.S., Kim, E.S., Bae, S.S., Lim, J.K., Matsumi, R., Lebedinsky, A.V., Sokolova, T.G., Kozhevnikova, D.A., Cha, S.-S., *et al.* (2010). Formate driven growth coupled with H_2 production. Nature. *67*, 352–355.

Kochetkova, T.V., Rusanov, I.I., Pimenov, N.V., Kolganova, N.V., Lebedinsky, A.V., Bonch-Osmolovskaya, E.A., and Sokolova T.G. (2011). Anaerobic transformation of carbon monoxide by microbial communities of Kamchatka hot springs. Extemophiles *15*, 319–325.

Könneke, M., Bernhard, A.E., de la Torre, J.R., Wolker, C.B., Waterbury, J.B., and Stahl, D.A. (2005). Isolation of an autotrophic ammonia-oxidizing marine archaeon. Nature *437*, 543–546.

Kozina, I.V., Kublanov, I.V., Kolganova, T.V., Chernyh, N.A., and Bonch-Osmolovskaya, E.A. (2010). *Caldanaerobacter uzonensis* sp. nov., an anaerobic, thermophilic, heterotrophic bacterium isolated from a hot spring. Int. J. Syst. Evol. Microbiol. *60*, 1372–1375.

Kuenen, J.G. (2008). Anammox bacteria: from discovery to application. Nat. Rev. Microbiol. *6*, 320–326.

Kumar, S., and Nussinov, R. (2001). How do thermophilic proteins deal with heat? Cell. Mol. Life Sci., *58*, 1216–1233.

Lebedeva, E.V., Alawi, M., Fiencke, C., Namsaraev, B., Bock, E., and Spieck, E. (2005). Moderately thermophilic nitrifying bacteria from a hot spring of the Baikal rift zone. FEMS Microbiol. Ecol. *54*, 297–306.

Lee, H.S., Kang, S.G., Bae, S.S., Lim, J.K., Cho, Y., Kim, Y.J., Jeon, J.H., Cha, S.S., Kwon, K.K., Kim, H.T., *et al.* (2008). The complete genome sequence of *Thermococcus onnurineus* NA1 reveals a mixed heterotrophic and carboxydotrophic metabolism. J. Bacteriol. *190*, 7491–7499.

Limauro, D., Cannio, R., Fiorentino, G., Rossi, M., and Bartolucci, S. (2001). Identification and molecular characterization of an endoglucanase gene, *celS*, from the extremely thermophilic archaeon *Sulfolobus solfataricus*. Extremophiles *5*, 213–219.

López-Garcia, P., Duperron, S., Philippot, P., Foriel, J., Susini, J., and Moreira, D. (2003). Bacterial diversity in hydrothermal sediment and epsilonproteobacterial dominance in experimental microcolonizers at the Mid-Atlantic Ridge. Environ. Microbiol. *5*, 961–976.

Lovley, D.R. (2006). Bug juice: harvesting electricity with microorganisms. Nat. Rev. Microbiol. *4*, 497–508.

Mardanov, A.V., Ravin, N.V., Svetlitchnyi, V.A., Beletsky, A.V., Miroshnichenko, M.L., Bonch-Osmolovskaya, E.A., and Skryabin, K.G. (2009). Metabolic versatility and indigenous origin of the archaeon *Thermococcus sibiricus*, isolated from a Siberian oil reservoir, as revealed by genome analysis. Appl. Environ. Microbiol. 75, 4580–4588.

Mardanov A.V., Ravin, N.V., Svetlichny, V.A., Beletsky, A.V., Prokofeva, M.I., Bonch-Osmolovskaya, E.A., and Skryabin, K.G. (2010). The genome sequence of crenarchaeon *Acidilobus saccharovorans* supports a new order *Acidilobales* and suggests its important ecological role in terrestrial acidic hot springs. Appl. Environ. Microbiol. *76*, 5652–5657.

Martin, W., Baross J., Kelley D., and Russell, M.J. (2008). Hydrothermal vents and the origin of life. Nature Revs. Microbiol. *6*, 805–814.

Miroshnichenko, M.L., Rainey, F.A., Rhode, M., and Bonch-Osmolovskaya, E.A. (1998). *Hippea maritima* gen. nov., sp. nov., a new genus of thermophilic, sulfur-reducing bacterium from submarine hot vents. Int. J. Syst. Bacteriol. *49*, 1033–1038.

Miroshnichenko, M.L., Hippe, H., Stackebrandt, E., Kostrikina, N.A., Chernyh, N.A., Jeanthon, C., Nazina, T.N., Belyaev, S.S., and Bonch-Osmolovskaya, E.A. (2001). Isolation and characterization of *Thermococcus sibiricus* sp. nov. from a Western Siberia high-temperature oil reservoir. Extremophiles *5*, 85–91.

Miroshnichenko, M.L., Lebedinsky, A.V., Chernyh, N.A., Tourova, T.P., Kolganova, T.V., Spring, S., and Bonch-Osmolovskaya, E.A. (2009). *Caldimicrobium rimae* gen. nov., sp. nov., an extremely thermophilic, facultatively lithoautotrophic, anaerobic bacterium from the Uzon Caldera, Kamchatka. Int. J. Syst. Evol. Microbiol. *59*, 1040–1044.

Mori, K., Sunamura, M., Yanagawa, K., Ishibashi, J., Miyoshi, Y., Iino, T., Suzuki, K., and Urate, T. (2008). First cultivation and ecological investigation of a bacterium affiliated with the candidate phylum OP5 from hot springs. Appl. Environ. Microbiol. *74*, 6223–6229.

Nunoura, T., Hirayama, H., Takai, H., Oida, H., Nishi, S., Shimanuora, S., Suzuki, Y., Inagaki, Y., Takai, K., Nealson, K., and Horikoshi, K. (2003). Genetic and functional properties of uncultivated thermophilic crenarchaeotes from a subsurface gold mine as revealed by analysis of genome fragments. Environ. Microbiol. *7*, 1967–1984.

Orphan, V. J. C., House, H., Hinrichs, K.-U., McKeegan, K.D., and DeLong E.F. (2001). Methane-consuming archaea revealed by directly coupled isotopic and phylogenetic analysis. Science *293*, 484–487.

Pearson, A., Huang, Z., Inglass, A.E., Romanek, C.S., Wiegel, J., Freeman, K.H., Smittenberg, R.H., and Zhang, C.L. (2004). Nonmarine crenarchaeol in Nevada hot springs. Appl. Environ. Microbiol. *70*, 5229–5237.

Pearson A., Pi V., Zhao, W., Li, W., Li, Y.-L., Inskeep, W., Perevalova, A.A., Romanek, C.S., Li, S., and Zhang, C.L. (2008). Factors controlling the distribution of archaeal lipids in terrestrial hot springs. Appl. Environ. Microbiol. *74*, 3523–3532.

Perevalova, A.A., Svetlichny, V.A., Kublanov, I.V., Chernyh, N.A., Kostrikina, N.A., N.A. Tourova, T.P., Kuznetsov, B.B., and Bonch-Osmolovskaya, E.A. (2005). *Desulfurococcus fermentans* sp. nov., a novel hyperthermophilic archaeon from a Kamchatka hot spring, and emended description of the genus *Desulfurococcus*. Int. J. Syst. Evol. Microbiol. *55*, 995–999.

Pimenov, N.V., and Bonch-Osmolovskaya, E.A. (2006). *In situ* activities in thermal environments. In: Methods in Microbiology, Vol. 14, Rainey, F., and Oren, A., eds. (Elsevier, Amsterdam), pp. 29–53.

Pol, A., Heijmans K., Harhangi, H.R., Tedesco, D., Jetten, M.S.M., and Op den Camp, H.J.M. (2007). Methanotrophy below pH 1 by a new *Verrucomicrobia* species. Nature *450*, 870–874.

Prokofeva, M.I., Kostrikina, N.A., Kolganova, T.V., Tourova, T.P., Lysenko, A.M., Lebedinsky, A.V., and Bonch-Osmolovskaya, E.A. (2009). Isolation of the anaerobic thermoacidophilic crenarchaeote *Acidilobus saccharovorans* sp. nov. and proposal of *Acidilobales* ord. nov., including *Acidilobaceae* fam. nov. and *Caldisphaeraceae* fam. nov. Int. J. Syst. Evol. Microbiol. *59*, 3116–3122.

Rattray, J.E., Vossenberg, J., Jaeschke, A., Hopmans, E.C., Wakeham, S.G., Lavik, G., Kuypers, M.M.M., Strous, M., Jettem, M.S.M., Schouten, S., *et al.* (2010). Impact of temperature on ladderane lipid distribution in anammox bacteria. Appl. Environ. Microbiol. *76*, 1596–1603.

Reigstad, L.J., Richter, A., Daims, H., Ulrich, T., Schwark, L., and Schleper, C. (2008). Nitrification in terrestrial hot springs of Iceland and Kamchatka. FEMS Microbiol. Ecol. *64*, 167–164.

Reysenbach, A.-L., and Shock, E. (2002). Merging genomes with geochemistry of hydrothermal ecosystems. Science *296*, 1077–1082.

Reysenbach, A.-L., Ehringer, M., and Hershberger, K. (2000). Microbial diversity at 83 degrees C in Calcite springs, Yellowstone National Park: another environment where the Aquificales and 'Korarchaeota' coexist. Extremophiles *4*, 61–67.

Reysenbach, A.-L., Liu, Y., Banta, A.B., Beveridge, T.J., Kirshtein, J.D., Schouten, S., Tivey, M.K., and Von Damm, K.L. (2006). A ubiquitous thermoacidophilic archaeon from deep-sea hydrothermal vents. Nature *442*, 444–447.

Saiki, T., Kobayashi, Y., Kawagoe, K., and Beppu, T. (1985). *Dictyoglomus thermophilum* gen. nov., sp. nov., a chemoorganotrophic, anaerobic, thermophilic bacterium. Int. J. Syst. Bacteriol. *35*, 253–259.

Segerer, A., Neuner, A., Kristiansson, J.K., and Stetter, K.O. (1986). *Acidianus infernus* gen. nov., sp. nov., and *Acidianus brierleyi* comb. nov.: facultatively aerobic, extremely acidophilic thermophilic sulfur-metabolizing archaebacteria. Int. J. Syst. Bacteriol. *36*, 559–564.

Semrau, J.D., Dispirito, A.A., and Murrell, J.C. (2008). Life in the extreme: thermoacidophilic methanotrophy. Trends Microbiol. *16*, 190–193.

Schleper, C., Jurgens, G., and Januscheit, M. (2005). Genomic studies of uncultivated archaea. Nat. Rev. Microbiol. *3*, 479–488.

Schönheit, P., and Schäfer, T. (1995). Metabolism of hyperthermophiles. World J. Microbiol. Biotechnol. *11*, 26–57.

Schouten, S., Wakeharm, S.G., Hopmans, E.S., and Damste, J.S.S. (2003). Biogeochemical evidence that thermophilic archaea mediate the anaerobic oxidation of methane. Appl. Environ. Microbiol. *69*, 1680–1686.

Schramm, A., DeBeer, D., Wagner, M., and Amann, R. (1998). *Nitrosospira* sp. and *Nitrospira* sp. as dominant populations in a nitrifying fluidized bed reactor: identification and activity *in situ*. Appl. Environ. Microbiol. *64*, 3480–3485.

Shestakova, N.M., Ivoilov, V.S., Tourova, T.P., Belyaev, S.S., Poltaraus, A.B., and Nazina, T.N. (2011). Syntrophic acetate degradation to methane in a high-temperature petroleum reservoir: culture-based and 16S rRNA genes characterization. In: Applied Microbiology and Molecular Biology in Oil Field Systems, Part 2. Whitby, C., and Skovhus, T.L., eds. (Springer): 45–53.

Shock, E.L., Holland, M., Meyer-Dombard, D.R., and Amend, J.P. (2005). Geochemical sources of energy for microbial metabolism in hydrothermal ecosystems: Obsidian Pool, Yellowstone National Park. In: Geothermal Biology and Geochemistry in Yellowstone National Park. Inskeep, W.P, and McDermott, T.R., eds. (Thermal Biology Institute, Bozeman, Montana), pp. 95–110.

Sokolova, T.G., Jeanthon, C., Kostrikina, N.A., Chernyh, N.A., Lebedinsky, A.V., Stackebrandt, E., and Bonch-Osmolovskaya, E.A. (2004). The first evidence of anaerobic CO oxidation coupled with H_2 production by a hyperthermophilic archaeon isolated from a deep-sea hydrothermal vent. Extremophiles *8*, 317–332.

Sokolova, T.G., Kostrikina, N.A., Chernyh, N.A., Tourova, T.P., and Bonch-Osmolovskaya, E.A. (2005). *Thermincola carboxydophila* gen.nov., sp.nov., a new anaerobic carboxydotrophic hydrogenogenic bacterium from a hot spring of Lake Baikal area. Int. J. Syst. Environ. Microbiol., *55*, 2069–2073.

Sokolova, T.G., Henstra, A.-M., Sipma, J., Parshina, S.N., Stams, A.J.M., and Lebedinsky, A.V. (2009). Diversity and ecophysiological features of thermophilic carboxydotrophic anaerobes. FEMS Microbiol. Ecol. *68*, 131–141.

Stams, A.J.M., and Plugge, C.M. (2009). Electron transfer in syntrophic communities of anaerobic bacteria and archaea. Nat. Rev. Microbiol. 7, 568–577.

Stetter, K.O. (1988). *Archaeoglobus fulgidus* gen. nov., sp. nov., a new taxon of extremely thermophilic *Archaebacteria*. Syst. Appl. Microbiol. 10, 172–173.

Stetter, K.O. (1996). Hyperthermophilic procaryotes. FEMS Microbiol. Rev. *18*, 149–158.

Stetter, K.O., König, H., and Stackebrandt, E. (1983). *Pyrodictium* gen. nov., a new genus of submarine disc-shaped sulphur reducing archaebacteria growing optimally at 105°C. Syst. Appl. Microbiol. *4*, 535–551.

Strous, M., and Jetten, M.S.M. (2004). The anaerobic oxidation of methane and ammonium. Ann. Rev. Microbiol. *58*, 99–117.

Subbotina, I.V., Chernyh, N.A., Sokolova, T.G., Kublanov, I.V., Bonch-Osmolovskaya, E.A., and Lebedinsky, A.V. (2003). Oligonucleotide probes for the detection of representatives of the genus *Thermoanaerobacter*. Microbiology (Ehgl. Transl. of Mikrobiologiya) 72, 374–382.

Svetlichny, V.A., and Svetlichnaya, T.P. (1988). *Dictyoglomus turgidus* sp. nov., a new extremely thermophilic eubacterium isolated from hot springs of the Uzon volcano caldera. Mikrobiologiya *57*, 435–441 [In Russian].

Svetlichny, V.A., Sokolova, T.G., Gerhardt, M., Ringpfei, M., Kostrikina, N.A., and Zavarzin, G.A. (1991). *Carboxydothermus hydrogenoformans* gen. nov., sp. nov., a CO-utilizing thermophilic anaerobic bacterium from hydrothermal environments of Kunashir Island. Syst. Appl. Microbiol. *14*, 254–260.

Takai, K., and Horikoshi, K. (1999). Genetic diversity of Archaea in deep-sea hydrothermal vent environments. Genetics *152*, 1285–1297.

Takai, K., Komatsu, T., Inagaki, F., and Horikoshi, K. (2001). Distribution of Archaea in a black smoker chimney structure. Appl. Environ. Microbiol. *67*, 3618–3629.

Takai, K., Nakamura, K., Miyazaki, J., Hirayama, H., Nakagawa, S., Nuonura, T., Horikoshi, K. (2008). Cell proliferation at 122°C and isotopically heavy CH_4 production by a hyperthermophilic methanogen under high-pressure cultivation. Proc. Natl. Acad. Sci. U.S.A. *105*, 10949–10954

Teske, A., Hinrichs, K.-U., Edgecomb, V., de Vera Gomes, A., Kyscha, D., Sylva, S.P., Sogin, M.L., and Jannasch, H.W. (2002). Microbial diversity of hydrothermal sediment in Guaymas Basin: evidence for anaerobic methanotrophic communities. Appl. Environ. Microbiol. *68*, 1994–2007.

Thamdrup, B., Finster, K., Hansen, J. W., and Bak, F. (1993). Bacterial disproportionation of elemental sulfur coupled to chemical reduction of iron or manganese. Appl. Environ. Microbiol. *59*, 101–108.

Thauer, R.K., Jungermann, K., and Decker, K. (1977). Energy conservation in chemotrophic anaerobic bacteria. Bacteriol. Rev. *41*, 100–180.

de la Torre, J.R., Walker, C.B., Ingalls, A.E., Könneke. M., and Stahl, D.A. (2008). Cultivation of ammonia oxidizing archaeon synthesizing crenarchaeol. Environ. Microbiol. *10*, 810–818.

Treusch, A., Leninger, S., Schuster, S., Kletzin, A., and Klenk, H.-P. (2005). Novel genes for nitrite reductase and Amo-related proteins indicate a role of uncultured mesophilic crenarchaeota in nitrogen cycling. Environ. Microbiol. 7, 1985–1995.

Tsubota, J., Eshinimaev, B.T., Khmelenina, V.N., and Trotsenko, Y.A. (2005). *Methylothermus thermalis* gen. nov., sp. nov., a novel moderately thermophilic obligate methanotroph from a hot spring in Japan. Int. J. Syst. Evol. Microbiol. *55*, 1877–1884.

Veille, C., and Zeikus, G.J. (2001). Hyperthermophilic enzymes: sources, uses and molecular mechanisms for thermostability. Microbiol. Mol. Biol. Rev. *65*, 1–43.

Windman, T., Zolotova, N., Schwandner, F., and Shock, E.L. (2007). Formate as an energy source for microbial metabolism in chemosynthetic zones of hydrothermal ecosystems. Astrobiology 7, 873–890

Wrighton, K.C., Agbo, P., Warnecke, F., Weber, K.A., Brodie, E.L., DeSantis, T.Z., Hugenholz, P., Andersen, G.L., and Goates, J.D. (2008). A novel ecological role of the *Frimicutes* identified in the thermophilic microbial fuel cells. ISME J. 2, 1146–1156.

Zavarzina, D.G., Sokolova, T.G., Tourova, T.P., Chernyh, N.A., Kostrikina, N.A., and Bonch-Osmolovskaya, E.A. (2006). *Thermincola ferriacetica* sp. nov., a new anaerobic, thermophilic, facultatively chemolithoautotrophic bacterium capable of dissimilatory Fe(III) reduction. Extremophiles *11*, 1–7.

Zhang, C.L., Pearson, A., Li, Y.-L., Mills, G., and Wiegel J. (2006). Thermophilic temperature optimum for crenarchaeol synthesis and implication for archaeal evolution. Appl. Environ. Microbiol. 72, 4419–4422.

Zhang, C.L., Ye, Q., Huang, Z., Li, W.J., Chen, J., Song, Z., Zhao, W., Bagwell, C., Inskeep, W.P., Ross, C., *et al.* (2008). Global occurrence of archaeal amoA genes in terrestrial hot springs. Appl. Environ. Microbiol. 74, 6417–6226.

Zillig, W., Stetter, K.O., Schaefer, W., Janekovic, D., Wunderl, S., Holz, I., and Palm, P. (1981). *Thermoproteales:* a novel type of extremely thermoacidophilic anaerobic archaebacteria isolated from Icelandic solfataras. Zentralbl. Mikrobiol. Parasitenkd. Infektionskr. Hyg. Abt. 1 Orig. C2, 205–227.

Zillig, W., Stetter, K.O., Prangishvili, D., Schaefer, W., Wunderl, S., Janekovich, D., Holz, I., and Palm, P. (1982). *Desulfurococcaceae,* the second family of the extremely thermophilic, anaerobic, sulfur-respiring *Thermoproteales.* Zentralbl. Bakteriol. Parasitenkd. Infektionskr. Hyg. Abt. 1 Orig. C3, 304–317.

Zillig, W., Holz, I., Janekovic D., Schäfer, W., and Reiter, W.D. (1983a). The archaebacterium *Thermococcus celer* represents a novel genus within the thermophilic branch of the archaebacteria. Syst. Appl. Microbiol. *4*, 88–94.

Zillig, W., Gierl, A., Schreiber, G., Wunderl, S., Janekovic, D., Stetter, K.O., and Klenk, H.P. (1983b). The archaebacterium *Thermofilum pendens* represents a novel genus of the thermophilic, anaerobic sulfur-respiring *Thermoproteales.* Syst. Appl. Microbiol. *4*, 79–87.

Zinder, S.H., Sowers, K.R., and Ferry, J.G. (1985). *Methanosarcina thermophila* sp. nov., a thermophilic, acetotrophic, methane-producing bacterium. Int. J. Syst. Bacteriol. 35, 522–523.

Cellulolytic Microorganisms from Thermal Environments

7

Tatiana A. Vishnivetskaya, Babu Raman, Tommy J. Phelps, Mircea Podar and James G. Elkins

Abstract

Conversion of lignocellulosic biomass to liquid fuels using biological processes offers a potential solution to partially offset the world's dependence on fossil fuels for energy. In nature, decomposition of organic plant biomass is brought about by the combined action of several interacting microorganisms existing in complex communities. Bioprospecting in natural environments with high cellulolytic activity (for example, thermal springs) may yield novel cellulolytic microorganisms and enzymes with elevated rates of biomass hydrolysis for use in industrial biofuel production. In this chapter, various cellulose-degrading microorganisms (in particular, thermophilic anaerobic bacteria), their hydrolytic enzymes, and recent developments in the application of biomass fermentations for production of sustainable bioenergy are reviewed. In this context, results from ongoing research at the Oak Ridge National Laboratory in the isolation and subsequent phylogenetic and metabolic characterization of thermophilic, anaerobic, cellulolytic bacteria from the hot springs of Yellowstone National Park are presented.

Introduction

Each year photosynthetic fixation of CO_2 yields over 4 billion tons of plant biomass (Detroy and Stjulian, 1983; Beguin and Aubert, 1994; Leschine, 1995). Depending on plant species and cell type, plant cell walls typically consist of between 35% to 50% cellulose, 20% to 35% hemicellulose, and 10% to 25% lignin on a dry weight basis (Saha, 2004; Somerville, 2006). In fact, cellulose is the most abundant organic biopolymer in terrestrial environments. Crops and other sources of biomass can be used for the production of fuel and energy, as they represent stored solar energy that is converted from sunlight during the fixation of carbon dioxide into organic compounds such as polysaccharides. Therefore, plant biomass is a substantial natural resource for renewable energy. There are several different routes for harvesting the energy stored in plant biomass:

1. Plant material (e.g. wood) can be burned directly to produce energy, adequate for domestic heating and cooking.
2. Organic material, including cellulosic biomass, can be converted to synthesis gas (a mixture, primarily of carbon monoxide and hydrogen) through the process of thermal gasification, which can then be converted to biofuels either chemically by Fischer-Tropsch synthesis or biologically using microbial processes (Inderwildi *et al.*, 2008; Valdez-Vazquez and Poggi-Varaldo, 2009).
3. Plant material in the form of animal manure can be decomposed to the flammable gas methane through methanogenesis (das Neves *et al.*, 2009).
4. Plant biomass can be enzymatically hydrolysed and the resulting sugars fermented to produce liquid ethanol, which can be used as a transportation fuel either directly or in combination with gasoline (Lynd *et al.*, 2002); and lastly
5. Plant material can be fermented by obligate or facultative anaerobes to higher alcohols

such as butanol and propanol, or converted to other chemicals such as acetone, diacetyl, etc.

Butanol could also serve as a direct replacement for gasoline due to its favourable physical properties (Garncarek and Kociolek-Balawejder, 2009).

In light of the growing concern regarding the long-term supply of fossil fuels, applied research aimed at unlocking the energy stored in plant biomass for anthropogenic use has increased (Lynd *et al.*, 2002; Saha, 2004). However, the recalcitrance of lignocellulosic materials to enzymatic hydrolysis still poses the largest technological hurdle in developing a photosynthetic-based bioenergy economy. Plant biomass is made in part from cellulose, a homopolymer consisting of a linear chain of linked d-glucose units. As a natural substrate, cellulose is insoluble in water (Chang *et al.*, 2006). Cellulose is associated with other plant polymers such as hemicelluloses, lignin, pectin, and proteins (Somerville, 2006). This interwoven, complex association hinders the ability of hydrolytic enzymes (e.g. cellulases) to act on cellulose fibrils, thereby inhibiting degradation. In order to improve enzyme accessibility and aid hydrolysis, plant biomass is usually pretreated using a combination of heat, pressure, and harsh chemicals to remove hemicellulose and proteins, and partially hydrolyse the crystalline cellulose (Wyman, 1999). Pretreated biomass is then hydrolysed using a cocktail of enzymes (for example, from the fungus *Trichoderma reesei*) and the liberated sugars are anaerobically converted to ethanol using fermentative microorganisms, such as *Escherichia coli*, *Zymomonas mobilis* or yeast. Along with readily fermented hexoses, cellulosic biomass also contains pentose sugars such as l-arabinose and d-xylose, derived from the hemicellulose component of plant cell walls. Industrial production of bioethanol from lignocellulolytic material is therefore dependent on the availability of robust microorganisms that are capable of fermenting both C_6 and C_5 carbohydrates present in the feedstock. For example, some wild-type bacteria (*Z. mobilis*) and yeast (*Saccharomyces cerevisiae*) are able to rapidly and efficiently convert glucose (C_6) substrates into ethanol but cannot ferment xylose (Dien *et al.*, 2003). These and other organisms have been engineered to obtain strains that are capable of fermenting both hexoses and pentoses to produce ethanol (Yanase *et al.*, 2007; Bettiga *et al.*, 2009).

Traditional ethanol production in a simultaneous saccharification and co-fermentation (SSCF) process configuration requires a dedicated step for cellulase production (Lynd *et al.*, 2005). An alternative process under development for the conversion of lignocellulosic biomass is termed consolidated bioprocessing (CBP). It involves the use of either native cellulolytic organisms (e.g. *Clostridium thermocellum*) or recombinant organisms which have been engineered to produce cellulolytic and other hydrolytic enzymes (e.g. yeast). This allows for the integration of several biologically mediated processes, specifically saccharolytic enzyme production, enzymatic hydrolysis of pretreated plant biomass carbohydrates, and fermentation of hexose and pentose sugars, into a single step process. Eliminating or minimizing the need for external addition of expensive cellulases offers significant economical advantages for the CBP process over the conventional SSCF. Cost comparison analysis of different bioprocess configurations has highlighted the potential impact of the CBP process in establishing an economically viable and sustainable ethanol industry. (Lynd *et al.*, 2002, 2005). Thermophilic and extreme thermophilic cellulolytic fermentative organisms potentially offer additional economical advantages in an industrial bioethanol production scheme. For example, fermentation processes operating at elevated incubation temperatures (80°C or higher) may allow for partial distillation of ethanol to occur. Removal of ethanol from the fermentation broth reduces its inhibitory effects on cell growth and metabolism, consequently leading to increased ethanol productivity (Hamilton-Brehm *et al.*, 2010). Natural environments such as thermal hot springs with decaying plant material (discussed in the ensuing section) are a rich resource of thermophilic cellulolytic organisms with potential use in high-temperature processes. For example, bioprospecting in these locations could lead to the discovery of novel microbial species or communities that are capable of both degrading plant polymers and fermenting the suite of sugars present in lignocellulosic biomass to ethanol.

Natural environments as a source of cellulose-degrading microorganisms

Cellulose-degrading microbial communities contribute significantly to the global carbon cycle, partly by playing an essential role in biomass degradation and turnover. Bacterial and fungal degradation of cellulose occurs through the action of extracellular enzymes that are either cell-free or associated with the outer cell envelope. Microbially mediated cellulose depolymerization provides a virtually endless resource of easily digestible metabolites and products, including soluble sugars, organic acids, alcohols, CO_2, and H_2 (Leschine, 1995). These compounds are readily utilized as carbon and energy sources by other, non-cellulolytic commensal members of the microbial community inhabiting the cellulolytic environment. CO_2 released into the atmosphere by microbes is converted back to sugars and further to lignocellulosic biomass during photosynthesis and plant growth (Leschine, 1995; Desvaux *et al.*, 2001; Doi, 2008).

Cellulose decomposition occurs in soils, sediments, aquatic environments as well as environments impacted by human activities or associated with animals (sewage sludge, compost piles, wood-chip piles, paper mills, wood processing plants). Soil and sediments are a huge reservoir of carbon in the form of dead and recalcitrant plant biomass, where cellulose is degraded either aerobically or anaerobically. Aerobic cellulose decomposition may be achieved through the activities of a single species or consortia of mesophilic fungi and bacteria. For instance, the cellulose and hemicellulose components of wood are completely degraded and metabolized by single fungal species of either *Phlebia radiate* (white-rot fungi), *T. reesei* or *Penicillium* species (Busto *et al.*, 1996; Fujii and Shintoh, 2006; Cho *et al.*, 2009). Mesophilic aerobic cellulolytic bacteria, such as *Bacillus pumilus*, *Cellulomonas* spp., *Cellvibrio gilvus*, *Pseudomonas fluorescens*, and *Streptomyces* spp., are often found in soils, on plant materials, in humus, animal faeces, and leaf litter (Deshpande *et al.*, 2009; Dinis *et al.*, 2009).

While most of the cellulose in nature is degraded aerobically, 5–10% is degraded anaerobically by cellulose-fermenting microorganisms (Leschine, 1995). Anaerobic cellulose decomposition, which requires mixed populations, benefits from being performed by a community of physiologically diverse microorganisms (Leschine, 1995). Anaerobes are metabolically diverse and can perform various fermentations and respirations, employing varied electron acceptors (e.g. carbon dioxide, inorganic sulfur compounds, and inorganic nitrogen compounds). A typical ecosystem undergoing anaerobic degradation of lignocellulosic plant matter by microbial communities is schematically represented in Fig. 7.1. In the absence of oxygen, cellulose is decomposed to acetate, H_2 and CO_2, which are then able to undergo methanogenesis through a complex microbial food chain (Leschine, 1995). Cellulose is initially hydrolysed through the action of extracellular enzymes of cellulolytic bacteria. The released sugar monomers and oligomers are fermented by cellulolytic and other saccharolytic microorganisms with production of CO_2, H_2, organic acids (e.g. acetate, propionate, butyrate), and alcohols (e.g. ethanol). Exogenous H_2 is immediately consumed by methanogens or acetogens, which use H_2 to reduce CO_2 to, respectively, CH_4 or acetate. Acetate can then serve as a terminal electron acceptor and the predominant end products of the methanogenesis appear to be CH_4, CO_2, and H_2O (Fig. 7.1). Cellulose-decomposing microbial communities comprising several symbiotic anaerobic bacteria are found in the gut flora of herbivorous mammals (e.g. the rumen of cattle and sheep), leaf eating birds (e.g. hoatzin, the only known representative), and plant eating insects (e.g. termites, grubs, larvae). In addition to anaerobic bacteria and archaea, the microbial communities in the intestinal tract of large herbivorous mammals also contain a number of anaerobic fungi, which are classified into six genera, namely *Anaeromyces, Caecomyces, Cyllamyces, Neocallimastix, Orpinomyces* and *Pyromyces* (Nicholson *et al.*, 2005).

While the majority of common natural habitats where cellulose degradation occurs (e.g. soils, sediments, compost and even gut environments) are temperate or even low-temperature niches, thermal (55–95°C) aquatic environments that contain decaying plant material are also a good source of lignocellulose-degrading microorganisms (see

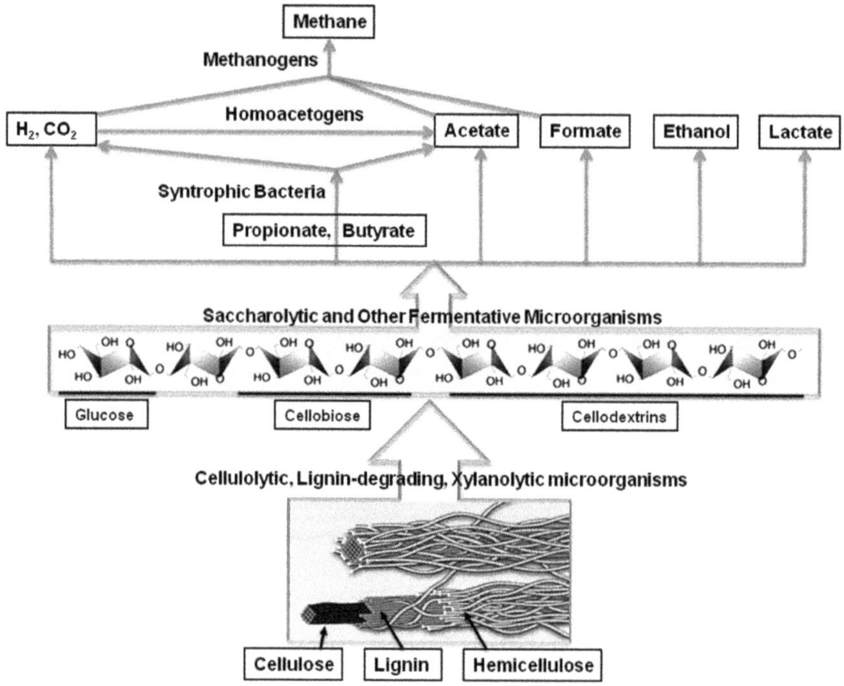

Figure 7.1 Schematic representation of anaerobic degradation of lignocellulosic plant biomass by microbial communities in soils and freshwater sediments. Cellulolytic, xylanolytic and ligninolytic microbes degrade the major polysaccharide components of plant cell walls to monomers and oligomers. Commensal microorganisms assimilate the soluble sugars and other end products (ethanol and lactate usually do not accumulate), which are eventually converted to methane and carbon dioxide.

Table 7.1). The range of substrates these thermophilic microorganisms can utilize depends on the temperature, biotope, and inorganic chemistry of the aqueous phase. For example, many anaerobic thermophilic bacteria of the order Thermotogales do not utilize crystalline cellulose as a substrate, but are capable of fermenting simple and complex sugars to H_2, CO_2, and acetate (Blumer-Schuette *et al.*, 2008). The ability to utilize crystalline cellulose as well as hemicellulose as a growth substrate is limited to terrestrial thermophilic bacteria of the phylum Firmicutes with growth temperature optima around 70°C. Cellulolytic, anaerobic thermophiles have been isolated from terrestrial hot springs of Kamchatka (Far East Russia), Iceland, New Zealand, Nagano Prefecture in Japan and the Great Artesian Basin in Australia. (Taya *et al.*, 1988; Hudson *et al.*, 1990; Svetlichnyi *et al.*, 1990; Bredholt *et al.*, 1995; Mladenovska *et al.*, 1995; Miroshnichenko *et al.*, 2008; Ogg and Patel, 2009). The cellulolytic bacteria isolated to date belong to the order Clostridiales, including the extremely thermophilic *Anaerocellum thermophilum*, *Caldicellulosiruptor kristjanssonii*, *Caldicellulosiruptor kronotskyensis*, *Caldicellulosiruptor hydrothermalis*, *Caldicellulosiruptor lactoaceticus*, and the thermophilic *Caloramator australicus*, *Clostridium thermocellum* and *Thermoanaerobacter thermohydrosulfuricus* (Freier et al., 1988; Svetlichnyi *et al.*, 1990; Mladenovska *et al.*, 1995; Koskinen *et al.*, 2008a; Miroshnichenko *et al.*, 2008; Ogg and Patel, 2009).

One of the most abundant distributions of thermal sites is within Yellowstone National Park (YNP), located in north-western USA. Bacterial and archaeal diversity within the different YNP hot springs is well documented, and several novel thermophilic microorganisms have been isolated (Hugenholtz *et al.*, 1998; Reysenbach *et al.*, 2000; Meyer-Dombard *et al.*, 2005; Nakagawa

Table 7.1 Thermophilic microorganisms capable of degrading lignocellulosic biomass

Species	Temperature range (optimum), °C	pH range (optimum)	Fermentation end-products[1]	Original source	Reference
Aerobic					
Acidothermus cellulolyticus	37–70 (55)	3.5–7.0 (5.0)	NA	Hot springs, Norris Geyser basin, YNP	Mohagheghi *et al.* (1986)
Caldibacillus cellulovorans	(68)	(7.0)	NA	Artificial compost heap	Bergquist *et al.* (1999)
Rhodothermus marinus	54–77 (65)	(7.0)	NA	Submarine alkaline hot spring, Iceland	Alfredsson *et al.* (1988), Hreggvidsson *et al.* (1996)
Anaerobic					
Anaerocellum thermophilum[2] DSMZ 6725	40–83 (75)	(7.2)	Lac, Ac, H_2, CO_2	Valley of Geysers, Kamchatka, Russia	Svetlichnyi *et al.* (1990); Yang *et al.* (2010)
Caldicellulosiruptor lactoaceticus	50–78 (68)	5.8–8.2 (7.0)	Lac, Ac, Et, H_2, CO_2	Alkaline hot spring, Hveragerdi, Iceland	Mladenovska *et al.* (1995)
Caldicellulosiruptor saccharolyticus	45–80 (70)	5.5–8.0 (7.0)	Ac, H_2, CO_2	Hot spring, Taupo-Rotorua region, New Zealand	Rainey *et al.* (1994)
Caldicellulosiruptor kristjanssonii	45–82 (78)	5.8–8.0 (7.0)	Lac, Ac, Et, H_2, CO_2	Slightly alkaline hot spring, Iceland	Bredholt *et al.* (1999)
Caldicellulosiruptor owensensis	50–80 (75)	5.5–9.0 (7.5)	Lac, Ac, Et, H_2, CO_2	Dry sediment, Owens Lake, California, USA	Huang *et al.* (1998)
Caldicellulosiruptor kronotskyensis 2002[3]	45–82 (70)	6.0–8.0 (7.0)	Lac, Ac,	Terrestrial hot springs, Kamchatka, Russia	Miroshnichenko *et al.* (2008)
Caldicellulosiruptor hydrothermalis 108[3]	50–80 (70)	6.0–8.0 (7.0)	Lac, Ac	Terrestrial hot springs, Kamchatka, Russia	Miroshnichenko *et al.* (2008)
Clostridium stercorarium	(65)	(7.3)	Et, Ac	Compost heap	Madden (1983)
Clostridium thermocellum JW20	37–69 (60)	6.1–7.5 (7.0)	Et, Ac, Lac, H_2, CO_2	Cotton bale, Louisiana, USA	Freier *et al.* (1988)
Desulfurococcus fermentans (crenarchaeote)	63–89 (81)	4.8–6.8 (6.0)	Ac, H_2, CO_2	Uzon caldera, Kamchatka, Russia	Perevalova *et al.* (2005)
Fervidobacterium islandicum	50–80 (65)	6.0–8.0 (7.2)	Lac, Ac, Et, H_2, CO_2	Thermal spring, Varma River, Hveragerdi, Iceland	Huber *et al.* (1990)
Spirochaeta thermophila	40–73 (67)	509–7.7 (7.0)	Ac, CO_2, H_2, Lac	Saline hot spring, Shiashkotan, Russia	Aksenova *et al.* (1992)
Thermoanaerobacter cellulolyticus	50–85 (75)	6.0 – 9.5 (8.1)	Ac, H_2 CO_2, Lac	Alkaline hot spring, Nagano Prefecture, Japan	Bredholt *et al.* (1995)
Thermotoga neapolitana	5 –90 (80)	5.5 – 9.0 (7.0)	NR	Submarine thermal vent, Naples	Jannasch *et al.* (1988), Sunna *et al.* (1997)

[1]End-products are those reported and abbreviated as follows: Lac, lactate; Ac, acetate; Et, ethanol. [2]*Anaerocellum thermophilum* was recently reclassified as *Caldicellulosiruptor bescii* (Svetlichnyi *et al.*, 1990; Yang *et al.*, 2010). [3]Type strain. NA, not applicable; NR, not reported.

et al., 2005). For example, two studies describe the isolation of mesophilic and thermophilic xylan-degrading bacteria from YNP. In the first, several mesophilic bacteria of the genus *Clostridium* showing high xylanolytic activity were isolated from agricultural and forest soil samples and water-sediment samples from acidic springs. This included the *Clostridium acetobutylicum* isolate 3BYR, which exhibited very high extracellular xylanase, xylosidase, alpha-l-arabinofuranosidase and alpha-glucuronidase activities, and utilized up to 80% of oat spelt xylan as a carbon source during growth (Trudeau *et al.*, 1992). In the second study, thermophilic, anaerobic, xylanolytic strains of *Thermoanaerobacterium xylanolyticum* sp. nov. and *Thermoanaerobacterium saccharolyticum* sp. nov. were isolated from Frying Pan Springs. These organisms are non-cellulolytic and grow chemoorganotrophically at 60°C and pH 6.0, fermenting xylan and starch to acetate, ethanol, lactate, CO_2, and H_2 (Lee *et al.*, 1993). compared with the above xylan-degrading microbes, studies dedicated to isolation of cellulolytic bacteria from YNP are scarce. Cellulolytic organisms isolated from YNP include the anaerobic thermophile *C. thermocellum* strain YS (Bayer *et al.*, 1983) from hot spring soil samples and the aerobic, acidophilic thermophile, *Acidothermus cellulolyticus*, from acidic hot springs (Mohagheghi *et al.*, 1986). We hypothesized that the anaerobic thermal springs of YNP probably contain a wide variety of cellulose-degrading microbial communities and focused our efforts on enriching these communities on natural pretreated biomass (see later), with the eventual aim being the isolation of novel cellulolytic microorganisms with high rates of lignocellulosic biomass utilization.

Enzyme systems of cellulolytic microorganisms

Plant cell wall composition

Plant cell walls are composed of three major polysaccharide polymers, namely, cellulose, hemicellulose and pectin. Cellulose is a stable unbranched homopolymer with linear chains of β-1,4-linked glucose units stacked tightly in parallel into microfibrils via strong inter- and intra-chain hydrogen bonding. The structurally rigid cellulose microfibrils are largely crystalline in nature, with semi-crystalline and amorphous regions in between (Chang *et al.*, 1981). Hemicellulose and pectin, on the other hand, are heterogeneous branched polymers and are chemically and structurally more complex than cellulose. Pectin is a family of galacturonic acid-rich polysaccharides including homogalacturonan, xylogalacturonan, and rhamnogalacturonan I and II polymers (Mohnen, 2008). Hemicellulose consists primarily of xylan or other backbone structures (e.g. galacto-glucomannan, arabinogalactan, lichenins), classified based on their monomeric composition. In addition, hemicellulose sugars are often acetylated or methylated and also linked with lignin, which is another heterogeneous plant polymer of linked aromatic monomers (Knox, 2008). To counter the chemical and structural complexity of plant cell wall polysaccharides, cellulolytic organisms have evolved enzyme systems with a wide array of cellulases and hemicellulases for hydrolysis and degradation of plant biomass (Schwarz *et al.*, 2004; Dashtban *et al.*, 2009).

Plant cell wall-degrading enzymes

Plant biomass-degrading enzymes were originally classified based on the substrates they hydrolysed and their mode of action. More recently, cellulases and other carbohydrate active enzymes (CAZymes; www.cazy.org) have been classified based on their sequence homology, structural fold and mechanistic themes (Henrissat and Davies, 1997; Cantarel *et al.*, 2009). As of September 2010, the CAZy database classification included 118 glycoside hydrolase (GH), 16 carbohydrate esterase (CE), 22 polysaccharide lyase (PL) and 61 carbohydrate binding module (CBM) families. Cellulases hydrolyse the β-1,4-glucosidic bonds in cellulose polymers either internally (endo-acting), or at the reducing/non-reducing ends (exo-acting) of the polysaccharide chain. Owing to a high degree of crystallinity and structural stability of the cellulose polymer, the combined synergistic action of several cellulases with differing specificities (e.g. glycoside hydrolases belonging to families GH48 and GH9 in bacteria, based on CAZy classification) is needed for its degradation (Wilson, 2009).

In contrast, hemicellulose degradation requires the action of several types of enzymes that act upon bonds between different carbohydrate moieties or the bonds between carbohydrate and non-carbohydrate moieties in this chemically heterogeneous plant polymer. For example, while xylanases (e.g. GH10, GH11 family enzymes) and mannanases (e.g. GH26) cleave the main backbone, other hemicellulases such as arabinofuranosidase (e.g. GH43) and glucuronidase cleave the bonds between backbone sugars and side-chain substituent sugars (Shallom and Shoham, 2003). Carbohydrate esterases are another group of enzymes that cleave the ester bond between the hemicellulose sugars and acetyl group (e.g. acetyl esterases belonging to the CE2, CE3 families) or lignin monomeric compounds such as ferulic and *p*-coumaric acids (e.g. feruloyl esterases belonging to the CE1 family) (Wong, 2006). Lastly, the polysaccharide lyase family of enzymes is active against glycosidic bonds present in anionic polysaccharides such as the pectin component of the plant cell wall (Michaud *et al.*, 2003).

Modular nature of enzymes

Carbohydrate-active enzymes, especially in anaerobic bacteria, have a modular structure. For example, cellulases and hemicellulases consist of several structurally and functionally distinct units called *domains* or *modules* (Gilkes *et al.*, 1991; Schwarz *et al.*, 2004). Apart from the catalytic module(s) which cleave the glycosidic bond, the additional non-catalytic domains participate and contribute to the overall hydrolytic activity of the enzyme. For example, the carbohydrate binding modules (CBM) help in targeting the catalytic module, and in attachment of the free enzymes, to the insoluble plant polysaccharide substrates (Shoseyov *et al.*, 2006). Other non-catalytic modules include the immunoglobulin-like (Ig), fibronectin type-III (Fn3), dockerins involved in cellulosome assembly and the surface-layer homology (SLH) module (Doi and Tamaru, 2001). SLH modules interact with cell wall polysaccharides and aid in the attachment of free enzymes or cellulosome anchor proteins to the cell surface (e.g. *C. thermocellum*; see discussion below). Recent evidence suggests that apart from the modular composition, their arrangement on the polypeptide chain also has a bearing on the role these accessory modules play in the overall functional activity of a given enzyme (Bayer *et al.*, 2004).

Complexed and non-complexed enzyme systems

Cellulolytic organisms have evolved to produce a host of enzymes for degrading plant cell wall polysaccharides. Broadly, two types of enzyme systems have been observed in cellulolytic microorganisms, namely the *non-complexed* and *complexed* systems (Bayer *et al.*, 2006; Doi, 2008). Aerobic fungi (e.g. *T. reesei*), aerobic thermophilic bacteria (e.g. *Acidothermus cellulolyticus*, *Thermobifida fusca*) and anaerobic, extremely thermophilic bacteria (e.g. *Caldicellulosiruptor saccharolyticus*, *A. thermophilum*) contain the non-complexed enzyme system, in which several individual secreted endo- and exo-glucanases act synergistically in degradation of plant cell wall cellulose (Svetlichnyi *et al.*, 1990; Rainey *et al.*, 1994). Complexed enzyme systems are found in anaerobic fungi (e.g. *Anaeromyces*, *Caecomyces*, *Cyllamyces*, *Neocallimastix*, *Orpinomyces*, and *Pyromyces*), and both mesophilic (e.g. *Bacteroidetes cellulosolvens*, *Clostridium cellulovorans*, *Clostridium cellulolyticum*, *Ruminococcus flavefaciens*) and thermophilic bacteria (*C. thermocellum*) (Ali *et al.*, 1995b; Ljungdahl, 2008). These microorganisms form large extracellular enzyme complexes on the cell surface, termed the cellulosome, consisting of a non-catalytic scaffolding protein(s) with many bound catalytic components exhibiting cellulolytic, hemicellulolytic and other hydrolytic activities (Bayer *et al.*, 1998; Miron *et al.*, 2001; Doi and Kosugi, 2004; Xu *et al.*, 2004; Fendri *et al.*, 2009).

Cellulosomal architecture

In cellulosomes, a structural component called scaffoldin contains binding sites (cohesins) that interact with the docking modules (dockerins) borne by the enzymatic units. The highly specific and stable cohesin-dockerin type interaction mediates the assembly of the catalytic components onto the scaffoldin (Type I interaction) and also the attachment of the scaffoldin itself (with bound subunits) onto the cell surface via anchor proteins

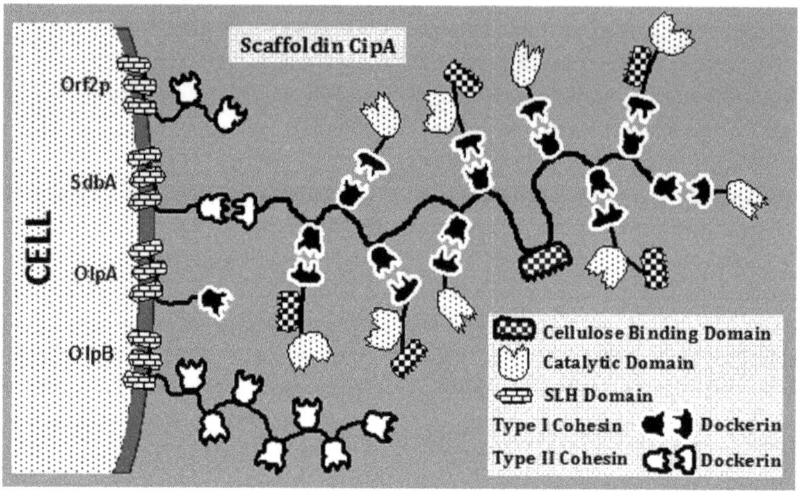

Figure 7.2 Simplified schematic representation of *Clostridium thermocellum* cellulosomal architecture. The cellulosome backbone scaffoldin protein CipA contains 9 type-I cohesins decorated with catalytic units bearing the type-I dockerin. CipA also contains a type-II dockerin which interacts with the type-II cohesins of the anchor proteins (e.g. SdbA, OlpB, Orf2p), which in turn are anchored to the bacterial cell surface via interaction of the SLH domains with the cell wall-associated polysaccharides. Also shown is another cell surface anchor protein, OlpA, containing a type-I cohesin. CipA and many catalytic units also contain cellulose (and other carbohydrate) binding domains which aid in attachment of the cellulosome complex to various plant cell wall components. Figure adapted from Raman *et al.* (2009).

(Type II interaction) (Beguin and Lemaire, 1996; Belaich *et al.*, 1997; Doi and Tamaru, 2001). A simplified schematic of the cellulosomal architecture in *C. thermocellum* is shown in Fig. 7.2; further details about this system are provided in a later section. More complex architectures involving multiple adapter scaffoldin proteins in cell wall anchoring has also been reported (Xu *et al.*, 2004; Rincon *et al.*, 2005). In cellulolytic organisms with complexed enzyme systems, the scaffoldin component usually contains a carbohydrate binding module (CBM) which aids in substrate recognition and binding. Thus the cellulosome complex provides for both proximity-based synergy of catalytic units via their assemblage on the scaffoldin, as well as targeting-based synergy between microbe and biomass via CBM-mediated formation of the cell-cellulosome-substrate tri-complex (Lu *et al.*, 2006). However, the mechanism(s) of microbial adhesion to cellulose and its relative importance in substrate utilization in cellulolytic organisms is not clearly understood and requires further research (Bayer *et al.*, 1998; Demain *et al.*, 2005).

Cellulolytic enzymes of anaerobic thermophilic bacteria

In this section, we will focus on the enzyme systems of three anaerobic cellulolytic bacteria that are either thermophilic (*C. thermocellum*, abbreviated *Cthe*) or extremely thermophilic (*A. thermophilum*, *Athe* and *Ca. saccharolyticus*, *Csac*), and whose genomes have been sequenced (http://genome.ornl.gov/microbial/cthe/; van de Werken *et al.*, 2008; Kataeva *et al.*, 2009). Table 7.2 compares the distribution of glycoside hydrolases, esterases, lyases and carbohydrate binding modules across these three organisms. Genome-level comparison reveals a similar total number of glycoside hydrolase domains encoded in *Cthe*, *Csac* and *Athe* (Table 7.2). Complete cellulose degradation requires the combined action of exo-acting cellobiohydrolases and endoglucanases (Warren, 1996). While all three organisms encode GH48 family bacterial reducing-end cellobiohydrolase, *Cthe* has a higher distribution of other cellulose-specific hydrolytic enzymes (e.g. belonging to the GH9 and GH8 families) as compared with *Csac* and *Athe* (Table 7.2). These observations are consistent with *C. thermocellum* being known as one of the most efficient cellulose-degrading microorganisms (Lynd and Zhang, 2002). In comparison, the extreme thermophiles display a wider distribution of glycoside hydrolase domains across several different families, including several GH13 amylases and other α-glucan acting enzymes (e.g. belonging to the GH4, GH29 and GH65 families), reflecting their broader substrate specificity (Table 7.2).

All three organisms encode enzymes for hydrolysing the non-cellulosic components of the plant cell wall, namely, hemicellulose and pectin. In addition to xylanases (e.g. belonging to the GH10 and GH11 families), Cthe contains more mannanase (GH26) and arabinofuranosidase

Table 7.2 Comparison of glycoside hydrolase (GH), carbohydrate esterase (CE), polysaccharide lyase (PL) and carbohydrate binding module (CBM) family domain distribution across the genomes of three anaerobic cellulolytic bacteria, the thermophilic *Clostridium thermocellum* ATCC 27,405(*Cthe*) and extremely thermophilic *Caldicellulosiruptor saccharolyticus* DSM 8903 (*Csac*) and *Anaerocellum thermophilum* DSM 6725 (*Athe*)

Domain Family	*C. thermocellum*	*Ca. saccharolyticus*	*A. thermophilum*
GH	# Domains		
1	2	1	1
2	1	4	4
3	2	2	1
4		2	2
5	12	6	6
8	1		
9	16	2	2
10	6	6	5
11	1		1
13	2	7	7
15	1	1	1
16	2	1	
18	3		
20		1	
23	2	2	2
26	3	1	
28		2	2
29		1	1
30		1	1
31		1	1
36		1	1
39	1	2	2
42		1	1
43	6	4	3
44	1	1	1
48	2	1	3
51	1	1	1
53	1		
65		2	2
67		1	1
74	1	1	1
78		1	
81	1		
88		1	
94	3	2	2
105		1	1
Total	71	61	56

Domain Family	*C. thermocellum*	*Ca. saccharolyticus*	*A. thermophilum*
CBM	# Domains		
3	23	10	16
4	7		
6	11	1	1
9	2		
11	1		
13	2		
16	3		
20		1	2
22	5	5	5
25	2		
28		1	1
30	1		
32	1	1	
34	1		
35	7		
36			1
41		1	1
42	4		
44	1		
48	1	3	3
50	2		
54	1	1	
Total	75	24	30

Domain Family	*C. thermocellum*	*Ca. saccharolyticus*	*A. thermophilum*
CE	# Domains		
1	3	1	
2	1		
3	2		
4	3	3	2
7	1	1	1
8	1		
9	1	1	1
12	2		1
NC	1		
Total	15	6	5

PL	# Domains		
1	2		
3			1
9	1		1
11	1		1
NC		1	
Total	4	1	3

(GH43) domains and carbohydrate esterases (e.g. CE1, CE4) than *Csac* and *Athe*. Among pectin-degrading enzymes, both *Cthe* and *Athe* contain polysaccharide lyases (e.g. PL9, PL11) which attack the pectin backbone, and esterases (e.g. CE8, CE12) which cleave the side-chains, while *Csac* contains glycoside hydrolases (e.g. belonging to the GH28 and GH105 families) that are active against pectin. Consistent with the biomass-degrading capacity encoded in their genomes, *Ca. saccharolyticus* and *A. thermocellum* can grow on cellulose, xylan, pectin and a range of other biomass-derived sugar monomers and oligomers (de Vrije *et al.*, 2009; Vanfossen *et al.*, 2009; Yang *et al.*, 2009a). In contrast, *C. thermocellum* can only grow on cellulose and the derived cello-oligosaccharides (Demain *et al.*, 2005). Therefore, the repertoire of enzymes encoded by *C. thermocellum* for hemicellulose and pectin degradation is only for unwrapping and exposing of its preferred substrate of cellulose, which is intertwined with these polysaccharides in plant cell walls.

The evolution of *C. thermocellum* as a voracious cellulose degrader is further highlighted by the preferential distribution of cellulose-specific carbohydrate binding modules in its genome. *Cthe* contains an abundance of CBMs (75 CBM domains from 18 families), including 23 cellulose-binding CBM3 family domains, as compared with *Csac* and *Athe* (Table 7.2). *Cthe* also has several domains (e.g. from the CBM4, CBM6, CBM35 families) that bind to various hemicellulose components, including xylan, glucomannan, and arabinoxylan. In contrast, Csac and Athe have representative domains in CBM families (e.g. CBM20, CBM48) that bind to starch, glycogen, etc., concordant with the presence of many α-glucan acting enzymes in their genomes, as discussed earlier.

While most of the CAZymes in *C. thermocellum* contain a dockerin and are cell-bound as part of the cellulosome, the organism also has a small number of cell-free enzymes including a GH48 family enzyme (Zverlov *et al.*, 2005). In comparison, *A. thermophilum* and *Ca. saccharolyticus* do not produce cellulosomes and contain cell-free, non-complexed enzyme systems. Table 7.3 lists all CAZymes from *Cthe, Athe* and *Csac* that have at least two domains belonging to the GH, CE, PL or CBM families. In *Cthe,* cellulosomal enzymes are bound to a scaffoldin subunit that contains a family 3a CBM and aids in the collective binding of the complex to crystalline cellulose. Yet, more than half of the cellulosomal catalytic units also contain CBM domain(s) of related substrate specificity, in addition to the dockerin module (Table 7.3). For example, several GH9 cellulosomal cellulases contain a family 3c CBM that is thought to play a 'helper' role, acting in concert with the catalytic domain, in transiently binding and feeding a loose non-crystalline cellulose chain into the active site of the enzyme (Gal *et al.*, 1997; Irwin *et al.*, 1998).

Cthe has several multifunctional enzymes that contain catalytic domains with complementing activities oriented towards hemicellulose degradation, potentially to counter the structural and chemical complexity of this heterogeneous polysaccharide. Examples include GH10 and GH11 xylanases paired with esterases belonging to the CE1 (XynZ, *Cthe*_1963) and CE4 (XynA, *Cthe*_2972) families, respectively. Some of the cellulosomal enzymes display a more complex modular architecture, as in the example of a GH9 cellulase paired with GH44 and CBM30, CBM44 domains (CelJ, *Cthe*_0624). Non-cellulosomal multi-modular enzymes in *Cthe* contain multiple repeated domains from the same CBM family, as in the case of GH16 (LicA, *Cthe*_2809) paired with 4 CBM4 domains (Table 7.3). A similar trend is also observed in the cell-free enzymes of *Athe* and *Csac,* which contain multiple CBM3 domains.

Although *Athe* and *Csac* have fewer multi-functional enzymes than *Cthe,* genome analysis reveals a higher order of modular architecture of enzymes in these extremely thermophilic organisms. Several enzymes contain two catalytic domains and multiple CBM domains, in some cases with differing substrate specificities (see Table 7.3). Examples of multi-domain enzymes sharing similar architecture in the two organisms include: (i) GH5 catalytic domain paired with a GH44 (*Athe*_1859 and *Csac*_1077, ManA) along with two CBM3 domains (Saul *et al.*, 1990; Gibbs *et al.*, 1992); and (ii) two GH43 arabinofuranosidase domains appended with a family 6 and -22 CBM domain (*Athe*_0182 and *Csac*_2411,

Table 7.3 Multimodular enzymes of *C. thermocellum* (cellulosomal and non-cellulosomal; *Cthe*), *A. thermophilum* (non-cellulosomal; *Athe*) and *Ca. saccharolyticus* (non-cellulosomal; *Csac*) containing at least two domains belonging to the GH, CE, PL or CBM families. Enzymes containing at least three domains are highlighted in grey.

C. thermocellum : Cellulosomal Enzymes

	Locus	Modular Architecture
	Cthe_0015	CBM42-Doc1-GH43
	Cthe_0043	GH9-CBM3-Doc1
	Cthe_0246	Doc1-CBM35-PL11
	Cthe_0412	CBM4-Ig-GH9-Doc1
	Cthe_0413	CBM4-Ig-GH9-CBM3-Doc1
	Cthe_0433	GH9-CBM3-Doc1
	Cthe_0543	GH9-CBM3-Doc1
	Cthe_0578	GH9-CBM3-Doc1
*	Cthe_0624	CBM30-Ig-GH9-GH44-Doc1-CBM44
	Cthe_0625	GH9-CBM3-Doc1
	Cthe_0661	GH43-CBM13-Doc1
	Cthe_0745	GH9-CBM3-Doc1
*	Cthe_0797	GH5-Doc1-CE2
^	Cthe_0798	CE3-CE3-Doc1
	Cthe_0821	GH5-CBM32-Doc1
*	Cthe_0912	CBM22-GH10-CBM22-Doc1-CE1
	Cthe_1271	GH43-CBM6-CBM6-Doc1
*	Cthe_1472	GH26-GH5-CBM11-Doc1
	Cthe_1838	CBM22-GH10-Doc1
*	Cthe_1963	CE1-CBM6-Doc1-GH10
	Cthe_2137	GH39-CBM35-CBM35-Doc1
	Cthe_2138	CBM42-GH43-Doc1
*	Cthe_2139	GH5-CBM42-GH43-Doc1
	Cthe_2147	CBM3-GH5-Doc1
*	Cthe_2179	PL1-Doc1-CBM35-PL9
	Cthe_2193	GH5-CBM6-Fn3-CBM13-Doc1
	Cthe_2194	CE1-CBM6-Doc1
	Cthe_2196	GH43-CBM6-Doc1
	Cthe_2197	GH2-CBM6-Doc1
	Cthe_2360	GH9-CBM3-Doc1
	Cthe_2590	CBM22-GH10-Doc1
	Cthe_2760	GH9-CBM3-Doc1
	Cthe_2761	GH9-CBM3-Doc1
	Cthe_2811	GH26-CBM35-Doc1
	Cthe_2950	PL1-Doc1-CBM35
*	Cthe_2972	GH11-CBM6-Doc1-CE4
	Cthe_3012	GH5-CBM6-Doc1
^	Cthe_3141	CE12-Doc1-CBM35-CE12

C. thermocellum : Non-cellulosomal Enzymes

Locus	Modular Architecture
Cthe_0040	GH9-CBM3-CBM3
Cthe_0071	GH48-CBM3
Cthe_0795	GH13-CBM34
Cthe_1257	CBM3-CBM4
Cthe_1800	CBM50-CBM50-Unk-GH18
Cthe_2191	CBM48-GH13
Cthe_2809	SLH-GH16-CBM4-CBM4-CBM4-CBM4

A. thermophilum

	Locus	Modular Architecture
	Athe_0089	GH11-CBM36
^	Athe_0182	GH43-CBM22-GH43-CBM6
	Athe_0183	CBM22-CBM22-GH10
	Athe_0558	CBM48-GH13
	Athe_0594	GH5-CBM28-SLH
	Athe_0609	CBM41-CBM48-GH13-CBM20
	Athe_0618	CBM22-CBM22-GH10
	Athe_1853	PL11-CBM3
*	Athe_1857	GH10-CBM3-CBM3-GH48
*	Athe_1859	GH5-CBM3-CBM3-GH44
*	Athe_1860	GH74-CBM3-CBM3-GH48
*	Athe_1865	GH9-CBM3-CBM3-CBM3-GH5
^	Athe_1866	GH5-CBM3-CBM3-CBM3-GH5
*	Athe_1867	GH9-CBM3-CBM3-CBM3-GH48

Ca. saccharolyticus

	Locus	Modular Architecture
	Csac_0671	CBM48-GH13
	Csac_0678	GH5-CBM28-SLH
	Csac_0689	CBM41-CBM48-GH13-CBM20
	Csac_0696	CBM22-CBM22-GH10
	Csac_0784	CBM48-GH13
*	Csac_1076	GH9-CBM3-CBM3-CBM3-GH48
*	Csac_1077	GH5-CBM3-CBM3-GH44
*	Csac_1078	GH10-CBM3-GH5
	Csac_1079	GH9-CBM3-CBM3-CBM3
	Csac_1085	GH74-CBM3
	Csac_2410	CBM22-CBM22-GH10
^	Csac_2411	GH43-CBM22-GH43-CBM6

Entries marked with ^ and * contain two catalytic domains of the same or different families, respectively.

Module classification according to the carbohydrate active enzyme (CAZy) database (www.cazy.org; Cantarel *et al.*, 2009); GH, glycoside hydrolase; CE, carbohydrate esterase; PL, polysaccharide lyase; CBM, carbohydrate binding module; Doc1, dockerin type I; Ig, immunoglobulin-like fold; Fn3, fibronectin III; SLH, surface layer homology domain; Unk, module of unknown function.

Information gathered from the carbohydrate active enzymes database, as of September 2010 (www.cazy.org; Cantarel *et al.*, 2009). NC, not classified.

XynF). Of particular interest is a very similar cellulase encoded by *Athe* and *Csac* that contains GH9 cellulase and GH48 cellobiohydrolase family catalytic domains along with three CBM3 domains (*Athe*_1867 and *Csac*_1076, CelA). This organization suggests synergism between these modules for concerted action on crystalline cellulose (Teo *et al.*, 1995; Zverlov *et al.*, 1998). There is no evidence for a similar enzyme with paired cellulase domains in the cellulosome producing *Cthe*.

It is possible that organisms with free, non-complexed enzyme systems evolved to encode multiple catalytic domains with complementing activities in a single polypeptide chain to compensate for the lack of 'proximity effect' achieved in a complexed enzyme system, such as the cellulosome. The multifunctional enzymes in *Athe* and *Csac* are also encoded in contiguous genomic regions (*Athe*_1857–60, 1865–67; *Csac*_1076–79) of potential operon structure, probably for their coordinated expression to synergistically degrade plant biomass. Diversity in the modular structure of enzymes in cellulolytic organisms probably reflects the complexity of the substrate chemistry encountered in plant cell walls, and is potentially important in the survival and persistence of these bacteria in their natural environment. The synergism promoted by multi-domain enzymes suggests an intermediate level in evolution between simple free enzymes (containing only one catalytic domain) and cellulosomal systems in which several enzymes with diverse catalytic functions are complexed together.

Among thermophilic bacteria, the cellulosome system of the anaerobic Gram-positive bacterium *C. thermocellum* has been extensively studied (Bayer *et al.*, 2008; Ding *et al.*, 2008). The primary scaffoldin CipA contains nine type-I cohesins and one type II dockerin. CipA is anchored to the cell surface via three well-characterized anchor proteins, SdbA, Orf2p and OlpB, containing 1, 2 and 7 type-II cohesins respectively [see Fig. 7.2, (Bayer *et al.*, 1998)]. The *C. thermocellum* genome encodes more than 70 type-I dockerin-containing components with diverse hydrolytic activities and specificities (Zverlov *et al.*, 2005). In general, cellulosome-containing organisms have an overabundance of dockerin-containing subunits compared with the number of cohesins on the scaffoldin. As a consequence, It is probable that multiple cellulosomal subpopulations exist, even within an organism. Indeed, chromatographic separation of cellulosomes from *C. papyrosolvens*, *C. thermocellum* and *C. cellulovorans* identified distinct cellulosomal subpopulations with differing subunit composition and enzyme activities in these organisms (Ali *et al.*, 1995a; Pohlschroder *et al.*, 1995; Han *et al.*, 2005). These findings suggest that the process of cellulosome assembly is not an entirely random process and the cellulosome composition is probably regulated to make the organism better suited for attacking specific growth substrates.

In order to understand the substrate-related changes in the subunit composition of the global cellulosome population in *C. thermocellum* ATCC 27405, we isolated cellulosomes from cultures grown on a variety of model substrates (cellobiose, amorphous cellulose, crystalline cellulose) and dilute-acid pretreated switchgrass. Using quantitative proteomics methods (multidimensional LC-MS/MS and ^{15}N-metabolic labelling) we comprehensively measured the relative changes in levels of cellulosomal subunit proteins on a per-scaffoldin basis. In total, 80% of all dockerin containing proteins (59/73) and all of the cohesin-module containing components in *C. thermocellum* were detected in the cellulosomal preparations, including many subunits not reported previously (Raman *et al.*, 2009). Cellulosomes from amorphous cellulose, cellobiose and pretreated switchgrass-grown cultures displayed the most distinct differences in composition as compared with cellulosomes from crystalline cellulose-grown cultures (Raman *et al.*, 2009). Several cellulosomal components showed differential protein abundance in response to the presence of non-cellulose plant biomass components in the growth medium. It is important to understand the relative contribution and importance of the various enzymatic components in plant cell wall degradation. Such knowledge would help in engineering a strain that is best suited to grow on a specific feedstock of interest, and also for constructing non-native designer cellulosomes with tailored enzyme composition in organisms for industrial ethanol production.

Cellulolytic fermentations in production of sustainable energy

Sustainable energy can be defined as the provision of energy for our current needs without compromising the ability of future generations to meet their energy needs. Sustainable energy sources include all renewable resources, such as solar, wind, hydro, geothermal and tidal power, as well as biofuels derived from plant biomass. Biofuels such as hydrogen and ethanol can be produced from renewable plant biomass or organic waste materials using microbial fermentations. For example, Karakashev and colleagues have enriched strictly anaerobic, extremely thermophilic H_2 producing bacteria affiliated with the genera *Bacillus* and *Clostridium* in a lab-scale digester treating household solid organic wastes at 80°C and pH 8.1 (Karakashev *et al.*, 2009). The mixed culture was able to utilize glucose, galactose, mannose, xylose, arabinose, maltose, sucrose, pyruvate and glycerol as carbon sources. Growth on glucose produced acetate, CO_2 and H_2 at a maximal rate of 1.9 mol H_2/mol glucose (Karakashev *et al.*, 2009). Bacterial species isolated from natural sources and which could be used in hydrogen and ethanol production are listed in Table 7.1.

Fermentation by pure cultures

Most cellulolytic bacteria produce acetate and H_2 as the main fermentation end products during growth on cellulose-, hemicellulose- and pectin-containing biomass, with varying amounts of ethanol, lactic acid, and CO_2. For example, *Ca. saccharolyticus* ferments cellulosic biomass to H_2 (3.8 mol/mol glucose), acetate, and lactate as major end products, and ethanol and CO_2 as minor fermentation products (Mladenovska *et al.*, 1995; van de Werken *et al.*, 2008; Ivanova *et al.*, 2009). *Ca. lactoaceticus* strain 6A ferments cellobiose (4 g/l; approximately 22 mM equivalents of glucose) to 17.8 mM H_2, 19.3 mM lactate, 10.3 mM acetate, 9.8 mM CO_2, and 0.3 mM ethanol. Other species of the genus *Caldicellulosiruptor*, such as *Ca. kristjanssonii* strain 177R1B(T), produce 8 mM acetate/g/l Avicel, H_2 and CO_2 as major fermentation products, together with small amounts of lactic acid and 0.55 mM ethanol/g/l Avicel (Bredholt *et al.*, 1999). Attributes such as a broad substrate specificity, high hydrogen-producing capacity, and ability to co-utilize glucose and xylose make bacteria of the genus *Caldicellulosiruptor* attractive candidates for the production of renewable bioenergy in the form of hydrogen (van Niel *et al.*, 2002; Kadar *et al.*, 2003; O-Thong *et al.*, 2008; van de Werken *et al.*, 2008; Vanfossen *et al.*, 2009).

Thermophilic bacteria of the genera *Thermoanaerobacter* and *Thermoanaerobacterium* are also known to produce considerable amounts of ethanol from soluble sugars and other polymeric substrates (e.g. starch) (Lee *et al.*, 1993; Koskinen *et al.*, 2008b). Hence, these bacteria are attractive candidates for the production of biofuels from lignocellulosic biomass. However, the known representatives from these genera are non-cellulolytic, although they can grow chemoorganotrophically by utilizing xylan and starch, and also on a number of di- and monosaccharides, including glucose and xylose. For example, a thermophilic (60°C and pH 6.0), anaerobic strain of the xylan-degrading bacterium, *Thermoanaerobacter ethanolicus* (similar to strains 39E and JW200) has been isolated from Frying Pan Springs in YNP (Lee *et al.*, 1993). *T. ethanolicus* JW200 produced approximately 5 g/l of ethanol from 20 g/l of sucrose, while *Thermoanaerobacter* strain 65-2 produced 6.5 g/l of ethanol from 40 g/l of molasses (Lee *et al.*, 1993; Koskinen *et al.*, 2008b). A mutant strain of *T. ethanolicus* (39E H8) has been adapted to grow in the presence of up to 8% (v/v) ethanol, compared with less than 1.5% (v/v) alcohol tolerance for the wild-type strain (39E). The ethanol tolerant mutant strain had an increased percentage of transmembrane long-chain C-30 fatty acids and lacked the primary alcohol dehydrogenase (ADH), which is probably involved in ethanol consumption, whereas the secondary ADH functioned primarily to produce ethanol (Burdette *et al.*, 2002). Along with ethanol, the fermentation products included acetate, lactate, H_2 and CO_2.

Fermentation by co-cultures

There are a few studies, reviewed below, on the development of an anaerobic fermentation system for the production of H_2 by a community of extremely thermophilic organisms, including *Caldicellulosiruptor* spp. Such a reactor was tested for H_2 production from hydrolysed paper sludge, which is a solid industrial waste containing mainly cellulose, by *Ca. saccharolyticus* and *Thermotoga elfii* (Kadar *et al.*, 2003). The maximal volumetric H_2 production rate from paper sludge hydrolysate was 5 to 6 mmol/l/h, which was lower than the production rate in media with glucose, xylose, or a combination of these sugars (9–11 mmol/l/h), possibly due to the presence of inhibiting components in paper sludge hydrolysate (Kadar *et al.*, 2004). In culture during sugar fermentation, *Ca. saccharolyticus* grown on sucrose and *T. elfii* grown on glucose yielded 3.3 mol of H_2 and 2 mol of acetate per molC6 sugar unit. The H_2 production level was about 83% of the theoretical maximum. *Ca. saccharolyticus* and *T. elfii* reached maximum cell densities of 1.1×10^9 and 0.8×10^9 cells/ml, and their maximum H_2 production rates were 11.7 and 5.1 mmol/g dry weight/h, respectively (van Niel *et al.*, 2002). Co-culture of *Ca. saccharolyticus* and *Thermotoga neapolitana* simultaneously and completely utilized all pentoses, hexoses and oligomeric saccharides present in miscanthus hydrolysate (a total sugar concentration up to 17 g/l) with production of 2.9–3.4 mol H_2/mol hexose, corresponding to 74 to 85% of the theoretical yield (de Vrije *et al.*, 2009). O-Thong *et al.* (1998) compared microbial communities in H_2 producing reactors (70°C), fed either with xylose or lignocellulosic hydrolysate, and identified the presence of *Thermoanaerobacterium* spp. only in reactors fed with the hydrolysate. Other major groups of hydrogen producers detected included *Caldicellulosiruptor* spp., other Firmicutes, and archaea (O-Thong *et al.*, 2008).

It has been suggested that a co-culture composed of cellulolytic and saccharolytic bacteria could convert cellulosic materials to ethanol more effectively (Erbeznik *et al.*, 1997). For example, the co-culture of *T. ethanolicus* JW200 with *C. thermocellum* JW20 allows for the utilization of the wide range of substrates present in plant biomass, including xylose, that is not otherwise utilized by *C. thermocellum* (Erbeznik *et al.*, 1997). These strains, isolated from a Louisiana cotton bale, showed cellulose degradation and subsequent growth at 60°C and pH 6.1–7.5 in the presence of as much as 10% w/v cellobiose (300 mM), 6% glucose (333 mM), 5% xylose (333 mM), and 6% fructose (333 mM). The fermentation products upon cellulose degradation were ethanol (0.92 μmol/ml), acetate, lactate, H_2 (1.4 μmol/ml), and CO_2 (Freier *et al.*, 1988). The latter study elucidated the effect of externally added H_2 in increasing the formation of ethanol during growth on cellulose or cellobiose (Freier *et al.*, 1988). Another example of a co-culture involves an extremely thermophilic cellulolytic Gram-negative anaerobe strain H173 isolated from a New Zealand hot spring and an ethanologenic extreme thermophile, *Thermoanaerobacter wiegelii* strain Rt8.B1. Strain H173 could completely degrade up to 10 g/l of Avicel at 70°C and pH 7.0 (Hudson *et al.*, 1990). Two growth phases have been observed for this bacterium – an initial phase during which a cellulase complex was produced and carbohydrates were fermented to form acetic and lactic acids, followed by a second phase during which the cellular metabolism stopped but the cellulase complex continued to actively break down cellulose (with accumulation of glucose in the medium). When strain H173 was co-cultured with *T. wiegelii* strain Rt8.B1, glucose was fermented to ethanol (12.7 mM) and acetate (11.9 mM), and no reducing sugars accumulated in the culture medium (Hudson *et al.*, 1990; Cook *et al.*, 1996).

Another enrichment, dominated by bacteria closely affiliated with *T. thermohydrosulfuricus*, produced 1.21 molethanol/mol glucose, and 0.68 molH_2/mol glucose at 78°C (Koskinen *et al.*, 2008a). Even higher ethanol production by a co-culture of *Clostridium uzonii* AK15 and *Thermoanaerobacterium aciditolerans* AK17 has been reported (Koskinen *et al.*, 2008b). In the co-culture, *C. uzonii* AK15 was the main hydrogen producer (up to 1.9 molH_2/mol xylose, 33% of the theoretical maximum), while *T. aciditolerans* AK17 produced ethanol from both glucose and xylose, up to 1.6 molethanol/mol glucose (80%) and 1.1 molethanol/mol xylose (66%), respectively. The hydrogen yields by *T. aciditolerans* AK17 were significantly lower, 1.2 molH_2/mol glucose (30% of theoretical maximum) and 1.0 molH_2/mol xylose (30%). Stable and continuous ethanol and hydrogen co-production was achieved in a continuous-flow 60°C bioreactor by a co-culture of *C. uzonii* AK15 and *T. aciditolerans* AK17, with an ethanol yield of 1.35 mol/mol glucose and a H_2 yield of 0.08 mol/mol glucose at a production rate of 6.1 mmol/l/h. *T. aciditolerans* AK17 showed high ethanol tolerance, up to 4% (v/v), and over time the AK17 became the dominant bacterium in the bioreactor (Koskinen *et al.*, 2008b). Since bacteria of the genus *Thermoanaerobacter* cannot degrade cellulose, in co-culture they survive by cross-feeding on soluble carbohydrates (e.g. glucose, cellobiose, and small cello-oligomers) released through the action of the cellulase complex of cellulolytic bacteria. However, if the same culture is subcultured on glucose, cellobiose, or xylose, which are growth substrates for *Thermoanaerobacter* spp., this stimulates the growth of the *Thermoanaerobacter* spp. cell population and accounts for the apparent ability of cellulolytic bacteria to utilize xylose (Erbeznik *et al.*, 1997).

Thermophilic cellulolytic bacteria from hot springs of Yellowstone National Park

This section describes research conducted at the Oak Ridge National Laboratory (ORNL, Oak Ridge, TN, USA) within the framework of the BioEnergy Science Centre (BESC), one of the three Bioenergy Research Centres sponsored by the United States Department of Energy. To our knowledge, the hot springs located within Yellowstone National Park (YNP) have not been extensively studied with respect to isolation of cellulolytic bacteria. Hence, our initial research efforts have been directed to the enrichment, identification and isolation of thermophilic microorganisms that are capable of degrading lignocellulosic biomass, from samples collected at select sites within YNP.

Sampling sites and anaerobic batch enrichments

A total of 134 samples consisting of spring water, sediments, microbial mats, and plant material (100 ml each) were collected from 7 different

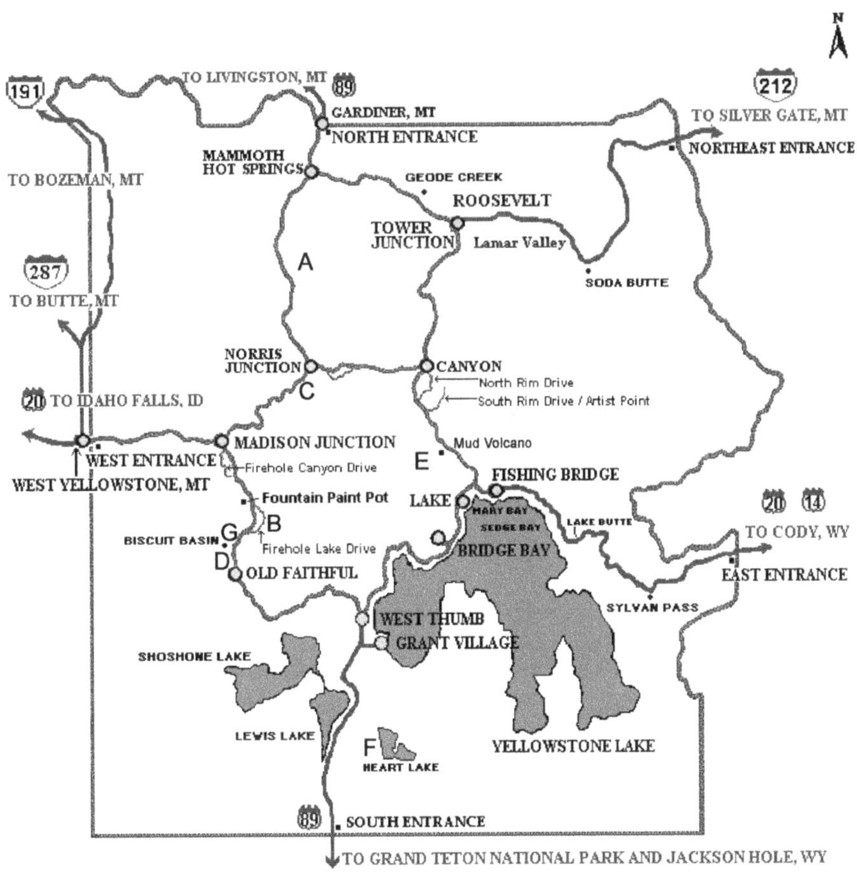

Figure 7.3 Locations (A through G) of sampling sites are shown on a map of Yellowstone National Park (WY, USA). The primary anaerobic enrichments established from these sites are indicated in Table 7.4.

locations (10 sites) within YNP (Fig. 7.3). Samples varied in their ambient temperature, from 50 to 93.7°C, and pH, from 3 to 9.5 (Table 7.4 and Fig. 7.4A). Immediately after collection, samples were reduced by adding a few crystals of sodium dithionite and 1 ml of 2.5% cysteine HCl, pH 7.2. The samples were kept at room temperature during transportation to ORNL. In the laboratory, enrichment cultures inoculated with 2% (v/v) of an environmental sample were initiated using 10 ml of a fermentative cellulolytic anaerobes (FCA) liquid growth medium at pH 7.0 (Hamilton-Brehm *et al.*, 2010), containing either dilute-acid pretreated switchgrass or *Populus* shavings (0.1% w/v) as the sole carbon and energy source. These primary anaerobic enrichments, incubated at temperatures ranging from 55 to 85°C in increments of 5°C, were screened for turbidity, pH changes, gas production, and end product formation. Tubes positive for growth were transferred to 50 ml of fresh FCA medium (N_2:CO_2 headspace; 80%:20%) and subcultured at least three consecutive times (Table 7.4).

Physiological characteristics of batch cellulolytic enrichments

Enrichment cultures inoculated with samples collected from Obsidian Pool, and thermal features around the Witch Creek/Heart Lake, and White Creek areas (Fig. 7.3) displayed gas production and visible turbidity upon repeated transfer in medium containing pretreated switchgrass. Stable enrichments from these sites were selected for further analysis using the same FCA medium with 0.2% (w/v) of dilute-acid pretreated substrates. End-products formed during growth on pretreated switchgrass included H_2, CO_2, acetate, and traces of ethanol. Acetate was the major end-product with concentrations reaching up to 0.73, 0.24, 0.18 g/l for the Obsidian Pool, Heart Lake, and White Creek enrichment cultures, respectively, after 120 hours of incubation at 75°C. Among the three, the Obsidian Pool enrichments displayed the fastest growth on pretreated switchgrass, and hydrolysed approximately 50% of the starting pretreated biomass within 24 hours. The Obsidian Pool enrichment also hydrolysed crystalline

Table 7.4 Anaerobic batch enrichments established in this study using dilute-acid pretreated switchgrass or *Populus* as sole carbon and energy source

			Enrichments – growth[2]	
Location[1]	Site	Environmental conditions	Switchgrass (*Panicum virgatum*) (60°C/80°C)	Populus *(Populus deltoides)* (60°C/80°C)
A	Amphitheater Stream	50–74.3°C, pH 3	–/++	–/–
B	Octopus Spring	80.6°C, pH 7.5	+++/+++	++/++
	White Creek area	73–92.1°C, pH 7	+/–	+/–
	Five Sisters	69.5–85°C, pH 8.2–8.5	–/+	–/+
	Fire Hole Spring	78°C, pH 7.5–8	–/+	–/–
C	Artist Paintpots	55.5–93.7°C, pH 6–9.5	++/–	+/+
D	Black Sands Basin	59–87°C, pH 7–9	–/+	–/–
E	Obsidian Pool area	61.9–90°C, pH 5.5	++/++++	–/++++
F	Witch Creek area	74–92°C, pH 7–8.5	+++/++	–/–
G	Rabbit Creek	64.7–92°C, pH 4–9	+++/+++	++/++

[1]Location of sites is shown on YNP map, Fig. 7.3.

[2]Growth in liquid enrichments was characterized based on turbidity using a scale from + (poor growth) to ++++ (very good growth) and – (no growth).

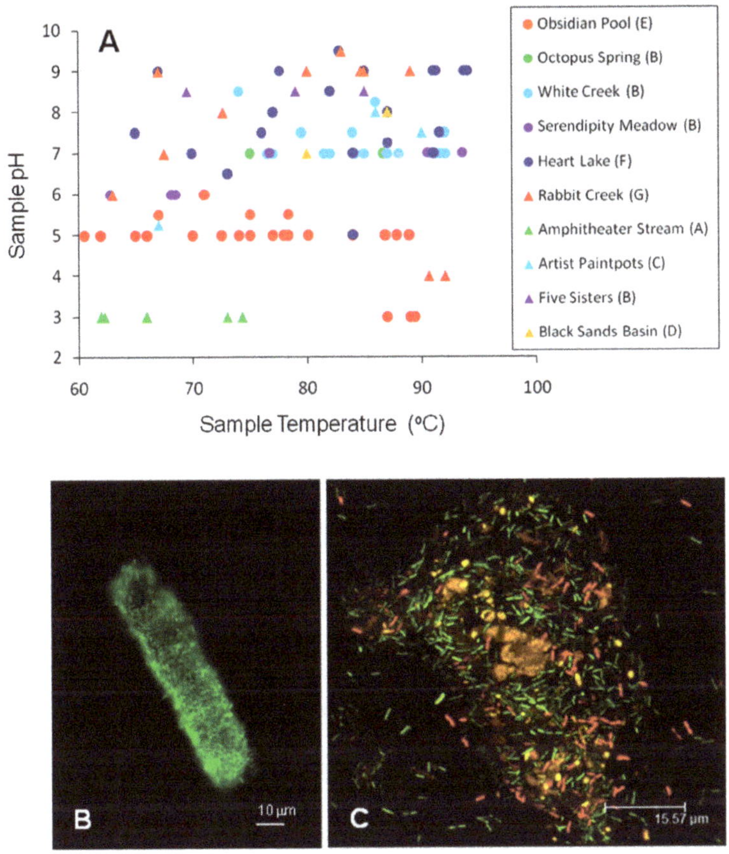

Figure 7.4 (A) Temperature and pH values of samples collected across various sites within YNP. (B) Epifluorescent microscopy imaging of SYTO9-stained cells shows attachment of bacterial cells to particles of switchgrass used in enrichment cultures. (C) Laser confocal microscopy visualization of Live/Dead dye-stained bacterial cells confirmed the predominance of live cells (green) attached to the insoluble substrate (dead cells are red).

cellulose (Avicel), Birchwood xylan and pre-treated *Populus,* but the overall cell densities were lower during growth on *Populus* as compared with either switchgrass or Avicel.

Staining the cultures with SYTO9 (Invitrogen, Carlsbad, CA) and imaging via epifluorescent or laser confocal microscopy revealed cells attached to particles of insoluble biomass (Fig. 7.4B,C). The cell morphology and overall cell densities in the established enrichments were examined using phase contrast and epifluorescent microscopy (Axioskop 2 plus, Zeiss). Enrichments were classified on the basis of observed morphotypes into cultures with high variety of morphotypes (≥ 3) and those with a prevalence of a particular morphotype (e.g. rods or cocci). Enrichment cultures with stable growth, multiple morphotypes and a cell density of ~10^8 cells/ml were selected for isolation attempts. An isolation method using flow cytometry was developed in order to circumvent plating on solid or semisolid media and for isolation of cells grown on insoluble biomass substrates (Elkins *et al.*, 2008). The cultures were chilled before sorting and oxygen exposure was minimized to preserve cell viability. Cells from enrichment cultures were dislodged from particles of insoluble biomass by shaking and sorted with an InFlux Cell Sorting Flow Cytometer (Cytopeia, Seattle, WA, USA) into 48-well plates. Each individual well in the plate, containing

0.6% (w/v) of pretreated switchgrass in 0.5 ml of reduced FCA medium, was inoculated with a single cell and the plate was incubated under anaerobic conditions at the respective enrichment culture growth temperature. Positive growth in the wells was detected using a pH indicator for acid production and was confirmed by microscopy. Wells displaying growth were considered putative isolates and were subcultured into Balch tubes containing fresh FCA medium with 0.1–0.5% (w/v) dilute-acid-pretreated switchgrass or Avicel. Successfully recovered cultures displaying growth on cellulosic substrates were resorted (and the above procedure repeated) and verified for purity by microscopy and sequencing of multiple PCR amplified 16S rRNA gene recombinants. A total of 155 putative isolates were recovered from multiple isolation attempts.

Community analyses of batch enrichments

The total community genomic DNA (cgDNA) was extracted using the MoBio PowerSoil™ kit (MoBio Laboratories, Inc., Carlsbad, CA). The 16S rRNA gene fragments were amplified using 8f and 1492r *Bacteria*-specific primers (Lane, 1991) and cloned using standard procedures. The plasmids from positive transformants were amplified using Rolling Circle Amplification (RCA) (TempliPhi process, Illustra™ GE Healthcare, UK). Subsequently, a total of 615 recombinants from enrichment cultures differing in growth substrate and temperature were sequenced (Vishnivetskaya *et al.*, 2009). Phylogenetic analysis of the 16S rRNA genes (Fig. 7.5) showed that the majority of the sequences were affiliated with the phylum Firmicutes (80.9% of clones), followed by Dictyoglomi (12.9%), Nitrospira (3.4%), and others (2.7%). The bacteria of the phylum Firmicutes fell into five groups. The most abundant group (54.4%) was represented by *Caldicellulosiruptor* spp., followed by *Caloramator* spp. (11.5%), *Thermoanaerobacter* spp. (8.8%), *Thermovenabulum ferriorganovorum* (4.1%), and *Clostridium xylanovorans* (2.0%). Similar results were obtained for enrichment cultures established using water from the hot springs of Uzon Caldera, Kamchatka, with ambient temperatures from 68 to 87°C and pH from 4.1 to 7.0 (Kublanov *et al.*, 2009). The latter study showed, using DGGE analysis of 16S rRNA gene fragments, that the bacterial communities developing on carbohydrates (cellulose, carboxymethyl cellulose, chitin, or agarose) contained the genera *Caldicellulosiruptor* and *Dictyoglomus*.

Dynamic changes in microbial community structure during continuous enrichment

We established a continuous-flow anaerobic enrichment culture to monitor the temporal changes in structure and composition of a thermophilic cellulolytic community. The enrichment culture was inoculated with 20 ml of spring water collected from the peripheral areas of Obsidian Pool and was maintained at a constant temperature of 65°C by using a jacketed vessel connected to a recirculating water/glycerol bath. FCA medium supplemented with 10 mM MOPS buffer (pH 7.0) was continuously added to the vessel at 8.3 ml/h. The culture initially contained 1 litre of FCA medium with 1% (w/v) Avicel as the carbon source and was supplemented with additional carbon every 7 days by adding 50 ml of nitrogen sparged FCA medium containing 4 g of Avicel. Bacterial growth was monitored by direct microscopic cell counts. Cell density increased from an initial 5.9×10^6 cells/ml to 6.3×10^8 cells/ml within two days, then decreased to 1.8×10^8 cells/ml after another 2 days, after which it remained essentially stable for the next 38 days (Fig. 7.6A). Acetate production by the thermophilic community, measured using HPLC, followed the growth curve with a maximum concentration of 5.33 mM on day 6, and then remained in the range of 4.39–3.70 mM from day 8 to day 21 (Fig. 7.6A). The pH remained circumneutral due to the continual addition of buffered FCA medium. At defined time points suspended Avicel and cell biomass were collected (Fig. 7.6A) and total community genomic DNA was extracted, amplified and cloned as described above. Sequences were aligned against the most similar sequences in the RDP (http://rdp.cme.msu.edu) and NCBI (http://www.ncbi.nlm.nih.gov/blast/Blast.cgi) databases.

Changes in the bacterial community composition at different time points are shown in Fig. 7.6B. After 48 hours of incubation, along with

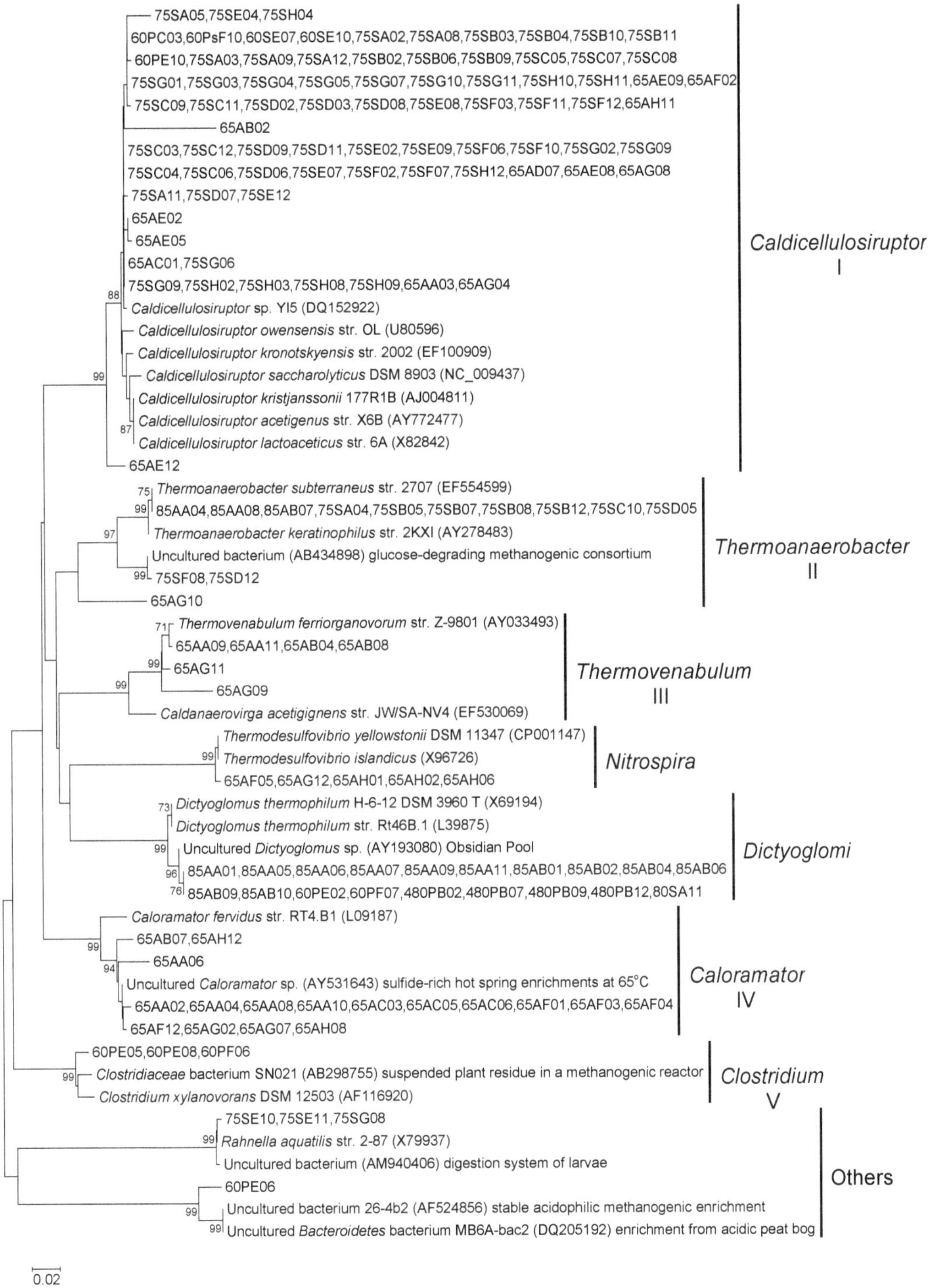

Figure 7.5 Phylogenetic analysis of environmental clones derived from batch cellulolytic enrichments. The tree was produced by a neighbour-joining method. Bootstrap values were based on 1000 replicates. The scale bar represents 0.02 changes per nucleotide position. Sequence names were designated as follows: first number, enrichment growth temperature (60°C, 65°C, 70°C, 75°C, 80°C or 85°C), followed by a letter indicating the carbon source (S, switchgrass; P, *Populus*; A, Avicel), followed by the specific location of the clone on the sequencing plate.

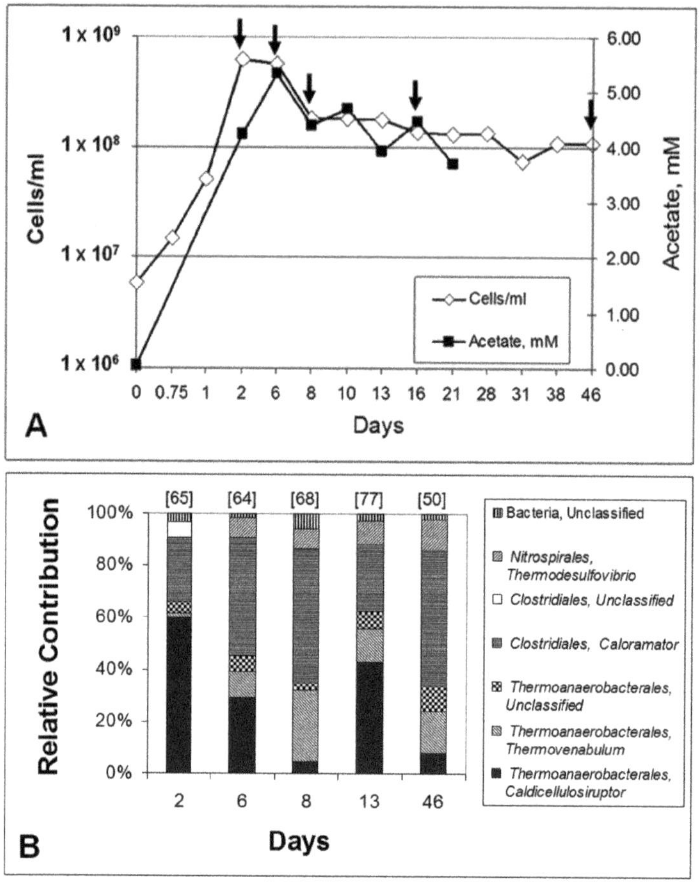

Figure 7.6 (A) Cell density and acetate production during continuous growth of Avicel enrichment of Obsidian Pool sample at 65°C for 46 days. Arrows indicate the time points at which samples used for clone library construction were collected. (B) Bacterial community composition as determined by 16S rRNA gene Sanger sequencing. The number of sequences obtained from each sample is given in parentheses. The bacterial taxonomy at order and genus levels is given on the right side of the figure.

unclassified bacteria (<80% homology, RDP database), two orders of the class Clostridia in the phylum Firmicutes, namely Thermoanaerobacterales (66.2% of total) and Clostridiales (30.8%), were detected. About 60.0% of Thermoanaerobacterales recombinants were affiliated with the genus *Caldicellulosiruptor*, which comprises hydrogen- and acetate-producing cellulolytic organisms. One of the sequences was 91% (687/751) identical to *T. ferriorganovorum*, a novel thermophilic, anaerobic bacterium isolated from a hydrothermal feature in the Uzon caldera on the Kamchatka peninsula (Zavarzina *et al.*, 2002). *T. ferriorganovorum* is capable of molecular hydrogen utilization in the presence of Fe(III) and an organic electron donor. The remaining 4.6% of recombinants were unclassified. Among Clostridiales, unclassified bacteria comprised 6.2%, while 24.6% recombinants were affiliated with the genus *Caloramator*. Another member of the genus *Caloramator, C. australicus* RC3T, a strictly anaerobic thermophilic bacterium, has been isolated from microbial mats colonizing thermal waters of the Great Artesian Basin, Australia (Ogg and Patel, 2009). *C. australicus* requires yeast extract for growth on energy sources such as cellobiose, cellulose, xylan and glucose (Ogg and Patel, 2009), with ethanol and acetate as the end products of glucose fermentation. In the presence of yeast extract, *C. australicus* was able to reduce iron (III), manganese (IV) and elemental sulfur, although there was no utilization of molecular hydrogen and/or carbon dioxide in the presence or absence of iron (III).

At subsequent time points (6–46 days) we observed the disappearance of unclassified Clostridiales and an increase in the relative contribution of the genus *Caloramator* (from 24.6% to 26.0–52.0%), while the relative abundance of the genus *Caldicellulosiruptor* decreased from 60.0% to < 42.9%. Starting from day 6 we observed the appearance of sequences (7.8–12.0%) which were 99% identical to Nitrospira isolated from methanogenic sludge – *Thermodesulfovibrio yellowstonii* DSM 11,347 (also named as strain ATCC 51303) and *Thermodesulfovibrio islandicus* TGL-LS1, both obligately anaerobic, thermophilic, sulfate-reducing bacteria (Sekiguchi *et al.*, 2006). Nitrospira-like bacteria are known to be nitrite oxidizers and they tend to be slow-growing organisms. The presence of bacteria of the phylum Nitrospirae in Obsidian Pool sediment has been reported previously (Hugenholtz *et al.*, 1998).

Strain isolation and phylogenetic identification

A total of 155 strains were isolated from established anaerobic thermophilic (60°C, 75°C and 85°C) batch enrichments by sorting with a flow cytometer (see description earlier). Detailed 16S rRNA-based phylogenetic analysis of 81 strains (Fig. 7.7) showed that 57% and 43% of the total isolates were close relatives of, respectively, *Caldicellulosiruptor* spp. and *Thermoanaerobacter* spp. Isolates identified as *Caldicellulosiruptor* spp. were closely related to *Caldicellulosiruptor owensensis* strain OL (ATCC 700167) (97.2–99.3% sequence identity) and to *Caldicellulosiruptor* sp. strain Y15 (96.8–99.3% sequence identity). Isolates affiliated with the genus *Thermoanaerobacter* were 96.9–98.8% identical to *Thermoanaerobacter tengcongensis* strain MB4 (JCM 11007). *Ca. owensensis*

Figure 7.7 Phylogenetic analysis of single cell isolates obtained upon sorting of batch cellulolytic enrichments by flow cytometry. The scale bar represents 0.02 changes per nucleotide position. Isolate names were designated and the phylogenetic tree was constructed as described in the legend to Fig. 7.5.

and *T. tengcongensis,* are both anaerobic, extremely thermophilic, rod-shaped non-spore-forming bacteria. *Ca. owensensis* was isolated from a sediment sample from temperate Owens Lake, California, while *T. tengcongensis* was isolated from a Chinese hot spring. Both organisms grow at temperatures between 50°C and 80°C, with an optimum at 75°C, and in a pH range of 5.5–9.0 with an optimum around pH 7.5. However, the two organisms differ in their ability to utilize cellulose. *Ca. owensensis* can utilize cellulose, xylan, starch, pectin, glucose, xylose, and other biomass sugars with a doubling time of 7.3 h during growth on glucose. The end products of glucose fermentation are acetate, lactate, H_2, CO_2, and traces of ethanol (Huang *et al.*, 1998). *T. tengcongensis* cannot utilize cellulose but can grow on other substrates, including starch, cellobiose, glucose, galactose, and mannose. The main end products from glucose fermentation are acetate (1.0 molacetate/mol glucose), CO_2 (1.5 mol), ethanol (0.7 mol), and H_2 (0.3 mol), with growth inhibition by hydrogen accumulation (Xue *et al.*, 2001). The *Caldicellulosiruptor* sp. strain OB47 isolated in our study was able to generate up to 37 mM acetate and 2 mM ethanol with no detectable lactate during the course of pretreated switchgrass fermentation (Hamilton-Brehm *et al.*, 2010). Screening of the other bacterial strains isolated from our enrichment cultures (e.g. members of the genera *Thermoanaerobacter* and *Caloramator*) for elevated cellulolytic and saccharolytic activity is in progress. Comparing the metabolic characteristics of various cellulolytic microorganisms will provide evidence whether the isolates derived from YNP enrichments are attractive candidates for further research and development towards their use in biofuel production.

Future trends

Continually rising crude oil prices and environmental concerns have revived interest in renewable sources of sustainable energy such as biofuels. While the first-generation ethanol industry depended on starch or sugars derived from maize grain, sugarcane and other crops, and thus potentially competed with the food supply, second-generation ethanol derived from cellulosic biomass promises to deliver a renewable biofuel from non-food feedstock. However, many challenges remain in establishing an economically viable and sustainable lignocellulosic bioethanol industry, and these need to be addressed. With regards to biomass, the challenges include engineering transgenic plants to increase feedstock productivity and decrease biomass recalcitrance without loss of structural integrity and tolerance traits. With regards to cellulolytic microorganisms, challenges include engineering robust organisms that can degrade plant polymers at high rates and convert all biomass-derived sugars to the fuel of interest at high yields, while being tolerant to pretreatment-derived inhibitors, fermentation products and other process conditions (Blaschek and White, 1995; Lynd *et al.*, 2002; Keller and Zengler, 2004; Himmel *et al.*, 2007).

Bioethanol production relies on the availability of individual organisms or consortia with high capability for degradation of plant polymers and broad substrate specificity for co-utilization of a wide range of sugars (C_5 and C_6) present in lignocellulosic biomass. Strains should be capable of converting concentrated feedstock (e.g. ≥ 50 g of carbohydrate per litre) at high rates with maximal productivity (Lynd and Zhang, 2002; Zhang *et al.*, 2007). At the same time, organisms should also be robust enough to function adequately under conditions typically encountered in an industrial environment. This would be achieved either by genetic modification of ethanologenic strains that are currently available or through the discovery of novel cellulolytic microorganisms.

Natural ecosystems offer a wealth of resources for bioprospecting, and our group will continue to tap into the thermal features within Yellowstone National Park to isolate novel organisms with high lignocellulolytic potential. The application of the novel flow cytometry-based technique for isolating bacterial strains from enrichment cultures, as described here, will allow a screening of isolates for robustness, fast growth and high cellulolytic activity (Elkins *et al.*, 2008). High-throughput approaches are needed for large-scale screening of naturally isolated strains, as well as strains engineered for high product yields and other desired process characteristics. Many of the challenges could be efficiently tackled through the successful application of biotechnological techniques.

Several tools are currently being developed for improving bioenergy production systems. Systems biology analysis using 'omics' tools probably plays a key role in the development of robust organisms tailored for industrial biofuel production. High-throughput genome sequencing probably highlights the cellulolytic and metabolic potential of newly isolated organisms for degradation and utilization of lignocellulosic biomass-derived polymers and sugars. Functional genomics employing proteomics, transcriptomics, metabolomics and other 'omics' methods probably reveal the active or inactive enzymatic and metabolic networks in these organisms. Development of novel technologies is necessary for identifying factor(s) that are limiting bacterial growth and performance, particularly under conditions that mimic biofuel production processes. These technologies will probably guide both bioprocess development and organism development with pointers to relevant gene targets for genetic manipulation. In order to engineer strains with desired attributes for biofuel production, it is necessary to develop practical and reproducible genetic tools that permit gene expression and gene knockout in a variety of cellulolytic and ethanologenic microorganisms. Currently, many cellulolytic organisms, in particular thermophilic anaerobes, are not easily engineered through the use of traditional microbial genetic systems, posing a challenge and opportunity for biotechnology tool development (Blaschek and White, 1995; Lynd *et al.*, 2002).

Web resources

There are a number of web sites that discuss various aspects of cellulolytic microorganisms, bioenergy and biofuel development:

A list of cellulolytic bacteria described in the literature as continuously hydrolysing crystalline cellulose substrates like Avicel or filter paper can be found at the web site of the Technical University of Munich (http://www.weihenstephan.de/mbiotec/cellmo.htm).

The genome sequences of cellulolytic thermophilic bacteria are available through the Integrated Microbial Genomes (IMG) system (http://img.jgi.doe.gov/cgi-bin/pub/main.cgi?page=home) and NCBI (http://www.ncbi.nlm.nih.gov/Genomes/) web pages. IMG serves as a community resource for comparative analysis and annotation of all publicly available genomes from three domains of life in a uniquely integrated context. NCBI provides several genomic biology tools and resources, including organism-specific pages that include links to many web sites and databases relevant to that species.

The description of the families of structurally related catalytic and carbohydrate-binding modules (or functional domains) of enzymes that degrade, modify, or create glycosidic bonds can be found at the CAZy database (http://www.cazy.org/). This website also provides links to commercially available carbohydrate-active enzymes (CAE) using CAZy nomenclature.

The BioEnergy Science Centre (BESC) web site (http://bioenergycenter.org/) introduces a person to the mission of the United States Department of Energy in developing alternative fuel solutions such as converting plant biomass into biofuels. BESC is a multi-institutional partnership that is connecting the world's leading scientific minds and resources for finding bioenergy solutions. The goal of BESC is to help develop viable, plentiful and clean alternative biofuel sources for generations to come.

Knowledge Base (http://besckb.ornl.gov) is another web site that provides integrated data and information representing systems biology analysis of key plants, microbes, and molecules within BESC. It is a resource for exploring metabolic networks, regulatory pathways, signalling maps, and gene expression, with links to primary genome sequence annotation and underlying experimental data, for several plant and bacterial species of importance in biofuels production.

The Food and Agriculture Organization (FAO) draws attention to high energy prices, environmental degradation, sustainability of current energy systems and the competition of food crops versus energy crops (http://www.fao.org/bioenergy/home/en/).

A list of worldwide organizations, conferences, papers and other information related to petroleum, bioenergy, natural gas, biomass, fossil-fuel and solar energy is provided at http://www.petrolpump.co.in/energy-resources-directory/b.htm.

Acknowledgements

This work was supported by the BioEnergy Science Centre (BESC), which is a US Department of Energy (DOE) Bioenergy Research Centre supported by the Office of Biological and Environmental Research in the DOE Office of Science, Oak Ridge National Laboratory. Oak Ridge National Laboratory is managed by UT-Battelle, LLC, for the US DOE under contract DE-AC05-00OR22725. We thank the Yellowstone National Park Service for coordinating and allowing sampling under permit #YELL-2008-SCI-5714. Microscopic images were collected by Dr. Jennifer L. Morrell-Falvey. We thank Sue Carroll, Steve Allman, and Marilyn Kerley for technical support.

References

Aksenova, H.Y., Rainey, F.A., Janssen, P.H., Zavarzin, G.A., and Morgan, H.W. (1992). *Spirochaeta thermophila* sp. nov., an obligately anaerobic, polysaccharolytic, extremely thermophilic bacterium. Int. J. Syst. Bacteriol. *42*, 175–177.

Alfredsson, G.A., Kristjansson, J.K., Hjorleifsdottir, S., and Stetter, K.O. (1988). *Rhodothermus marinus*, gen. nov, sp. nov., a thermophilic, halophilic bacterium from submarine hot springs in Iceland. J. Gen. Microbiol. *134*, 299–306.

Ali, B.R.S., Romaniec, M.P.M., Hazlewood, G.P., and Freedman, R.B. (1995a). Characterization of the subunits in an apparently homogeneous subpopulation of *Clostridium thermocellum* cellulosomes. Enzyme Microbial Technol. *17*, 705–711.

Ali, B.R.S., Zhou, L.Q., Graves, F.M., Freedman, R.B., Black, G.W., Gilbert, H.J., and Hazlewood, G.P. (1995b). Cellulases and hemicellulases of the anaerobic fungus *Piromyces* constitute a multiprotein cellulose-binding complex and are encoded by multigene families. FEMS Microbiol. Lett. *125*, 15–21.

Bayer, E.A., Kenig, R., and Lamed, R. (1983). Adherence of *Clostridium thermocellum* to cellulose. J. Bacteriol. *156*, 818–827.

Bayer, E.A., Shimon, L.J.W., Shoham, Y., and Lamed, R. (1998). Cellulosomes – structure and ultrastructure. J. Structural Biol. *124*, 221–234.

Bayer, E.A., Belaich, J.P., Shoham, Y., and Lamed, R. (2004). The cellulosomes: Multienzyme machines for degradation of plant cell wall polysaccharides. Annu. Rev. Microbiol. *58*, 521–554.

Bayer, E.A., Shoham, Y., and Lamed, R. (2006). Cellulose-decomposing bacteria and their enzyme systems. In The Prokaryotes, Dworkin, M., Falkow, S., Rosenberg, E., Schleifer, K.–H., Stackerbrandt, E., eds (Springer, New York), pp. 578–617.

Bayer, E.A., Lamed, R., White, B.A., and Flint, H.J. (2008). From cellulosomes to cellulosomics. Chemical Record *8*, 364–377.

Beguin, P., and Aubert, J.P. (1994). The biological degradation of cellulose. FEMS Microbiol. Rev. *13*, 25–58.

Beguin, P., and Lemaire, M. (1996). The cellulosome: An exocellular, multiprotein complex specialized in cellulose degradation. Crit. Rev. Biochem. Mol. Biol. *31*, 201–236.

Belaich, J.P., Tardif, C., Belaich, A., and Gaudin, C. (1997). The cellulolytic system of *Clostridium cellulolyticum*. J. Biotechnol. *57*, 3–14.

Bergquist, P.L., Gibbs, M.D., Morris, D.D., Te'o, V.S., Saul, D.J., and Moran, H.W. (1999). Molecular diversity of thermophilic cellulolytic and hemicellulolytic bacteria. FEMS Microbiol. Ecol. *28*, 99–110.

Bettiga, M., Bengtsson, O., Hahn-Hagerdal, B., and Gorwa-Grauslund, M.F. (2009). Arabinose and xylose fermentation by recombinant *Saccharomyces cerevisiae* expressing a fungal pentose utilization pathway. Microb. Cell. Fact. *8*, 12–40.

Blaschek, H.P., and White, B.A. (1995). Genetic systems development in the *Clostridia*. FEMS Microbiol. Rev. *17*, 349–356.

Blumer-Schuette, S.E., Kataeva, I., Westpheling, J., Adams, M.W.W., and Kelly, R.M. (2008). Extremely thermophilic microorganisms for biomass conversion: status and prospects. Curr. Opin. Biotechnol. *19*, 210–217.

Bredholt, S., Mathrani, I.M., and Ahring, B.K. (1995). Extremely thermophilic cellulolytic anaerobes from Icelandic hot springs. Antonie Van Leeuwenhoek *68*, 263–271.

Bredholt, S., Sonne-Hansen, J., Nielsen, P., Mathrani, I.M., and Ahring, B.K. (1999). *Caldicellulosiruptor kristjanssonii* sp. nov., a cellulolytic extremely thermophilic, anaerobic bacterium. Int. J. Syst. Bacteriol. *49*, 991–996.

Burdette, D.S., Jung, S.H., Shen, G.J., Hollingsworth, R.I., and Zeikus, J.G. (2002). Physiological function of alcohol dehydrogenases and long-chain (C-30) fatty acids in alcohol tolerance of *Thermoanaerobacter ethanolicus*. Appl. Environ. Microbiol. *68*, 1914–1918.

Busto, M.D., Ortega, N., and PerezMateos, M. (1996). Location, kinetics and stability of cellulases induced in *Trichoderma reesei* cultures. Bioresour. Technol. *57*, 187–192.

Cantarel, B.L., Coutinho, P.M., Rancurel, C., Bernard, T., Lombard, V., and Henrissat, B. (2009). The Carbohydrate-Active EnZymes database (CAZy): an expert resource for Glycogenomics. Nucleic Acids Res. *37*, D233-D238.

Chang, M.M., Chou, T.Y.C., and Tsao, G.T. (1981). Structure, pretreatment and hydrolysis of cellulose. In: Bioenergy, (Springer, Berlin/Heidelberg), pp. 15–42.

Cho, N.S., Hatakka, A.I., Rogalski, J., Cho, H.Y., and Ohga, S. (2009). Directional degradation of lignocellulose by *Phlebia radiata*. J. Fac. Agric. Kyushu Uni. *54*, 73–80.

Cook, G.M., Rainey, F.A., Patel, B.K.C., and Morgan, H.W. (1996). Characterization of a new obligately anaerobic thermophile, *Thermoanaerobacter wiegelii* sp. nov. Int. J. Syst. Bacteriol. *46*, 123–127.

Dashtban, M., Schraft, H., and Qin, W. (2009). Fungal bioconversion of lignocellulosic residues; opportunities and perspectives. Int. J. Biol. Sci. *5*, 578–595.

Demain, A.L., Newcomb, M., and Wu, J.H.D. (2005). Cellulase, clostridia, and ethanol. Microbiol. Mol. Biol. Rev. *69*, 124–154.

Deshpande, P., Nair, S., and Khedkar, S. (2009). Water hyacinth as carbon source for the production of cellulase by *Trichoderma reesei*. Appl. Biochem. Biotechnol. *158*, 552–560.

Desvaux, M., Guedon, E., and Petitdemange, H. (2001). Kinetics and metabolism of cellulose degradation at high substrate concentrations in steady-state continuous cultures of *Clostridium cellulolyticum* on a chemically defined medium. Appl. Environ. Microbiol. *67*, 3837–3845.

Detroy, R.W., and Stjulian, G. (1983). Biomass conversion – fermentation chemicals and fuels. CRC Crit. Rev. Microbiol. *10*, 203–228.

Dien, B.S., Cotta, M.A., and Jeffries, T.W. (2003). Bacteria engineered for fuel ethanol production: current status. Appl. Microbiol. Biotechnol. *63*, 258–266.

Ding, S.Y., Xu, Q., Crowley, M., Zeng, Y., Nimlos, M., Lamed, R., Bayer, E.A., and Himmel, M.E. (2008). A biophysical perspective on the cellulosome: new opportunities for biomass conversion. Curr. Opin. Biotechnol. *19*, 218–227.

Dinis, M.J., Bezerra, R.M.F., Nunes, F., Dias, A.A., Guedes, C.V., Ferreira, L.M.M., Cone, J.W., Marques, G.S.M., Barros, A.R.N., and Rodrigues, M.A.M. (2009). Modification of wheat straw lignin by solid state fermentation with white-rot fungi. Bioresour. Technol. *100*, 4829–4835.

Doi, R.H. (2008). Cellulases of mesophilic microorganisms: cellulosome and noncellulosome producers. Ann. N.Y. Acad. Sci. *1125*, 267–279.

Doi, R.H., and Kosugi, A. (2004). Cellulosomes: Plant-cell-wall-degrading enzyme complexes. Nat. Rev. Microbiol. *2*, 541–551.

Doi, R.H., and Tamaru, Y. (2001). The *Clostridium cellulovorans* cellulosome: An enzyme complex with plant cell wall degrading activity. Chem. Rec. *1*, 24–32.

Elkins, J.G., Vishnivetskaya, T.A., Podar, M., Phelps, T.J., and Keller, M. (2008). High-throughput isolation of extreme thermophiles that produce ethanol from switchgrass and *Populus*. Paper presented at: 30th Symposium on Biotechnology for Fuels and Chemicals (Society for Industrial Microbiology, New Orleans, LA).

Erbeznik, M., Jones, C.R., Dawson, K.A., and Strobel, H.J. (1997). *Clostridium thermocellum* JW20 (ATCC 31549) is a coculture with *Thermoanaerobacter ethanolicus*. Appl. Environ. Microbiol. *63*, 2949–2951.

Fendri, I., Tardif, C., Fierobe, H.P., Lignon, S., Valette, O., Pages, S., and Perret, S. (2009). The cellulosomes from *Clostridium cellulolyticum*. FEBS J. *276*, 3076–3086.

Freier, D., Mothershed, C.P., and Wiegel, J. (1988). Characterization of *Clostridium thermocellum* JW20. Appl. Environ. Microbiol. *54*, 204–211.

Fujii, K., and Shintoh, Y. (2006). Degradation of mikan (Japanese mandarin orange) peel by a novel *Penicillium* species with cellulolytic and pectinolytic activity. J. Appl. Microbiol. *101*, 1169–1176.

Gal, L., Gaudin, C., Belaich, A., Pages, S., Tardif, C., and Belaich, J.P. (1997). CelG from *Clostridium cellulolyticum*: a multidomain endoglucanase acting efficiently on crystalline cellulose. J. Bacteriol. *179*, 6595–6601.

Garncarek, Z., and Kociolek-Balawejder, E. (2009). Biobutanol. Perspectives of the production development. Przemysl Chemiczny *88*, 658–666.

Gibbs, M.D., Saul, D.J., Luthi, E., and Bergquist, P.L. (1992). The beta-mannanase from *Caldocellum saccharolyticum* is part of a multidomain enzyme. Appl. Environ. Microbiol. *58*, 3864–3867.

Gilkes, N.R., Henrissat, B., Kilburn, D.G., Miller, R.C., and Warren, R.A.J. (1991). Domains in microbial beta-1,4-glycanases – sequence conservation, function, and enzyme families. Microbiol. Rev. *55*, 303–315.

Hamilton-Brehm, S.D., Mosher, J.J., Vishnivetskaya, T., Podar, M., Carroll, S., Allman, S., Phelps, T.J., Keller, M., and Elkins, J.G. (2010). *Caldicellulosiruptor obsidiansis* sp. nov., an anaerobic, extremely thermophilic, cellulolytic bacterium isolated from Obsidian Pool, Yellowstone National Park. Appl. Environ. Microbiol. *76*, 1014–1020.

Han, S.O., Yukawa, H., Inui, M., and Doi, R.H. (2005). Effect of carbon source on the cellulosomal subpopulations of *Clostridium cellulovorans*. Microbiology *151*, 1491–1497.

Henrissat, B., and Davies, G. (1997). Structural and sequence-based classification of glycoside hydrolases. Curr. Opin. Struct. Biol. *7*, 637–644.

Himmel, M.E., Ding, S.Y., Johnson, D.K., Adney, W.S., Nimlos, M.R., Brady, J.W., and Foust, T.D. (2007). Biomass recalcitrance: Engineering plants and enzymes for biofuels production. Science *315*, 804–807.

Hreggvidsson, G.O., Kaiste, E., Holst, O., Eggertsson, G., Palsdottir, A., and Kristjansson, J.K. (1996). An extremely thermostable cellulase from the thermophilic eubacterium *Rhodothermus marinus*. Appl. Environ. Microbiol. *62*, 3047–3049.

Huang, C.Y., Patel, B.K., Mah, R.A., and Baresi, L. (1998). *Caldicellulosiruptor owensensis* sp. nov., an anaerobic, extremely thermophilic, xylanolytic bacterium. Int. J. Syst. Bacteriol. *48*, 91–97.

Huber, R., Woese, C.R., Langworthy, T.A., Kristjansson, J.K., and Stetter, K.O. (1990). *Fervidobacterium islandicum* sp. nov., a new extremely thermophilic eubacterium belonging to the Thermotogales. Arch. Microbiol. *154*, 105–111.

Hudson, J.A., Morgan, H.W., and Daniel, R.M. (1990). Cellulolytic properties of an extremely thermophilic anaerobe. Appl. Microbiol. Biotechnol. *33*, 687–691.

Hugenholtz, P., Pitulle, C., Hershberger, K.L., and Pace, N.R. (1998). Novel division level bacterial diversity in a Yellowstone hot spring. J. Bacteriol. *180*, 366–376.

Inderwildi, O., R., Jenkins, S., J., and King, D., A. (2008). Mechanistic studies of hydrocarbon combustion and synthesis on noble metals. Angewandte Chemie International Edition 47, 5253–5255.

Irwin, D., Shin, D.H., Zhang, S., Barr, B.K., Sakon, J., Karplus, P.A., and Wilson, D.B. (1998). Roles of the catalytic domain and two cellulose binding domains of *Thermomonospora fusca* E4 in cellulose hydrolysis. J. Bacteriol. *180*, 1709–1714.

Ivanova, G., Rakhely, G., and Kovacs, K.L. (2009). Thermophilic biohydrogen production from energy plants by *Caldicellulosiruptor saccharolyticus* and comparison with related studies. Int. J. Hydr. Energy *34*, 3659–3670.

Jannasch, H.W., Huber, R., Belkin, S., and Stetter, K.O. (1988). *Thermotoga neapolitana* sp. nov. of the extremely thermophilic, eubacterial genus *Thermotoga*. Arch. Microbiol. *150*, 103–104.

Kadar, Z., De Vrije, T., Budde, M.A.W., Szengyel, Z., Reczey, K., and Claassen, P.A.M. (2003). Hydrogen production from paper sludge hydrolysate. Appl. Biochem. Biotechnol. *105*, 557–566.

Kadar, Z., De Vrijek, T., van Noorden, G.E., Budde, M.A.W., Szengyel, Z., Reczey, K., and Claassen, P.A.M. (2004). Yields from glucose, xylose, and paper sludge hydrolysate during hydrogen production by the extreme thermophile *Caldicellulosiruptor saccharolyticus*. Appl. Biochem. Biotechnol. *113*, 497–508.

Karakashev, D., Kotay, S.M., Trably, E., and Angelidaki, I. (2009). A strict anaerobic extreme thermophilic hydrogen-producing culture enriched from digested household waste. J. Appl. Microbiol. *106*, 1041–1049.

Kataeva, I.A., Yang, S.J., Dam, P., Poole, F.L., Yin, Y., Zhou, F.F., Chou, W.C., Xu, Y., Goodwin, L., Sims, D.R., *et al.* (2009). Genome sequence of the anaerobic, thermophilic, and cellulolytic bacterium '*Anaerocellum thermophilum*' DSM 6725. J. Bacteriol. *191*, 3760–3761.

Keller, M., and Zengler, K. (2004). Tapping into microbial diversity. Nat. Rev. Microbiol. *2*, 141–150.

Knox, J.P. (2008). Revealing the structural and functional diversity of plant cell walls. Curr. Opin. Plant Biol. *11*, 308–313.

Koskinen, P.E.P., Beck, S.R., Orlygsson, J., and Puhakka, J.A. (2008a). Ethanol and hydrogen production by two thermophilic, anaerobic bacteria isolated from Icelandic geothermal areas. Biotechnol. Bioeng. *101*, 679–690.

Koskinen, P.E.P., Lay, C.H., Beck, S.R., Tolvanen, K.E.S., Kaksonen, A.H., Orlygsson, J., Lin, C.Y., and Puhakka, J.A. (2008b). Bioprospecting thermophilic microorganisms from Icelandic hot springs for hydrogen and ethanol production. Energy Fuels *22*, 134–140.

Kublanov, I.V., Perevalova, A.A., Slobodkina, G.B., Lebedinsky, A.V., Bidzhieva, S.K., Kolganova, T.V., Kaliberda, E.N., Rumsh, L.D., Haertle, T., and Bonch-Osmolovskaya, E.A. (2009). Biodiversity of thermophilic prokaryotes with hydrolytic activities in hot springs of Uzon Caldera, Kamchatka (Russia). Appl. Environ. Microbiol. *75*, 286–291.

Lane, D.J. (1991). 16S/23S rRNA sequencing. In Nucleic Acid Techniques in Bacterial Systematics, Stackebrandt, E., and Goodfellow, M., eds. (John Wiley & Sons, Chichester, New York), pp. 15–176.

Lee, Y.E., Jain, M.K., Lee, C.Y., Lowe, S.E., and Zeikus, J.G. (1993). Taxonomic distinction of saccharolytic thermophilic anaerobes – description of *Thermoanaerobacterium xylanolyticum* gen. nov, sp. nov, and *Thermoanaerobacterium saccharolyticum* gen. nov, sp. nov. Reclassification of *Thermoanaerobium brockii*, *Clostridium thermosulfurogenes*, and *Clostridium thermohydrosulfuricum* E100–69 as *Thermoanaerobacter brockii* comb. nov., *Thermoanaerobacterium thermosulfurigenes* comb. nov., and *Thermoanaerobacter thermohydrosulfuricus* comb. nov., respectively, and transfer of *Clostridium thermohydrosulfuricum* 39E to *Thermoanaerobacter ethanolicus*. Int. J. Syst. Bacteriol. *43*, 41–51.

Leschine, S.B. (1995). Cellulose degradation in anaerobic environments. Annu. Rev. Microbiol. *49*, 399–426.

Ljungdahl, L.G. (2008). The cellulase/hemicellulase system of the anaerobic fungus *Orpinomyces* PC-2 and aspects of its applied use. Ann. N.Y. Acad. Sci. *1125*, 308–321.

Lu, Y.P., Zhang, Y.H.P., and Lynd, L.R. (2006). Enzyme-microbe synergy during cellulose hydrolysis by *Clostridium thermocellum*. Proc. Natl. Acad. Sci. U.S.A. *103*, 16165–16169.

Lynd, L.R., and Zhang, Y.H. (2002). Quantitative determination of cellulase concentration as distinct from cell concentration in studies of microbial cellulose utilization: Analytical framework and methodological approach. Biotechnol. Bioeng. *77*, 467–475.

Lynd, L.R., Weimer, P.J., van Zyl, W.H., and Pretorius, I.S. (2002). Microbial cellulose utilization: Fundamentals and biotechnology. Microbiol. Mol. Biol. Rev. *66*, 506–577.

Lynd, L.R., van Zyl, W.H., McBride, J.E., and Laser, M. (2005). Consolidated bioprocessing of cellulosic biomass: an update. Curr. Opin. Biotechnol. *16*, 577–583.

Madden, R.H. (1983). Isolation and characterization of *Clostridium stercorarium* sp. nov., cellulolytic thermophile. Int. J. Syst. Bacteriol. *33*, 837–840.

Meyer-Dombard, D.R., Shock, E.L., and Amend, J.P. (2005). Archaeal and bacterial communities in geochemically diverse hot springs of Yellowstone National Park, USA. Geobiology *3*, 211–227.

Michaud, P., Da Costa, A., Courtois, B., and Courtois, J. (2003). Polysaccharide lyases: recent developments as biotechnological tools. Crit. Rev. Biotechnol. *23*, 233–266.

Miron, J., Ben-Ghedalla, D., and Morrison, M. (2001). Invited review: adhesion mechanisms of rumen cellulolytic bacteria. J. Dairy Sci. *84*, 1294–1309.

Miroshnichenko, M.L., Kublanov, I.V., Kostrikina, N.A., Tourova, T.P., Kolganova, T.V., Birkeland, N.K., and Bonch-Osmolovskaya, E.A. (2008). *Caldicellulosiruptor kronotskyensis* sp. nov. and

Caldicellulosiruptor hydrothermalis sp. nov., two extremely thermophilic, cellulolytic, anaerobic bacteria from Kamchatka thermal springs. Int. J. Syst. Evol. Microbiol. *58*, 1492–1496.

Mladenovska, Z., Mathrani, I.M., and Ahring, B.K. (1995). Isolation and characterization of *Caldicellulosiruptor lactoaceticus* sp. nov., an extremely thermophilic, cellulolytic, anaerobic bacterium. Arch. Microbiol. *163*, 223–230.

Mohagheghi, A., Grohmann, K., Himmel, M., Leighton, L., and Updegraff, D.M. (1986). Isolation and characterization of *Acidothermus cellulolyticus* cen. nov., sp. nov., a new genus of thermophilic, acidophilic, cellulolytic bacteria. Int. J. Syst. Bacteriol. *36*, 435–443.

Mohnen, D. (2008). Pectin structure and biosynthesis. Curr. Opin. Plant Biol. *11*, 266–277.

Nakagawa, S., Shtaih, Z., Banta, A., Beveridge, T.J., Sako, Y., and Reysenbach, A.L. (2005). *Sulfurihydrogenibium yellowstonense* sp nov., an extremely thermophilic, facultatively heterotrophic, sulfur-oxidizing bacterium from Yellowstone National Park, and emended descriptions of the genus *Sulfurihydrogenibium, Sulfurihydrogenibium subterraneum* and *Sulfurihydrogenibium azorense.* Int. J. Syst. Evol. Microbiol. *55*, 2263–2268.

Nicholson, M.J., Theodorou, M.K., and Brookman, J.L. (2005). Molecular analysis of the anaerobic rumen fungus *Orpinomyces* – insights into an AT-rich genome. Microbiology *151*, 121–133.

van Niel, E.W.J., Budde, M.A.W., de Haas, G.G., van der Wal, F.J., Claasen, P.A.M., and Stams, A.J.M. (2002). Distinctive properties of high hydrogen producing extreme thermophiles, *Caldicellulosiruptor saccharolyticus* and *Thermotoga elfii.* Int. J. Hydr. Energy *27*, 1391–1398.

das Neves, L.C.M., Converti, A., and Penna, T.C.V. (2009). Biogas production: new trends for alternative energy sources in rural and urban zones. Chem. Eng. Technol. *32*, 1147–1153.

O-Thong, S., Prasertsan, P., Karakashev, D., and Angelidaki, I. (2008). Thermophilic fermentative hydrogen production by the newly isolated *Thermoanaerobacterium thermosaccharolyticum* PSU-2. Int. J. Hydr. Energy 33, 1204–1214.

Ogg, C.D., and Patel, B.K.C. (2009). *Caloramator australicus* sp. nov., a thermophilic, anaerobic bacterium from the Great Artesian Basin of Australia. Int. J. Syst. Evol. Microbiol. *59*, 95–101.

Perevalova, A.A., Svetlichny, V.A., Kublanov, I.V., Chernyh, N.A., Kostrikina, N.A., Tourova, T.P., Kuznetsov, B.B., and Bonch-Osmolovskaya, E.A. (2005). *Desulfurococcus fermentans* sp. nov., a novel hyperthermophilic archaeon from a Kamchatka hot spring, and emended description of the genus *Desulfurococcus.* Int. J. Syst. Evol. Microbiol. *55*, 995–999.

Pohlschroder, M., Canaleparola, E., and Leschine, S.B. (1995). Ultrastructural diversity of the cellulase complexes of *Clostridium papyrosolvens* C7. J. Bacteriol. *177*, 6625–6629.

Rainey, F.A., Donnison, A.M., Janssen, P.H., Saul, D., Rodrigo, A., Bergquist, P.L., Daniel, R.M., Stackebrandt, E., and Morgan, H.W. (1994). Description of *Caldicellulosiruptor saccharolyticus* gen. nov, sp. nov – an obligately anaerobic, extremely thermophilic, cellulolytic bacterium. FEMS Microbiol. Lett. *120*, 263–266.

Raman, B., Pan, C., Hurst, G.B., Rodriguez, M., Jr., McKeown, C.K., Lankford, P.K., Samatova, N.F., and Mielenz, J.R. (2009). Impact of pretreated Switchgrass and biomass carbohydrates on *Clostridium thermocellum* ATCC 27,405 cellulosome composition: a quantitative proteomic analysis. PLoS One *4*, e5271.

Reysenbach, A.L., Ehringer, H., and Hershberger, K. (2000). Microbial diversity at 83 degrees C in Calcite Springs, Yellowstone National Park: another environment where the *Aquificales* and '*Korarchaeota*' coexist. Extremophiles *4*, 61–67.

Rincon, M.T., Cepeljnik, T., Martin, J.C., Lamed, R., Barak, Y., Bayer, E.A., and Flint, H.J. (2005). Unconventional mode of attachment of the *Ruminococcus flavefaciens* cellulosome to the cell surface. J. Bacteriol. *187*, 7569–7578.

Saha, B.C. (2004). Lignocellulose biodegradation and applications in biotechnology. In Lignocellulose Biodegradation, Saha, B.C., and Hayashi, K., eds. (American Chemical Society, Washington, D.C.), pp. 2–34.

Saul, D.J., Williams, L.C., Grayling, R.A., Chamley, L.W., Love, D.R., and Bergquist, P.L. (1990). celB, a gene coding for a bifunctional cellulase from the extreme thermophile *Caldocellum saccharolyticum.* Appl. Environ. Microbiol. *56*, 3117–3124.

Schwarz, W.H., Zverlov, V.V., and Bahl, H. (2004). Extracellular glycosyl hydrolases from Clostridia. Adv. Appl. Microbiol. *56*, 215–261.

Sekiguchi, Y., Imachi, H., Susilorukmi, A., Muramatsu, M., Ohashi, A., Harada, H., Hanada, S., and Kamagata, Y. (2006). *Tepidanaerobacter syntrophicus* gen. nov., sp nov., an anaerobic, moderately thermophilic, syntrophic alcohol- and lactate-degrading bacterium isolated from thermophilic digested sludges. Int. J. Syst. Evol. Microbiol. *56*, 1621–1629.

Shallom, D., and Shoham, Y. (2003). Microbial hemicellulases. Curr. Opin. Microbiol. *6*, 219–228.

Shoseyov, O., Shani, Z., and Levy, I. (2006). Carbohydrate binding modules: Biochemical properties and novel applications. Microbiol. Mol. Biol. Rev. *70*, 283–295.

Somerville, C. (2006). Cellulose synthesis in higher plants. Annu. Rev. Cell. Dev. Biol. *22*, 53–78.

Sunna, A., Moracci, M., Rossi, M., and Antranikian, G. (1997). Glycosyl hydrolases form hyperthermophiles. Extremophiles *1*, 2–13.

Svetlichnyi, V.A., Svetlichnaya, T.P., Chernykh, N.A., and Zavarzin, G.A. (1990). *Anaerocellum thermophilum* gen. nov., sp. nov. – an extremely thermophilic cellulolytic eubacterium isolated from hot springs in the Valley of Geysers. Microbiology *59*, 598–604.

Taya, M., Hinoki, H., Yagi, T., and Kobayashi, T. (1988). Isolation and characterization of an extremely thermophilic, cellulolytic, anaerobic bacterium. Appl. Microbiol. Biotechnol. *29*, 474–479.

Teo, V.S.J., Saul, D.J., and Bergquist, P.L. (1995). CelA, another gene coding for a multidomain cellulase from the extreme thermophile *Caldocellum saccharolyticum*. Appl. Microbiol. Biotechnol. *43*, 291–296.

Trudeau, D.G., Bernier, R.L., Gannon, D.J., and Forsberg, C.W. (1992). Isolation of *Clostridium acetobutylicum* strains and the preliminary investigation of the hemicellulolytic activities of isolate 3byr. Can. J. Microbiol. *38*, 1120–1127.

Valdez-Vazquez, I., and Poggi-Varaldo, H.M. (2009). Hydrogen production by fermentative consortia. Renewable Sustainable Energy Rev. *13*, 1000–1013.

Vanfossen, A.L., Verhaart, M.R., Kengen, S.M., and Kelly, R.M. (2009). Carbohydrate utilization patterns for the extremely thermophilic bacterium *Caldicellulosiruptor saccharolyticus* reveal broad growth substrate preferences. Appl. Environ. Microbiol.

Vishnivetskaya, T.A., Podar, M., Phelps, T.J., Keller, M., and Elkins, J.G. (2009). Diversity analysis and cultivation of novel cellulolytic bacteria from Obsidian Pool, Yellowstone National Park. Paper presented at: 10th ASM General Meeting (ASM Press, Philadelphia, PA).

de Vrije, T., Bakker, R.R., Budde, M.A., Lai, M.H., Mars, A.E., and Claassen, P.A. (2009). Efficient hydrogen production from the lignocellulosic energy crop Miscanthus by the extreme thermophilic bacteria *Caldicellulosiruptor saccharolyticus* and *Thermotoga neapolitana*. Biotechnol. Biofuels 2, 12.

Warren, R.A.J. (1996). Microbial hydrolysis of polysaccharides. Annu. Rev. Microbiol. *50*, 183–212.

van de Werken, H.J.G., Verhaart, M.R.A., VanFossen, A.L., Willquist, K., Lewis, D.L., Nichols, J.D., Goorissen, H.P., Mongodin, E.F., Nelson, K.E., van Niel, E.W.J., *et al.* (2008). Hydrogenomics of the extremely thermophilic bacterium *Caldicellulosiruptor saccharolyticus*. Appl. Environ. Microbiol. *74*, 6720–6729.

Wilson, D.B. (2009). Cellulases and biofuels. Curr. Opin. Biotechnol. *20*, 295–299.

Wong, D.W.S. (2006). Feruloyl esterase – a key enzyme in biomass degradation. Appl. Biochem. Biotechnol. *133*, 87–112.

Wyman, C.E. (1999). Biomass ethanol: Technical progress, opportunities, and commercial challenges. Annu. Rev. Energy Environ. *24*, 189–226.

Xu, Q., Bayer, E.A., Goldman, M., Kenig, R., Shoham, Y., and Lamed, R. (2004). Architecture of the *Bacteroides cellulosolvens* cellulosome: Description of a cell surface-anchoring scaffoldin and a family 48 cellulase. J. Bacteriol. *186*, 968–977.

Xue, Y.F., Xu, Y., Liu, Y., Ma, Y.H., and Zhou, P.J. (2001). *Thermoanaerobacter tengcongensis* sp. nov., a novel anaerobic, saccharolytic, thermophilic bacterium isolated from a hot spring in Tengcong, China. Int. J. Syst. Evol. Microbiol. *51*, 1335–1341.

Yanase, H., Sato, D., Yamamoto, K., Matsuda, S., Yamamoto, S., and Okamoto, K. (2007). Genetic engineering of *Zymobacter palmae* for production of ethanol from xylose. Appl. Environ. Microbiol. *73*, 2592–2599.

Yang, S.J., Kataeva, I., Hamilton-Brehm, S.D., Engle, N.L., Tschaplinski, T.J., Doeppke, C., Davis, M., Westpheling, J., and Adams, M.W.W. (2009a). Efficient degradation of lignocellulosic plant biomass, without pretreatment, by the thermophilic anaerobe *Anaerocellum thermophilum* DSM 6725. Appl. Environ. Microbiol. *75*, 4762–4769.

Yang, S.J., Kataeva, I., Wiegel, J., Yin, Y., Dam, P., Xu, Y., Westpheling, J., and Adams, M.W. (2010). Reclassification of '*Anaerocellum thermophilum*' as *Caldicellulosiruptor bescii* strain DSM 6725T sp. nov. Int. J. Syst. Evol. Microbiol. *60*, 2011–2015.

Zavarzina, D.G., Tourova, T.P., Kuznetsov, B.B., Bonch-Osmolovskaya, E.A., and Slobodkin, A.I. (2002). *Thermovenabulum ferriorganovorum* gen. nov., sp nov., a novel thermophilic, anaerobic, endospore-forming bacterium. Int. J. Syst. Evol. Microbiol. *52*, 1737–1743.

Zhang, Y.H.P., Ding, S.Y., Mielenz, J.R., Cui, J.B., Elander, R.T., Laser, M., Himmel, M.E., McMillan, J.R., and Lynd, L.R. (2007). Fractionating recalcitrant lignocellulose at modest reaction conditions. Biotechnol. Bioeng. *97*, 214–223.

Zverlov, V., Mahr, S., Riedel, K., and Bronnenmeier, K. (1998). Properties and gene structure of a bifunctional cellulolytic enzyme (CelA) from the extreme thermophile *Anaerocellum thermophilum* with separate glycosyl hydrolase family 9 and 48 catalytic domains. Microbiology *144*, 457–465.

Zverlov, V.V., Kellermann, J., and Schwarz, W.H. (2005). Functional subgenomics of *Clostridium thermocellum* cellulosomal genes: Identification of the major catalytic components in the extracellular complex and detection of three new enzymes. Proteomics 5, 3646–3653.

8

Extreme to the Fourth Power! Oil-, High Temperature-, Salt- and Pressure-tolerant Microorganisms in Oil Reservoirs. What Secrets can they Reveal?

Hans Kristian Kotlar

Abstract

In the deep biosphere, extraordinary new types of microorganisms, sedimented or buried 200 – 500 million years ago, can be found. These organisms can be identified and characterized. The information obtained can be developed into novel tools for searching for new oil in sensitive regions like the Arctic, Antarctica and jungle areas.

Relatively few enzymes are used in large-scale industrial applications. Enzymes isolated from these extremophile/thermophile organisms might provide 'game changing' new possibilities. They may furnish incentives for the development of entirely new technical processes. These microbes provide opportunities for new technologies in second generation biofuel production. Several companies are working on alternative routes for the production of fuels using biomass as the raw source material.

Traditional heavy oil extraction methods have major difficulties in justifying their high energy usage, CO_2 emissions and soil and environment pollution. The first company implementing a large-scale process based on biotechnology principles in enhanced oil recovery will gain huge strategic and economic benefits.

The knowledge of this huge subsurface population of diverse microorganisms provides excellent opportunities for bioprospecting. There should be a multitude of spin-offs outside the oil industry. The world is desperately in need of new enzymes, new antibiotics, new immunosuppressant, new anticancer agents, etc. This chapter reviews just some of the areas we have been working on at Statoil. Hopefully some of these investigations could one day solve some of the problems we will face in the future. One day these extremophiles could be on the payroll of many different companies.

The deep biosphere

Different research organizations have been searching for the 'gold of the ocean' and have made important discoveries of organisms that can produce new compounds seen as potential antibiotics and anti-cancer medicines. They have been searching from the high mountain moors to the sediments of the ocean floor, and even in high temperature hydrothermal vents.

At Statoil we are digging even deeper. Deep down in oil reservoirs we are finding a whole new world of microorganisms that, until recently, we had very little knowledge of (Fig. 8.1). Deep in the subsurface there are microorganisms with unique properties. Some day they could be on the payroll of many different companies.

These microorganisms live at depths of between 2 and 4 km in the lithosphere, and their communities represent ancient phylotypes. The traditional understanding of the generation of oil reservoirs is through the sedimentation of river deltas from the Cambrian through to the Jurassic period, i.e. 500–200 million years ago, especially in the Jurassic. Microorganisms living throughout this period

Figure 8.1 Indigenous microorganisms originating from the extra deep subsurface. 2000–5000 m below the sea-bottom there is a new, unknown microflora that mankind has just recently started to explore.

might have been trapped in the different phases of sedimentation and survived in specialized niches. Their descendants are the microbes found in oil and gas reservoirs today.

A large number of cores and research samples are collected during exploration and initial development of a new oil field. Samples from production wells, seabed sediments and seeps, and pockmarks have been investigated for their microbial content. All of these samples have given a completely new insight into the abundance and types of organisms that can be found in the deep biosphere.

How can they be identified?

DNA is extracted from rock, water and oil samples and organisms identified by a traditional molecular genetic approach: the 16S rRNA PCR products obtained from the extracted DNA are cloned and/or sequenced, the data is BLASTed against available databases and the closest matching taxa are identified. DNA probes/primers specific for these taxa can then be synthesized (Fig. 8.2).

One point should be emphasized, however: A match can only be made if the organisms are found in the reference sequence database used. Many deep biosphere samples contain unknown microbes, i.e. they have never been identified before. These microorganisms could possibly be the more exciting ones. Another important consideration is that even a 100% match at the 16S rDNA level does not necessarily guarantee physiological or metabolic identity with the matching database strain. With these reservations in mind, it is a big surprise to find that many of the observed oil- and gas-associated microorganisms are very close to or similar to already sequenced organisms. This fact is of special interest, as these indigenous reservoir organisms are thought to have been separated and sealed off for many millions of years.

As already mentioned, using sequence information DNA probes for the recognition of these reservoir microorganisms can be synthesized (Fig. 8.2). Specific probes for selected organisms could be printed on microarray plates or developed into lab-on-a-chip systems. DNA extracted from environmental samples can be extracted, amplified and exposed to the microarray probes. A predefined pattern will indicate whether a drilling

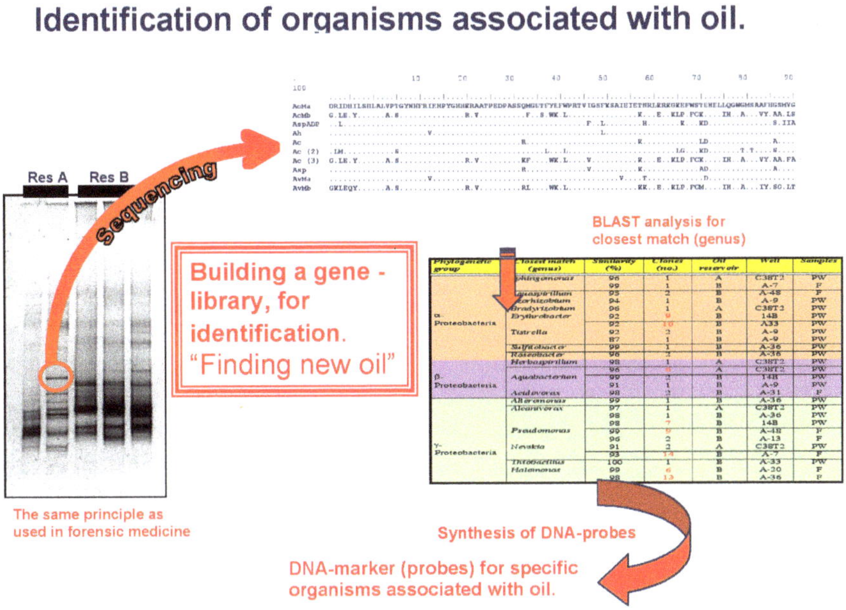

Figure 8.2 Concept for the generation of DNA-probes to identify oil-related microorganisms. DNA is extracted from reservoir samples (e.g. exploration cores, formation water, drill cuttings, seeps, sediments or others). 16S rDNA is amplified by PCR, cloned and/or sequenced, and BLASTed for the closest match at the genus level. Selected sequences are used to synthesize new DNA probes for the identification of oil-related microorganisms.

prospect contains hydrocarbons, i.e. gas and/or oil, or not (Fig. 8.3). This approach could provide a new tool in exploration of sensitive areas like the Arctic, Antarctica and jungle areas. In the future, traditional seismic logging might be abandoned in these areas for environmental reasons, so the use of non-invasive search methods will prove to be of great value.

A challenging question is whether these microbes are really indigenous or not. Samples extracted from exploration cores, formation water and oil, all originating from the same oil reservoir, basically identify the same organisms (although the formation water can show a higher biodiversity). In contrast, when seawater or outcrop core samples are subjected to exactly the same extraction and analytical protocols, none of these organisms are identified, indicating that the reservoir-associated microbes are indeed indigenous. Exploiting this difference in microbial ecosystem profiles provides a powerful new tool for the exploration and discovery of new hydrocarbon resources or reservoirs.

In addition, Statoil has used direct denaturating gradient gel electrophoresis (DGGE) patterns to discriminate between pockmark areas with a hydrocarbon (HC) gas flux from areas outside the pockmark sites. Pockmarks are found on the sea floor at different locations worldwide, and they are often connected to subsurface hydrocarbon-reservoirs or subsurface gas pockets. They are frequently found in both the North Sea and the Black Sea. At pockmarks the HC fluxes coming from the subsurface are strongest in the slope area of the pockmark. Thus, if microorganisms have adapted to metabolize the HC, this should be the place to search for differences with the surroundings of non-HC seeps. Bacterial and archaeal DGGE patterns of samples from an active pockmark site in the Norwegian Sea were analysed. Results from both domains yielded the same overall observation, i.e. distinct DGGE patterns of organisms in the slope of the pockmark (Fig. 8.4). Such findings prove the concept of developing tools for exploration and detection of new hydrocarbons by 'fingerprints' of the site-specific DNAs.

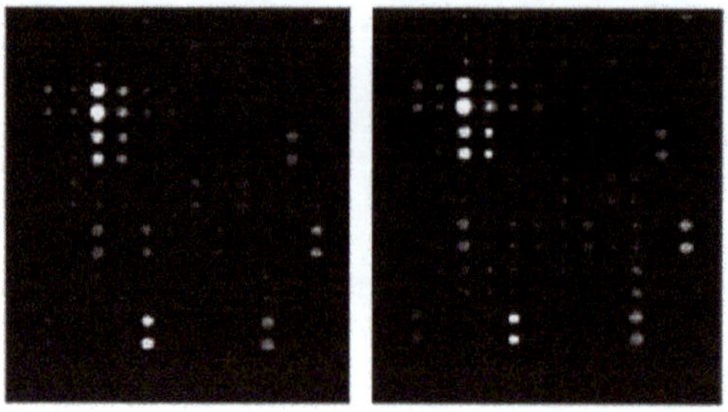

Figure 8.3 Microarray identification system of oil-related microorganisms. More than 20 different DNA probes were selected and printed onto a microarray plate. Light spots indicate a match between the probe and the microbes tested.

What are they and what are their special features?

Another surprising observation is the abundance of bacteria versus archaea in oil reservoirs and HC seeps. According to the general acceptance of the phylogenetic tree, one would perhaps have expected that the archaea should be predominant, given the ancient time of their appearance on Earth. However, in most oil reservoirs on the Norwegian Continental Shelf (NCS), bacterial species diversity is about five times higher than that of archaea, and they are also present in higher numbers (Table 8.1).

Thermophilic and hyperthermophilic organisms are characterized by their ability to live at elevated temperatures, i.e. 50°C –70 °C, and 70°C – 100°C, respectively. Most of the oil reservoirs on the NCS typically range from 70°C to 170°C. An investigation of high temperature reservoirs (Table 8.2) reveals microorganisms to be found in all except the two with the highest temperatures (125°C and 165°C). Several of the 16S rDNA PCR products from these wells have been cloned and sequenced (note: intellectual property issues prevent publication of the identity of these microorganisms).

The world record reported for living organisms is very close to 120°C. In the study shown in Table 8.2 there are no borderline reservoirs, i.e. with temperatures around 110–120°C. Some of the reservoirs tested are at around 100°C and displayed quite a high biological diversity (data not shown). The upper limits for biodegradation, given by most geochemists, is said to be at about 80°C. These new findings provide an adequate argument for an extension of the traditional boundary of biodegradation to at least 100°C.

As noted in the chapter title, these oil-associated

Table 8.1 Phylotypes found in different oil reservoirs and hydrocarbon seeps. This table includes five different North Sea oil reservoirs; however, more have been analysed. Clones only specified to phylum level

CLONES AND SEQUENCES FROM OIL RESERVOIRS AND HYDROCARBON SEEPS							
Environment	Designation	Clones (no.)		Sequences (no.)		No. different phylotypes	
		Bacteria	Archaea	Bacteria	Archaea	Bacteria	Archaea
Oil reservoirs North Sea	A	152	53	79	39	11	4
	B	288	17	98	14	11	1
	C	12	0	11	0	5	0
	G	12	9	7	8	4	2
	H	51	15	25	5	7	1
Marine HC seeps	I	18	5	7	4	3	2
	II	18	20	8	3	3	4
	III	13	6	3	2	3	0
	V	12	17	6	7	3	2
Terrestrial seeps	SA	110	110	34	30	8	A 2
	As	137	113	63	38	13	A 5

Table 8.2 Temperature as a selection criterion for life in oil reservoirs

Sample	PCR results	
	Bacteria	Archaea
Reservoir A washed oil, 165°C	–	–
Reservoir A oil pellets, 165°C	–	–
Reservoir B oil pellets, 98°C	+	+
Reservoir C, well K3 oil pellet (50% ethanol or oil with toluene) 93°C	+	+
Reservoir D, oil pellet (oil with toluene) 125°C	–	–

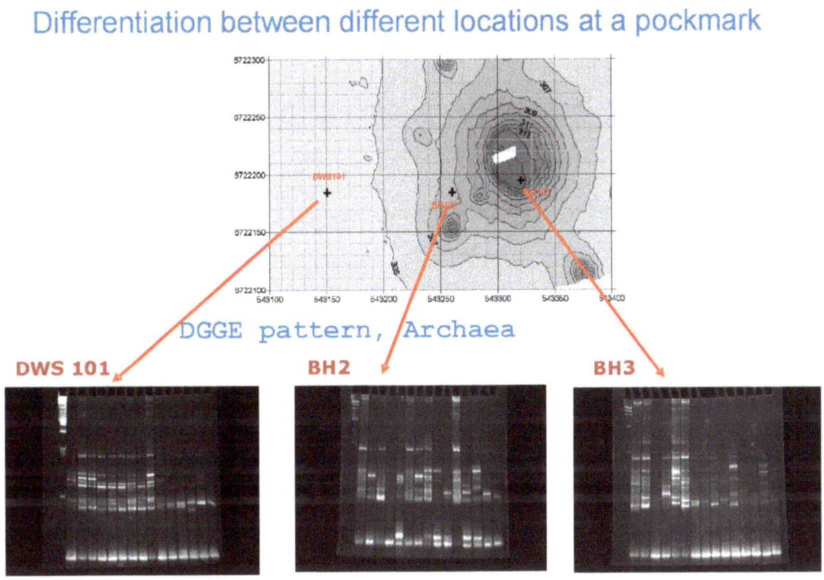

Figure 8.4 DGGE patterns as a basis for differentiation between HC and non-HC-containing areas of pockmarks. Sediment samples were taken by an ROV at three different locations at an active pockmark site in the Norwegian Sea. At each site a depth profile from a few cm to 3–5 metres below the ocean floor was collected, and archaeal DGGE patterns analysed. Samples outside (DWS 101), on the slope of (BH2), and in the midst of the pockmark (BH3) give distinctive patterns.

microorganisms are extremophiles to the fourth power: they tolerate a high load of hydrocarbons, extreme temperatures, high pressures and high salt concentrations. Inherently this will generate some very exciting properties. Some of the properties of cellular components (enzymes, proteins and nucleic acids) from these polyextremophilic microbes include (i) thermostability; (ii) stability against denaturing substances like detergents, organic solvents; (iii) stability against extreme pH conditions; (iv) suitability for fermentative processes; (v) significantly improved solubility of many substrates; and (vi) reduced contamination/undesired complications at higher temperatures.

One mechanism by which these organisms tolerate high temperatures, a broader range of pH and a higher load of organic compounds, is by increasing the amount of hydrophobic amino acids in their proteins. High salt tolerance results from conformational changes provided by amino acids that are stabilized by salt-bridges and/or ionic interactions.

Biotechnology industries are on a constant search for new enzymes with high stability and high conversion rates. Approximately 5000 different enzymes have so far been named. Nevertheless, enzymes used in biotechnological and industrial applications are still insufficient to meet all demands. Enzymes with enhanced stability under denaturating conditions are of particularly great value in biotechnological processing. Different strategies exist for achieving the desired biocatalytic activities. For example, the following approaches are used for the generation of thermostable enzymes: (a) introduce features for internal stabilization in thermophile proteins; (b) introduce internal thermostability in mesophile proteins by site-directed mutagenesis or directed evolution; and (c) develop thermostable proteins by directed evolution. However, there are multitudes of interacting properties that will contribute to the overall thermostability of an enzyme. Thus, *the best approach is to bioprospect for the enzyme directly in thermophilic microbes.*

Piezophilic organisms

Although the above-mentioned reservoir-derived organisms are extremophiles for different properties, their piezophilic dependency or nature is not normally taken into account. However, this is of special importance in cultivation of such microorganisms. In order to understand these organisms, one has to study them under their 'home conditions'. Many different features are attributed to high pressure, including, for example, increased solubility of potentially essential nutrients. In order to perform investigations on pressure responses, Statoil has built an apparatus that functions at reservoir conditions (Fig. 8.5). It is constructed to maintain pressurized samples from the oil field reservoirs, and contains six individual cells that can be kept at reservoir conditions or at different individual pressures. The results described below indicate the importance of culturing at high pressure.

Improved growth using high-pressure culture conditions

Higher partial pressure of components into the water phase might be beneficial to the growth of piezophiles. Some components of the oil have low solubility in the water phase at standard conditions, i.e. 1 bar and 20°C. This characteristic might be dramatically changed at higher temperatures and pressures. For example, if the reservoir pressure is above the bubble point of the HC phase, the

Figure 8.5 Pressure cells for the cultivation of microorganisms at reservoir conditions. Piezophilic reservoir microorganisms were collected in pressurized containers and transferred at the appropriate pressure (20–500 bars) to this pressurized cultivation cell. In the set-up shown, six individual cells can be maintained at reservoir conditions. Small aliquots can be withdrawn at different intervals from each individual cell and analysed for growth.

reservoir will have no discrete gas cap, i.e. the gas is completely dissolved into the HC phase. When the molar concentration of the gas is high and the pressure is high, the partial pressure of the different gases will be higher in the water-phase as well and the solubility of the gases into the water-phase is increased. In the oil reservoir, the HC phase and the water phase are in equilibrium with each other. The increased solubility into the water-phase of components originating from the oil phase might be essential and at lower pressure, rate-limiting for microbial growth. Other more basic effects of the high pressure are the piezophilic constraints on the differential change in volume, ΔV, of enzymes. Some enzymes will not work *in vivo* unless the protein is pressurized to such a degree that the conformation of the catalytic site is optimal for the bioconversion, i.e. a change in the entire volume of the enzyme molecule (Bartlett *et al.*, 1995).

Direct proof of the importance of simulating piezophilic 'home conditions' is shown in Fig. 8.6. After cultivation for 3 months and subsequent harvesting, enrichments in pressurized cells, showed a much higher diversity and abundance of organisms compared with non-pressurized controls. Direct extraction of DNA and subsequent analysis clearly indicated a higher diversity of bacteria as well as archaea in the high-pressure cultivation. This result was consistent in samples from four different oil wells.

How can they be used?

The direct use of microbial ecology surveys in exploration to find new oil fields has already been mentioned. Other biotechnological industrial approaches, discussed below, encompass both short-term and long-term possibilities.

New, smart increased oil recovery (IOR) and production methods

Discussion on whether peak oil has occurred or is still to come is a subject of debate amongst market analysts. However, there is a common agreement that all major existing HC resources have been detected. Thus, a future decline in oil production is to be expected. In this context, methods for increased oil recovery (IOR) from existing oil fields will receive increased emphasis from oil companies. On a world average, less than 50% of the oil present in an oil reservoir is recoverable by traditional extraction methods. Hopefully, this can be improved using new, smart, non-polluting and environmentally friendly methods. One such method is the use of specially designed microorganisms. Either naturally selected or genetically modified organisms (GMOs) could provide completely new options for IOR. In addition, microorganisms coupled with delivery systems will be a new way of thinking as opposed to conventional IOR.

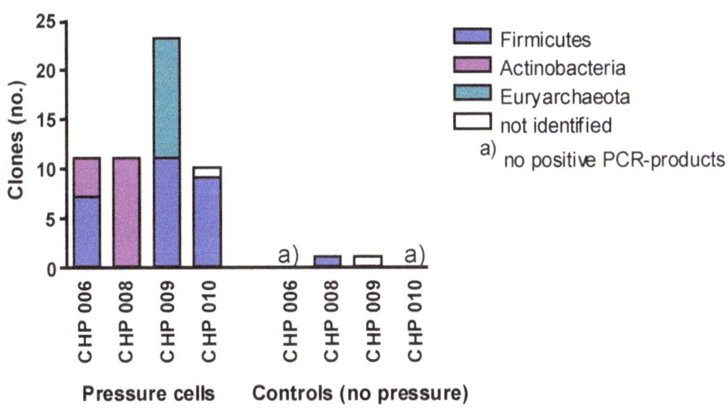

Figure 8.6 Pressure dependent cultivation of microorganisms from an oil reservoir. Four different reservoir samples were each cultured under two different pressure conditions (253 bar in a pressure cell, and 1 bar) at 85°C, and diversity subsequently assessed by 16S rRNA gene PCR, cloning and sequencing. Cultivation under high pressure was clearly an important factor in the successful growth of the microbes.

Preventative 'medication'

Oil wells can lose productivity by a variety of different mechanisms. Some of the major problems include scaling, asphaltene precipitation, and hydrate formation. Mixing of incompatible waters occurs when injection water, i.e. seawater, hits the formation water or brine. If the concentration of particular ion-pairs exceeds the solubility product, salt will precipitate. The most common salt precipitation, i.e. scaling, is by carbonates and sulfates. Precipitation of salt blocks the pore structure of the oil formation and can precipitate in the gravel packs as well as in the production tubing, thereby reducing oil production. Likewise, if the temperature and pressure conditions are outside the phase envelope for asphaltene dissolution, asphaltene will precipitate and block the near-well region or production tubing. If gas and water are mixed at the right temperature and pressure conditions, gas hydrates can form. The hydrates can also reduce or block oil production.

In order to mitigate these problems, the oil industry administers various types of inhibitors to hinder these precipitants from forming. Inhibitors can have different chemistries, but often consist of molecules that will hinder the nucleation of the salt crystals or interfere with crystal growth. For several years Statoil has had a patented version (US PA No 10/478776; PCT/GB02/02359; EU PA no 200301214) of preventative 'medication'. This invention is based on a hyperthermophilic host organism that contains a transfected gene that codes for the production of a scale inhibitor. The secreted version of this inhibitor acts as a continuous supply of the treatment chemical (Fig. 8.7). For hydrate inhibition, nature itself produces kinetic inhibitors like anti-freeze proteins and small alcohols for freezing point depression. By lowering the freezing point, the phase envelope

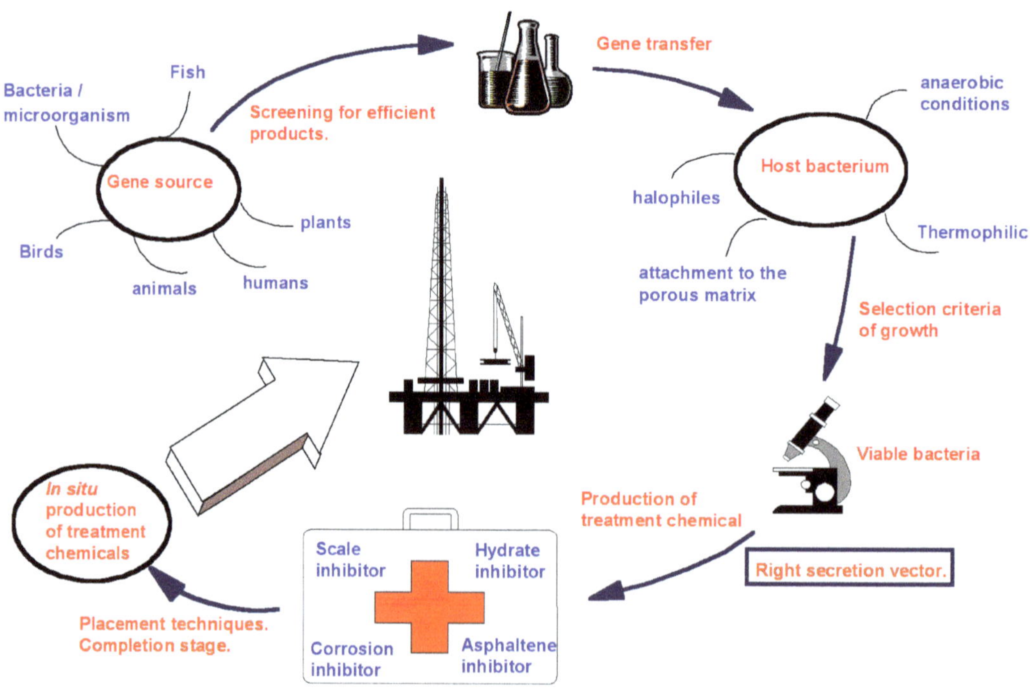

Figure 8.7 Preventative 'medication'. Flow diagram on the use of hyperthermophiles as host organism for the production of desired products for IOR. This concept could be expanded to a range of different treatment chemicals. It is anticipated as a 'doctor's bag' with remedial actions for many different oil production well 'diseases'.

can be maintained outside the envelope for hydrate formation. This approach can also be extended to organisms producing corrosion inhibitors, for example, surface-active proteins coating the pipeline.

Biofuels

Replacing fossil fuels with biofuels (e.g. biodiesel and bioalcohols) will reduce CO_2, CO, SO_2 and particulate atmospheric emissions. In order to be considered a biofuel, the fuel must contain over 80% renewable materials. Biofuels are produced from living organisms or from metabolic by-products (organic or food waste products). It is originally derived from the photosynthetic process and can therefore often be referred to as a solar energy source. Thus, polysaccharides, organic oils and animal fats can be looked upon as storage devices for solar energy. Different feedstocks of these types of organic substrates are used as the source to produce biofuels. The triglycerides in vegetable oils and fat are converted to fatty acid residues that are subsequently incorporated into biodiesel, and the various types of saccharides are fermented to bioalcohols for gasoline substitution.

Biodiesel and European Union (EU) regulations – EU directives and arable farmland

To an oil company, it is of great concern not to be embroiled in ethical dilemmas relating to competition with arable farmland and crops used for human food. Nonetheless, a huge amount of raw materials are needed to fulfil the EU regulation of 10% substitution of biodiesel into diesel fuels by 2010. A European parliament directive in 2003 established that 5.75% of all fuel sold in EU countries by 2010 should be biofuel. Later, EU leaders decided that by 2020, 20% of all EU energy consumption should be from renewable energies. The biomass and feedstock sources for biodiesel are diverse, among which are agricultural and forestry products. Use of agricultural areas for biomass-production could be competing with areas utilizable for food production.

Currently, the essential part of first generation biodiesel production (non-microbial) is transesterification of oils with alcohol and a catalyst. Second generation (2G) biodiesel, NExBTL, is produced by an energy-demanding hydrogenation (using abiotically produced H_2) of the vegetable fat or oil components. While vegetable oils have good ignition characteristics, other properties, such as their viscosity and quick oxidation, need to be reduced or altered to allow them to be used as fuel in modern diesel engines. Because second generation technology involves the administration of hydrogen to the feedstock mix, this process is suited for operation at a petroleum refinery, where hydrogen is generated as a part of its regular processes. Hydrogenation demands high energy and therefore high costs.

As an alternative to the above abiotic approach, biotechnology can be used in the pretreatment of feedstock before the hydrogenation step. By applying an approach that utilizes reservoir-derived microbes, more diverse feedstock sources or mixes of feedstock can be used, including waste products. By utilizing special enzymatic pathways in different microorganisms, bioconversion is feasible at several other stages of biodiesel production. The use of a bio-catalytic process should reduce the amount of hydrogen required for biodiesel production, making the production more energy and cost effective.

Chemical background of 1G biodiesel and development of biotechnology-derived 2G production

Animal fat contains triglycerides composed of fatty acids and glycerol. This fat cannot be used as biodiesel due to its high viscosity at ambient temperatures. The triglycerides must first be cleaved to produce free fatty acids, which are subsequently converted into fatty acid esters or into the corresponding *n*-alkanes.

In Norway there is a small amount of biodiesel production from fish oil and deep-fried fat waste in a conventional (i.e. non-biological) transesterification process, which is an endothermic reaction where the triglycerides are mixed with methanol/ethanol and a strong alkaline catalyst. The fatty acids are released from the glycerol and converted into methyl/ethyl esters, called biodiesel or fatty acid methyl esters (FAMEs) (Fig. 8.8A). Transesterification is an expensive and disadvantageous process due to its high-energy demand (heat) and use of toxic chemicals like methanol. Alternatively, using bacterial bioconversion, the same process would be less expensive and more environmentally friendly (Fig. 8.8B).

The above concept is patented by Statoil and implies the use of bacteria for biocatalytic conversion of triglycerides into biodiesel, as well as modification of the fatty acid chain length and biohydrogenation of unsaturated fatty acids to improve the winter properties of biodiesel [briefly, the general rule is that the longer the *n*-alkanes, the worse the winter properties, owing to the reduced wax appearance temperature (WAT)]. Verification of the concept of bacterial conversion of triglycerides to biodiesel was done in two steps (see below). Four types of fat/oil were tested as substrates: animal fat, soybean oil, refined salmon oil, and untreated fish/salmon oil.

1 The first approach was to modify the fatty acid chain length in experiments with an *Acinetobacter* sp. and mutants thereof. This species can degrade *n*-alkanes from decane ($C_{10}H_{22}$) to tetraoctane ($C_{40}H_{82}$) chain lengths. The fatty acid structure is similar to the *n*-alkane structure. By shortening the chain length the winter properties of the biodiesel are improved. To measure changes in oil properties after treatment with bacteria, the oil phase and the bacterial phase were harvested separately. The oil phase was checked for changes in triglyceride content by electron spray ionization mass spectroscopy (ESI-MS), chain length was analysed using gas chromatography (GC), and viscosity using a rheometer. *Acinetobacter* sp. gave encouraging results with fish oil, but its maximum temperature of growth is too low (about 30°C). Therefore, it was decided to continue the research using thermophilic organisms.

2 An alternative, second approach for environmentally friendly biodiesel production was investigated using microbes from the Statoil strain collection (the BTC collection). This collection includes both archaea and bacteria associated with oil reservoirs, and other terrestrial samples associated with proven effects on hydrocarbons. The BTC culture collection was screened for thermophilic bacteria

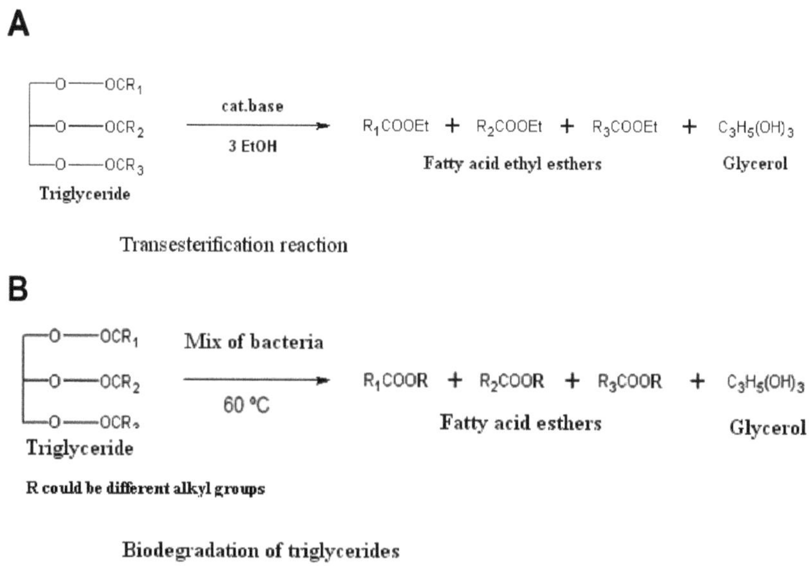

Figure 8.8 Conventional versus biocatalytic conversion of triglycerides. (A) Currently used fatty acid methyl esters (FAMEs) production. (B) Potential bacterial bioconversion.

having the ability to cleave triglycerides into fatty acid and glycerol at 60°C, using plant oils, waste fish oils and waste animal fat as substrate. Following culturing with various inocula in the presence of substrate(s), the oil phase and the 'water' phase were harvested separately. The oil phase was analysed by ESI-MS, detecting tri-, di- and monoglycerides in positive mode, and free fatty acids in negative mode. ESI-MS data were processed with a multivariate analytical software program to identify significant differences between the experimental samples and negative controls (no inoculum).

Initial data with different substrates revealed differences between the bio-converted (with thermophilic bacterial inocula) and negative control (no inoculum). These results are illustrated in Figs. 8.9 to 8.12. Note that different inocula were used for the various experimental results shown. Fig. 8.9 shows that different inocula were able to convert fish oil to fatty acids, with inoculum L019 displaying the most effective bioconversion when compared with the negative control. When soybean vegetable oil was used as the substrate for the biodiesel conversion (Fig. 8.10), ESI-MS showed significant changes in the triglycerides from the inoculated cultures compared with the triglyceride composition of the non-inoculated negative control. This result indicated that the triglycerides were modified by the bioinocula, with mix 2 (L004) and mix 3 (L018) giving the strongest responses.

A corresponding experiment using animal fat and four different inocula was also conducted (Fig. 8.11). A dramatic bioconverting action was observed with inocula A and C. Again, this experiment provided good evidence that the triglycerides had been changed by the inoculated microbes. In the final example (Fig. 8.12), refined salmon oil was used as the substrate. It is interesting to note the different behaviour of the biocatalysts, i.e. the various inocula, which generally grouped around

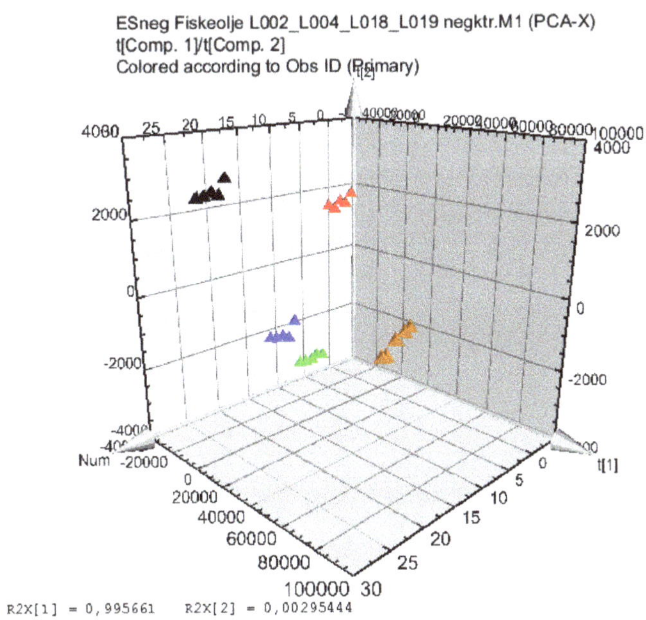

Figure 8.9 Bioconversion of fish oil by thermophilic bacterial inocula. 3D multivariate representation on negative ESI-MS, showing fatty acids present in the oil phase. Axis labelling: t1 = principal component 1, t2 = principal component 2. Coloured triangles identify the activity, i.e. the combined concentrations and types of fatty acids produced, by inocula L002 (red), L004 (green), L018 (blue) and L019 (orange). Black triangles represent the negative control without biocatalyst/inoculum. L019 displayed the most effective bioconversion.

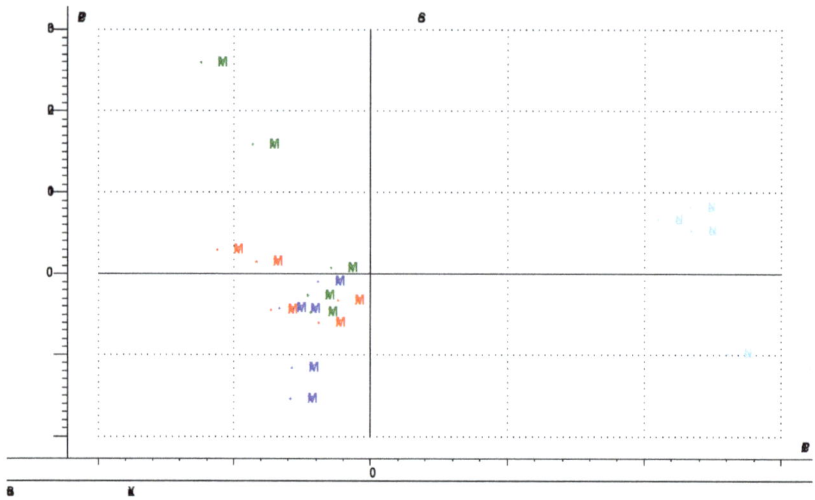

Figure 8.10 Bioconversion of soybean oil by different thermophilic bacterial inocula (mixes 1–3). Positive ESI-MS scores plot showing mono-, di- and triglycerides present in the oil phase. x-axis, principal component 1; y-axis, principal component 2. Inocula: mix 1, L002; mix 2, L004; and mix 3, L018. Negative control: no inoculum. The strongest responses were seen with mix 2 and mix 3.

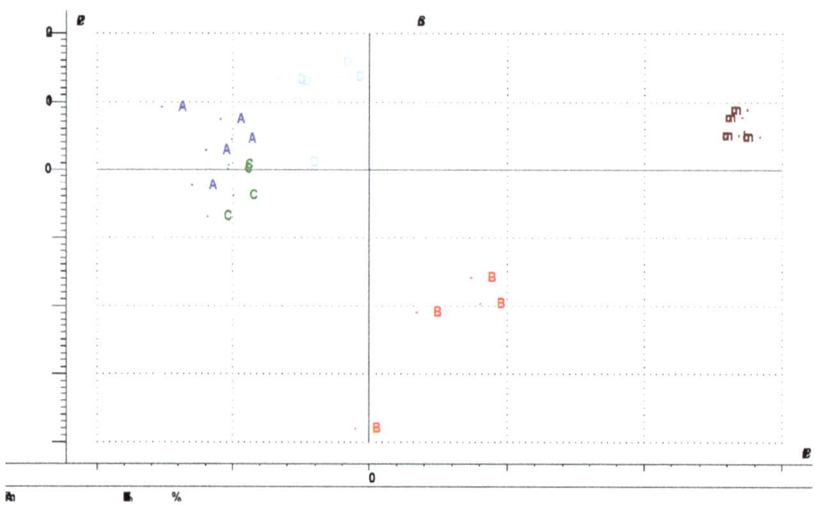

Figure 8.11 Bioconversion of animal fat by different thermophilic bacterial inocula (A, B, C, D). Positive ESI-MS scores plot showing mono-, di- and triglycerides present in the oil phase. Axis labelling as in Fig. 8.10. Negative control, no inoculum. The best bioconversion was achieved with inocula A and C.

the non-inoculated sample. The only exception was seen with inoculum D, which produced by far the best response.

The experiments described above show that Statoil has microorganisms in its strain collection that can tolerate and have the capability to bioconvert different substrates. This finding is of particular interest in relation to our aim to co-mingle substrates for an industrially feasible biocatalytic biodiesel process.

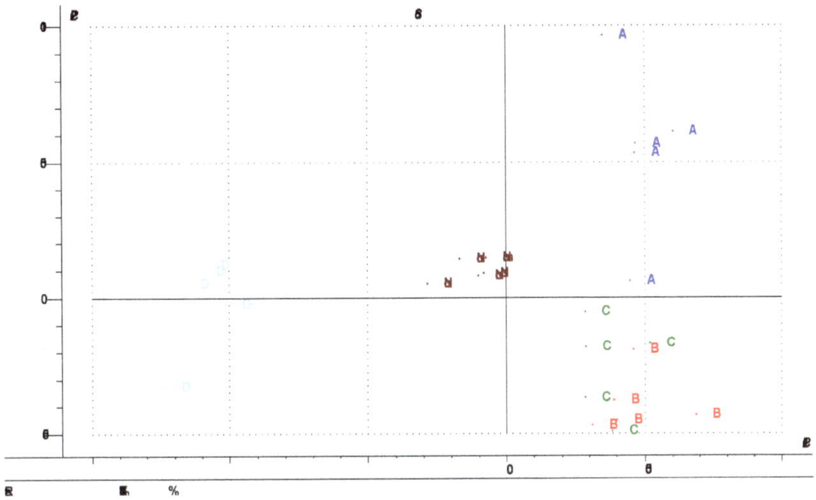

Figure 8.12 Bioconversion of refined salmon oil by different thermophilic bacterial inocula (A, B, C, D). Negative ESI-MS scores plot showing fatty acids present in the oil phase. Axis f the remaining original triglyceride peaks, as well as those of generated di- and mono-glycerides. This pattern indicates that fatty acids have been split off the triglycerides, but that on this particular occasion the reaction has not reached completion.

Commingling of raw materials for biodiesel production

To overcome the ethical aspects and costs associated with today's biodiesel production, an alternative method using thermophilic bacteria that produce fatty acids from triglycerides in aqua- and agricultural waste products has been investigated. In order to compensate for a deficiency in the amount of raw material, different types of biomass waste and co-mingling of raw materials has been assessed, including: (i) animal fats from slaughter houses (beef and pig tallow); (ii) plant oils from rapeseed; and (iii) fish oil waste.

Statoil has performed bioconversion experiments to identify new stabilized and thermotolerant species harbouring special enzyme systems. High thermotolerance is necessary because some animal fats will not melt until a temperature of 50–60°C is reached. The result of one such experiment is shown in Fig. 8.13. Results from replicate experiments like these indicate that this process could be a feasible and less energy intensive route for future biodiesel production.

Identification of the biocatalyst

Subsequent to the culturing experiments described in the previous section, three microbial mixes (#2, #4, #18) that alter the triglyceride/fatty acid composition of the substrates were identified. These mixes were isolated from oil well formation water and a number of other environments (e.g. Norwegian garden compost, a water treatment plant, asphalt samples from old roads, sediments and mud volcanoes). 16S rDNA sequencing data revealed the microbes responsible for the triglyceride conversions to be from the genus *Geobacillus*. The conversion experiments were successfully repeated with pure isolates of *Geobacillus* sp., in which they were able to change the triglyceride composition of soybean oil, animal fat and refined salmon oil. Further work with thermophilic bacteria will focus on 'real' compositions (as opposed to single substrate compositions) of commingled feedstock as substrate. This is particularly beneficial as biofuel production can be flexible to substrate availability.

Second generation (2G) alcohols

Another area of interest for Statoil is the production of bioalcohols. Of the alcohols, butanol

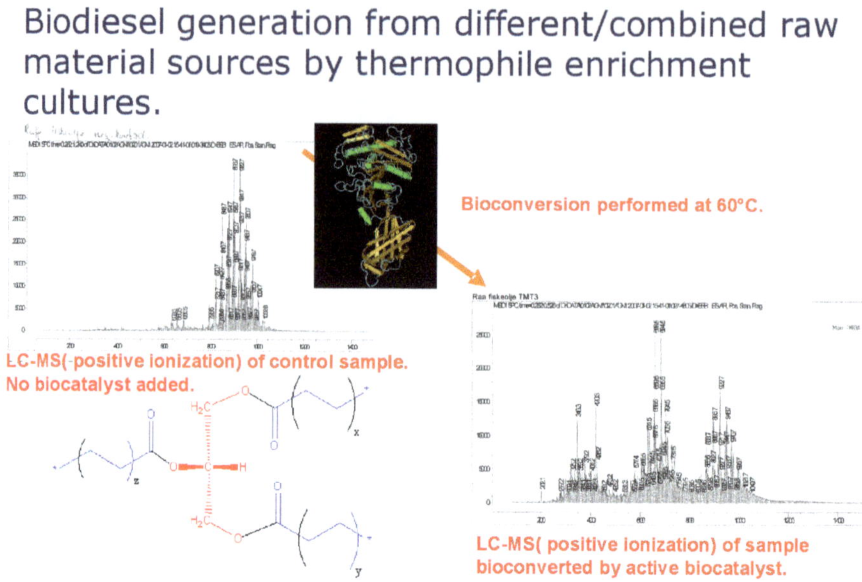

Figure 8.13 Biocoversion of a waste product (fish oil) into diesel. LC-MS spectrum of fish oil triglycerides before (top left) and after (bottom right) bioconversion. The post-biocatalysis spectrum shows some of the remaining original triglyceride peaks, as well as those of generated di- and mono-glycerides. This pattern indicates that fatty acids have been split off the triglycerides, but that on this particular occasion the reaction has not reached completion.

might be the most desirable as the energy density of the molecule is quite close to that of gasoline. Furthermore, butanol is less hygroscopic and less corrosive than ethanol. However, there are many considerations that come into play in making production of bioalcohols economically feasible. For example: what would be the best and most reliable source of raw material, i.e. the biomass, the amount of this type of biomass in the nearby geographical area, logistics and transportation of the biomass, what technology should be chosen, the necessary pretreatment process, fermentation technology and product purification and concentration. As emphasized, it is imperative to our company not to compete with the food chain. Thus, our search will be for raw materials for second generation, 2G, and beyond, biofuels.

The potential use of thermophilic biocatalytic processes has been launched by several research groups and companies around the world. The use of specially selected organisms that can convert cellulose, hemicellulose and lignin in lignocellulosic material to C_6 and C_5 sugars will be of great importance, as these can be bioconverted to ethanol, butanol or H_2 via the pentose-6-phosphate and Embden–Meyerhof metabolic pathways.

Several enzymes are involved in the generation of the alcohols, as shown in Fig. 8.14. If these enzymes could be optimized in a thermophilic organism, this process could have huge environmental and economic benefits. Today, the process incorporates the biocatalyst and fermentation performed by living organisms. Future processes might involve direct reactor technology with purified, GMO-transformed, optimized enzymes as a new concept of biocatalysts.

Bioconversion of heavy oil

Another area of interest is the search for microorganisms with one or more metabolic pathways for the bioconversion of heavy oil. Crude oil is defined in a number of categories, which also includes heavy oil (Table 8.3).

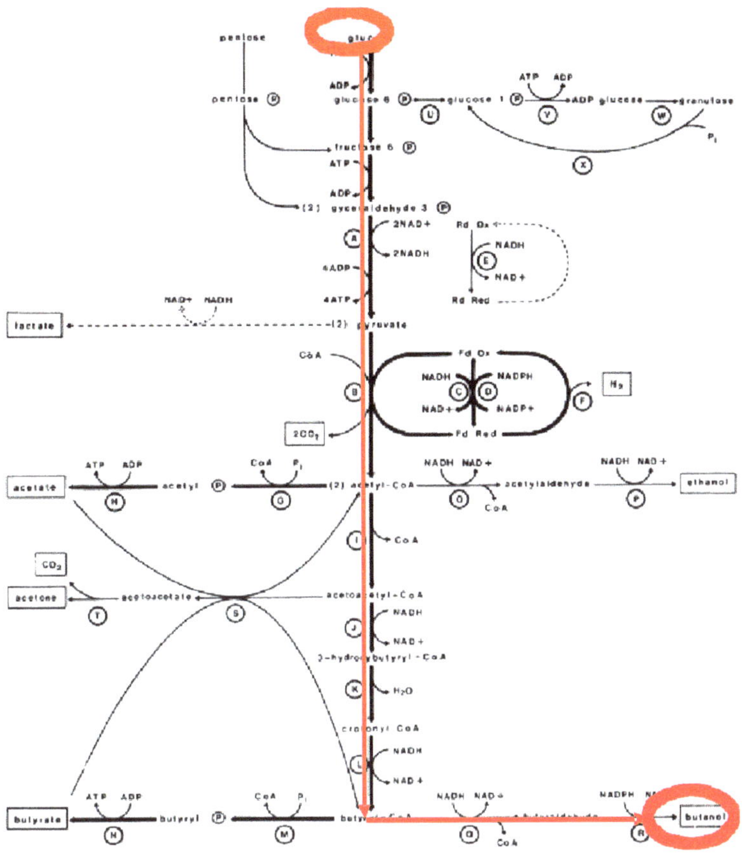

Figure 8.14 Metabolic pathway for the generation of butanol from sugars (adapted from Lee *et al.*, 2008). *The predominant organisms known to produce butanol are Clostridia* sp. Several research groups (including Statoil) have begun attempts to optimize butanol production by modifying selected enzymes in this metabolic pathway.

Table 8.3 Categories of crude oil

Crude oil categories	American Petroleum Institute gravity	Density (kg/m^3)
Light crude oil	>31 API	<870
Medium crude oil	31–22 API	870–920
Heavy crude oil	22–10 API	920–1000
Extra-heavy crude oil	<10 API	>1000

Heavy oil extraction

Two thirds of the world's extractable fossil fuels are categorized as heavy to extra heavy oil. The world's average recovery rate from these types of oil reservoirs is about 7%. Technologies that could boost these recoveries would have a tremendous economic impact. One of the major challenges, in this respect, is the huge mobility ratio between the water phase and the oil phase: oil viscosities can range from some tens of millipascal-seconds (mPa s) to 10^6 mPa s, while water is mostly less than 1 mPa s, depending on temperature. Today, different technologies have been developed to extract these oils, e.g. steam assisted gravity drainage (SAGD), vapour extraction (VAPEX) and cold heavy oil extraction with sand (CHOPS).

These are all high-cost, high-energy and high-emission technologies that are also associated with other environmental concerns. Combinations of conventional methods with biotechnological processes represent a completely novel approach to overcome many of these obstacles.

In this section and below are described Statoil's investigations focusing on the use of extremophile microorganisms as *in situ* biocatalysts for conversion of heavy oils. Experiments were designed to mimic reservoir conditions, with special emphasis on the biocatalytic processes involved in reducing the viscosity of the heavy oil components. These experiments will help address a major challenge from an oil production technology aspect: the control and the regulation of these *in situ* bioprocesses in the oil reservoir.

Two different prototype reservoir models, introducing radial flow, have been designed: (1) containing a central horizontal production well, and (2) containing a central vertical production well. Enrichment cultures from the Statoil BTC strain collection have been assayed for catalytic activities specific for the heavy constituents of various heavy oils. Mixed BTC culture inocula have been injected or soaked into the reservoir models. Depending on the type of oil, viscosities were reduced 2- to 3-fold. In experiments at reservoir-like conditions this gave between 2- and 3-fold increases in recovery. Different types of extremophiles were found to be involved in the bioconversion, and these were most probably thermophiles because the reactions were conducted at 55–60°C. Within 2–3 days a substantial conversion was observed.

Environmental impact

Development of a biologically based process, as opposed to today's conventional methods, would have major implications for the environmental issues raised against heavy oil extraction. A major concern of critics is whether the SAGD process can be justified or not, due to the enormous associated CO_2 emissions and high-energy demand.

There is a general tendency to consider heavy oil or extra heavy oil as the last step in an active ongoing biodegradation of conventional oil. Microorganisms degrade the *n*-alkanes and the light components, leaving the polyaromatic and heavy residues behind, as indicated from gas chromatograph (GC) patterns of these oils. However, there might be more to this than the usually considered descriptive models. There are oil reservoirs containing a high load of different microorganisms, and yet this oil is not characterized as biodegraded. In contrast, there are oils considered heavily biodegraded despite a very low biodiversity and bioactivity in the reservoir zones. Some of these discrepancies might be explained by different reservoir histories, i.e. infill and loading/maturation.

A biocatalytic process for heavy oil production might result in huge environmental benefits. The first practical technology to be implemented will provide a tremendous drive for further developments. A biotech process will provide very low CO_2 emissions, and less air and water pollution. Significantly lower usage of diluents or organic solvents and water in the process flow would help protect the environment. Without being categoric, there should be huge potential savings in CAPEX, as fewer boilers and heat exchangers will be needed. Furthermore, a biocatalysis based method might allow a simpler construction of the separator and process train, thereby reducing the overall process facility costs.

Results and discussion: Bioconversion

Active organisms from the BTC strain collection have been characterized at the genetic level. Fig. 8.15 shows the direct, physical effects of biocatalytic disintegration of oil. Further, it can be shown that this conversion lowers the viscosity. More interestingly, effects are also seen on the distinct bioconversion of special compounds that are constituents of the oil. This facet will be investigated further under the new technology platform of 'petroleomics'.

A standardized procedure has been developed for the identification of suitable biocatalysts/microbial consortia that will result in the desired bioconversion effects. A number of different consortia have been identified using this approach. Fig. 8.15 shows a typical experimental result.

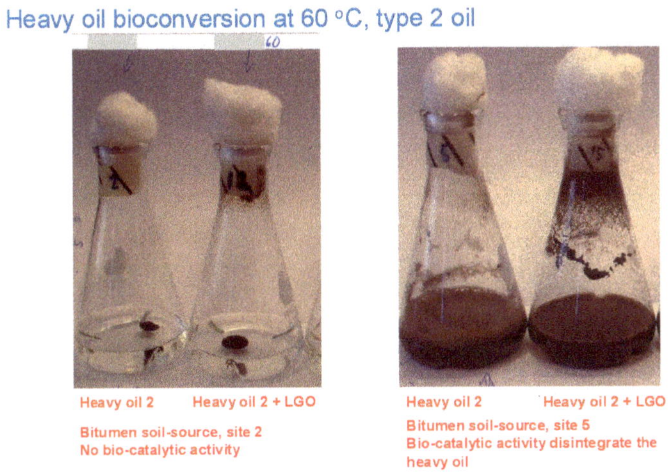

Figure 8.15 Inocula with heavy oil bioconversion activity. Left panel: shaker flasks containing a lump of asphaltfenic (type 2) heavy oil in mineral medium (MM). In addition, the flask on the right also contains light gas oil (LGO). An inoculum from a bitumen source (site 2) was introduced to both flasks. This inoculum did not show any biodegradative activity. Right panel: a corresponding set up, but using an inoculum from a different bitumen source (site 5). This inoculum actively bioconverted the substrate, and within 2–3 days the lump of heavy oil had completely disintegrated.

Results and discussion: reservoir extraction models

One thing is to prove the concept in a laboratory set-up; another is to make the concept work under reservoir conditions. In order to answer some of the basic questions raised in this regard, several sand packs and radial reservoir models have been constructed. Sand pack columns containing sterilized Zuata heavy oil are shown in Fig. 8.16. A special standardized procedure has been developed, and the experiments are run to mimic reservoir conditions as closely as possible.

Under traditional water-flood conditions (i.e. the forced flow of seawater through the reservoir), the viscosity of the Zuata oil at $100\,s^{-1}$ and 55°C is 20,251 mPa s, and the overall recovery at the end of the experiment is 39% of the *standard, total original oil in place* (STOOIP; Fig. 8.16). In contrast, when a mixed inoculum of microorganisms is introduced, the viscosity is lowered to 9258 mPa s and the recovery is boosted to 66% of STOOIP. In a third experiment, where a small amount of acetate was added to the water-flood, a viscosity of 12, 638 mPas were observed, but a total recovery of 73%. Whilst the viscosity was a little higher than for experiment 2, the recovery was equal or even higher. This result is probably due to the generation of more biosurfactants: the acetate helps/stimulates the microorganisms to produce more biosurfactant, giving better access to the substrates in the heavy oil.

Despite the promising results, it could still be argued that the sand pack experiments only provide a linear flow regime. In real life, in the reservoir, there is a radial flow pattern converging towards the well. Thus, experiments with and without the biocatalyst were also performed in a radial reservoir model (Fig. 8.17). The total oil recovery was found to be 68% and 53% of STOOIP for heavy and extra-heavy oil, respectively. Moreover, there was a markedly sharp reduction in viscosity in the bioconversion setup inoculated with a microbial consortium. The total overall recovery was a little lower in the radial compared with the sand pack model.

Experiments using both models have provided evidence that the concept of bioconversion of heavy oil is viable, even under reservoir conditions. The results indicate that the total oil viscosity is to be reduced and the recovery factor is to be increased.

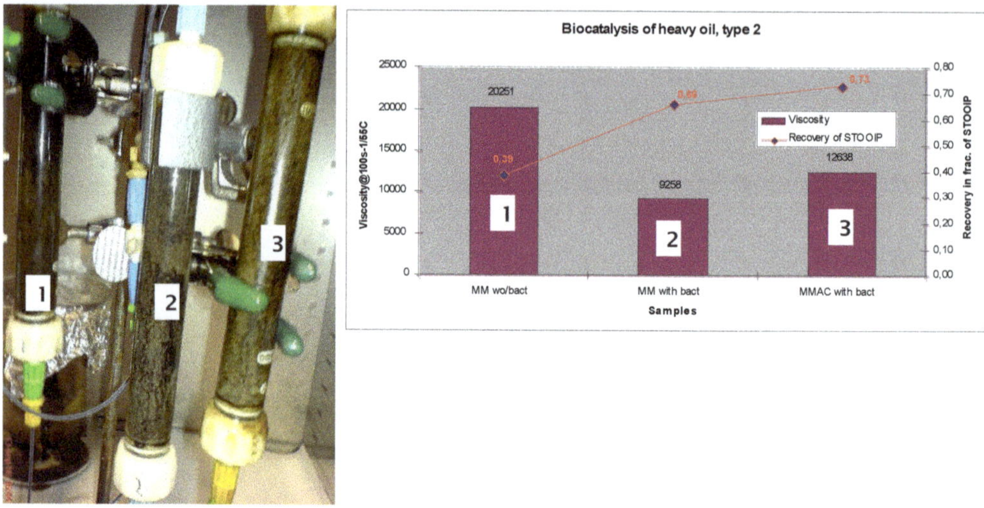

Figure 8.16 Sandpacks designed to mimic heavy oil reservoir conditions. Left: Column sandpacks were set up with no bio-inocula (column 1), active biocatalyst (2) and active biocatalyst + acetate (3). Right: Results – left ordinate, viscosity at 100 s^{-1} and 55°C; right ordinate, overall recovery (as a fraction) of the standard total original oil in place (STOOIP). The two sandpacks with an active biocatalyst (2 and 3) give an almost 50% reduction in viscosity, and almost a doubling in recovery, compared with the control (1).

Biocatalysis of heavy oil – Reservoir like models

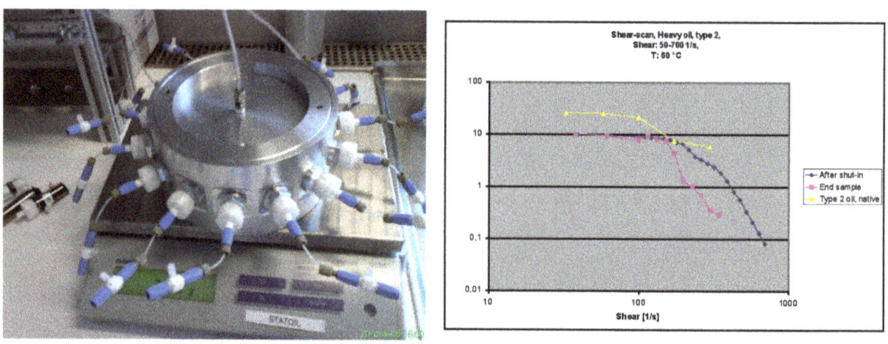

Figure 8.17 Heavy oil biocatalysis under reservoir-like conditions. Left: a radial flow reservoir model. The oil well is in the centre and the inflow is divided amongst various ports at the periphery. This model is run with mixed bio-inocula to deduce the effects of the biocatalyst (controls do not have the biocatalyst). Right: viscosity as a function of shear on type 2, heavy oil, at reservoir conditions. Yellow line – viscosity of native type 2 oil; pink line – end sample with biocatalyst; blue line – after shut-in, i.e. the control without biocatalyst.

Results and discussion: identification of oil fractions by thin layer chromatography (Iatroscan)

Whilst it is critical to observe the physical effects of bioconversion on oil, it is also important to understand some of the underlying mechanisms. In this approach we used saturated-, aromatic-, resins- and asphaltenic hydrocarbons (SARA) quantification of the heavy oil components using thin layer chromatography, TLC, coupled with a flame ionization detector (FID), instrumented by Iatroscan. This analysis compares native crude heavy oil with the bioconverted heavy oil.

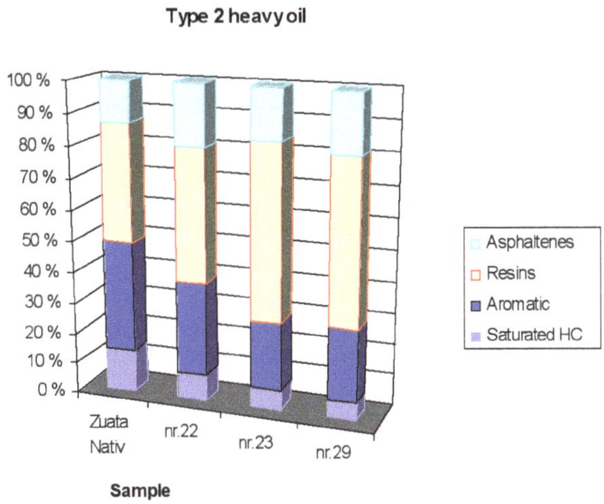

Figure 8.18 SARA quantification of components using Iatroscan. The first column shows the component distribution of the native Zuata oil. Sample no. 22, heavy oil after hot water extraction; sample nos. 23 and 29, heavy oil after incubation with biocatalyst for, respectively, approximately one and two weeks.

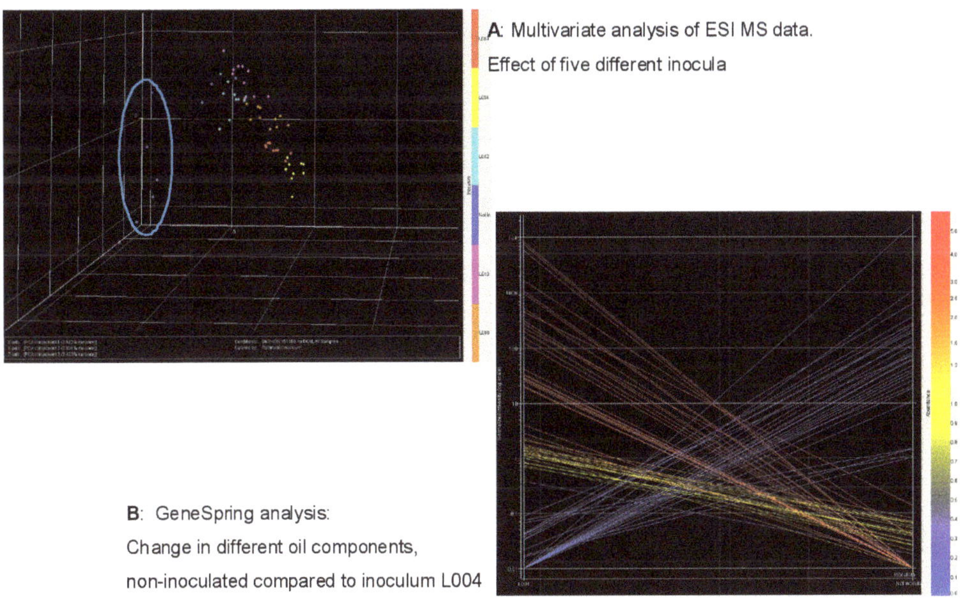

Figure 8.19 Changes in the composition of heavy oil upon bioconversion, assessed using multivariate analysis of ESI MS data and GeneSpring analysis. (A) The native heavy oil, not bioconverted (dark blue dots), groups into a specific cluster in the principal component analysis, PCA, plot. Note that all of the various inocula (different colours) resulted in a distinct bioconversion activity, discriminating each one from the control, uninoculated sample. The inocula used were L009, L004, L002, L010 and L006, all from the BCT strain collection. (B) On the right axis is the control, the non-inoculated Zuata heavy oil. On the left axis is the Zuata heavy oil following bioconversion by inoculum L004. The blue lines identify molecules whose concentration decreased from the non-inoculated control crude oil to the bioconverted oil. The yellow and red lines identify the product molecules generated as a consequence of the bioconversion. The red lines indicate a dramatic increase in new products in the bioconverted oil (left axis).

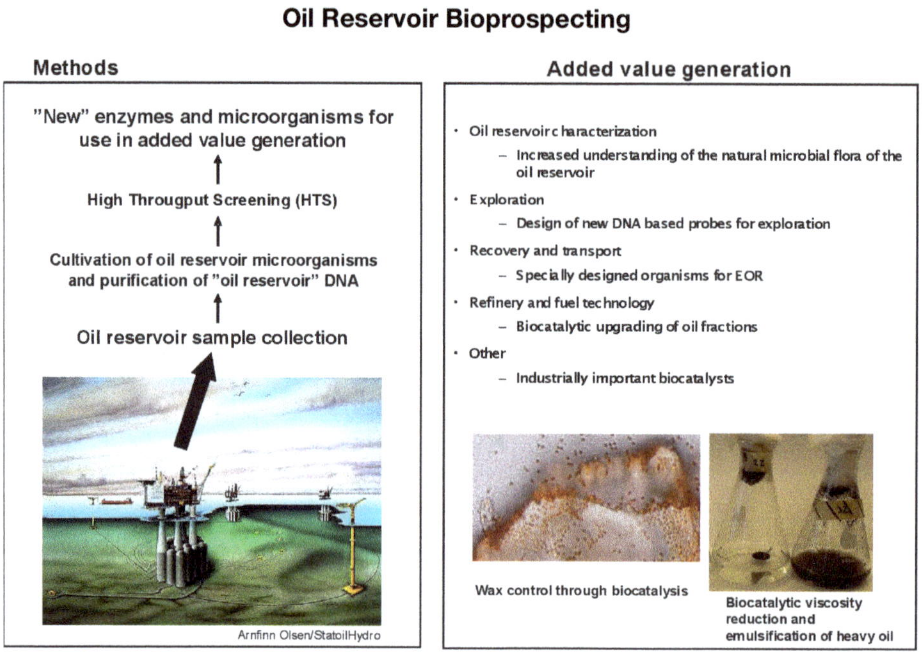

Figure 8.20 Bioprospecting for microorganisms used for value adding. Schematic for sample collection of oil reservoir-associated microorganisms, purification of DNA and high throughput screening of microbes that are potentially useful for value adding (examples of which are given in the right panel).

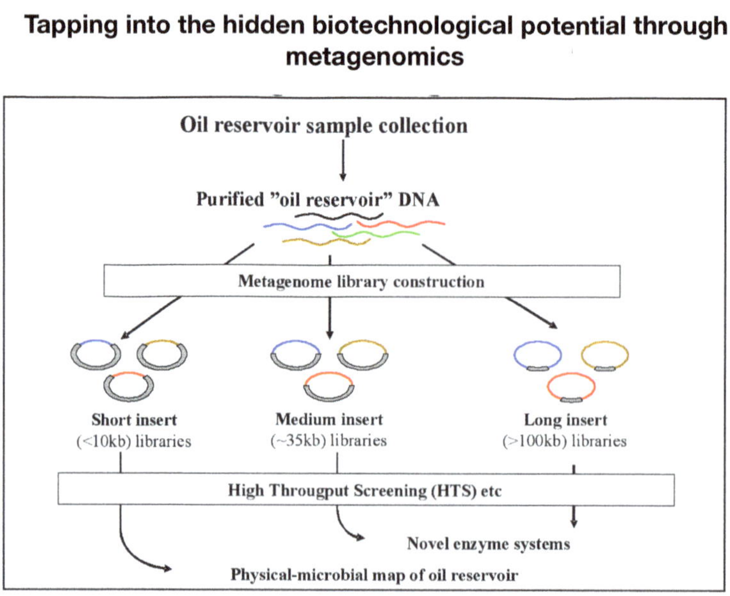

Figure 8.21 Tapping into the hidden potential provided by nature through metagenomics. Schematic view of new techniques in metagenome constructions. Today, more than 1181 organisms are fully sequenced and around 3342 partially sequenced. These numbers will escalate through the years, meaning that there is a tremendous data bank of information that can be tapped into.

Furthermore, it is important to compare with the appropriate controls. The hot water (55–60°C) extraction might alter some of the heavy oil fractions. The relevant control is necessary to discriminate changes caused by the hot water alone from those directly attributed by the bioconversion. Fig. 8.18 illustrates a typical SARA quantification result. Note that the first column indicates the different fractions found in native Zuata oil. As this is a relative measure of the different hydrocarbon fractions, it does not provide absolute quantitative values. Furthermore, the total constituents are always normalized to 100%. In sample 22 (oil sample after extraction with 55–60°C water), the results indicate that some saturated and aromatic hydrocarbons have been extracted from the native Zuata oil. However, as the total fractions are normalized to 100%, it does not directly imply that the asphaltenes fraction has been increased compared with native Zuata, but as the saturates and the aromatic fractions decreased, the relative fraction of asphaltenes has increased (but not its absolute value).

The actual effects of the bioconversion can be seen by comparing (Fig. 8.18) the difference between sample 22, after finishing the water flood, and sample 23 (heavy oil post biocatalyst inoculation/incubation and further flooding at 60°C). Specifically, saturates have decreased marginally, while there is a significant loss in the aromatic fraction; consequently the resin fraction is increased. Sample 29 was collected at the end of the experiment, approximately one week after sample 23 (i.e. after running the bio-inoculated radial reservoir model for 14 days); this oil was quite similar to the first bioconverted oil (sample 23).

The Iatroscan results are qualitative rather than quantitative. Thus, with the current set-up a change of greater than 5% in the various fractions is required for the difference to be considered significant. The heavier components of the oil are usually considered responsible for its high viscosity. However, the interactions of the different types of molecules might be the more influential part of this phenomenon. The important goal is to disrupt the molecular entanglements and interactions that give heavy oil its high visco-elasticity. In this type of bioconversion, there is little information on which molecule is the primary point of attack by the microorganisms. Aromatic compounds could be good candidates.

Results and discussion: 'petroleomics'

Oil is composed of more than 30,000 types of different molecules, many of which are very difficult to identify. The term 'petroleomics' (Hugley *et al.*, 2002) refers to the study of these constituents of oil. Because petroleum crude oil is one of the most complex organic mixtures in the world, to understand it at the molecular level will require very sophisticated instrumentation. For example, using a 9.4-tesla Fourier transform ion cyclotron resonance mass spectrophotometer (FT ICR MS; Hugley *et al.*, 2002), over 11,000 peaks have been identified in a single positive ionization (ESI) mass spectrum. Such an ultra-high-mass resolving power instrument would allow an understanding of the mechanisms behind the various types of bioconversions occurring in experiments such as those described in the above sections. The FT ICR MS could be used to identify the actual substrate(s) and end product(s), which might allow insights into the enzymes and metabolic pathways activated in the biocatalyst or inoculum. Unfortunately, very few researchers have access to such an instrument. The results described below have instead been obtained with a quadrupole time of flight (QTOF) MS instrument.

MS data obtained from a bioconversion experiment with Zuata heavy oil was analysed using a multivariate analytical tool (Fig. 8.19A). Ten consecutive samples were injected in each series, giving 10 replicate readings of each. Of particular interest was the fact that the bioconverted oil resulting from the activity of five different inocula (i.e. five different bioactive microbial consortia) clustered into individual groups with little overlap (Fig. 8.19A). This suggests that the different microbial consortia activated different metabolic pathways.

The inoculum with the highest activity was the consortium designated L004. This inoculum was analysed further (see Fig. 8.19B) – using multivariate analyses to compare with the non-inoculated control. The blue lines in Fig. 8.19 identify molecules, i.e. m/z ratios, whose concentration decreased from the non-inoculated control crude oil to the oil bioconverted by consortium L004. Thus, all of the 'blue molecules' were substrates in various enzymatic conversions. Note that many of the substrate molecules were almost completely bioconverted, being close to zero in the inoculated heavy oil. Conversely, product molecules (red lines, Fig. 8.19B) almost absent in the non-bioconverted oil rose dramatically following bioconversion. The yellow lines represent molecules with a small increase going from the native crude heavy oil to the bioconverted one (Fig. 8.19B).

The overall relative changes of some of these molecules are extraordinary. About ten different molecules showed a greater than 20,000-fold change (a 10-fold change represents a significance better than $P = 0.01$). Four molecules had a 100,000-fold change; the m/z ratio of each of these molecules is at the resolving power of the QTOF. With access to extensive data banks and analytical software it should be possible to identify some of the actual molecules. The metabolic pathways involved could then be postulated, which would provide an interesting insight into the mechanisms involved in the bioconversion of heavy oil.

Bioconversion of heavy oil – conclusion

The results described here provide strong evidence for the heavy oil bioconversion activities of several microbial consortia/inocula. Given the right conditions, some of these consortia have an extraordinary bioconverting capacity. Experiments performed in various reservoir-like models indicate that these consortia also provide a feasible oil extraction method under reservoir conditions. Both in sandpacks and radial flow models, residual oil has been 'attracted' by the microbial consortia and viscosity reduced and recovery substantially increased.

The more descriptive analytical tools like SARA quantification reveal a significant change in the overall composition of native heavy oil fractions following bioconversion. MS measurements and sophisticated analytical software seems to provide potential methods to elucidate the actual metabolic pathways that will be necessary for an efficient industrial bioconversion.

This new heavy oil bioconversion technology will be implemented in combination or in sequence with non-biologically based CHOPS, VAPEX and hot water extraction procedures. To make this approach work, even new strategies for reservoir sweep and drainage (i.e. technologies for covering as much as possible of the reservoir volume and attempting to drain it towards the production well) might have to be developed.

Bioprospecting – what nature provides

Microorganisms can be used at various stages throughout the value chain of an oil company. Future use may call for an expansion of the BTC strain collection with even more specialized organisms or GMOs. These organisms can be searched for in many different, extreme habitats around the world. So far, our search has been concentrated in oil reservoirs, mud volcanoes, seeps and areas associated with hydrocarbons. In many different environments, mother nature has provided a pre-selection of particularly well-suited organisms. In the oil industry there is a need for organisms optimized for oil recovery, thermophile *n*-alkane degradation or thermophile ring opening processes, as well as organisms involved in the production of biofuels. Of special interest will be organisms optimized for high temperature pretreatment of lignocelluloses, producing the perfect hydrolysate for further fermentation. Such an organism would be specialized in high yield fermentations producing bioalcohols or in the direct generation of bioenergy. The bioprospecting and value adding schematic shown in Fig. 8.20 is currently being pursued by Statoil using a metagenomic approach on samples from three different wells in the Norwegian Continental Shelf. Future use of the BTC strain collection and extremophile organisms might open a completely new world.

A more in-depth outline of our metagenomic research project is given in Fig. 8.21. Currently, DNA is extracted from three different North Sea wells. Metagenomic library construction and sequencing is under way. Preliminary results indicate some very exciting new findings. For example, there are variations in biodiversity between the three wells producing from different reservoir zones in the same oilfield. Furthermore, there are distinct variations in organisms present in the water phase versus organisms found in the oil or hydrocarbon phase. As we delve into more extreme areas like the Arctic and Antarctica, cold adapted psychrophilic organisms might prove to contain exciting new enzymes or metabolic pathways to explore for possible industrial processes, such as oil production in extremely cold climates. Such new findings may also possibly give a more extensive understanding of basic reservoir mechanisms and processes, provide insights into the reservoir microorganism gene-pool, and shed new knowledge on hydrocarbon generation.

Spin-offs outside the oil industry

Bioprospecting of the different variety of oil-associated microorganisms may provide spin-offs for many different applications: new antibiotics, new anti-cancer compounds, new metabolites and monomers, new stereospecific enzymes and a variety of thermotolerant new industrial enzymes. Such spin-offs could open up a completely new range of industrial applications. At this stage it is just too early to judge the implications of any future findings on possible beneficial spin-offs. However, the potential seems huge. The limiting factors might be boundaries set by our own overly narrow-minded imagination. Creativity and boldness to think in completely new terms will help to solve the problems of the future.

Summary and future trends

The use of extremophiles, such as thermophiles, in various biocatalytic processes has already provided a new wave of biotechnological applications in industry. With the completely new and so far unevaluated candidates discovered using the techniques described in this chapter, bioprocesses can be performed at temperatures and pressure conditions never before considered possible.

In the deep biosphere, extraordinary new types of organisms, sedimented or buried 200–500 million years ago, can be found. Through molecular genetic approaches that are nowadays even learnt by school students, these organisms can be identified and characterized. The information obtained can then be developed into new tools for searching for new oil in sensitive areas like the Arctic, Antarctica and jungle areas; even 'fit for the purpose' microarray and lab-on-a-chip systems can be developed as detection systems.

What are these bacterial and archaeal microbes? These organisms are extremophiles to the fourth power – they tolerate a high load of hydrocarbons, extreme temperatures, high pressures and high salt concentrations. Inherently, this will generate some very exciting properties. These organisms will also provide new knowledge about, and understanding of, various oil reservoir-associated mechanisms and characteristics. For example, will biodegradation of oil occur even at 100°C? What are the actions and the biological significance of these microbes in the reservoir, irrespective of whether they are grouped within the bacterial or archaeal domains?

Relatively few enzymes are used in large-scale industrial applications. Enzymes isolated from these extremophile/thermophile organisms might provide 'game changing' new possibilities. They may furnish new incentives for the development of entirely new processes where the partial pressure and concentration of certain compounds in the water phase can be dramatically changed. Another possibility is the use of selected new thermophilic organisms as hosts for the introduction of other desired properties: a new brand of specially designed hyperthermophilic GMOs can be constructed.

These new microbes provide opportunities for new technological processes in biofuel production. Second generation technology for biodiesel production might be implemented in the relatively near future. Several companies are also working on alternative routes for the production of jet fuels using biomass as the raw source material. Also, in the realm of second generation bioalcohols, potential technologies using thermophilic to hyperthermophilic conditions seem to provide the most promising solutions for rapid implementation.

Users of traditional heavy oil extraction methods have major difficulties in justifying the associated high use of energy, and pollution of soil and the environment. Furthermore, the emission profile is not very acceptable. The first company that implements a large-scale facility based on biotechnology principles will gain huge strategic benefits. Lastly, the knowledge of this huge subsurface population of diverse microorganisms provides excellent opportunities for bioprospecting, with a view to developing new solutions to the various technology gaps a company might have. There should be a multitude of spin-offs outside the oil industry. The world is desperately in need of new enzymes, new antibiotics, new immunosuppressants, new anticancer agents, etc.

The few suggestions mentioned above are just some of the areas we have been working on at Statoil. Other companies/research institutions might work on complementary areas, potentially leading to some very exciting new discoveries. Let us hope that some of these investigations will one day solve just a few of the problems and obstacles that we as humans will face in the future.

Acknowledgements

The author is indebted to colleagues at Statoil ASA and to the company in particular for permission to publish results. Many thanks to my Biotech-group: Anita Skarstad, Evy Mellemsæther, Hege Engan, Rebecca Kristensen, Amund Heggholmen and Christian Collin-Hansen. Furthermore, the contract work done by SINTEF organization employees Sidsel Markussen, Asgeir Winnberg, Trond Ellingensen, Odd Gunnar Brakstad, Mimmi Throne-Holst, Kolbjørn Zahlsen and Kristin Bonaunet is highly valued, as is the project work with the NTNU group – Svein Valla and Anna Lewin. Without the support of these groups, this work and the ideas within would not have been possible.

References

Bartlett, D.H., Kato, C., and Horikoshi, K. (1995). High pressure influences and protein expression. Res. Microbiol. *146*, 697–706.

Hugley, C.A., Rodgers, R.P., and Marshall, A.G. (2002). Resolution of 11,000 compositionally distinct components in a single electrospray ionization fourier transform ion cyclotron resonance mass spectrum of crude oil. Anal. Chem. *74*, 4145–4149.

Lee, S.Y., Park, J.H., Jang, S.H., Nielsen, L.K., Kim, J., and Jung, K.S. (2008). Fermentative Butanol Production by Clostridia. Biotechnol. Bioengin. *110*, 209–228

Hyperthermophiles: Metabolic Diversity and Biotechnological Applications

9

Kazem Kashefi

Abstract

The isolation and characterization of novel hyperthermophilic, microorganisms from modern hot environments have greatly increased our understanding of how microbes can live and thrive in such inhospitable environments. The finding that microorganisms have the ability to grow at these high temperature has implications for delimiting when and where life might have evolved on a hot, early Earth; the depth to which life exists in the Earth's subsurface; and the potential for life in hot, extraterrestrial environments. The study of hyperthermophilic microorganisms provides valuable insights into microbial respiration in a diversity of modern and ancient hydrothermal systems. In addition, it provides information about the fate of metals such as iron, uranium, technetium, and even gold. Reduction of these metals by hyperthermophiles provides, for example, a probable explanation for a number of geologically, environmentally and economically important ore deposits. This allows us to identify geological signatures for biological processes, something that may prove instrumental in our search for life on other planets. Finally, enzymes capable of functioning at high temperatures have a number of important applications in biomass conversion, in biotechnology, and in the pharmaceutical, food and cosmetic industries.

Introduction

Modern hot environments such as those found in hydrothermal systems are of great interest to the scientific community because they resemble those ancient environments where life first arose on Earth. They also share many characteristics with environments on other planets where life may have once existed, or still does (Kashefi and Lovley, 2003; Orange *et al.*, 2009). A wide diversity of organisms with optimum growth temperatures above 80°C, and known as hyperthermophiles, inhabit these environments. They have been isolated from a wide variety of extreme thermal habitats such as deep-sea hydrothermal vents, marine and volcanic environments, solfataricus fields, deep geothermally heated oil-containing stratifications, shallow, neutral hot springs, and hot industrial environments (Table 9.1). This geographic diversity is reflected in a wide diversity of morphological, physiological, and metabolic traits (Table 9.1). Most hyperthermophiles isolated to date belong to the third domain of life, the *Archaea* (Sakuraba *et al.*, 2004). Examples include the genera *Pyrodictium, Pyrobaculum, Thermoproteus, Desulfurococcus, Sulfolobus, Methanopyrus, Pyrococcus, Thermococcus, Methanococcus,* and *Archaeoglobus*. Hyperthermophilic bacteria also have been isolated, such as those represented by the genera *Geothermobacterium, Aquifex,* and *Thermotoga* (Table 9.1). To date, approximately 100 species of hyperthermophilic archaea and bacteria have been recovered in pure culture. These are listed in Table 9.1. These species are grouped into 39 genera and 10 orders and represent only a fraction of the estimated hyperthermophilic diversity on Earth.

Hyperthermophiles are very distinct in terms of their phylogeny as well as their physiological and biochemical properties. Based on 16S rRNA,

Table 9.1 Taxonomic description and characteristics of hyperthermophilic microorganisms

Genus	Species	Aerobe/ Anaerobe	Hetero-/ Autotroph	Temperature (°C) Range	Temperature (°C) Opt.	pH Range	pH Opt.	% (w/v) Salt Range	% (w/v) Salt Opt.	mol % G+C
Bacteria										
Aquifex	*aeolicus*	AE	A	58–95	85	5.5–8.0	6.5–7.0	1.0–5.0	3.0	43.4
Aquifex	*pyrophilus*	FAN	A	67–95	85	5.4–7.5	6.8	1.0–5.0	3.0	47.3
Geothermobacterium	*ferrireducens*	AN	A	65–100	85–90	N.D.	6.8–7.0	0.0–0.75	0.0–0.5	N.D.
Rhodothermus	*obamensis*	AE	H	50–85	80	5.5–9.0	7.0	1.0–5.0	3.0	66.6
Thermocrinis	*ruber*	AE	FA	44–89	80	7.0–8.5	N.D.	0.0–0.4	N.D.	47.2
Thermotoga	*maritima*	AN	FA	55–90	80	5.5–9.0	6.5	0.25–3.75	2.8	46
Thermotoga	*neapolitana*	AN	H	55–90	80	5.5–9.0	7.0	0.25–6.0	2.0	41
Thermotoga	*naphthophila*	AN	H	48–86	80	5.4–9.0	7.0	0.1–6.0	1.0	46.1

Growth Substrates					
Carbon Sources & Electron Donors		Electron Acceptors	Morphology & Size	Habitat	References
CO_2	Hydrogen, Sulfide	Oxygen	Rods, 0.4–0.5 x 2–6 µm	Shallow marine hydrothermal vent; Vulcano, Italy	Deckert *et al.*, 1998; Warren *et al.*, 1998; Chatelet *et al.*, 1999; Huber and Stetter, 1999; Nubel *et al.*, 2000
CO_2	Hydrogen, Sulfur, Thiosulfate	Nitrate, Oxygen	Rods, 0.4–0.5 x 2–6 µm	Submarine vent; Kolbeinsey Ridge, Iceland	Burggraf, 1992; Huber *et al.*, 1992; Huber and Stetter, 1999
CO_2	Hydrogen	Poorly crystalline Fe(III) oxide	Rods, 0.5 x 1.0–1.2 µm	Hot sediment from Obsidian Pool; Yellowstone National Park, WY, United States	Kashefi *et al.*, 2001; Kashefi *et al.*, 2002
Amino acids (Alanine, Arginine, Aspartate, Glycine, Proline, Serine, Threonine), Casamino acids, Casein, Cellobiose, Galactose, Glucose, Maltose, Mannitol, Peptone, Sorbitol Starch, Sucrose, Yeast extract	Lactate, Pyruvate	-	Rods, 4–10 x 0.5–1.0 µm	Effluent water and sediments from a hydrothermal vent in Tachibana Bay; Nagasaki Prefecture, Japan	Sako *et al.*, 1996
CO_2	Hydrogen, Formamide, Formate, Sulfur, Thiosulfate	Oxygen	Rods, 1–3 x 0.4 µm	Pink filaments obtained from Octopus Spring in the White Creek area; Yellowstone National Park, WY, United States	Huber *et al.*, 1998
CO_2, Cellulose, Galactose, Glucose, Glycogen, Lactose, Maltose, Mannan, Raffinose, Ribose, Salicin, Starch, Sucrose, Xylan, Xylose, Yeast extract, Cell extracts of *Archaea* or *Bacteria*	Hydrogen	Poorly crystalline Fe(III) oxide, Ferric citrate, AQDS, Au(III)	Rods, 1.5–11 x 0.6 µm	Geothermally heated sea floor; the Azores and Italy	Huber *et al.*, 1986; Vargas *et al.*, 1998; Lovley *et al.*, 2000; Kashefi *et al.*, 2001
Galactose, Glucose, Glycogen, Lactose, Maltose, Ribose, Starch, Sucrose, Yeast extract. Weak growth on Peptone and Tryptone.	Hydrogen	Sulfur, Thiosulfate	Rods, 1.5–11 x 0.6 µm	Lucrino hot spring; Naples, Italy	Jannasch *et al.*, 1988
Arabinose, Fructose, Galactose, Glucose, Lactose, Maltose, Mannitol, ribose, Starch, Sucrose, Peptone, Yeast extract	-	Sulfur (reduces specific growth rate and final cell yield), Thiosulfate	Rods, 2–7 x 0.8–1.2 µm	Formation waters of production fluid in the Kubiki oil reservoir; Niigata, Japan	Takahata *et al.*, 2001

Table 9.1 continued

Genus	Species	Aerobe/ Anaerobe	Hetero-/ Autotroph	Temperature (°C) Range	Temperature (°C) Opt.	pH Range	pH Opt.	% (w/v) Salt Range	% (w/v) Salt Opt.	mol % G+C
Thermotoga	*petrophila*	AN	H	47–88	80	5.2–9.0	7.0	0.1–5.5	1.0	46.8
Archaea										
Acidianus	*ambivalens*	FAN	A	Up to 87	80	1.0–3.5	N.D.	N.D.	N.D.	N.D.
Acidianus	*infernus*	FAN	A	65–96	85–90	1.0–5.5	2.0	0.1–4.0	0.2	31
Acidilobus	*aceticus*	AN	H	60–92	85	2.0–6.0	3.8	N.D.	0.0	53.8
Aeropyrum	*camini*	AE	H	70–97	85	6.5–8.8	8.0	2.2–5.3	3.5	54.4
Aeropyrum	*pernix*	AE	H	70–100	90–95	5.0–9.0	7.0	1.8–7.0	3.5	56
Archaeo-globus	*fulgidus*	AN	FA	60–95	83	6.4–7.4	6.9	0.5–4.5	1.8	46

Growth Substrates					
Carbon Sources & Electron Donors		Electron Acceptors	Morphology & Size	Habitat	References
Cellulose, Arabinose, Cellulose, Fructose, Glucose, Lactose, Maltose, Ribose, Sucrose, Starch, Peptone, Yeast extract	-	Sulfur (reduces specific growth rate and final cell yield)	Rods, 2–7 x 0.7–1.0 µm	Formation waters of production fluid in the Kubiki oil reservoir; Niigata, Japan	Takahata *et al.*, 2001
CO_2	Hydrogen, Sulfur	Oxygen, Sulfur	Cocci,	Icelandic isolates from geographically distant solfataric fields; Worldwide	Zillig *et al.*, 1985; Zillig *et al.*, 1986; Fuchs *et al.*, 1996; Kletzen *et al.*, 1999
CO_2, Alanine, Arabinose, Beef extract, Casamino acids, Celate, Citrate, Galactose, Glucose, Glycerol, Glycin, Lactose, Lulose, Maltose, Mannitol, Melibiose, Peptone, Raffinose, Ribose, Starch, Sucrose, Yeast extract, Xylose	Hydrogen, Sulfur, Acetate, Butyrate, Valerate	Oxygen, Sulfur	Cocci, 0.5–2 µm diameter	Hot water, mud and marine sediments at geothermal springs in Naples and Vulcano; Naples, Italy	Segerer *et al.*, 1986
Beef extract, Soya extract, Starch, Yeast extract	Hydrogen, Sulfides, Sulfur	Sulfur	Cocci, 1–2 µm diameter	Solfataric field near the Moutnovsky volcano; Kamchatka, Russia	Prokofeva *et al.*, 2000; Dworkin *et al.*, 2006
Arginine, Asparagine, Asparate, Casamino acids, Casein, Citrate, Cellobiose, Chitin, Fructose, Galactose, Gelatin, Glucose, Glutamate, Glycerol, Lactose, Malate, Maltose, Myo-inositol, Phenylalanine, Proline, Serine, Sorbitol, Starch, Sucrose, Trehalose, Tryptone, Valine, Yeast extract, Xylose	Acetate, Pyruvate	Sulfur, Thiosulfate	Cocci, 2.1–11.2 µm diameter	Deep-sea hydrothermal vent chimney, depth 1385 m; Suiyo Seamount, Izu-Bonin Arc, Japan	Nakagawa *et al.*, 2004
Compound proteins	Thiosulfate (stimulates growth)	Oxygen	Cocci, 0.8–1.0 µm diameter	Hot sediments and venting waters at a coastal solfataric vent; Kodakara-Jima Island, Japan	Sako *et al.*, 1996; Kawarabayasi *et al.*, 1999
CO_2, Casamino acids, Casein, Gelatin, Glucose, Peptone, Starch, Cell extracts of *Archaea* or *Bacteria*, Meat extract, Yeast extract	Hydrogen, Formamide, Formate, DL-Lactate, Pyruvate	AQDS, Ferric Citrate, Sulfate, Sulfite, Thiosulfate	Cocci, 0.3–1.0 µm diameter	Geothermally heated sea floor; Vulcano Italy	Stetter *et al.*, 1987; Stetter, 1988; Beeder *et al.*, 1994; Klenk *et al.*, 1997; Vargas *et al.*, 1998; Lovley *et al.*, 2000; Hartzell and Reed, 2006

Table 9.1 continued

Genus	Species	Aerobe/ Anaerobe	Hetero-/ Autotroph	Temperature (°C)		pH		% (w/v) Salt		mol % G+C
				Range	Opt.	Range	Opt.	Range	Opt.	
Archaeoglobus	*lithotrophicus*	AN	A	55–87	80	6.0–8.5	6.0	0.6–4.8	N.D.	40
Archaeoglobus	*profundus*	AN	A	65–90	82	4.5–7.5	6.0	0.9–3.6	1.8	41
Archaeoglobus	*veneficus*	AN	FA	65–80	80	6.5–8.0	7.0	0.5–4.0	2.0	45
Caldariella	*acidophila*	AE	FA	63–89	87	3.0–4.5	N.D.	N.D.	0.0	N.D.
Caldivirga	*maquilingensis*	FAN	H	60–92	85	2.3–5.9	3.7–4.2	0.0–0.75	0.5	50
Caldococcus (Reclassified from *Thermococcus*)	*litoralis*	AN	H	55–100	88	5.9–7.0	6.4	1.0–6.0	2.5	57.1
Caldococcus	*'noboribetus'*	AN	H	70–96	92	1.5–4	3	N.D.	0.01	57.5
Desulfurococcus	*amylolyticus*	AN	H	68–97	90–92	5.7–7.5	6.4	0.0–1.12	0.12	41.2

Growth Substrates					
Carbon Sources & Electron Donors		Electron Acceptors	Morphology & Size	Habitat	References
CO_2	Hydrogen, Acetate, Formate, Pyruvate	Sulfate, Thiosulfate	Cocci, 0.4–1.2 µm diameter	Well-head samples upstream of topsides processing facilities at the Thistle oil platform; North Sea, Germany	Stetter *et al.*, 1993; Vorholt *et al.*, 1995; Hartzell and Reed, 2006
Acetate, Crude oil, Lactate, Meat extract, Peptone, Pyruvate, Yeast extract	Hydrogen (required)	Sulfate, Sulfite, Sulfur, Thiosulfate	Cocci, 1.3 µm diameter	Deep-sea hydrothermal system in Guaymas Basin; Gulf of California, United States	Burggaraf *et al.*, 1990; Hartzell and Reed, 2006
CO_2	Hydrogen, Acetate, Ethanol, Formate, Fumarate, Glucose, Isopropanol, Pyruvate	Sulfite, Thiosulfate	Cocci, 0.5–1.2 µm diameter	Black smoker wall, "Snake Pit" site; Mid Atlantic Ridge, Iceland	Huber *et al.*, 1997; Hartzell and Reed, 2006;
CO_2, Glucose, Lactose, Maltose, Sucrose, Xylose	Fe(II), Sulfur	Oxygen	N.D.	Solfatara, steam heated acidic waterhole; Pisciarelli, Italy	De Rosa *et al.*, 1974; 1975; Siling *et al.*, 1980; De Rosa *et al.*, 1984; Grogan, 1989
Beef extract, Gelatin, Glycogen, Peptone, Tryptone, Yeast extract	-	Sulfate, Sulfur, Thiosulfate	Rods, 3–20 x 0.4–0.7 µm	Hot spring water, solfataric soils and mud from a hot spring: Mud Spring, Mt. Makiling; Luzon, Laguna, Philppines	Itoh *et al.*, 1999
Amino acids, Compound proteins	-	Sulfur	Cocci, 1–2 µm diameter	Coastal hot volcanic vents of the Hot Beach; Kunashir Island, Japan	Neuner *et al.*, 1990; **Reclassified by** Kostyukova *et al.*, 1999
N.D.	N.D.	Sulfur	Cocci, 1–3 µm diameter	Noboribetsu hot spring; Hokkaido, Japan	Aoshima *et al.*, 1996; Aoshima and Oshima, 1997
Amino acids, Compound proteins, Glycogen, Pectin, Starch	-	Sulfur	Cocci, 0.7–1.5 µm diameter	Thermal spring of Kamchatka and Kunashir; Kunashir, Japan	Bonch-Osmolovskaya *et al.*, 1988

Table 9.1 continued

Genus	Species	Aerobe/ Anaerobe	Hetero-/ Autotroph	Temperature (°C)		pH		% (w/v) Salt		mol % G+C
				Range	Opt.	Range	Opt.	Range	Opt.	
Desulfurococcus	*mobilis*	AN	H	55–97	85	4.5–7.0	6.0	N.D.	0.0	50
Desulfurococcus	*mucosus*	AN	H	55–97	85	2.2–6.5	6.0	N.D.	0 .0	52
Ferroglobus	*placidus*	AN	FA	65–95	85	6.0–8.5	7.0	0.5–4.5	1.9	43
Geoglobus	*ahangari*	AN	FA	65–95	88	5.0–7.6	7.0	0.9–3.8	1.9	58.7
Hyperthermus	*butylicus*	AN	H	77–108	95–106	N.D.	7.0	0–2.9	1.7	55.6
Ignicoccus	*hospitalis*	AN	A	73–98	90	4.5–7.0	5.5	0.5–5.5	1.4	56
Ignicoccus	*islandicus*	AN	A	70–98	90	3.8–6.5	5.8	0.3–5.5	2.0	41
Ignicoccus	*pacificus*	AN	A	75–98	90	4.5–7.0	6.0	1.0–5.0	2.0	45
Ignisphaera	*aggregans*	AN	H	85–98	92–95	5.4–7.0	6.4	0.0–0.5	N.D.	52.9
Methanocaldococcus (Reclassified from *Methanococcus*)	fervens	AN	A	48–92	85	5.5–7.6	6.5	0.5–5.0	3.0	33
Methanocaldococcus	*indicus*	AN	A	50–86	85	5.5–6.7	6.5	1.5–5.0	3.0	30.7

Growth Substrates					
Carbon Sources & Electron Donors		Electron Acceptors	Morphology & Size	Habitat	References
Complex organic compounds	-	Sulfur	Cocci, 0.5–1.0 µm diameter	Solfataric fields, Hveravellir; Askja, Iceland	Zillig *et al.*, 1982; Dworkin *et al.*, 2006
Complex organic compounds	-	Sulfur	Cocci, 0.3–2.0 µm diameter	Solfataric fields, Hveravellir; Askja, Iceland	Zillig *et al.*, 1982; Dworkin *et al.*, 2006
CO_2, (Acetate, Pyruvate and Yeast extract stimulate growth)	Hydrogen, Aromatics, Fe(II), Sulfide	Fe(III) oxide, Nitrate, Thiosulfate	Cocci, 0.7–1.3 µm diameter	Shallow marine hydrothermal system; Vulcano, Italy	Hafenbradl *et al.*, 1996; Tor *et al.*, 2001
CO_2, Yeast extract	Hydrogen, Acetate, Alanine, Cysteine, Ethanol, Fumarate, Lactate, Malate, Palmitate, Pyruvate, Serine, Stearate	AQDS, Au(III), FC, Fe(III)	Cocci, 0.3–0.5 µm diameter	Sediments from Guaymas Basin; Gulf of California, United States	Kashefi *et al.*, 2002
Compound proteins	-	Sulfur	Spheres, 1.5 µm diameter	Azores seafloor; Sao Miguel, Portugal	Zillig *et al.*, 1990; Dworkin *et al.*, 2006
CO_2	Hydrogen	Sulfur	Cocci, 1–6 µm diameter	Isolated from the Kolbeinsey Ridge, to the north of Iceland	Paper *et al.*, 2007
CO_2 (Gelatin, Meat extract and Tryptone stimulate growth)	Hydrogen	Sulfur	Cocci, 1.2–3 µm diameter	Submarine (103–106 m) sandy sediments and venting water from Kolbeinsey Ridge, north of Iceland	Huber *et al.*, 2000
CO_2 (Gelatin, Meat extract and Tryptone stimulate growth)	Hydrogen	Sulfur	Cocci, 1–2 µm diameter	Black smoker samples at East Pacific Rise (9 N, 104 W) in the Pacific Ocean (2500 m); Mid-Oceanic Ridge, Pacific Ocean	Huber *et al.*, 2000
β-cyclodextrin, Galactose, Glucose, Glycogen, Konjac Glucomannan, Lactose, Maltose, Mannose, Typticase peptone	-	Cystine, Nitrate, Nitrite, Sulfite, Sulfur, Thioglycolate, Thiosulfate	Cocci, 1–1.5 µm diameter	Hot Springs; New Zealand and Yellowstone National Park, WY, United States	Niederberger *et al.*, 2006
CO_2 (Casamino acids, Trypticase and Yeast extract stimulate growth)	Hydrogen	(Selenate or Tungstate stimulate growth)	Cocci, 1–2 µm diameter	Deep-sea hydrothermal vent core from Guaymas Basin (2003 m); Gulf of California, United States	Zhao *et al.*, 1988; Jeanthon *et al.*, 1999; Whitman *et al.*, 2001
CO_2 (Yeast extract stimulates growth)	Hydrogen	(Selenate or tungstate stimulate growth)	Cocci. 1–3 µm diameter	Isolated from a deep-sea hydrothermal chimney sample collected on the Central Indian Ridge at a depth of 2420 m; Saint Paul and Amsterdam	L'Haridon *et al.*, 2003

Table 9.1 continued

Genus	Species	Aerobe/ Anaerobe	Hetero-/ Autotroph	Temperature (°C)		pH		% (w/v) Salt		mol % G+C
				Range	Opt.	Range	Opt.	Range	Opt.	
Methanocaldococcus (Reclassified from *Methanococcus*)	*infernus*	AN	A	55–91	85	5.25–7.0	6.5	1.3–5.6	2.5	33
Methanocaldococcus (Reclassified from *Methanococcus*)	*jannaschii*	AN	A	50–85	85	5.2–7.0	6.0	1.0–5.0	3.0	31
Methanocaldococcus (Reclassified from *Methanococcus*)	*vulcanius*	AN	A	49–89	80	5.25–7.0	6.5	0.63–5.6	2.5	31
Methanopyrus	*kandleri*	AN	A	84–110	98	5.5–7.0	6.5	0.2–4.0	2.0	59
Methanothermus	*fervidus*	AN	A	65–97	83	4.0–7.0	6.5	N.D.	1.2	33
Methanothermus	*sociabilis*	AN	A	55–97	88	5.5–7.5	6.5	N.D.	0.0	33
Methanotorris (Reclassified from *Methanococcus*)	*igneus*	AN	A	45–91	88	5.0–7.5	5.7	0.9–5.4	1.8	31
Palaeococcus	*ferrophilus*	AN	H	60–88	83	4.0–8.0	6.0	2.0–7.3	4.3	53.5
Palaeococcus	*helgesonii*	AN	H	45–85	80	5.0–8.0	6.5	0.5–6.0	2.8	42.5
Pyrobaculum	*aerophilum*	FAN	FA	75–104	100	5.8–9.0	7.0	0.0–3.6	1.5	52
Pyrobaculum	*arsenaticum*	AN	FA	68–100	95	N.D.	6.5–7.0	0.0–3.0	N.D.	58.3

Growth Substrates					
Carbon Sources & Electron Donors		Electron Acceptors	Morphology & Size	Habitat	References
CO_2 (Casamino acids, Trypticase and Yeast extract stimulate growth)	Hydrogen	Sulfur (Selenate or Tungstate stimulate growth)	Cocci, 1–3 µm diameter	Deep-sea hydrothermal vent chimney at the Mid-Atlantic Ridge (14 45' N); Iceland	Jeanthon *et al.*, 1998; Whitman *et al.*, 2001
CO_2	Hydrogen	(Selenate or Tungstate stimulate growth)	Cocci, 1.5 µm diameter	Sedimentary material at the base of a 'white smoker' vent in the East Pacific Rise (2600 m); Mid-Oceanic Ridge, Pacific Ocean	Jones *et al.*, 1983; Boone *et al.*, 1993; Bult *et al.*, 1996; Whitman *et al.*, 2001
CO_2 (Casamino acids, Trypticase and Yeast extract stimulate growth)	Hydrogen (Selenate or Tungstate stimulate growth)	Sulfur	Cocci, 1–3 µm diameter	Deep-sea hydrothermal vent chimney on the East Pacific Rise; Mid-Oceanic Ridge, Pacific Ocean	Jeanthon *et al.*, 1999; Whitman *et al.*, 2001
CO_2 (Yeast extract stimulates growth)	Hydrogen	AQDS, Sulfur	Rods, 2–14 x 0.5 µm	Guaymas Basin; Gulf of California, United States	Huber *et al.*, 1989; Kurr *et al.*, 1991; Lovley et al, 2000; Slesarev *et al.*, 2002
CO_2 (Yeast extract (required)	Hydrogen	CO_2	Rods, 1–3 x 0.3–0.4 µm	Hot spring, solfataric waterhole; Kerlingarfjoll, Iceland	Stetter *et al.*, 1981
CO_2	Hydrogen	CO_2	Rods, 3–5 x 0.3–0.4 µm	Hot spring, solfataric mudhole; Kerlingarfjoll, Iceland	Lauerer *et al.*, 1986; Garcia, 1990
CO_2	Hydrogen	CO_2 (Elemental Sulfur or Sulfide are required as Sulfur sources)	Cocci, 1.3–1.8 µm diameter	Shallow (106 m) submarine hydrothermal vent at the Kolbeinsey rise on the Mid Atlantic Ridge, Iceland	Burggraf *et al.*, 1990; Whitman *et al.*, 2001
Casein, Peptone, Sulfur, Tryptone, Yeast extract	Fe(II) (required)	Nitrate	Cocci, 0.5–2.0 µm diameter	Black smoker sample from Myojin Knoll, at 1338 m; Ogasawara-Bonin Arc, Japan	Takai *et al.*, 2000
Beef extract, Casein, Peptone, Tryptone, Yeast extract	-	Sulfur	Cocci, 0.6–1.5 µm diameter	Shallow geothermal well; Vulcano Island, Italy	Amend *et al.*, 2003
CO_2, Casamino acids, Gelatin, Meat extract, Peptone, Tryptone, Yeast extract	Hydrogen, Acetate, Propionate, Thiosulfate	AQDS, Arsenate, FC, Fe(III) oxide, Nitrate, Nitrite, Oxygen, Selenate, Selenite	Rods, 3–8 x 0.6 µm	Marine water holes in the tidal zone of Maronti Beach; Ischia, Italy	Volkl *et al.*, 1993; Huber *et al.*, 2000; Kashefi & Lovley, 2000
CO_2	Hydrogen	Arsenate, Selenate, Sulfur, Thiosulfate	Rods, 3–7 x 0.7 µm	Pisciarelli solfatara hot spring sample; Naples, Italy	Huber *et al.*, 2000

Table 9.1 continued

Genus	Species	Aerobe/ Anaerobe	Hetero-/ Autotroph	Temperature (°C)		pH		% (w/v) Salt		mol % G+C
				Range	Opt.	Range	Opt.	Range	Opt.	
Pyrobaculum	*calidifontis*	FAN	H	75–100	90–95	5.5–7.5	6.0	0.0–0.8	0.0	51
Pyrobaculum	*islandicum*	AN	FA	74–102	100	5.0–7.0	6.0	0.0–0.8	N.D.	45
Pyrobaculum	*oguniense*	FAN	H	70–97	90–94	5.4–7.4	6.3–7.0	0.0–1.5	0.0	48
Pyrobaculum	*organotrophum*	AN	H	78–102	102	5.0–7.0	6.0	0.0–0.5	N.D.	46
Pyrococcus	*abyssi*	AN	FA	67–102	96	4.0–8.5	6.8	1.5–4.5	3.0	45
Pyrococcus	*endeavori*	AN	H	55–110	90	4.0–8.0	5.0–7.0	1.0–6.4	2.7	55
Pyrococcus	*furiosus*	AN	FA	70–103	100	5.0–9.0	7.0	0.5–5.0	2.0	38
Pyrococcus	*glycovorans*	AN	H	75–104	95	2.5–9.5	7.5	2.0–6.0	3.0	47
Pyrococcus	*horikoshii*	AN	H	80–102	98	5.0–8.0	7.0	1.0–5.0	2.4	44
Pyrococcus	*woesei*	AN	H	95.0–104.8	100–103	6.0–6.5	N.D.	2.0–3.0	3.0	37.5

Growth Substrates			Morphology & Size	Habitat	References
Carbon Sources & Electron Donors		Electron Acceptors			
Casamino acids, Nutrient broth, Peptone, Tryptone, Yeast extract	Hydrogen, Thiosulfate	Oxygen, Nitrate (Anaerobically)	Rod, 1.5–10 x 0.5–1.0 µm	A terrestrial hot spring; Philippines	Amo *et al.*, 2002
CO_2, Cell homogenates of *Archaea* or *Bacteria*, Meat extract, Peptone, Yeast extract	Hydrogen	AQDS, Au(III), Cr(VI), FC, Fe(III) oxide, L-Cysteine, Glutathione (oxidized), Mn(IV), Sulfite, Sulfur, Tc(VII), Thiosulfate, U(VI)	Rods, 1.5–8 µm long, 0.5–0.6 µm wide	Boiling solfataric and geothermal waters; Hvergerthi, Iceland	Huber *et al.*, 1987; Kashefi & Lovley, 2000; Lovley *et al.*, 2000; Kashefi *et al.*, 2001
Trypticase peptone, Yeast extract		L-Cysteine, Glutathione (oxidized), Oxygen, Sulfur, Thiosulfate	Rods, 2–10 µm long, 0.6–1.0 µm wide	Water samples of a terrestrial hot spring (Tuetate); Oguni-cho, Kamamoto Prefecture, Japan	Sako *et al.*, 2001
Cell homogenates of *Archaea* or *Bacteria*, Meat extract, Peptone, Yeast extract	Hydrogen	Sulfur, L-Cysteine, Glutathione (oxidized)	Rods, 0.5–1.8 µm long, 0.5 µm wide	Boiling solfataric and geothermal waters; Iceland, and the Azores, Italy	Huber *et al.*, 1987
CO_2, Gelatin, Tryptone	Hydrogen (inhibits growth except in presence of Sulfur), Amino acids, Pyruvate, Volatile fatty acids	Cysteine, Polysulfide, Sulfur	Cocci, 0.8–2.0 µm diameter	Hot fluid from an active chimney; North Fiji Basin (SW Pacific)	Erauso *et al.*, 1993
Peptone, Starch, Yeast extract	Hydrogen, Amino acids	Sulfur	Cocci	Flange from the hydrothermal vent system; Juan de Fuca Ridge, WA, United States	Pledger and Baross, 1991
Casamino acids, Casein, Maltose, Peptone, Starch, Tryptone, Yeast extract	Hydrogen	AQDS, Au(III), FC, Sulfur	Cocci, 0.8–1.5 µm diameter	Geothermally heated marine sediments at the beach; Porto di Levante,Vulcano, Italy	Fiala and Stetter, 1986; Lovley *et al.*, 2000; Kashefi *et al.*, 2001
Glucose, Maltose, Peptone, Starch, Yeast extract	Hydrogen, Amino acids	Sulfur	Cocci, 0.5–1.5 µm diameter	13 N East Pacific Rise (EPR), Totem site, active deep-sea (2550 m) vent, chimney wall pieces and alvinellid tubes in EPR; Mid-Oceanic Ridge, Pacific Ocean	Barbier *et al.*, 1999
Beef extract, Casamino acids, Casein, Peptone, Yeast extract	Hydrogen, Amino acids	Sulfur	Cocci, 0.8–2.0 µm diameter	Okinawa Trough vents at 27 33'N, 126 56'E at depth of 1395 m; East China Sea	Gonzalez *et al.*, 1998; Kawarabayasi *et al.*, 1998
Peptone, Yeast extract	Hydrogen	Sulfur	Sphere, 0.5–2.0 µm diameter	Marine solfataras at the beach; Vulcano, Italy	Zillig *et al.*, 1987

Table 9.1 continued

Genus	Species	Aerobe/ Anaerobe	Hetero-/ Autotroph	Temperature (°C) Range	Temperature (°C) Opt.	pH Range	pH Opt.	% (w/v) Salt Range	% (w/v) Salt Opt.	mol % G+C
Pyrodictium	*abyssi*	AN	FA	80–110	97	4.7–7.1	5.5	0.7–4.2	2.0	60
Pyrodictium	*brockii*	AN	A	80–110	105	4.5–7.2	5.5	0.2–12.0	1.5	51.5–56.6
Pyrodictium	*occultum*	AN	A	80–110	105	4.5–7.2	5.5	0.2–12.0	1.5	62
Pyrolobus	*fumarii*	FAN	A	90–113	106	4.0–6.5	5.5	1.0–4.0	1.7	53
Staphylothermus	*hellenicus*	AN	H	70–90	85	4.5–7.0	6.0	2.0–8.0	4.0	38
Staphylothermus	*marinus*	AN	H	65–98	92	4.5–8.5	6.5	1.0–3.5	1.5	35
Stetteria	*hydrogenophila*	AN	FA	68–102	95	4.5–7.0	6.0	0.5–6.0	2.0–3.5	65
Stygiolobus	*azoricus*	AN	A	57–89	80	1.0–5.5	2.5–3.0	N.D.	N.D.	38
Sulfolobus	*islandicus*	AE	H	N.D.	80	3.0–3.5	N.D.	0.0	0.0	35.3
Sulfolobus	*shibatae*	AE	FA	65–86	81	2.5–4.5	3.0	N.D.	N.D.	35
Sulfolobus	*solfataricus*	AE	FA	50–87	85	2.0–5.5	4.5	N.D.	N.D.	35
Sulfolobus	*tokodaii*	AE	H	70–85	80	2.0–5.0	2.5–3.0	N.D.	N.D.	32.8

Growth Substrates					
Carbon Sources & Electron Donors		Electron Acceptors	Morphology & Size	Habitat	References
CO_2, Carbohydrates, Cell extracts of *Archaea* or *Bacteria*	Hydrogen, Acetate, Formate	AQDS, FC, Sulfur, Thiosulfate	Cocci, 1.0–2.0 µm diameter	"Black smoker" in Guaymas Basin; Gulf of California, USA; and Kolbenisey Ridge, Iceland	Pley *et al.*, 1991; Erauso *et al.*, 1993; Lovley *et al.*, 2000
CO_2	Hydrogen	Sulfur	Cocci, 0.3–2.5 µm diameter	Hot sea water, Vulcano, Italy	Pley *et al.*, 1991; Stetter, *et al.*, 1983
CO_2	Hydrogen	Sulfur	Cocci, 0.3–2.5 µm diameter	Hot sea water, Vulcano, Italy	Pley *et al.*, 1991; Stetter, *et al.*, 1983
CO_2	Hydrogen	Nitrate, Oxygen, Thiosulfate	Cocci, 0.7–2.5 µm diameter	Black smoker wall, TAG site; Mid Atlantic Ridge	Blochl *et al.*, 1997
Cell extracts of *Archaea* or *Bacteria*, Meat extract, Peptone, Tryptone, Yeast extract	-	Sulfur	Cocci, 0.8–1.3 µm diameter	Geothermal marine sediments (9.4 m); Palaeochori Bay in Milos, Greece	Arab *et al.*, 2000
Cell extracts of *Archaea* or *Bacteria*, Meat extract, Peptone, Tryptone, Yeast extract	Formate	Sulfur	Cocci; 0.5–1 µm diameter, and up to 15 µm diameter with high (0.2%) yeast extract concentrations	11N EPR Smoker; Vulcano, Italy	Fiala *et al.*, 1986
CO_2 (stimulates growth), Casamino acids, Cell-free extract of *Archaea* or *Bacteria*, Peptone, Tryptone, Yeast extract	Hydrogen	Sulfur, Thiosulfate	Cocci, 0.5–1.5 µm diameter	Shallow submarine hydrothermal system at Paleohori Bay; Milos Greece	Jochimsen *et al.*, 1997
Yeast extract (stimulates growth)	Hydrogen	Sulfur	Cocci, 0.5–1.8 µm diameter	Hot water, mud, and soil at acidic geothermal springs; Sao Miguel Island, Azores, Italy	Segerer *et al.*, 1991
Tryptone, Yeast extract	Hydrogen	Oxygen	N.D.	Various solfataric fields; Iceland	Zillig *et al.*, 1994; Keeley *et al.*, 1996; Zillig *et al.*, 1998; Stedmen *et al.*, 2000
Starch, Sugars, Tryptone, Yeast extract	Hydrogen, Sulfur	Oxygen	Cocci, 0.7–1.5 µm diameter	Mud pots; Kiushu Island, Japan	Grogan, 1989; Grogan *et al.*, 1990; Huber *et al.*, 1992
Casamino acids, Glutamic acid, Sugars, Tryptone, Yeast extract	Hydrogen, Sulfur,	Oxygen	Polyhedrons, 0.8–2.0 µm diameter	Solfatara; Pisciarelli, Italy	Zillig *et al.*, 1980; Huber *et al.*, 1992
Amino acids, Casamino acids, Sugars, Tryptone, Yeast extract	Sulfur	Oxygen	Cocci, 0.5–0.8 µm diameter	Beppu Hot Spring in the geothermal area; Kyushu Island, Japan	Kawarabayasi *et al.*, 2001; Suzuki *et al.*, 2002

Table 9.1 continued

Genus	Species	Aerobe/ Anaerobe	Hetero-/ Autotroph	Temperature (°C) Range	Temperature (°C) Opt.	pH Range	pH Opt.	% (w/v) Salt Range	% (w/v) Salt Opt.	mol % G+C
Sulfolobus	*yangmingensis*	AE	H	65–95	80	2.0–6.0	4.0	-	-	42
Sulfophobo-coccus	*zilligii*	AN	H	70–95	85	6.5–8.5	7.5	0.0–0.2	-	54.7
Sulfurisphaera	*ohwakuensis*	FAN	H	63–92	84	1.0–5.0	2.0	-	-	32.9
Thermococcus	*acidaminovorans*	AN	H	56–93	85	5.0–9.5	9.0	1.0–6.0	2.0–3.0	49
Thermococcus	*aegaeicus*	AN	H	50–95	90	4.5–7.5	6.0	0.5–6.5	2.0	45
Thermococcus	*aggregans*	AN	H	60–94	88	5.6–7.9	7.0	1.0–3.0	2.5	42
Thermococcus	*alcaliphilus*	AN	H	56–90	85	6.5–10.5	9.0	1.0–6.0	2.0–3.0	42
Thermococcus	*barophilus*	AN	H	48–100	85	4.5–9.5	7.0	1.0–4.0	2.0–3.0	37.1
Thermococcus	*barosii*	AN	H	60–92	82.5	4.0–9.0	6.5–7.5	1.0–4.0	2.0	60.0
Thermococcus	*celer*	AN	FA	N.D.-93	88	5.7–7.2	5.8	3.5-N.D.	3.8–4.0	56.6

Growth Substrates					
Carbon Sources & Electron Donors		Electron Acceptors	Morphology & Size	Habitat	References
Amino acids (except cysteine), Sugars (D-arabinose, D-fructose, D-galactose, D-glucose, D-mannose, D-ribose, D-xylose, L-rhamnose, cellobiose, raffinose, sorbitol, lactose, maltose, sucrose, trehalose), Yeast extract	Sulfide, Sulfur, Thiosulfate	Oxygen	Lobed, spherical shaped cells, 0.8–1.3 µm diameter	Longford valley, hot spring vent; Southern Yang-Ming National Park, Taiwan	Jan *et al.*, 1999
Yeast extract	-	-	Cocci, 3–5 µm diameter	Hot alkaline springs; Iceland	Hensel *et al.*, 1997
Peptone, Yeast extract	-	-	Cocci; 1.2–1.5 µm diameter	Muddy water of an acidic hot spring of Ohwaku Valley; Hakone, Japan	Kurosawa *et al.*, 1998
Amino acids, Casamino acids, Peptone, Yeast extract	-	Polysulfide or Sulfur (stimulate growth)	Cocci	Shallow marine sediments, Vulcano, Italy	Dirmeier *et al.*, 1998; Arab *et al.*, 2000
Complex organics, Peptone, Tryptone, Yeast extract	-	Sulfur (stimulates growth)	Cocci, 0.8–1.5 µm diameter	Shallow marine sediments at Palaeochori Bay; Milos, Greece	Arab *et al.*, 2000
Casein, Maltose, Starch, Trypticase, Yeast extract	-	Sulfur (stimulates growth)	Cocci, 1.0–1.5 µm diameter	Marine sediments from Guaymas Basin; Gulf of California, United States	Canganella *et al.*, 1998
Casein, Meat extract, Peptone, Tryptone, Yeast extract	-	Polysulfide or Sulfur (stimulate growth)	Cocci, 0.8–1.3 µm diameter	Shallow marine hydrothermal system; Vulcano Island, Italy	Keller *et al.*, 1995
Amino acids, Compound proteins	Pyruvate (weak growth)	Sulfur (stimulates growth)	Cocci, 0.8–2.0 µm diameter	Chimney wall, under 40 Mpa hydrostatic pressure, Snakepit (3550 m); Mid-Atlantic Ridge, Iceland	Marteinsson *et al.*, 1999
Compound proteins, Oligo-maltose	-	Sulfur (required)	Cocci	Rock fragments from a flange formation from the Juan de Fuca Ridge, East Pacific Rise; Mid-Oceanic Ridge, Pacific Ocean	Duffaud *et al.*, 1998
CO_2 (with Hydrogen as the electron donor and AQDS as the electron acceptor), Compound proteins	Hydrogen (with AQDS as the electron acceptor and CO_2 as the source of carbon)	AQDS, Sulfur (stimulates growth)	Cocci, 1 µm diameter	Solfataric marine water holes; Vulcano, Italy	Zilig *et al.*, 1983; Lovley *et al.*, 2000

Table 9.1 continued

Genus	Species	Aerobe/ Anaerobe	Hetero-/ Autotroph	Temperature (°C)		pH		% (w/v) Salt		mol % G+C
				Range	Opt.	Range	Opt.	Range	Opt.	
Thermococcus	*chitonophagus*	AN	H	60–93	85	3.5–9.0	6.7	0.8–8.0	2.0	46.5
Thermococcus	*fumicolans*	AN	H	73–103	85	4.0–9.5	8.0	0.6–4.0	1.3–2.6	54–55
Thermococcus	*gorgonarius*	AN	H	68–95	80–88	5.8–8.5	6.5–7.2	1.0–5.0	2.0–3.5	50.6
Thermococcus	*guaymasensis*	AN	H	56–90	88	5.6–8.1	7.2	1.0–5.0	3.0	46
Thermococcus	*hydrothermalis*	AN	H	55–100	80–90	3.5–9.5	5.5–6.5	1.3–5.2	2.0–2.6	58
Thermococcus	*kodakaraensis*	AN	H	60–100	85	5.0–9.0	6.5	0.0–5.0	3.0	52
Thermococcus	*litoralis*	AN	H	65–95	88	6.2–8.5	7.2	1.8–6.5	2.5	38
Thermococcus	*pacificus*	AN	H	70–95	80–88	6.0–8.0	6.5	1.0–6.0	2.0–3.5	53.3
Thermococcus	*peptonophilus*	AN	H	60–100	85	4.0–8.0	6.0	1.0–5.0	3.0	52
Thermococcus	*profundus*	AN	H	50–90	80	4.5–8.5	7.5	1.0–6.0	2.0–4.0	52.5
Thermococcus	*siculi*	AN	H	50–93	85	5.0–9.0	7.0	1.0–4.0	2.0	55.8

Growth Substrates					
Carbon Sources & Electron Donors		Electron Acceptors	Morphology & Size	Habitat	References
Chitin, Meat peptone, Peptone, Yeast extract	-	Sulfur (stimulates growth)	Cocci, 1.2–2.5 µm diameter	Deep sea hydrothermal vent enviroment (depth: 2600 m) at 20 50' N, 109 06'W at Guaymas Basin; Gulf of California	Huber *et al.*, 1995
Amino acids (mixture), Compound proteins, Maltose	Pyruvate	Sulfur (stimulates growth)	Cocci, 0.8–2.0 µm diameter	An active chimney wall fragment obtained at a hydrothermal site; Northern Fiji Basin, South Pacific Ocean	Godfroy *et al.*, 1996
Compound proteins	Pyruvate (weak growth)	Sulfur (required)	Cocci, 0.3–1.2 µm diameter	Sand from shore; Whale Island, New Zealand	Miroshnichenko *et al.*, 1998
Casein, Maltose, Starch, Trypticase, Yeast extract	-	Sulfur (stimulates growth)	Cocci, 1.0–3.0 µm diameter	Marine sediments from Guaymas Basin; Gulf of California, United States	Canganella *et al.*, 1998
Amino acids (mixture), Compound proteins, Maltose	Pyruvate	Sulfur (stimulates growth)	Cocci, 0.8–2.0 µm diameter	An active chimney wall fragment obtained at a hydrothermal site on EPR (21 N); Mid-Oceanic-Ridge, Pacific Ocean	Godfroy *et al.*, 1997
Peptone, Tryptone, Yeast extract	-	Sulfur	Cocci, 1–2 µm diameter	Solfatara, sediments, water; Kagoshima, Japan	Morikawa *et al.*, 1994; Nakatari *et al.*, 2000
Casein, Meat extract, Peptone, Tryptone, Yeast extract	Pyruvate	Sulfur (stimulates growth)	Cocci, 0.5–3.0 µm diameter	Shallow submarine solfataras at Lucrino; Naples and Porto di Levante, Vulcano, Italy	Belkin and Jannasch, 1985; Neuner *et al.*, 1990; Kostyukova *et al.*, 1999
Compound proteins, Starch	-	Sulfur (required)	Cocci, 0.7–1.0 µm diameter	Bottom deposits; Bay of Plenty, New Zealand	Miroshnichenko *et al.*, 1998
Beef extract, Non-hydrolyzed casein, Peptone, Tryptone, Yeast extract	-	Sulfur (stimulates growth)	Cocci, 0.7–2.0 µm diameter	Marine hydrothermal vents; Izu-Bonin Arc and South Mariana Trough, Japan	Gonzalez *et al.*, 1995
Peptone, Tryptone, Yeast extract, Casein, Gelatin, Pyruvate, Maltose, Starch	-	Sulfur (required)	Cocci, 1–2 µm	A deep-sea hydrothermal vent; Iheya Small Ridge and Mid Okinawa Trough, East China Sea	Kobayashi *et al.*, 1994
Casein, Peptone, Tryptone, Yeast extract	-	Cysteine or Sulfur (stimulate growth)	Cocci, 0.8–2 µm diameter	Hydrothermal vent fluids collected; Mid-Okinawa Trough, East China Sea	Grote *et al.*, 1999

Table 9.1 continued

Genus	Species	Aerobe/ Anaerobe	Hetero-/ Autotroph	Temperature (°C) Range	Temperature (°C) Opt.	pH Range	pH Opt.	% (w/v) Salt Range	% (w/v) Salt Opt.	mol % G+C
Thermococcus	*thioreducens*	AN	H	55–94	83–85	5.0–8.5	7.0	1.0–5.0	3.0	52.9
Thermococcus	*waiotapuensis*	AN	H	60–90	85	5.0–8.0	7.0	1.0–13.9	5.4	50.4
Thermodiscus	*maritimus*	AN	FA	75–98	90	5.0–7.0	5.5	1.0–4.0	2.0	49
Thermofilum	*pendens*	AN	H	70–100	85–90	2.8–6.7	5.0–6.0	0.1–2.0	N.D.	57.6
Thermoproteus	*neutrophilus*	AN	FA	70–97	85	5.0–7.5	6.5	N.D.	N.D.	N.D.
Thermoproteus	*tenax*	AN	FA	70 - 96	88	2.5–6.0	5.5	N.D.	N.D.	55.5
Thermoproteus	*uzoninesis*	AN	H	74–102	90	4.6–6.8	5.6	N.D.	N.D.	56.5
Thermosphaera	*aggregans*	AN	H	67–90	85	5.0–7.0	6.5	0.0–0.7	0.0	46
Vulcanisaeta	*distributa*	AN	H	70–92	90	3.5–5.6	4.5	0.0–1.0	≤ 1.0	43.9–46.2
Vulcanisaeta	*souniana*	AN	H	65–89	85	3.5–5.0	4.5	0.0–1.25	≤ 1.25	44.9–45.4

A, Autotroph; AE, Aerobic; AN, Anaerobic; AQDS, Anthraquinone-2,6-Disulfonate; Au, Auric ("Gold"); Cr, Chromium; FA, Facultative Aerobe; FAN, Facultative Anaerobe; FC, Ferric Citrate; Fe(III) oxide, Iron (III) oxide or Ferric oxide; H, Heterotroph; Mn, Manganese; N.D., Not Determined; Tc, Technetium; U, Uranium

Growth Substrates					
Carbon Sources & Electron Donors		Electron Acceptors	Morphology & Size	Habitat	References
Bacto-tryptone, Casamino acids, Peptone, Yeast extract	-	Sulfur	Cocci, 0.7–1.8 µm diameter	Isolated from 'black smoker' chimney material from the Rainbow hydrothermal vent site; Mid-Atlantic Ridge, Iceland	Pikuta *et al.*, 2007
Amino Acids, Compound proteins, Maltose, Starch	Pyruvate	Sulfur	Cocci, 0.7–1.5 µm diameter	Isolated from a freshwater hot spring in the Lake Taupo area; North Island, New Zealand	Gonzalez *et al.*, 1999
CO_2, Yeast extract	Hydrogen	Sulfur	Disk-shaped, 0.3–3 µm in diameter and 0.1–0.2 µm thickness	Hot sea water; Vulcano, Italy	Fischer *et al.*, 1983; Stetter *et al.*, 1990
Compound proteins, Polar lipids	Hydrogen	Sulfur	Rods, 1–100 µm long, 0.17–0.35 µm wide	Solfataric hot springs; Hveravellir, Iceland	Zilig *et al.*, 1983
CO_2, Acetate	Hydrogen	Sulfur	Rods, 1–40 µm long, 0.4 µm wide	Hot spring; Kerlingarfjoll, Iceland	Fischer *et al.*, 1983
CO_2, Amino acids, fumarate, Glucose, Glycogen, Starch, Yeast extract (stimulates growth)	Hydrogen	Sulfur, Malate	Rods, 1–80 µm long, 0.3 µm wide	Mud hole; Krafla, Iceland	Fischer *et al.*, 1983; Zilig *et al.*, 1983
Compound proteins	-	Sulfur (stimulates growth)	Rods, 1–20 µm long, 0.3–0.4 µm wide	Hot springs, soil and geothermal waters; Uzon caldera, Kamchatka Peninsula, Russia	Bonch-Osmolovskaya *et al.*, 1990
Amino acids, Gelatin, Glucose, Peptone, Yeast Extract, Xylan (heat-treated)	-	N.D.	Cocci, 0.2–0.8 µm diameter	Obsidian Pool; Yellowstone National Park, WY, United States	Huber *et al.*, 1998
Beef extract, Casamino acids, Gelatin, Maltose, Peptone, Starch, Yeast extract	Malate	Sulfur, Thiosulfate	Rods, 0.4–0.6 x 3–7 µm	Hot springs, Ohwakudani, Kanagawa	Itoh *et al.*, 2002
Beef extract, Casamino acids, Gelatin, Maltose, Peptone, Starch, Yeast extract	Malate	Sulfur, Thiosulfate	Rods, 0.4–0.6 x 3–7 µm	Hot springs, Ohwakudani, Kanagawa	Itoh *et al.*, 2002

hyperthermophilic bacteria and archaea represent the deepest phylogenetic branches and have the shortest evolutionary distances to the root of the tree of life (Blochl *et al.*, 1995; Baumann *et al.*, 1998; Stetter, 2006b). As a result, they are regarded as the most ancient forms of life and closely related to the 'Last Universal Common Ancestor' (LUCA) of all extant life. Hyperthermophiles have evolved very little from these ancestral organisms since their estimated appearance on Earth 3.5 billion years ago (Madigan *et al.*, 2009). As a result, modern hyperthermophiles serve as model systems to investigate the biochemical evolution of primary metabolic pathways in mesophilic organisms, especially so for hyperthermophilic archaea, which are more closely related to higher life forms in the *Eukarya* domain.

In addition to their evolutionary significance, interest in hyperthermophiles has been stimulated by the biotechnological potential of their metabolic flexibility and heat resistance mechanisms (Vieille and Zeikus, 2001; Turner *et al.*, 2007). Hyperthermophilic organisms are metabolically versatile, as evidenced by their ability to reduce a variety of metals such as iron, uranium, technetium (Kashefi and Lovley, 2000; Kashefi *et al.*, 2008a), and gold (Kashefi *et al.*, 2001). This type of metabolism, if properly harnessed, could aid in the bioremediation of radioactive, toxic and heavy metals at high temperatures. Lignocellulose biomass also is efficiently scavenged in hot environments such as hydrothermal habitats, which requires a wide variety of glycan-degrading enzymes with potential for applications in biomass processing at high temperatures (Chou *et al.*, 2008). Bioremediation of lignin-containing complex natural wastes and industrial effluents discharged at high temperature is also a perfect target for hyperthermophiles and their enzymes (Hasan *et al.*, 2006). The final products of such degradative processes are environmentally friendly substances such as hydrogen and ethanol, which can be recovered as value-added products. Hyperthermophiles are of interest for the biogeneration of hydrogen gas, because yields are generally higher than in mesophiles. However, volumetric productivities are low.

The metabolic versatility observed in these organisms also probably reflects the presence of unique metabolic pathways. Recent studies have revealed some unusual metabolic processes in hyperthermophilic archaea, including a modified Embden-Meyerhof pathway (the most common form of glycolysis) not described in the bacterial or eukaryotic domains. Studies of this pathway in the model hyperthermophilic archaeon *Pyrococcus furiosus* showed several novel metabolic enzymes such as ADP-dependent glucokinase (Dorr *et al.*, 2003), ADP-dependent phosphofructokinase (Tuininga *et al.*, 1999; Kengen, *et al.*, 2001; Pikuta *et al.*, 2007) glyceraldehyde-3-phosphate ferredoxin oxidoreductase (Mukund and Adams, 1995; Hagedoorn *et al.*, 1999), formaldehyde ferredoxin oxidoreductase (Hu *et al.*, 1999; Roy *et al.*, 2001), phosphoenolpyruvate synthase (Hutchins *et al.*, 2001; Sakuraba *et al.*, 2001, 2004), pyruvate–ferredoxin oxidoreductase (Mai and Adams, 1994; Smith *et al.*, 1994; Ma *et al.*, 1997), and ADP-forming acetyl-CoA synthetase. These unique pathways may evoke metabolisms at an early stage of evolution (Sakuraba and Ohshima, 2002) but also show promise for the discovery of novel enzymes of biotechnological potential. Thus, the study of hyperthermophilic microorganisms has large implications for biotechnological applications, as well as providing a potential window into our evolutionary history.

Biodiversity of hyperthermophiles

Much of what is known about the diversity of hyperthermophilic microbes comes from both non-culture based studies of hyperthermophilic environments (Reysenbach *et al.*, 2000a,b; Takai *et al.*, 2001, 2004; Hoek *et al.*, 2003; Sokolova *et al.*, 2004; Nercessian *et al.*, 2005a,b; Prokofeva *et al.*, 2005; Biddle *et al.*, 2006; Brazelton *et al.*, 2006; Huber *et al.*, 2007; Dahle *et al.*, 2008; Reigstad *et al.*, 2009; Ver Eecke *et al.*, 2009), and from the recovery of microorganisms in pure culture (Table 9.1). A complete list of cultured hyperthermophiles is presented in Table 9.1. These presently cultured strains represent only 1% or less of the existing diversity estimated by non-culture based techniques such as 16S rRNA gene sequencing (Amann *et al.*, 1994; Barns *et al.*, 1994). Much biodiversity remains untapped. Isolation approaches are mostly limited by the difficulty in reproducing the extreme conditions of hot environments in the laboratory. Hyperthermophiles grow optimally above 80°C and, in some cases, growth is restricted at temperatures above 80–90°C (Stetter, 2006a). Generally, they can be preserved at ambient temperatures in the same way as mesophilic organisms are preserved at low temperatures for a prolonged time. The thermal tolerance of hyperthermophilic microorganisms indicates that they have cellular machinery such as proteins, genetic material, transcription and translation system etc. that is functional at high or extreme temperatures. Hyperthermophilic enzymes, for example, are inactive at moderate temperatures, and regain their activity when temperatures increase (Koutsopoulos *et al.*, 2005). Thus, temperature ranges for growth are critical parameters for successful isolations. Other conditions must be taken into account. For example, most hyperthermophiles have been isolated at atmospheric pressure, but in some cases altering the pressure may prove to be more fruitful (Blochl *et al.*, 1997; Takai *et al.*, 2008). In addition, careful selection of nutrients has proven key to the recovery of novel hyperthermophiles with unique metabolisms (Huber *et al.*, 1986, 1987, 2002; Stetter, 1988; Volkl *et al.*, 1993; Hafenbradl *et al.*, 1996; Blochl *et al.*, 1997; Kashefi *et al.*, 2002a,b; Kashefi and Lovley, 2003; Takai *et al.*, 2008). Examples of this approach are provided later in this chapter.

By the early 1990s, most of the reported hyperthermophilic diversity resulted from microbes isolated from shallow waters or continental areas, with exceptions such as *Sulfolobus marinus, Thermodiscus maritimus* and *Methanopyrus kandleri*, which were isolated from geothermally heated coastal environments (Fiala *et al.*, 1986; Kurr *et al.*, 1991; Huber and Stetter, 1992). A large hyperthermophilic diversity is found in deep-sea hot environments, most of it concentrated around hydrothermal vents. Water surrounding these vents is characterized by low pH, high concentrations of dissolved gases (hydrogen sulfide, methane, carbon monoxide, carbon dioxide, and hydrogen), and minerals (silicon, iron, manganese, zinc, etc.). The vent fluids are also a rich source of heavy metals, which precipitate at high temperature (350–400°C) when they mix with cold oxygenated seawater (average temperature about 2°C) (Tivey and Delaney, 1986; Tivey *et al.*, 1999). These precipitates accumulate to form mineral constructions known as 'chimneys', 'black smokers', or hydrothermal vents. These structures are inhabited by microbial communities with unique physiological characteristics, and with a primarily chemolithotrophic metabolism (Prieur, 1997; Corre *et al.*, 2001; Kelley *et al.*, 2002). Among these are methanogens [e.g. *Methanocaldococcus* (previously known as *Methanococcus*), and *Methanopyrus*], sulfate reducers (*Archaeoglobus*), facultative autotrophs and heterotrophs (*Pyrococcus, Thermococcus, Desulfurococcus, Staphylothermus, Thermodiscus*, and *Pyrodictium*). Phylogenetic analysis of microbial diversity from a hydrothermal sulfide chimney sample from the Main Endeavor vent field of the Juan de Fuca Ridge showed that most of the organisms were autotrophic or heterotrophic, and belonged to the orders Desulfurococcales, Thermoproteales, Archaeoglobales, and Thermococcales (Zhou *et al.*, 2009). These organisms utilized a wide range of substrates, such as sulfate (SO_4^{2-}), thiosulfate ($S_2O_3^{2-}$) or elemental sulfur (S^0), with hydrogen sulfide (H_2S) as their end product (Zhou *et al.*, 2009). All were strict anaerobes and grew at temperatures ranging from 48°C to 110°C, with optimal growth at 82–98°C and doubling times from 20 (*Methanococcus* AG86) to 270 min (*Staphylothermus marinus*). Some hyperthermophiles are considered fast growing organisms, as they have a doubling time of less than an hour and are very well adapted to the extreme temperatures existing '*in situ*'. However, it has been observed that the temperature at a hydrothermal vent is variable (Chevaldonne *et al.*, 1991). Thus, hyperthermophiles must tolerate transient periods of cold (Erauso *et al.*, 1993). Oxygen intrusions also are possible, especially in cold temperatures, which increase the solubility of oxygen. Hence the fact that some hyperthermophilic microorganisms are aerotolerant or respire oxygen. The genus *Aeropyrum*, for example, is composed of obligate aerophiles (Sako *et al.*, 1996), while species such as the archaeon *Pyrobaculum aerophilum* and the bacterium *Aquifex pyrophilus* (Stetter, 1996) are microaerophiles and grow at partial oxygen pressures up to 5%.

Most of the hyperthermophilic diversity identified to date belongs to the domain *Archaea*. Recognized in the late 1970s by Carl Woese and George Fox (Woese and Fox, 1977), the domain *Archaea* consist of four phyla: Crenarchaeota, Euryarchaeota, Korarchaeota and Nanoarchaeota. Hyperthermophilic sulfur-dependent microorganisms usually belong to the phylum Crenarchaeota, whereas the phylum Euryarchaeota includes some methanogens and most halophiles (Lebedinsky *et al.*, 2007; Madigan *et al.*, 2009). The phylum Korarchaeota was first identified from environmental rRNA sequences from Obsidian Pool in Yellowstone National Park (Barns *et al.*, 1996), and later from a diversity of marine and terrestrial hot environments (Reigstad *et al.*, 2009). Although no members of the Korarchaeota have been recovered in pure culture, 16S rRNA analyses suggest that this group branches off very deeply within the *Archaea* (Huber *et al.*, 2006c). The availability of some hyperthermophilic archaea in pure culture has allowed the complete genome sequence of many representative species to be determined. For example, Euryarchaeota such as *Pyrococcus horikoshii* (Kawarabayasi *et al.*, 1998), *Pyrococcus abyssi* (Cohen *et al.*, 2003), *Pyrococcus furiosus* (Robb et al., 2001), *Archaeoglobus fulgidus* (Klenk *et al.*, 1997), *Methanococcus jannashii* (Bult *et al.*, 1996), *Methanobacterium thermoautotrophicum* (Smith *et al.*, 1997), and *Thermococcus kodakaraensis* strain KOD1 (Fukui *et al.*, 2005); the Crenarchaeotes *Pyrobaculum aerophilum* (Fitz-Gibbon *et al.*, 2002), *Desulfurococcus kamchatkensis* (Ravin *et al.*, 2009), *Ignicoccus hospitalis* (Podar *et al.*, 2008), and *Staphylothermus marinus* (Anderson *et al.*, 2009); and *Nanoarchaeum equitans* (Waters *et al.*, 2003) from the phylum Nanoarchaeota.

The most frequently isolated hyperthermophilic archaea belong to the order Thermococcales in the euryarchaeal lineage (Bertoldo and Antranikian, 2006). The order Thermococcales is divided into three genera: *Thermococcus*, *Pyrococcus*, and *Palaeococcus*. All are anaerobic heterotrophs and utilize peptides as their carbon and energy source. Their growth temperatures vary by species, lying between 70 and 105°C, with optima ranging from 80 to 100°C. These organisms exhibit increased growth in the presence of elemental sulfur and produce a large amount of hydrogen sulfide. When grown in the absence of elemental sulfur, cells produce hydrogen as an electron sink, which severely limits growth unless removed. Of the three genera, *Pyrococcus* has been studied most extensively in terms of its physiology and enzymology, especially the enzymes responsible for peptide and carbohydrate metabolism (Adams *et al.*, 2001). Thermococcales are geographically diverse and have mostly been isolated from both deep and shallow heterotrophic ecosystems. Exceptions include *Thermococcus sibiricus*, isolated from a high temperature oil reservoir (Miroshnichenko *et al.*, 2001), and *T. zilligii* (Ronimus *et al.*, 1997) and *T. waiotapuensis* (Gonzalez *et al.*, 1999), found in a continental hot spring.

The smallest microorganism isolated to date, *Nanoarchaeum equitans* (Huber *et al.*, 2002, 2006c; Jahn *et al.*, 2008) is a hyperthermophilic archaeon. This nano-sized (400 nm in diameter) hyperthermophile, isolated from a submarine hot vent off the coast of Iceland, was assigned to a new phylum, the Nanoarchaeota (Huber *et al.*, 2002). Phylogenetic analyses based on SSU rRNA gene placed this phylum as the deepest branch of the *Archaea*, and at a large distance from the Euryarchaeota, Crenarchaeota, and Korarchaeota branches (Huber *et al.*, 2002, 2006c). However, comparative analyses of the ribosomal proteins from several sequenced archaeal genomes raised doubts about the phylogenetic placement of *N. equitans* as a separate, deeply branching phylum (Brochier *et al.*, 2005). Furthermore, a similarity survey of all open reading frames in the genome of *N. equitans* led to the surprising finding that most of the closest hits were with Euryarchaeota homologues, mostly in the order Thermococcales (Brochier *et al.*, 2005). This was further supported by phylogenetic analyses of a subset of ribosomal proteins (Brochier *et al.*, 2005). These findings strongly suggest that *N. equitans* may be a member of the euryarchaeal lineage, most closely related to Thermococcales, rather than a separate, early divergent phylum. *N. equitans* has the smallest known genome (0.5 Mb) among all archaea (Waters *et al.*, 2003; Randau *et al.*, 2005). The genome lacks key metabolic pathways for the synthesis of nucleotides, amino acids, lipids and cofactors and the organism's ability to produce its own ATP is questionable (Waters *et al.*, 2003; Burghardt *et al.*, 2008). With no apparent biosynthetic and catabolic capacity, *N. equitans* obtains metabolites and energy by growing and dividing at the surface of a crenarchaeal chemoautotrophic host, Ignicoccus *hospitalis*, which couples the oxidation of hydrogen to the reduction of sulfur (Huber *et al.*, 2002). This metabolic dependence is obligate and, as a result, *N. equitans* cannot be cultivated independently, suggestive of an obligate symbiotic, or even parasitic, relationship (Waters *et al.*, 2003; Randau *et al.*, 2005). Molecular analyses indicate that the Nanoarchaeota phylum is widely distributed in extremophilic environments (Hohn *et al.*, 2002; Casanueva *et al.*, 2008).

Hyperthermophilic members of the domain *Bacteria* span five genera (*Aquifex, Geothermobacterium, Rhodothermus, Thermocrinis* and *Thermotoga*), and include nine species available in pure culture. They have been isolated from a diversity of terrestrial and marine hydrothermal environments (Table 9.1). Regardless of the small number of pure culture representatives characterized to date, their physiology is diverse (Table 9.1). Their optimum growth temperatures range from 80–90°C. Their pH and salt (NaCl) optima range from 6.5 to 8.5 and from 0% to 3.0% (w/v), respectively. They are capable of autotrophic, facultative autotrophic and heterotrophic growth under aerobic (*A. aeolicus, R. obamesis and T. ruber*), facultatively anaerobic (*A. pyrophilus*), and strictly anaerobic (*G. ferrireducens, T. maritima, T. neapolitana, T. naphthophila and T. petrophila*) conditions. They can gain energy for growth by coupling the reduction of O_2, NO_3^- (*T. ruber*), Fe(III) oxide (*G. ferrireducens* and *T. maritima*), Fe(III) citrate (*T. maritima*), anthraquinone-2,6-disulfonate (AQDS) (*T. maritima*), S^0 and $S_2O_3^{2-}$ (*T. neapolitana, T. naphthophila* and *T. petrophila*), to the oxidation of a variety of organic and inorganic electron donors, and by using a wide diversity of organic and inorganic carbon sources. With an optimum growth temperature of 80–90°C, and temperature maxima of 100°C, the iron reducing *G. ferrireducens* is the most thermophilic member of the domain *Bacteria* (Kashefi *et al.*, 2002a).

Metabolic diversity of hyperthermophiles

With a limited number of cultured isolates available, the first metabolic classification of hyperthermophiles was restricted to sulfate reducers, methanogens and S^0-dependent microorganisms, with the majority of isolates belonging to the latter group. Oxygen nitrate, arsenate, humic acid, smectite and Fe(III) respiration were subsequently recognized as novel hyperthermophiles capable of utilizing these electron acceptors to gain energy for growth were isolated. Of all the diverse metabolic capacities described, the ability to reduce Fe(III) and humic acids is a conserved trait amongst all of the hyperthermophilic microorganisms tested thus far. The significance of Fe(III) and humic acid-reduction at hyperthermophilic temperatures is discussed in detail later in this chapter.

With the exception of homoacetogenesis and photogenesis, all other forms of metabolism that are present in mesophiles are also found in hyperthermophilic microorganisms. Although homoacetogenesis has not been reported at hyperthermophilic temperatures, it has been reported to take place at psychrophilic (Kotsyurbenko *et al.*, 2001; Nozhevnikova *et al.*, 2003), mesophilic (Nie *et al.*, 2007) and thermophilic (Siriwongrungson *et al.*, 2007) conditions. To date, there are no hyperthermophilic homoacetogens in pure culture. Also, owing to the absence of (sun)-light in the deep hot biosphere, hyperthermophilic microorganisms cannot use photosynthesis to gain energy for growth, but instead grow by chemosynthesis. During chemosynthetic growth, energy is obtained from a series of oxidation-reduction reactions during which a reduced carbon, or inorganic, compound is oxidized and a (usually inorganic) terminal electron acceptor is reduced. For example, geotonic sources of H_2 can serve as the electron donor for the reduction of the often abudant environmental sources of iron and/or sulfurous compounds as electron acceptors. At hyperthermophilic temperatures oxygen respiration may not appear to be significant because of the low solubility of oxygen at higher temperatures. However, some hyperthermophilic species, such as *P. calidifontis* (Amo *et al.*, 2002) and *P. aerophilum* (Volkl *et al.*, 1993), can utilize

O_2 as an electron acceptor under aerobic and microaerophilic conditions, respectively.

Similar to their mesophilic and thermophilic counterparts, hyperthermophiles can derive energy for growth under chemolithoautotrophic or (obligate)-heterotrophic conditions (Table 9.1). Chemolithoautotrophic metabolisms couple the reduction of S^0, SO_4^{2-}, NO_3^- (nitrate), NO_2^- (nitrite), CO_2 (carbon dioxide), As(V) [arsenate], Fe(III) [ferric iron], Mn(IV) [manganese], humic acids, or smectite to the oxidation of H_2 (or other inorganic electron donors, such as Fe^{2+}, S^0, S^{2-}), using CO_2 as the sole source of carbon (Table 9.1). Under (obligate)-heterotrophic growth conditions some hyperthermophiles can use a variety of organic substrates as carbon and electron sources. Some are also capable of fermentative growth. Such metabolic diversity reflects an adaptive response that enables hyperthermophilies to thrive in their natural habitats (e.g. terrestrial hot springs, deep-sea hydrothermal vents, and the deep hot biosphere).

Electron acceptors and modes of respiration

In hydrothermal environments a variety of hyperthermophilic microorganisms are responsible for the decomposition of organic material, using a diversity of electron acceptors. As a result, they drive many of the chemical reactions that occur in these environments. SO_4^{2-}, $S_2O_3^{2-}$ and SO_3^{2-} (sulfite), NO_3^-, NO_2^-, CO_2, As(V), Mn(IV), humic acids, Fe(III), and smectite are some of the major electron acceptors utilized in anaerobic respiration by hyperthermophiles.

Oxygen respiration

The hydrothermal environments in which hyperthermophiles thrive are usually oxygen free. This fact is mainly due to the low solubility of O_2 at elevated temperatures and the abundance of reducing gases, which help in maintaining anoxic conditions throughout these environments. Nevertheless, O_2 respiration has been observed in a phylogenetically diverse group of hyperthermophiles. Their presence is largely influenced by their accessibility to O_2 in their natural habitats. Terrestrial hot springs (where atmospheric O_2 can easily diffuse) and coastal solfataric vents are examples of environments in which oxygen is accessible. Availability of dissolved O_2 in ocean water is inversely proportional to the depth of a hydrothermal system, and therefore, is more relevant to the shallow regions of such systems.

The ability to respire O_2 has been reported in hyperthermophilic archaeal and bacterial species belonging to the genera *Acidianus, Aquifex, Pyrobaculum, Sulfolobus, Thermocrinis* and *Pyrolobus*. These species can obtain energy for growth by coupling the oxidation of hydrogen to the reduction of oxygen, with water as the end product, by the following reaction (eqn 9.1, also known as the Knallgas reaction; Reysenbach and Shock, 2002):

$$O_2 + 2H_2 \rightarrow 2H_2O \qquad (9.1)$$

(in *Acidianus, Aquifex, Pyrobaculum, Sulfolobus, Thermocrinis* and *Pyrolobus*).

In addition to H_2, members of the genera *Acidianus, Aquifex*, and *Sulfolobus* can grow by oxidizing inorganic electron donors such as S^0, H_2S and $S_2O_3^{2-}$ (eqns 9.2–94, respectively) under microaerophilic conditions, with sulfuric acid as the end product:

$$3O_2 + 2S^0 + 2H_2O \rightarrow 2H_2SO_4 \qquad (9.2)$$

$$2O_2 + H_2S \rightarrow H_2SO_4 \qquad (9.3)$$

$$2O_2 + H_2S_2O_3 + H_2O \rightarrow 2H_2SO_4 \qquad (9.4)$$

(in *Acidianus, Aquifex* and *Sulfolobus*).

Metal-leaching, thermoacidophilic *Acidianus infernus* can grow by coupling the reduction of O_2 to the oxidation of the sulfides in sulfidic ores such as pyrite (FeS_2), chalcopyrite ($CuFeS_2$), chalcocite (Cu_2S) and sphalerite [a zinc iron sulfide ($ZnFeS_2$) mineral with variable Fe content]. As a result, the heavy metal ions (Fe^{3+}, Cu^{2+} and Zn^{2+}) are dissolved by leaching out of the sulfidic ores due to the production of sulfuric acid (eqns 9.5–9.8):

$$15O_2 + 4FeS_2 + 2\,H_2O \rightarrow 8\,SO_4^{2-} + 4\,H^+ + 4Fe^{3+} \qquad (9.5)$$

$$17O_2 + 4CuFeS_2 + 10H_2O \rightarrow 4Fe(OH)_3 + 8SO_4^{2-} + 8H^+ + 4Cu^{2+} \quad (9.6)$$

$$5O_2 + 2Cu_2S + 4H^+ \rightarrow 2SO_4^{2-} + 2H_2O + 4Cu^{2+} \quad (9.7)$$

$$17O_2 + 4ZnFeS_2 + 10H_2O \rightarrow 4Fe(OH)_3 + 8SO_4^{2-} + 8H^+ + 4Zn^{2+} \quad (9.8)$$

(in *Acidianus infernus*).

Sako *et al.* (1996) isolated the hyperthermophilic archaeon, *Aeropyrum pernix* strain K1 from a coastal solfataric vent at Kodakara-Jima Island (Kyushu, Japan). With an optimum growth temperature of 90–95°C (T_{max} = 100°C), and belonging to the phylum Crenarchaeota, *A. pernix* was the first strictly aerobic hyperthemophile isolated. During growth with O_2 as a terminal electron acceptor, *A. pernix* gained energy by utilizing a wide range of complex proteinaceous compounds (represented as CH_2O in eqn 9.9) as substrates:

$$O_2 + CH_2O \rightarrow H_2O + CO_2 \quad (9.9)$$

(in *Aeropyrum pernix*).

Amo *et al.* (2002) isolated a novel facultatively aerobic, heterotrophic hyperthermophilic archaeon from a terrestrial hot spring in the Philippines. Belonging to the genus *Pyrobaculum*, the new isolate was named *P. caldifontis* (strain VA1). Under anaerobic conditions *P. caldifontis* grew with NO_3^- as a terminal electron acceptor. Under aerobic conditions, however, O_2 was used as the terminal electron acceptor. *P. caldifontis* was able to grow optimally at 90–95°C with vigorous shaking under atmospheric air. $S_2O_3^{2-}$ was reported to enhance growth under aerobic culture conditions (Amo *et al.*, 2002).

Nitrate respiration

Nitrate reduction in hydrothermal environments was originally considered to be controlled by abiotic rather than biotic processes. This belief was mainly due to the instability and abiotic reduction of NO_3^- to ammonium (NH_4^+) under pyrite forming conditions, and at elevated temperatures (Blochl *et al.*, 1997). However, this view is no longer valid. The ability to gain energy for growth by the reduction of NO_3^- (as a terminal electron acceptor), via NO_2^- under anaerobic conditions has been observed in a handful of hyperthermophilic archaea and bacteria, including *Aquifex pyrophilus* (Huber *et al.*, 1992), *Ferroglobus placidus* (Hafenbradl *et al.*, 1996), *Pyrobaculum aerophilum* (Volkl *et al.*, 1993), *P. calidifontis* (Amo *et al.*, 2002), and *Pyrolobus fumarii* (Blochl *et al.*, 1997). *A. pyrophilus, P. aerophilum* and *P. calidifontis* reduce NO_3^- to NO_2^-, NO (nitric oxide), N_2O (nitrous oxide) and finally to N_2 (Volkl *et al.*, 1993), using the enzymes nitrate reductase (egn 9.10), nitrite reductase (eqn 9.11), nitric oxide reductase (eqn 9.12) and nitrous oxide reductase (eqn 9.13), respectively. This metabolic pathway is reminiscent of nitrate assimilation (or denitrification) by mesophilic nitrate reducers, following the energy-yielding reaction summarized by eqn 9.14:

$$NO_3^- + 2H^+ + 2e^- \rightarrow NO_2^- + H_2O$$

(catalysed by nitrate reductase) (9.10)

$$NO_2^- + 2H^+ + e^- \rightarrow NO + H_2$$

(catalysed by nitrite reductase) (9.11)

$$2NO + 2H^+ + 2e^- \rightarrow N_2O + H_2O$$

(catalysed by nitric oxide reductase) (9.12)

$$N_2O + 2H^+ + 2e^- \rightarrow N_2 + H_2O$$

(catalysed by nitrous oxide reductase) (9.13)

$$2HNO_3 + 5H_2 \rightarrow N_2 + 6H_2O \quad (9.14)$$

(in *A. pyrophilus, P. aerophilum* and *P. calidifontis*).

Thus far, *A. pyrophilus* is the only known hyperthermophilic bacterium that gains energy for growth by the reduction of NO_3^- to N_2 via NO_2^-, using inorganic electron donors such as H_2, S^0 or $S_2O_3^{2-}$ (Huber *et al.*, 1992) (eqns 9.14–16, respectively):

$$6HNO_3 + 5S^0 + 2H_2O \rightarrow 3N_2 + 5H_2SO_4 \quad (9.15)$$

(in *A. pyrophilus*).

$$8HNO_3 + 5H_2S_2O_3 + H_2O \rightarrow 4N_2 + 10H_2SO_4 \quad (9.16)$$

(in *A. pyrophilus* and *P. aerophilum*).

P. aerophilum can also gain energy for growth by coupling the reduction of NO_3^- and NO_2^- (as terminal electron acceptors) to the oxidation of inorganic (i.e. H_2 or $S_2O_3^{2-}$, eqns 9.14 and 9.16, respectively) or organic and proteinaceous substrates (represented as CH_2O in eqn 9.17) as electron donors (Table 9.1; Volkl *et al.*, 1993; Huber *et al.*, 1998, 2006b):

$$4HNO_3 + 5CH_2O \rightarrow 5CO_2 + 7H_2O + 2N_2 \quad (9.17)$$

(in *P. aerophilum* and *P. calidifontis*).

P. calidifontis was also reported to grow by the reduction of NO_3^- (as terminal electron acceptor), using a variety of complex carbon compounds (Amo *et al.*, 2002). Under anaerobic conditions and with tryptone as a growth substrate, *P. calidifontis* reduced NO_3^- to N_2 (eqn 9.17), although N_2O could not be detected by gas chromatography (Amo *et al.*, 2002), which could be due to the rapid reduction of N_2O to N_2 by *P. calidifontis*.

F. placidus, a hyperthermophilic archaeon, can grow by the reduction of NO_3^- to NO, coupled to the oxidation of H_2 (eqns 9.10 and 9.11), Fe(II) (eqns 9.18–9.20) or H_2S (eqns 9.23 and 9.24) as electron donors (Hafenbradl *et al.*, 1996):

$$HNO_3 + 2FeCO_3 + 5H_2O \rightarrow HNO_2 + 2Fe(OH)_3 + 2H_2CO_3 \quad (9.18)$$

$$HNO_2 + FeCO_3 + 2H_2O \rightarrow NO + Fe(OH)_3 + H_2CO_3 \quad (9.19)$$

The overall energy-yielding reaction is summarized in eqn 9.20:

$$HNO_3 + 3FeCO_3 + 7H_2O \rightarrow NO + 3Fe(OH)_3 + 3H_2CO_3 \quad (9.20)$$

(in *F. placidus*).

$$2HNO_3 + H_2S \rightarrow 2H_2O + 2NO_2 + S^0 \quad (9.21)$$

NO_2 is then reduced to NO by nitrite reductase. The overall reaction is summarized in eqn 9.22:

$$2HNO_3 + 3H_2S \rightarrow 4H_2O + 2NO + 3S^0 \quad (9.22)$$

(in *F. placidus*).

P. fumarii is the only known hyperthermophile capable of reducing NO_3^- as a terminal electron acceptor to NH_4^+ via NO_2^- under anaerobic conditions [(similar to dissimilatory nitrate reduction by mesophilic and thermophilic microorganisms) (Fig. 9.1)]. During this process NO_3^- is first reduced to NO_2^- (using nitrite reductase, eqn 9.11) by *P. fumarii*. NO_2^- is then reduced to ammonium (NH_4^+), with ammonium hydroxide as the end product of nitrate reduction (eqn 9.23):

$$HNO_3 + 4H_2 \rightarrow NH_4OH + 2H_2O \quad (9.23)$$

(in *P. fumarii*).

A summary of the putative pathways used by hyperthermophilic nitrate reducers is shown in Fig. 9.1.

Respiration of arsenate, selenate and selenite

Arsenate reduction at hyperthermophilic temperatures has only been observed in the *Pyrobaculum* species *P. arsenaticum* and *P. aerophilum*. Huber *et al.* (2000) described *P. arsenaticum*, isolated from a hot spring at Pisciarelli Solfarata (Naples, Italy), as the first hyperthermophilic microorganism able to grow either organotrophically (eqn 9.24) or chemolithoautotrophically (eqn 9.25) by utilizing arsenate (AsO_4^{3-}) as a terminal electron acceptor under anaerobic conditions. It was also reported that *P. arsenaticum* could grow with selenate (SeO_4^{2-}) as an electron acceptor, under similar organotrophic culture conditions (eqn 9.26; Huber *et al.*, 2000b, 2006b). Arsenite (AsO_3^{3-}) and selenite (SeO_3^{2-}) were the end products of AsO_4^{3-} and SeO_4^{2-} reduction, respectively, under the culture conditions tested. *P. aerophilum* was the only other member of the genus *Pyrobaculum* able to grow organotrophically and autotrophically (with H_2 as the electron donor), under strict anaerobic conditions with AsO_4^{3-}, SeO_4^{2-} and SeO_3^{2-} (eqns 9.24–9.28 respectively), as electron acceptors. Elemental selenium (Se^0) was the end product of SeO_3^{2-} reduction by *P. aerophilum* (Huber *et al.*, 2000b, 2006b):

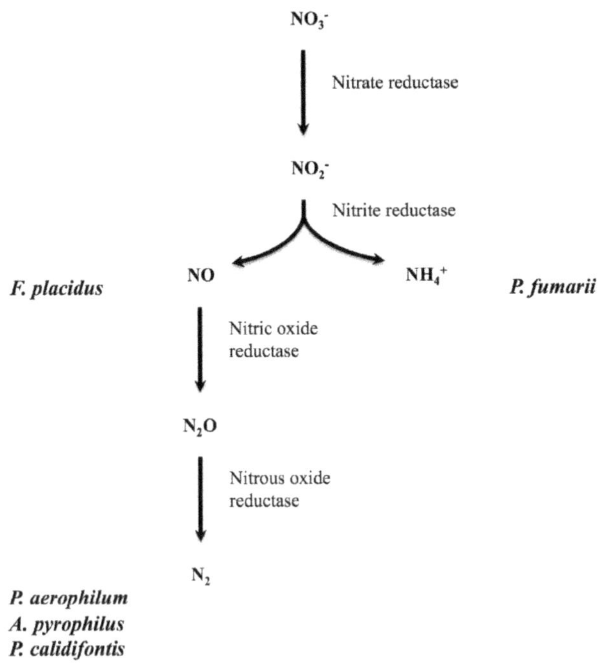

Figure 9.1 The pathway and enzymes involved in nitrate reduction in various hyperthermophiles. The end product of nitrate reduction for *P. fumarii* is NH_4^+. *P. aerophilum*, *A. pyrophilus* and *P. calidifontis* reduce NO_3^- to N_2. For *F. placidus* the pathway ends at NO.

$$2HAsO_4^{2-} + CH_2O + 4H^+ \rightarrow 2HAsO_2 + 3H_2O + CO_2 \quad (9.24)$$

$$HAsO_4^{2-} + H_2 + 2H^+ \rightarrow HAsO_2 + 2H_2O \quad (9.25)$$

$$2SeO_4^{2-} + CH_2O \rightarrow 2SeO_3^{2-} + H_2O + CO_2 \quad (9.26)$$

(in *P. arsenaticum* and *P. aerophilum*).

CH_2O in the above reactions represents organic compounds.

$$SeO_4^{2-} + H_2 \rightarrow SeO_3^{2-} + H_2O \quad (9.27)$$

$$SeO_3^{2-} + 2H_2 + 2H^+ \rightarrow Se^0 + 3H_2O \quad (9.28)$$

(in *P. aerophilum*).

Dissimilatory reduction of sulfate and other sulfurous compounds

Sulfate (SO_4^{2-}) is the second most abundant anion in seawater after chloride. Reduction of SO_4^{2-} to hydrogen sulfide (H_2S) is a process that is generally referred to as (dissimilatory) sulfate-reduction, or 'sulfate respiration'. During this process, sulfate is used as an electron acceptor to oxidize inorganic and/or organic matter and provide the cell with energy for vital activities. Sulfate-reduction at hyperthermophilic temperatures is conducted mainly by species in the genus *Archaeoglobus* (class Archaeoglobi, phylum Euryarchaeota) (Table 9.1). The genus *Archaeoglobus* is represented by six (hyper)thermophilic, sulfate-reducing, archaeal species. *Archaeoglobus fulgidus* VC-16, isolated from a shallow submarine hydrothermal vent of Volcano, Italy, was the first described sulfate-reducing hyperthermophile (Stetter, 1988). It grows optimally at 83°C under strict anaerobic, sulfate-reducing conditions. Three other members of the *Archaeoglobus* genus, *A. profundus* (Burggraf *et al.*, 1990), *A. veneficus* (Huber *et al.*, 1997) and *A. lithotrophicus* (Stetter *et al.*, 1993) are all hyperthermophiles which have also been isolated from marine hydrothermal habitats (Table 9.1). To date, the recently isolated *A. solfaticallidus* (Steinsbu *et al.*, 2010) and *A. infectus* (Mori *et al.*, 2008) are the only known thermophilic (T_{opt} = 75°C) members of the genus *Archaeoglobus* (not included in Table 9.1 and not discussed here).

All six species in the genus *Archaeoglobus* are strict anaerobes, and therefore, are unable to grow in the presence of dissolved oxygen. *Archaeoglobus* species are also unable to use S^0 (Hartzell and Reed, 2006). Instead, they obtain energy through the reduction of sulfate to H_2S. As a result they play an important role in the accumulation of reduced sulfur compounds in their natural habitats. Their activity is often recognized by the smell of H_2S, and also by the blackening of sediments due to the formation of insoluble iron(II) sulfide (FeS). At cooler temperatures H_2S and FeS react with O_2, which can then be converted to Fe(III) and S^0. Hyperthermophilic sulfate reducers are typically heterotrophic. *A. fulgidus* is capable of gaining energy for growth by anaerobic oxidation of organic acids such as lactate (eqn 9.29) or pyruvate under sulfate-reducing conditions, with H_2S and CO_2 as the end products:

$$3H_2SO_4 + 2C_3H_6O_3 \rightarrow 6H_2O + 3H_2S + 6CO_2 \quad (9.29)$$

(in *A. fulgidus*).

A. fulgidus and *A. lithotrophicus* can also grow fermentatively in the absence of sulfate (Hartzell and Reed, 2006). Under strict anaerobic conditions, using thiosulfate as the electron acceptor, *A. fulgidus* is also capable of oxidizing short- and long-chain (up to C_{18}) fatty acids (e.g. stearate; eqn 9.30) and alkenes (hexadecane; eqn 9.31; Khelifi *et al.*, 2010):

$$13H_2SO_4 + C_{18}H_{36}O_2 \rightarrow 18H_2O + 13H_2S + 18CO_2 \quad (9.30)$$

$$12H_2S_2O_3 + C_{16}H_{32} \rightarrow 4H_2O + 24H_2S + 16CO_2 \quad (9.31)$$

(in *A. fulgidus*).

In addition, *A. fulgidus* and *A. lithotrophicus* are also capable of autotrophic growth by reduction of SO_4^{2-}, $S_2O_3^{2-}$ and SO_3^{2-} to H_2S, with the concomitant oxidation of H_2 (eqns 9.32–9.34, respectively; Stetter, 1988):

$$H_2SO_4 + 4H_2 \rightarrow 4H_2O + H_2S \quad (9.32)$$

$$H_2S_2O_3 + 4H_2 \rightarrow 3H_2O + 2H_2S \quad (9.33)$$

$$H_2SO_3 + 3H_2 \rightarrow 3H_2O + H_2S \quad (9.34)$$

(in *A. fulgidus* and *A. lithotrophicus*).

It should be mentioned that in addition to *Archaeoglobus* spp., members of the genera *Pyrobaculum* (*P. arsenaticum*, *P. islandicum* and *P. oguniense*), *Pyrodictium* (*P. abyssi*), *Pyrolobus* (*P. fumarii*) and *Stetteria* (*S. hydrogenophila*), are also capable of reducing $S_2O_3^{2-}$ (Table 9.1). Similarly, sulfite respiration has also been observed in members of the genera *Pyrobaculum*, *Pyrodictium* and *Pyrolobus* (Table 9.1). H_2S is the end product of anaerobic $S_2O_3^{2-}$ and SO_3^{2-} reduction by these hyperthermophiles.

A. fulgidus can utilize a wide variety of carbon sources, including formate, formamide, lactate, pyruvate, glucose, starch, casamino acids, peptone, gelatine and casein (this growth substrate versatility suggests that *A. fulgidus* is a rich source of heat stable enzymes for commercial applications – see the biotechnology section later in this chapter). In contrast, *A. profundus* shows an absolute requirement for H_2 and acetate for growth (Burggraf *et al.*, 1990). *A. fulgidus* is also recognized as the hyperthermophilic archaeal counterpart of the mesophilic model sulfate reducers, such as the bacterium *Desulfovibrio vulgaris*. C-type cytochromes are important electron carriers in the sulfate reduction pathways in both model systems. For example, comparative studies indicated that the characterized C-cytochromes of *A. fulgidus* and *A. profundus* resembled the *Desulfovibrio* C_3-cytochromes in spectral properties and molecular weight (Kashefi and Hartzell, unpublished data).

A. fulgidus, like sulfate-reducing bacteria (SRB, e.g. *Desulfovibrio* spp.), contains enzymes involved in sulfate reduction, including adenosine triphosphate sulfurylase (ATP sulfurylase), adenylylsulfate reductase and bisulfate reductase (Speich and Truper, 1988; Dahl *et al.*, 1990; Lampreia *et al.*, 1991). Whilst this metabolic pathway is similar to that found in SRBs, *A. fulgidus* differs from the bacterial sulfate reducers in that it employs a modified acetyl-CoA/carbon monoxide dehydrogenase pathway to oxidize lactate to CO_2 (Fig. 9.2). Furthermore, *A. fulgidus* produces many novel cofactors (including deazaflavin F_{420}, methanofuran and tetrahydromethanopterin) and enzymes (e.g. formyl-H_4 MPT:MFR formyltransferase, methyl-H_4 MPT cyclohydrolase, methylene-H_4 MPT:F_{420} oxidoreductase) that are used by methanogenic archaea to reduce CO_2 to CH_4 (Ferry, 1992). In *Archaeoglobus*, the methanogenic cofactors serve as C_1 carriers for the oxidation of lactate to CO_2, producing the same intermediates formed during CO_2 reduction to CH_4 in methanogens (Fig. 9.2; Möller-Zinkhan and Thauer, 1990; Hartzell and Reed, 2006).

Interestingly, because *A. fulgidus* produces most of the enzymes and cofactors necessary for methane production it generates trace amounts of CH_4 (~0.1 μmol/ml) during growth (Möller-Zinkhan and Thauer, 1990; Hartzell and Reed, 2006). However, the methanogenic enzymes methyl-Co M reductase, the enzyme that catalyses the terminal step of methanogenesis, and F_{430}, the nickel tetrapyrrole bound to methyl-CoM reductase, have not been detected in *A. fulgidus* (Möller-Zinkhan and Thauer, 1990; Hartzell and Reed, 2006). The similarities of *Archaeoglobus* metabolic pathways to those of methanogens supports the 16S rRNA data that *A. fulgidus* is the 'evolutionary missing link' between methanogenic and sulfur-metabolizing archaea (Möller-Zinkhan and Thauer, 1990; Gorris *et al.*, 1991; Hartzell and Reed, 2006).

Archaeoglobus species are ubiquitous, having been isolated from a wide variety of geographically distinct hydrothermal habitats, including geothermally heated oil reservoirs. Members of the genus *Archaeoglobus* are unable to utilize elemental sulfur. Instead they gain energy for growth by coupling the oxidation of lactate, pyruvate or H_2 as electron donors to the reduction of sulfate, thiosulfate or sulfite as electron acceptors, with H_2S as the end product of sulfate reduction. Generation of H_2S by these hyperthermophiles may explain the formation of 'sour oil' (H_2S in geothermally heated oil wells; Stetter and Huber, 2000; Hartzell and Reed, 2006; Zhang *et al.*, 2011), which can cause the spoiling of crude oil deposits and may result in the loss of billions of dollars of this precious resource (see Chapter 8). A basic understanding of the physiology and metabolism of these sulfate reducers may help in discovering efficient ways to block the 'oil souring' capabilities of these organisms. Because *A. fulgidus* (in particular) can utilize a wide variety of substrates, tolerate oxygen and grow at high temperatures (83–85°C), it is a particularly hardy and abundant member of its genus. It is ironic that the physiological properties that make *A. fulgidus* the menace of secondary oil recovery (Stetter *et al.*, 1993; Stetter and Huber, 2000; Hartzell and Reed, 2006; Khelifi *et al.*, 2010; Zhang *et al.*, 2011) may also place this organism in a central role in global carbon and sulfur cycles.

Respiration of elemental sulfur (S^0)

The oxidation and reduction of S^0 is a common characteristic of a wide diversity of microorganisms inhabiting terrestrial hot springs and marine hydrothermal environments. This important energy-yielding reaction is arguably one of the most ancient forms of respiration (Fischer *et al.*, 1983). The majority of S^0-dependent hyperthermophiles are obligate heterotrophs growing under strictly anaerobic conditions, with peptides as the primary source of carbon (Adams, 1999). S^0 respiration can be performed by chemolithoautotrophic hyperthermophiles, including members of the genera *Acidianus, Aquifex, Ignicoccus, Metallosphaera, Pyrobaculum, Pyrodictium, Staphylothermus, Stetteria, Stygiolobus, Sulfolobus* and *Thermoproteus* (Table 9.1; Stetter *et al.*, 1983; Garrity *et al.*, 2005; Huber *et al.*, 2006b; Huber and Prangishvili, 2006; Huber and Stetter, 2006a; Stetter, 2006b). Members of the genus *Acidianus* are facultative anaerobes; under anaerobic conditions they conserve energy for growth by coupling the oxidation of H_2 (electron donor) to the reduction of S^0 (electron acceptor). H_2S is the end product of S^0 reduction by *A. ambivalens* under anaerobic conditions (eqn 9.35) (Kletzin, 1994), as well as by *I. islandicus* and *I. pacificus* (Huber *et al.*, 2000a; Huber *et al.*, 2006a), *P. islandicum* (Selig and Schonheit, 1994), *P. brockii* (Pihl *et al.*, 1992), *P. abyssi* (Dirmeier *et al.*, 1998), *S. azoricus*

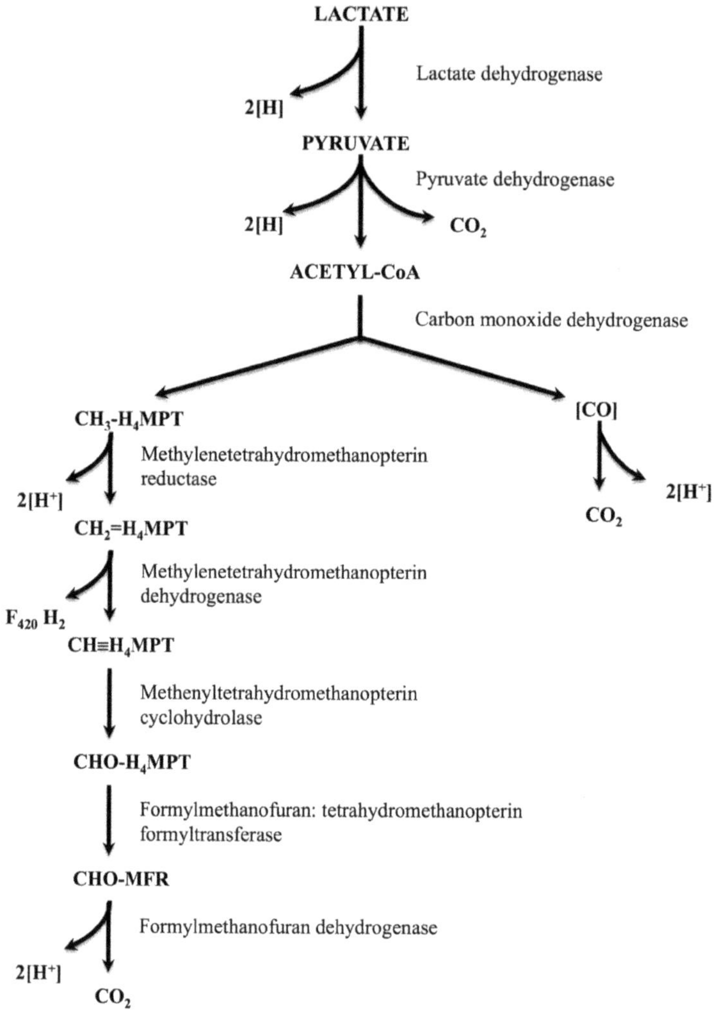

Figure 9.2 Pathway for oxidation of lactate to CO_2 in *A. fulgidus* (adapted from Moller-Zinkhan *et al.*, 1989). CH_3-H_4MPT, methyl-tetrahydromethanopterin; $CH_2=H_4$MPT, methylene-tetrahydromethanopterin; $CH\equiv H_4$MPT, methenyl-tetrahydromethanopterin; CHO-H_4MPT, formyl-tetrahydromethanopterin; CHO-MFR, formylmethanofuran. With kind permission of Springer Science+Business Media.

(Segerer *et al.*, 1991; Berg *et al.*, 2010) and *T. tenax* (Selig and Schonheit, 1994):

$$S^0 + H_2 \rightarrow H_2S \tag{9.35}$$

(in *Acidianus, Aquifex, Ignicoccus, Metallosphaera, Pyrobaculum, Pyrodictium, Staphylothermus, Stetteria, Stygiolobus, Sulfolobus* and *Thermoproteus* spp.).

Under aerobic conditions, S^0 is used as the electron donor. *Acidianus, Aquifex, Metallosphaera* and *Sulfolobus* species grow by coupling the oxidation of S^0 to the reduction of O_2 (Kletzin *et al.*, 2004). As a result, S^0 is oxidized to SO_4^{2-} (eqn 9.36), via SO_3^{2-} and $S_2O_3^{2-}$, in a pathway involving membrane-bound and soluble enzymes (Segerer *et al.*, 1986; Kletzin *et al.*, 2004):

$$2S^0 + 3O_2 + 2H_2O \rightarrow 2H_2SO_4 \quad (9.36)$$

(in *Acidianus, Aquifex, Metallosphaera* and *Sulfolobus* spp.).

Dissimilatory oxidation and reduction of S^0 in (hyper)thermophilic archaea has been reviewed in detail by Kletzin *et al.* (2004).

Methanogenesis

Regardless of their growth temperature optima, all methanogens belong to the phylum Euryarchaeota. They are capable of autotrophic growth under strictly anaerobic conditions by using H_2 as an electron donor and CO_2 as carbon source and/or terminal electron acceptor (Table 9.1). Currently there are 10 species of methanogens belonging to hyperthermophilic genera of the order Methanococcales, available in pure cultures, which have been isolated from a diversity of marine hydrothermal environments (Table 9.1). These species were originally represented by 3 genera: *Methanococcus, Methanopyrus* and *Methanothermus*. However, armed with new information on the phenotypic characteristics of methanogens (Keswani *et al.*, 1996), the order Methanococcales was revisited and reclassified based on a polyphasic taxonomic characterization (Whitman *et al.*, 2001). As a result, the hyperthermophilic genus *Methanococcus* was eliminated and the creation of two new genera, *Methanocaldococcus* and *Methanotorris*, was proposed. The proposed classification of the order Methanococcales by Whitman *et al.* (2001) greatly improved the phylogenetic affiliation of the (hyper)thermophilic species within this order and their kinship within their represented genera (Takai *et al.*, 2004).

Hyperthermophilic methanogens are represented by species capable of growth at very high temperatures (T_{opt} ranging from 80–98°C; Table 9.1). *Methanopyrus kandleri* has amongst the highest growth temperature for known hyperthermophiles [T_{opt} = 98°C and T_{max} = 110°C at about 2 bar = 206 kilopascal (kPa) = 0.002 megapascal (MPa)]. Takai *et al.* (2008) reported that a newly isolated strain of *M. kandleri* strain 116 with a T_{max} of 116°C (at 0.4 MPa pressure), was able to grow as high as 122°C when tested under a high *in situ* pressure of 20 MPa. Peptides, selenate, and tungstate enhance the growth of *Methanocaldococcus* species, *Methanocaldococcus fervens, Methanocaldococcus infernus, Methanocaldococcus jannaschii* and *Methanocaldococcus vulcanius* (Table 9.1). In their description of *M. fervidus*, Stetter *et al.* (1981) reported that yeast extract was required for growth.

M. jannaschii (Bult *et al.*, 1996), *Methanobacterium thermoautotrophicum* (Smith *et al.*, 1997) and *Methanopyrus kandleri* (Slesarev *et al.*, 2002) each represent a different genus. However, comparative analysis of their genome sequences suggest that the genes involved in methanogenesis are identical among these species (eqn 9.37):

$$CO_2 + 4H_2 \rightarrow 2H_2O + CH_4 \quad (9.37)$$

(in *Methanocaldococcus, Methanopyrus, Methanothermus* and *Methanotorris* spp.).

Fe(III) respiration in modern and ancient hot environments

It has been suggested that both ancient and modern hot environments provide conditions that are favourable for the activity of Fe(III)-reducing microorganisms. Geological evidence suggests that, with the exception of inorganic carbon, Fe(III) was the most abundant potential electron acceptor for anaerobic respiration as life was first evolving on early Earth. A variety of studies using different approaches have hinted that Fe(III) reduction was one of the first, if not the first, kind of microbial respiration (Walker, 1987; Lovley, 1991; de Duve, 1995a,b, 1998; Russell *et al.*, 1998; Vargas *et al.*, 1998; Russell and Hall, 2002; Kashefi and Lovley, 2003; Kashefi *et al.*, 2004; Lovley *et al.*, 2004). Hydrogen and Fe(III) are thought to have accumulated on the pre-biotic Earth as the result of high-intensity UV irradiation of the Archaean seas, hydrolysing the abundant dissolved Fe(II) to Fe(III) and hydrogen (Cairns-Smith *et al.*, 1992). Hydrogen was probably lost to the atmosphere, while Fe(III) precipitated as insoluble Fe(III) oxides (Russell and Hall, 2002). Hydrothermal fluids may have also contributed to Fe(III) accumulation (Cairns-Smith *et al.*, 1992). Modern hydrothermal vents, for example, release large quantities of Fe(II) that, once mixed with the cold, oxygenated waters, is oxidized to Fe(III) and precipitated to nearby vent sediments (Jannasch, 1995; Karl, 1995; Pichler and Veizer, 1999). Sediment temperatures are generally elevated due to the emerging super-heated water, and geotectonic sources of hydrogen could have been available, making these sediments a probable place for early life (Walker, 1980).

The combination of high temperatures, high concentrations of Fe(III), and availability of hydrogen represent ideal conditions for the growth and activity of hyperthermophilic microorganisms capable of coupling the reduction of Fe(III) to the oxidation of hydrogen (Kashefi *et al.*, 2002b). Thus, environmental conditions on the early Earth may have been ideal to support the development of a hydrogen oxidizing, Fe(III)-reducing biological system. This form of metabolism has been hypothesized to be the first protometabolism that eventually led to the evolution of life (Russell *et al.*, 1998). Similar conditions favouring the growth of hyperthermophilic Fe(III) reducers are found around terrestrial hot springs (Brock *et al.*, 1976; Kashefi *et al.*, 2002a).

Although a geological record for this earliest proposed respiration has yet to be recognized, there is strong geological evidence in the form of the Precambrian Banded Iron Formations that Fe(III) reduction was an important form of respiration slightly later in the Earth's history (Baur *et al.*, 1985; Walker, 1984, 1987). Microbial reduction of Fe(III) generates ultrafine-grained minerals, such as magnetite, and large quantities of such magnetite have been recovered from regions 5.5–6.7 km below the Earth's surface (Gold, 1992). This discovery eventually led to the proposal that a deep hot biosphere may actually exist on Earth because of the activity of Fe(III)-reducing hyperthermophilic microorganisms, and this hot biosphere may in fact be similar to hot biospheres below the surface of other planets (Gold, 1992). When magnetite with a similar morphology, size and magnetic properties to that produced by Fe(III)-reducing microorganisms was identified in Martian meteorite ALH84001, it provided evidence for some, that similar forms of life may have existed on Mars (McKay *et al.*, 1996).

Fe(III) respiration is a metabolic trait of most hyperthermophiles

Fe(III) respiration is one of the most recently recognized forms of microbial respiration (Balashova and Zavarzin, 1980; Lovley and Phillips, 1986; Lovley *et al.*, 1987). Previously doubted to be energetically viable, Fe(III) respiration was first shown to yield enough energy to support cell growth in certain mesophilic bacteria, when coupled to the oxidation of organic compounds or hydrogen (Balashova and Zavarzin, 1980). Fe(III) respiration was for long an unsuspected respiratory capability of microbes living at elevated temperatures (85–100°C), based on early reports that indicated that these organisms could not reduce Fe(III) (Stetter, 1996). In a pioneering study by Vargas *et al.* (Vargas *et al.*, 1998), a wide phylogenetic and physiological range of hyperthermophilic archaea and bacteria were tested for their ability to oxidize hydrogen with the reduction of Fe(III). All tested strains were capable of reducing Fe(III), suggesting that this ability may be widespread in hyperthermophiles (Vargas *et al.*, 1998). Two of the hyperthermophilic microbes tested, the archaeon *Pyrobaculum islandicum* and the bacterium *Thermotoga maritima*, were also found to conserve energy to support growth by coupling the reduction of Fe(III) to the oxidation of hydrogen, a previously unknown metabolic capability of hyperthermophiles. Growth of *T. maritima* by Fe(III) respiration was particularly surprising because this organism was previously reported to conserve energy for growth only by fermentation (Stetter, 1996). *P. islandicum* also was the first hyperthermophilic microorganism shown to convert large quantities of Fe(III) oxide to magnetite, a highly magnetic iron oxide (Fe_3O_4) of mixed oxidation state [two parts Fe(III) and one part Fe(II)], using H_2 as the terminal electron donor (Kashefi and Lovley, 2000; eqn 9.40). This magnetite was similar in morphology to the magnetite recovered from the deep, hot biosphere, thus providing a reasonable explanation for the accumulation of large quantities of magnetite in ancient formations. The microaerophilic, nitrate-reducing archaeon *P. aerophilum* (a close relative of *P. islandicum*) also grew by coupling hydrogen oxidation to Fe(III) reduction (Kashefi and Lovley,

2000). This observation is of interest because the availability of the genome sequences of *P. aerophilum*, *T. maritima* and *P. islandicum* opens new opportunities for understanding the biochemical basis of Fe(III) respiration at high temperatures.

The above studies not only suggest that Fe(III) reduction may be a highly conserved metabolic trait shared by most hyperthermophiles, they also demonstrate that dissimilatory Fe(III) reduction can take place in modern hot biospheres. Furthermore, if deeply branching, hyperthermophilic microorganisms provide suitable models for early forms of microbial respiration (Pace, 1991; Holm, 1992; de Duve, 1995a; Barns and Nierzwicki-Bauer, 1997), then some of the earliest microorganisms may have had the ability to reduce Fe(III) (Vargas *et al.*, 1998; Kashefi and Lovley, 2000). This hypothesis would be consistent with geological evidence indicating that hydrogen and abiotically formed Fe(III) oxides were likely to be abundant on the pre-biotic Earth, which may have favoured the growth of microorganisms able to couple hydrogen oxidation to the reduction of Fe(III). It has been suggested that life may have evolved from inorganic membranes that catalyse the oxidation of hydrogen, and couple this to Fe(III) reduction (Russell *et al.*, 1998). This idea is compatible with the proposal that Fe(III) reduction appeared early in microbial evolution, and strongly suggests that the Last Universal Common Ancestor of all modern organisms had the ability to oxidize hydrogen using Fe(III) as an electron acceptor.

Hyperthermophilic reduction of humic acids and other extracellular quinones

Studies of the metabolism of hyperthermophilic Fe(III) reducers revealed yet one more unsuspected metabolic capability: that of humic acid reduction. Humic substances are complex, quinone-like organic compounds derived from terrestrial plant material or microbial sources. In mesophilic environments, microbially derived humics can serve as electron acceptors for humic-reducing microorganisms (Scott *et al.*, 1998) and, even at low concentrations, they can greatly stimulate Fe(III) reduction by acting as electron shuttles between the cell and insoluble Fe(III) oxides (Nevin and Lovley, 2000). Humic-like substances are also thought to be present in hot environments (Nissenbaum *et al.*, 1975; Shock, 1996), with the high biological productivity associated with many hydrothermal systems proposed to facilitate the formation of microbially based humic-like material (Scott *et al.*, 1998). Initially, it was not known whether humics could serve as electron acceptors or electron shuttles for hyperthermophilic Fe(III) reducers, as shown in mesophilic systems. However, the humic acids analogue, anthraquinone 2,6-disulfonate (AQDS) has been shown to be readily reduced to anthrahydroquinone-2,6-disulfonate (AHQDS) at 100°C by the model hyperthermophilic Fe(III) reducer *P. islandicum* (Lovley *et al.*, 2000). AQDS can also serve as an electron shuttle that greatly stimulates the reduction of Fe(III) and crystalline Fe(III) oxides such as goethite and hematite by *P. islandicum*, even at low, catalytic concentrations (Lovley *et al.*, 2000). The ability to reduce humics was found to be highly conserved among deeply branching hyperthermophilic microorganisms, suggesting that, in addition to Fe(III), early organisms on Earth may have also had the ability to reduce extracellular quinones (Lovley *et al.*, 2000). Humic-like substances could have accumulated on the pre-biotic Earth as a result of large-scale condensation of dissolved organic matter in the pre-biotic hydrosphere and, thus, could have been available, along with Fe(III), as potential electron acceptors in early forms of respiration (Nissenbaum *et al.*, 1975). Furthermore, the high levels of UV radiation on the pre-biotic Earth that are considered to have led to the accumulation of Fe(III) may also have resulted in the accumulation of quinones (Lovley *et al.*, 2000).

Reduction of toxic, radioactive and precious metals by hyperthermophilic Fe(III) reducers

The study of hyperthermophilic Fe(III)-reduction may also provide information about the fate of other metals in hot environments, including the formation of economically important ore deposits. The potential for hyperthermophiles to reduce metals other than Fe(III) was examined in detail with *P. islandicum* as the model organism (Kashefi and Lovley, 2000). *P. islandicum* coupled the oxidation of hydrogen to the enzymatic reduction of toxic and radioactive metals such as U(VI) [uranium], Tc(VII) [technetium], Cr(VI) [chromium], Co(III) [cobalt], and Mn(IV) (Kashefi and Lovley, 2000). These results were the first to demonstrate that microorganisms could contribute to the speciation of these metals in hydrothermal environments, and are of particular interest in terms of uranium biogeochemistry. It should be mentioned that the role of microbial U(VI) reduction in uranium deposition at moderate temperatures had been previously recognized (Lovley *et al.*, 1991). However, studies with *P. islandicum* demonstrated that microbially catalysed deposition of uranium was possible at ca. 100°C (Kashefi and Lovley, 2000; Kashefi *et al.*, 2008a). This finding provided a plausible explanation for the origin of many sandstone-type uranium deposits (Hostetler and Garrels, 1962), and the naturally generated Oklo nuclear reactor in Gabon (Brookins, 1990). From a practical point of view, this discovery highlights the potential of hyperthermophilic U(VI) reducers, and their enzymes, in the remediation of hot, metal-contaminated environments or waste streams (Kashefi and Lovley, 2000; Kashefi *et al.*, 2008a).

When hydrogen was used as the electron donor, *P. islandicum* and other hyperthermophiles were also found to reduce the precious metal gold, another geologically and economically significant metal (Kashefi *et al.*, 2001). Soluble ionic gold, Au(III), was rapidly reduced to the metallic state, Au(0), and precipitated, when added to cell suspensions of various hyperthermophiles previously cultured with Fe(III) as the electron acceptor. The metallic gold precipitate was extracellular, and deposited on the cell surface. Interestingly, at about the same time a gold-laced portion of a hydrothermal vent sulfide chimney was recovered by an Australian team from a depth of two kilometres in the Bismarck Sea (News, 2000). This recovery suggests that the Au(III) transported in hydrothermal fluids as a chloride complex ($AuCl_4^-$) may be microbially reduced to Au(0) (Kashefi *et al.*, 2001). This process of 'microbial alchemy', while of great economical potential, also offers a plausible explanation for the presence of Au(0) deposits in some hot sediments.

Metabolic versatility of hyperthermophilic Fe(III) reducers

The use of Fe(III) as an electron acceptor for the characterization of known hyperthermophiles or for the enrichment and isolation of novel Fe(III) reducers at high temperatures led to the discovery of unsuspected metabolic processes. For example, using a novel strategy with Fe(III) oxide as the sole electron acceptor in solidified media, several hyperthermophilic microorganisms capable of autotrophic growth with hydrogen as the electron donor and Fe(III) oxide as the sole electron acceptor were recovered (Kashefi *et al.*, 2002a; Kashefi *et al.*, 2002b; Kashefi and Lovley, 2003). Magnetite (Fe_3O_4) was the end product of $Fe(OH)_3$ reduction (eqn 9.38):

$$6Fe(OH)_3 + H_2 \rightarrow 2Fe_3O_4 + 10H_2O \qquad (9.38)$$

(in *Ferroglobus placidus, Geoglobus ahangari* and *Geothermobacterium ferrireducens, Pyrobaculum aerophilum, P. islandicum*, 'strain 121', and *Thermotoga maritima*).

Of particular interest was a novel isolate from the Guaymas Basin in the Gulf of California, *Geoglobus ahangari* (Kashefi *et al.*, 2002b). *G. ahangari* was the first documented hyperthermophilic microorganism capable of conserving energy for growth by coupling Fe(III) reduction to the oxidation of acetate (eqn 9.39) and long chain fatty acids such as palmitate ($C_{16}H_{32}O_2$) and stearate ($C_{18}H_{36}O_2$), which are a significant component of organic matter in many anaerobic environments (Kashefi *et al.*, 2002b). The oxidation of palmitate and stearate by *G. ahangari* proceeds as according to the reactions shown in eqns 40 and 41, respectively, with magnetite (Fe_3O_4) and CO_2 as the end products:

$$CH_3COOH + 8Fe^{3+} + 2H_2O \rightarrow 2CO_2 + 8Fe^{2+} + 8H^+ \qquad (9.39)$$

(in *G. ahangari* and *F. placidus*).

$$C_{16}H_{32}O_2 + 92Fe^{3+} + 30H_2O \rightarrow 16CO_2 + 92Fe^{2+} + 92H^+ \qquad (9.40)$$

$$C_{18}H_{36}O_2 + 104Fe^{3+} + 34H_2O \rightarrow 18CO_2 + 104Fe^{2+} + 104H^+ \quad (9.41)$$
(in *G. ahangari*).

G. ahangari was also the first documented hyperthermophilic microorganism capable of completely oxidizing acetate to carbon dioxide, coupled to Fe(III) reduction (Tor *et al.*, 2001). Until the discovery of *G. ahangari*, acetate was not thought to be used by hyperthermophilic microbes, despite it being the most abundant organic acid in most hot environments (Slobodkin *et al.*, 1999). It seems probable that as the diversity of hyperthermophiles is further explored, hyperthermophiles capable of anaerobically oxidizing acetate with electron acceptors other than Fe(III) will also be discovered.

The ability to completely oxidize aromatic compounds, such as benzoate (C_6H_5COOH) and phenol (C_6H_5OH) by a hyperthermophilic microorganism was first documented in *Ferroglobus placidus* (Tor and Lovley, 2001). Neither benzoate nor phenol was previously known to be microbially degraded in thermal ecosystems. Furthermore, *F. placidus* can oxidize aromatic compounds specifically with Fe(III) as an electron acceptor (eqns 9.42 and 9.43; Tor and Lovley, 2001). The finding that a hyperthermophile could oxidize organic acids and aromatic compounds suggested, for the first time, that the complete oxidation of complex organic matter back to carbon dioxide was possible in hot microbial ecosystems (Tor *et al.*, 2001; Tor and Lovley, 2001; Kashefi *et al.*, 2002b):

$$C_6H_5COOH + 30Fe^{3+} + 12H_2O \rightarrow 7CO_2 + 30Fe^{2+} + 30H^+ \quad (9.42)$$

$$C_6H_5OH + 28Fe^{3+} + 11H_2O \rightarrow 6CO_2 + 28Fe^{2+} + 28H^+ \quad (9.43)$$
(in *F. placidus*).

Furthermore, recent studies have shown that a wide range of hyperthermophiles are able to conserve energy to support growth by coupling the oxidation of molecular hydrogen to the reduction of structural Fe(III) in smectite, a dominant mineral in the clay fraction in sedimentary environments (Kashefi *et al.*, 2008b).

Recovering novel hyperthermophiles with Fe(III)

The use of Fe(III) as an electron acceptor has enabled the enrichment, isolation, and recovery in pure culture of some organisms previously considered unculturable. The 16S rDNA sequence of a novel Fe(III) reducer, *Geothermobacterium ferrireducens* (Kashefi *et al.*, 2002a), is closely related to 16S rDNA sequences abundant in Obsidian Pool, Yellowstone National Park (Hugenholtz *et al.*, 1998). The physiology of these 'as-yet-uncultured' microorganisms was in doubt because of the lack of close relatives in culture. Characterization of the metabolic capabilities of *G. ferrireducens* revealed that this organism grows exclusively with hydrogen as the electron donor and Fe(III) oxide as electron acceptor, thereby providing new insights into the metabolic processes that may predominate in Obsidian Pool (Kashefi *et al.*, 2002a). In addition, all previously described hyperthermophilic and mesophilic Fe(III) reducers, including *P. islandicum* and *T. maritima*, required complex organic carbons such as yeast extract for growth, with hydrogen and Fe(III) as the electron donor and acceptor, respectively. *G. ferrireducens* and *G. ahangari*, were the first hyperthermophilic microorganisms shown to grow autotrophically with hydrogen (donor) and Fe(III) (acceptor). This finding also demonstrated that life at high temperatures could be sustained with only abiotically formed hydrogen and available Fe(III), with little or no organic carbon (Kashefi *et al.*, 2002a,b).

'Strain 121', arguably one of the most famous hyperthermophiles isolated to date, is a novel species in the novel genus '*Geogemma*' ('jewel from Earth'), family *Pyrodictiaceae* (Kashefi and Lovley, 2003). This strain, originally isolated from the 'Finn' hydrothermal vent in the Pacific Ocean, was named for its ability to grow (exclusively with Fe(III)) at 121°C, the standard temperature of sterilization procedures. It can even tolerate temperatures as high as 130°C for up to 2 hours (Kashefi and Lovley, 2003). The ability to grow at 121°C raises the upper temperature limit for the growth of any known organism. Two other members of the genus '*Geogemma*' have been isolated from different hydrothermal vents, one in the Pacific Ocean and the other in the Indian Ocean. Like 'strain 121', they grow with hydrogen and

Fe(III), but they display differences in electron donor use, possibly reflecting specific adaptations to their particular ecosystem (Kashefi, unpublished results).

The possibility of culturing so-called 'as-yet-uncultured' hyperthermophiles with Fe(III) as the sole electron acceptor suggests that those hyperthermophilic microorganisms that have not been successfully isolated with alternative electron acceptors (i.e. sulfur forms, nitrate, oxygen, or carbon dioxide) may actually be Fe(III) reducers. Taken together, these findings have provided new insights into the biogeochemistry of hot environments as well as the evolution of microbial respiratory systems.

Biotechnological applications of hyperthermophiles

Hyperthermophilic microorganisms are very attractive for industrial and biotechnological applications because of their metabolic flexibility and highly thermostable enzymes. Some of these applications are discussed below.

Hyperthermophilic enzymes

Because hyperthermophiles grow optimally at temperatures above 80°C, their enzymes are extremely resistant to heat denaturation, as well as being highly stable against protein denaturants such as detergents, organic solvents and chemicals. Although inactive at moderate temperatures, hyperthermophilic enzymes regain their activity as temperature increases (Koutsopoulos *et al.*, 2005). These unique properties make them excellent candidates for industrial applications.

Applications in molecular biology

Hyperthermophilic enzymes such as DNA polymerases and other DNA-modifying enzymes are used in recombinant DNA techniques and gene-based diagnostics (Cowan, 1995). The production of recombinant DNA polymerase *Taq* (Chien *et al.*, 1976) from the thermophilic bacterium *Thermus aquaticus* (Brock and Freeze, 1969) revolutionized molecular biology by replacing the mesophilic *E. coli* DNA polymerase I Klenow fragment in the Polymerase Chain Reaction (PCR). *Taq* is stable at the high temperatures required for the denaturation and annealing of DNA during PCR (Saiki *et al.*, 1988). When voided of 3′–5′ proofreading exonuclease activity, it has a high processivity, though low fidelity. Proofreading DNA polymerases such as Pfu (Lundberg *et al.*, 1991) and Vent (Mattila *et al.*, 1991; Zheng *et al.*, 2001) (isolated from *P. furiosus* and *Thermococcus litoralis*, respectively) are widely used in high-fidelity PCR amplification. Another hyperthermophilic DNA polymerase, KOD, isolated from *Thermococcus kodakaraensis* is one of the most efficient, most accurate thermostable DNA polymerases commercially available (Takagi *et al.*, 1997). Other hyperthermophilic enzymes with applications in molecular biology include DNA ligases (e.g. the *Pfu* DNA ligase from *P. furiosus*), proteases (e.g. Protease S from *P. furiosus*), aminopeptidases (e.g. the methionine and the pyroglutamate aminopeptidases from *P. furiosus*) and alkaline phosphatases (e.g. the alkaline phosphatase from *Thermotoga neapolitana*) (Vieille and Zeikus, 2001, and references therein).

Applications in biomass conversion

Thermophilic and hyperthermophilic enzymes are used for industrial starch processes that involve the hydrolysis of starch into sugar syrups at high temperatures (liquefaction) prior to saccharification (Vieille and Zeikus, 2001). The starch granules are gelatinized at temperatures above 105°C and low pH and partially hydrolysed by thermostable α-amylases at 95°C for several hours. The control of temperature and pH during this first step is critical. Optimum gelatinization requires that temperatures do not drop below 105°C. However, currently available α-amylases are inactivated above this temperature. Thus, hyperthermophilic α-amylases with optimal activities at the optimal gelatinization temperature and pH would be desirable. For example, the α-amylase from *P. furiosus* has an optimal activity of 105 °C and pH of 5.5–6.0 (Dong *et al.*, 1997), and is stable above 105°C.

Following liquefaction, the sugar syrups are enzymatically treated to convert high molecular weight saccharides into glucose or maltose (saccharification step). The enzymes involved in this step are β-amylases, glucoamylases, α-glucosidases, and pullunanases. All commercially available enzymes for the saccharification of starch syrups are thermophilic and only marginally stable at 60°C. For this reason, the temperature of the starch syrup needs to be dropped to 60°C after liquefaction, which increases the processing cost, slows down the reaction, and increases both the viscosity of the syrup (and, therefore, pumping costs) and the risk of contamination. Hyperthermophilic saccharification enzymes could overcome these disadvantages and greatly improve the saccharification process.

The high-glucose syrup produced from the bioprocessing of starch can be converted into high-fructose maize syrup (HFCS) by the isomerization of glucose into fructose, a reaction catalysed by immobilized xylose isomerases (Vieille and Zeikus, 2001). The direction of the isomerization reaction is controlled by temperature: as it increases, so does the isomerization of glucose into fructose. The xylose isomerases available to date are only moderately stable at 60°C. For this reason, the large-scale production of HFCS is run in packed-bed bioreactors at 58–60°C. At these temperatures only 42% of the glucose in the syrup is isomerized to fructose. Increasing temperature above 60°C shifts the isomerization rate towards fructose but also promotes side reactions at alkaline pHs. For this reason, hyperthermophilic xylose isomerase enzymes with excellent stability at 60°C and low pHs are desirable.

A major challenge in the production of biofuels from lignocellulose is the conversion of biomass into fermentable sugars, which currently requires a chemical pretreatment at low pHs or high temperatures and an enzymatic saccharification step at moderate pH and temperature (Lynd *et al.*, 2008). Many hyperthermophiles are capable of producing enzymes involved in the hydrolysis of cellulose and hemicellulose carbohydrates (reviewed in Blumer-Schuette *et al.*, 2008), yet the ability of hyperthermophiles to grow on crystalline cellulose or hemicellulose has never been demonstrated. Most hyperthermophiles lack primary cellulases required for growth on crystalline cellulose. Endoglucanases, which hydrolyse cellulose chains at internal points, and cellobiohydrolyases, which split the cellobiose disaccharides into glucose monomers, have been identified and characterized in hyperthermophilic bacteria (Vieille and Zeikus, 2001, and references therein). Species in the order *Thermococcales*, though unable to grow on crystalline cellulose, produce glycoside hydrolases and/or have genomes encoding various glucosidases and glucanases (Kawarabayasi *et al.*, 1998; Maeder *et al.*, 1999; Lecompte *et al.*, 2001; Kaper *et al.*, 2002; Cohen *et al.*, 2003; Fukui *et al.*, 2005). Endoglucanase enzymes from some *Thermococcales* have also been characterized (Gueguen *et al.*, 1997; Bauer *et al.*, 1999; Ando *et al.*, 2002; Kashima *et al.*, 2005). Hyperthermophilic cellulases also show promise for the production of ethanol from maize starch, as they can be added to the liquefaction and saccharification steps to improve yields (Lewis, 1996). Hyperthermophilic xylanases stable at neutral or alkaline pHs could be used during paper pulping to minimize both the amount of chlorine and/or chlorine dioxide currently used in the high-temperature bleaching process, and the resulting polluted waste streams generated and discharged (Tolan, 1996).

Other applications

Hyperthermophilic enzymes find many applications in the pharmaceutical, food and cosmetic industry. Cyclomaltodextrin glucanotransferases, for example, catalyse the synthesis of cyclodextrins used to remove undesirable tastes or odours, bind hydrophobic substances for increased water solubility, stabilize volatile substances, or protect compounds from side reactions (reviewed in Vieille and Zeikus, 2001). Hyperthermophilic enzymes also show promise for the development of improved or novel industrial-scale chemical syntheses of food additives (Yagasaki and Hashimoto, 2008). Hyperthermophilic hydrogenases also are of interest for applications as hydrogen catalysts. However, this application requires both heat and oxygen tolerance. Interestingly, almost all hydrogenases characterized to date are very sensitive to and become irreversibly inactivated by oxygen (Jenney and Adams, 2008). Recently, however, the hydrogenase of the novel hyperthermophilic archaeal isolate, *Aeropyrum camini*, was characterized and found to retain 75% of its activity during prolonged exposure to air at high temperatures (Nishimura and Sako, 2009). This finding highlights the potential of hyperthermophilic hydrogenases for applications in biological generation of hydrogen. Other hyperthermophilic enzymes that could potentially be compatible with industrial needs are pectinolytic enzymes such as methylesterases and depolymerases. These enzymes are used in many processes in the food industry to control viscosity and colour in fruit-derived juices and wine-making (Vieille and Zeikus, 2001). Enzymes from hyperthermophiles also serve as great model systems to investigate the basis of protein thermostability. As new approaches to protein engineering and evolution are investigated, the rational design of proteins with customized heat tolerance may be possible in the near future.

Conclusions

The majority of hyperthermophiles available in pure culture belong to the domain *Archaea.* Their optimum growth temperatures range from 80°C to 107°C, the highest temperature maximum is reported to be 122°C under 20 MPa (= 20,000 kPa = 200 bar) pressure (Takai *et al.*, 2008), and they are unable to grow below 60°C (Mesbah and Wiegel, 2008). Currently, there are about 90 species and 34 genera of hyperthermophilic archaea, and about nine species of hyperthermophilic bacteria in pure culture.

The archaea are a unique group of organisms that have been shown by 16S rRNA analysis to represent a separate biological domain (Woese and Fox, 1977; Woese *et al.*, 1991). Within this domain, the organisms are subdivided into three main groups according to their habitat and substrate utilization: (a) methane producers, (b) extreme halophiles and (c) hyperthermophiles. The last group is further subdivided into four smaller groups. The first group comprises mainly sulfur-metabolizing organisms that either oxidize or reduce elemental sulfur. The second group consists of the methanogens. The third group belongs to the genus *Archaeoglobus,* which contains all hyperthermophilic sulfate-reducing species of the domain *Archaea.* The last group harbours the iron reducers. Iron reduction has been found to be a major form of respiration in hyperthermophilic ecosystems.

Despite the low solubility of oxygen at elevated temperatures and the abundance of reducing gases, which help in maintaining anoxic conditions in hydrothermal environments, oxygen respiration has been observed in a phylogenetically diverse group of hyperthermophiles. Only a handful of hyperthermophilic archaea and bacteria possess the ability to gain energy for growth by nitrate (NO_3^-) reduction under anaerobic conditions. These hyperthermophiles use NO_3^- as a terminal electron acceptor, which is first reduced to nitrite (NO_2^-). NO_2^- can be further reduced either to ionic nitrogenous oxides (NO and N_2O) and nitrogen gas (N_2), or directly to ammonium (NH_4^+). Unlike their sulfate-reducing counterparts, hyperthermophilic nitrate reducers are phylogenetically diverse and have the ability to utilize a variety of electron acceptors (Slobodkin *et al.*, 1999). Arsenate and selenate reduction at hyperthermophilic temperatures has only been observed in the *Pyrobaculum* species *P. arsenaticum* and *P. aerophilum,* and only under strict anaerobic conditions. Selenite reduction has only been detected in *P. aerophilum.*

Members of the genera *Archaeoglobus* are believed to be the only known hyperthermophilic sulfate reducers. Reduction of sulfurous compounds is the universal electron-accepting pathway in the metabolism of sulfate reducing archaea (von Jan *et al.*, 2010). Generation of H_2S by these hyperthermophiles may explain the formation of 'sour oil' in geothermally heated oil wells (Stetter and Huber, 2000; Hartzell and Reed, 2006; Zhang *et al.*, 2011).

Recent studies suggest that the ability to couple the reduction of Fe(III) to the oxidation of molecular hydrogen is a common characteristic of hyperthermophilic microorganisms. Enrichment and isolation with Fe(III) oxide as the electron acceptor has proven to be a successful strategy for isolating previously uncultured hyperthermophiles, many of which are obligate Fe(III) reducers. Utilization of Fe(III) minerals as terminal electron acceptors has been instrumental in isolating as-yet-uncultured organisms from hot environments, expanding the known physiological capabilities of hyperthermophiles, especially in the range of organic compounds that can be oxidized at high temperatures, and also in determining the upper temperature limit for life. The study of hyperthermophilic Fe(III)-reducing microorganisms may aid in the understanding of the fate of other metals in hot environments, including the formation of economically important ore deposits. Fe(III) respiration also provides a persuasive analysis and interpretation for the massive formation of magnetite banded-iron formations in the Precambrian.

It is suggested that life began in a hot, anaerobic, Fe(III)- and H_2-rich environment. It has also been suggested that the physiological properties of the present hyperthermophiles most closely related to the Last Universal Common Ancestor(s) could help us understand the physiology of the early forms of life. If the above statements are correct, it is therefore not unreasonable to conclude that Fe(III) reduction was the first form of microbial respiration. Finally, hyperthermophilic microorganisms have significant importance in biotechnological and industrial applications, as they can serve as excellent sources of thermostable biomolecules while exhibiting flexible metabolic pathways. Furthermore, they are capable of utilizing a wide range of synthetic and natural products from their surroundings, in addition to producing renewable biofuels such as hydrogen and ethanol.

Hyperthermophiles represent the deepest phylogenetic branches of the tree of life (Blochl *et al.*, 1995; Baumann *et al.*, 1998; Stetter, 2006b). Hence, they are regarded as the most ancient known forms of life. Furthermore, they inhabit environments that share many similarities with ancient hot environments where life first arose on Earth (Kashefi and Lovley, 2003; Orange *et al.*, 2009). Thus, modern hyperthermophiles may have evolved very little from the ancestral organisms that are estimated to have appeared on Earth 3.5 billion years ago (Madigan *et al.*, 2009). Despite their evolutionary significance, only a small fraction (< 1%) of the estimated existing microbial diversity has been recovered, and therefore, studied in pure culture. This greatly limits evolutionary studies aimed at understanding the metabolic and physiological characteristics of the Last Universal Common Ancestor. Clearly, recovering additional modern descendants of ancient hyperthermophiles in pure culture is instrumental to these studies. To accomplish this, laboratory conditions used during the isolation process must recreate the conditions of temperature and electron donors and acceptors thought to have been present when the first forms of life arose on Earth.

Web resources

Microbial life – educational resources (http://serc.carleton.edu/microbelife/index.html)

Microbial life in extremely hot environments, created by Heather Beal, Montana State University (http://serc.carleton.edu/msu/heather.html)

The Astrobiology Web: Your Online Guide to the Living Universe™

Life in Extreme Environments (http://www.astrobiology.com/extreme.html)

New Millennium Observatory (NeMO) (http://www.pmel.noaa.gov/vents/nemo/index.html)

NeMO Dive! Dive to an active submarine volcano! (http://www.pmel.noaa.gov/vents/nemo/dive.html)

npr" New Discoveries in Deep-Sea Biodiversity (http://www.npr.org/templates/story/story.php?storyId=15233498)

University of Washington School of Oceanography, The Four Vent Fields (http://www.ocean.washington.edu/hydrothermalvents/fourfields.html)

Dive and Discover Expeditions to the Seafloor (http://www.divediscover.whoi.edu/expedition1/glossary.html#B; http://www.divediscover.whoi.edu/expedition7/index.html)

Extrasolar Planets and the Search for Life (http://www.astro.wisc.edu/~townsend/static.php?ref=diploma)

Voyage to the Deep, Mid-Ocean Ridge (http://www.ceoe.udel.edu/deepsea/level-2/geology/ridge.html)

Inferno 'flame' vent

http://www.pmel.noaa.gov/vents/nemo/explorer/ashes/inferno_flame.html

Microbiology: the beginning, Extremophiles (http://www.theguardians.com/Microbiology/gm_mbm04.htm)

Todar's Online Textbook of Bacteriology (by Kenneth Todar, 2009): The Procaryotes (http://www.textbookofbacteriology.net/procaryotes.html)

Methane on Mars, Brian Shiro (http://www.scribd.com/doc/10494415/Methane-on-Mars)

NASA Astrobiology Institute

Looking For Life (http://nai.arc.nasa.gov; http://passporttoknowledge.com/life)

Acknowledgements

I would like to thank Gemma Reguera for critical reading of this chapter and helpful insights. This work was supported by Michigan State University's AgBioResearch, a Strategic Partnership Grant from the MSU Research Foundation, and DOE Great Lakes Bioenergy Research Centre (DOE BER Office of Science DE-FC02-07ER64494).

References

Adams, M.W.W. (1999). The biochemical diversity of life near and above 100°C in marine environments. J. App. Microbiol. Symp. Suppl. *85*, 108S–117S.

Adams, M.W.W., Holden, J.F., Menon, A.L., Schut, G.J., Grunden, A.M., Hou, C., Hutchins, A.M., Jenney, F.E., Kim, C., Ma, K.S., *et al.* (2001). Key role for sulfur in peptide metabolism and in regulation of three hydrogenases in the hyperthermophilic archaeon *Pyrococcus furiosus*. J. Bacteriol. *183*, 716–724.

Amann, R., Ludwig, W., and Schleifer, K.-H. (1994). Identification of uncultured bacteria: A challenging task for molecular taxonomists. ASM News *60*, 360–365.

Amend, J.P., Meyer-Dombard, D.R., Sheth, S.N., Zolotova, N., and Amend, A.C. (2003). *Palaeococcus heigesonii* sp nov., a facultatively anaerobic, hyperthermophilic archaeon from a geothermal well on Vulcano Island, Italy. Arch. Microbiol. *179*, 349–401.

Amo, T., Paje, M.L., Inagaki, A., Ezaki, S., Atomi, H., and Imanaka, T. (2002). *Pyrobaculum calidifontis* sp. nov., a novel hyperthermophilic archaeon that grows in atmospheric air. Archaea *1*, 113–121.

Anderson, I.J., Dharmarajan, L., Rodriguez, J., Hooper, S., Porat, I., Ulrich, L.E., Elkins, J.G., Mavromatis, K., Sun, H., Land, M., *et al.* (2009). The complete genome sequence of *Staphylothermus marinus* reveals differences in sulfur metabolism among heterotrophic Crenarchaeota. BMC Genomics, *145*, doi: 10.1186/1471-2164-10-145.

Ando, S., Ishida, H., Kosugi, Y., and Ishikawa, K. (2002). Hyperthermostable endoglucanase from *Pyrococcus horikoshii*. Appl. Environ. Microbiol. *68*, 430–433.

Balashova, V.V., and Zavarzin, G.A. (1980). Anaerobic reduction of ferric iron by hydrogen bacteria. Microbiology *48*, 635–639.

Barns, S.M., and Nierzwicki-Bauer, S.A. (1997). Microbial diversity in ocean, surface, and subsurface environments. Rev. Mineral *35*, 35–79.

Barns, S.M., Fundyga, R.E., Jeffries, M.W., and Pace, N.R. (1994). Remarkable archaeal diversity detected in a Yellowstone National Park hot spring environment. Proc. Natl. Acad. Sci. U.S.A. *91*, 1609–1613.

Barns, S.M., Delwiche, C.F., Palmer, J.D., and Pace, N.R. (1996). Perspectives on archaeal diversity, thermophily and monophyly from environmental rRNA sequences. Proc. Natl. Acad. Sci. U.S.A. *93*, 9188–9193.

Bauer, M.W., Driskill, L.E., Callen, W., Snead, M.A., Mathur, E.J., and Kelly, R.M. (1999). An endoglucanase, EglA, from the hyperthermophilic archaeon *Pyrococcus furiosus* hydrolyzes β-1,4 bonds in mixed-linkage (1 → 3), (1 → 4)-β-d-glucans and cellulose. J. Bacteriol. *181*, 284–290.

Baumann, B., Judex, M., Huber, H., and Wirth, R. (1998). Estimation of genome sizes of hyperthermophiles. Extremophiles *2*, 101–108.

Baur, M.E., Hayes, J.M., Studley, S.A., and Walter, M.R. (1985). Millimeter-scale variations of stable isotope abundances in carbonates from banded iron-formations in the Hamersley Group of Western Australia. Econ. Geol. *80*, 270–282.

BBC News (2000). Seafloor 'chimney' recovered (Available at: http://news.bbc.co.uk/2/hi/science/nature/743165.stm).

Berg, I.A., Ramos-Vera, W.H., Petri, A., Huber, H., and Fuchs, G. (2010). Study of the distribution of autotrophic CO_2 fixation cycles in Crenarchaeota. Microbiology *156*, 256–269.

Bertoldo, C., and Antranikian, G. (2006). The Order Thermococcales. In The prokaryotes—a handbook on the biology of bacteria, M. Dworkin, S. Falkow, H. Rosenberg, K.-H. Schleifer, and E. Stackebrandt, eds. (New York, Springer), pp. 69–81.

Biddle, J.F., Lipp, J.S., Lever, M.A., Lloyd, K.G., Sorensen, K.B., Anderson, R., Fredricks, H.F., Elvert, M., Kelly, T.J., Schrag, D.P., *et al.* (2006). Heterotrophic Archaea dominate sedimentary subsurface ecosystems off Peru. Proc. Natl. Acad. Sci. U.S.A. *103*, 3846–3851.

Blochl, E., Burggraf, S., Fiala, G., Laurer, G., Huber, G., Huber, R., Rachel, R., Segerer, A., Stetter, K.O., and Volkl, P. (1995). Isolation, taxonomy, and phylogeny of hyperthermophilic microorganisms. World J. Microbiol. Biotechnol. *11*, 9–16.

Blochl, E., Rachel, R., Burggraf, S., Hafenbradl, D., Jannasch, H.W., and Stetter, K.O. (1997). *Pyrolobus fumarii*, gen. and sp. nov., represents a novel group of archaea, extending the upper temperature limit for life to 113°C. Extremophiles *1*, 14–21.

Blumer-Schuette, S.E., Kataeva, I., Westpheling, J., Adams, M.W.W., and Kelly, R.M. (2008). Extremely thermophilic microorganisms for biomass conversion: status and prospects. Curr. Opin. Chem. Biol. *19*, 210–217.

Bräsen, C., Schmidt, M., Grötzinger, J., and Schönheit, P. (2008). Reaction mechanism and structural model of ADP-forming Acetyl-CoA synthetase from the hyperthermophilic archaeon *Pyrococcus furiosus*: Evidence for a second active site histidine residue. J. Biol. Chem. *283*, 15409–15418.

Brazelton, W.J., Schrenk, M.O., Kelley, D.S., and Baross, J.A. (2006). Methane- and sulfur-metabolizing microbial communities dominate the Lost City Hydrothermal Field ecosystem. Appl. Environ. Microbiol. 72, 6257–6270.

Brochier, C., Gribaldo, S., Zivanovic, Y., Confalonieri, F., and Forterre, P. (2005). Nanoarchaea: representatives of a novel archaeal phylum or a fast-evolving euryarchaeal lineage related to Thermococcales? Genome Biol. *6*, R 42.

Brock, T.D., Cook, S., Petersen, S., and Mosser, J.L. (1976). Biogeochemistry and bacteriology of ferrous iron oxidation in geothermal habitats. Geochim. Cosmochim. Acta *40*, 493–500.

Brock, T.D., and Freeze, H. (1969). *Thermus aquaticus* gen. *n.* and sp. *n.* a non-sporulating extreme thermophile. J. Bacteriol. *98*, 289–297.

Brookins, D.G. (1990). Radionuclide behavior at the Oklo nuclear reactor, Gabon. Waste Manage *10*, 285–296.

Bult, C.J., White, O., Olsen, G.J., Zhou, L.X., Fleischmann, R.D., Sutton, G.G., Blake, J.A., FitzGerald, L.M., Clayton, R.A., Gocayne, J.D., *et al.* (1996). Complete genome sequence of the methanogenic archaeon, *Methanococcus jannaschii*. Science *273*, 1058–1073.

Burggraf, S., Jannasch, H.W., Nicolaus, B., and Stetter, K.O. (1990). *Archaeoglobus profundus* sp. nov., represents a new species within the sulfate-'reducing Archaebacteria. System. Appl. Microbiol. *13*, 24–28.

Burghardt, T., Saller, M., Gürster, S., Müller, D., Meyer, C., Jahn, U., Hochmuth, E., Deutzmann, R., Siedler, F., Babinger, P., *et al.* (2008). Insight into the proteome of the hyperthermophilic Crenarchaeon *Ignicoccus hospitalis*: the major cytosolic and membrane proteins. Arch. Microbiol. *190*, 379–394.

Cairns-Smith, A.G., Hall, A.J., and Russell, M.J. (1992). Mineral theories of the origin of life and an iron sulfide example. Origins of life and evolution of biospheres *22*, 161–180.

Casanueva, A., Galada, N., Baker, G.C., Grant, W.D., Heaphy, S., Jones, B., Yanhe, M., Ventosa, A., Blamey, J., and Cowan, D.A. (2008). Nanoarchaeal 16S rRNA gene sequences are widely dispersed in hyperthermophilic and mesophilic halophilic environments. Extremophiles *12*, 651–656.

Chevaldonne, P., Desbruyeres, D., and Le Haitre, M. (1991). Time-series of temperature from three deep-sea hydrothermal vent sites. Deep-Sea Res *38*, 1417–1430.

Chien, A., Edgar, D.B., and Trela, J.M. (1976). Deoxyribonucleic acid polymerase from the extreme thermophile *Thermus aquaticus*. J. Bacteriol. *27*, 1550–1557.

Chou, C.-J., Jenney, F.E., Adams, M.W.W., and Kelly, R.M. (2008). Hydrogenesis in hyperthermophilic microorganisms: Implications for biofuels. Metabol. Eng. *10*, 6394–6404.

Cohen, G.N., Barde, V., Flament, D., Galperin, M.D., Heilig, R., Lecompte, O., Poch, O., Prieur, D., Querellou, J., Ripp, R., *et al.* (2003). An integrated analysis of the genome of the hyperthermophilic archaeon *Pyrococcus abyssi*. Mol. Microbiol. *47*, 1495–1512.

Corre, E., Reysenbach, A.L., and Prieur, D. (2001). Epsilon-Proteobacterial diversity from a deep-sea hydrothermal vent on the Mid-Atlantic Ridge. FEMS Microbiol. Lett. *205*, 329–335.

Cowan, D.A. (1995). Hyperthermophilic enzymes: biochemistry and biotechnology. In Hydrothermal Vents and Processes, L.M. Parson, Walker, C.L., and Dixon, D.R., eds. (Geological Soc., London), pp. 351–364.

Dahl, C., Koch, H.-G., Keuken, O., and Truper, H.G. (1990). Purification and characterization of ATP sulfurylase from the extremely thermophilic archaebacterial sulfate-reducer, *Archaeoglobus fulgidus*. FEMS Microbiol. Lett. *67*, 27–32.

Dahle, H., Garshol, F., Madsen, M., and Birkeland, N.-K. (2008). Microbial community structure analysis of produced water from a high-temperature North Sea oil-field. Antonie van Leeuwenhoek 93, 37–49.

de Duve, C. (1995a). The beginnings of life on Earth. Amer. Sci. *83*, 428–437.

de Duve, C. (1995b). Vital Dust (Basic Books, New York).

de Duve, C. (1998). Possible starts for primitive life. Clues from present-day biology: the thioester world. In The Molecular Origins of Life: Assembling Pieces of the Puzzle, Brack, A., ed. (Cambridge University Press, Cambridge,), pp. 219–236.

Dirmeier, R., Keller, M., Frey, G., Huber, H., and Stetter, K.O. (1998). Purification and properties of an extremely thermostable membrane-bound sulfur-reducing complex from the hyperthermophilic *Pyrodictium abyssi*. Eur. J. Biochem. *252*, 486–491.

Dong, G., Vieille, C., Savchenko, A., and Zeikus, J.G. (1997). Cloning, sequencing, and expression of the gene encoding extracellular α-amylase from *Pyrococcus furiosus* and biochemical characterization of the recombinant enzyme. Appl. Environ. Microbiol. *63*, 3569–3576.

Dorr, C., Zaparty, M., Tjaden, B., Brinkmann, H., and Siebers, B. (2003). The hexokinase of the hyperthermophile Thermoproteus tenax. ATP-dependent hexokinases and ADP-dependent glucokinases, teo alternatives for glucose phosphorylation in Archaea. J. Biol. Chem. *278*, 18744–18753.

Erauso, G., Reysenbach, A.L., Godfroy, A., Meunier, J.R., Crump, B., Partensky, F., Baross, J.A., Marteinsson, V., Barbier, G., Pace, N.R., *et al.* (1993). *Pyrococcus abyssi* sp. nov.. a new hyperthermophilic archaeon isolated from a deep-sea hydrothermal vent. Arch. Microbiol. *160*, 338–349.

Ferry, J.G. (1992). Biochemistry of methanogenesis. Crit. Rev. Biochem. Mol. Biol. *27*, 473–503.

Fiala, G., Stetter, K.O., Jannasch, H.W., Langworthy, T.A., and Madon, J. (1986). *Staphylothermus marinus* sp. nov. represents a novel genus of extremely thermophilic submarine heterotrophic archaebacteria growing up to 98°C. Syst. Appl. Microbiol. *8*, 106–113.

Fischer, F., Zillig, W., Stetter, K.O., and Schreiber, G. (1983). Chemolithoautotrophic metabolism of anaerobic extremely thermophilic archaebacteria. Nature *301*, 511–513.

Fitz-Gibbon, S.T., Ladner, H., Kim, U.-J., Stetter, K.O., Simon, M.I., and Miller, J.H. (2002). Genome sequence of the hyperthermophilic crenarchaeon *Pyrobaculum aerophilum*. Proc. Natl. Acad. Sci. U.S.A. *99*, 984–989.

Fukui, T., Atomi, H., Kanai, T., Matsumi, R., Fujiwara, S., and Imanaka, T. (2005). Complete genome sequence of the hyperthermophilic archaeon *Thermococcus kodakaraensis* KOD1 and comparison with *Pyrococcus* genomes. Genome Res. *15*, 352–363.

Garrity, G.M., Bell, J.A., and Lilburn, T. (2005). The revised road map to the Manual. In Bergey's Manual of Systematic Bacteriology, Brenner, D.J., Krieg, N.R., and Staley, J.T., eds. (Springer, New York), pp. 159–187.

Gold, T. (1992). The deep, hot biosphere. Proc. Natl. Acad. Sci. USA. *89*, 6045–6049.

Gonzalez, J.M., Sheckells, D., Viebahn, M., Krupatkina, D., Borges, K.M., and Robb, F.T. (1999). *Thermococcus waiotapuensis* sp. nov., an extremely thermophilic archaeon isolated from a freshwater hot spring. Arch. Microbiol. *172*, 95–101.

Gorris, L.G., Voet, A.C., and van der Drift, C. (1991). Structural characteristics of methanogenic cofactors in the non-methanogenic archaebacterium *Archaeoglobus fulgidus*. Biofactors 3, 29–35.

Gueguen, Y., Voorhorst, W.G.B., Van Der Oost, J., and de Vos, W.M. (1997). Molecular and biochemical characterization of an endo-β-1,3-glucanase of the hyperthermophilic archaeon *Pyrococcus furiosus*. J. Biol. Chem. 272, 31258–31264.

Hafenbradl, D., Keller, M., Dirmeier, R., Rachel, R., Robnagel, P., Burggraf, S., Huber, H., and Stetter, K.O. (1996). *Ferroglobus placidus* gen. nov., sp. nov., a novel hyperthermophilic archaeum that oxidizes Fe^{+2} at neutral pH under anoxic conditions. Arch. Microbiol. *166*, 308–314.

Hagedoorn, P.L., Freije, J.R., and Hagen, W.R. (1999). *Pyrococcus furiosus* glyceraldehyde 3-phosphate oxidoreductase has comparable $W^{6+/5+}$ and $W^{5+/4+}$ reduction potentials and unusual [4Fe–2S] EPR properties. FEBS Lett. J. *462*, 66–70.

Hartzell, P., and Reed, D.W. (2006). The genus Archaeoglobus. In The Prokaryotes, Falkow, S., Rosenberg, E., Schleifer, K.-H., and Stackebrandt, E., eds. (Springer-Verlag, New York), pp. 82–100.

Hasan, F, Shah, A.A., and Hameed, A. (2006). Industrial applications of microbial lipases. Enzyme Microbial. Technol. *39*, 235–251.

Hoek, J., Banta, A., Hubler, F., and Reysenbach, A.-L. (2003). Microbial diversity of a sulphide spire located in the Edmond deep-sea hydrothermal vent field on the Central Indian Ridge. Geobiology *1*, 119–127.

Hohn, M.J., Hedlund, B.P., and Huber, H. (2002). Detection of 16S rDNA sequences representing the novel phylum 'Nanoarchaeota': indication for a wide distribution in high temperature biotopes. Syst. Appl. Microbiol. *25*, 551–554.

Holm, N.G. (1992). Why are hydrothermal systems proposed as plausible environments for the origin of life? Origins Life Evol. Biosphere *22*, 5–14.

Hostetler, P.B., and Garrels, R.M. (1962). Transportation and precipitation of uranium and vanadium at low temperatures with special reference to sandstone-type uranium. Econ. Geol. *57*, 137–167.

Hu, Y., Faham, S., Roy, R., Adams, M.W.W., and Rees, D.C. (1999). Formaldehyde ferredoxin oxidoreductase from *Pyrococcus furiosus*: the 1.85 A resolution crystal structure and its mechanistic implications. J. Mol. Biol. *286*, 899–914.

Huber, H., and Prangishvili, D. (2006). Solfolobales. In The Prokaryotes, Falkow, S., Rosenberg, E., Schleifer, K.-H., and Stackebrandt, E., eds. (Springer-Verlag, New York), p.959.

Huber, R., and Stetter, K.O. (1992). The Order Thermoproteales. In The Prokaryotes, Balows, A., Truper, H.G., Dworkin, M., Harder, W., and Schleifer, K.H., eds. (Springer-Verlag, New York), pp. 677–683.

Huber, H., and Stetter, K.O. (2006a). Desulfuroccales. In The Prokaryotes, Falkow, S., Rosenberg, E., Schleifer, K.-H., and Stackebrandt, E., eds. (Springer-Verlag, New York), p.959.

Huber, H., and Stetter, K.O. (2006b). Thermoplasmatales. In The Prokaryotes, Falkow, S., Rosenberg, E., Schleifer, K.-H., and Stackebrandt, E., eds. (Springer-Verlag, New York), p.959.

Huber, R., Langworthy, T.A., Konig, H., Thomm, M., Woese, C.R., Sleytr, U.B., and Stetter, K.O. (1986). *Thermotoga maritima* sp. nov. represents a new genus of unique extremely thermophilic eubacteria growing up to 90 °C. Arch. Microbiol. *144*, 324–333.

Huber, R., Kristjansson, J.K., and Stetter, K.O. (1987). *Pyrobaculum* gen. nov., a new genus of neutrophilic, rod-shaped archaebacteria form continental solfataras growing optimally at 100 °C. Arch. Microbiol. *149*, 95–101.

Huber, R., Wilharm, T., Huber, D., Trincone, A., Burggraf, S., Konig, H., Rachel, R., Rockinger, I., Fricke, H., and Stetter, K.O. (1992). *Aquifex pyrophilus* gen. nov., sp. nov., represents a novel group of marine hyperthermophilic hydrogen-oxidizing bacteria. Syst. Appl. Microbiol. *15*, 340–351.

Huber, H., Jannasch, H., Rachel, R., Fuchs, T., and Stetter, K.O. (1997). *Archaeoglobus veneficus* sp. nov., a novel facultative chemolithotrophic hyperthermophilic sulfite reducer, isolated from abyssal black smokers. Syst. Appl. Microbiol. *20*, 374–380.

Huber, R., Eder, W., Heldwein, S., Wanner, G., Huber, H., Rachel, R., and Stetter, K.O. (1998). *Thermocrinis ruber* gen. nov., sp. nov., a pink-filament-forming hyperthermophilic bacterium isolated from Yellowstone National Park. Environ. Microbiol. *64*, 3576–3583.

Huber, H., Burggraf, S., Mayer, T., Wyschkony, I., Rachel, R., and Stetter, K.O. (2000a). *Ignicoccus* gen. nov., a novel genus of hyperthermophilic, chemolithoautotrophic Archaea, represented by two new species, *Ignicoccus islandicus* sp. nov. and *Ignicoccus pacificus* sp. nov.. Int. J. Syst. Evol. Microbiol. *50*, 2093–2100.

Huber, R., Sacher, M., Vollmann, A., Huber, H., and Rose, D. (2000b). Respiration of arsenate and selenate by hyperthermophilic archaea. System. Appl. Microbiol. *23*, 305–314.

Huber, H., Hohn, M.J., Rachel, R., Fuchs, T., Wimmer, V.C., and Stetter, K.O. (2002). A new phylum of Archaea represented by a nanosized hyperthermophilic symbiont. Nature *417*, 63–67.

Huber, H., Hohn, M.J., Rachel, R., and Stetter, K.O. (2006a). Nanoarchaeota. In The Prokaryotes, Falkow, S., Rosenberg, E., Schleifer, K.-H., and Stackebrandt, E., eds. (Springer-Verlag, New York), p.959.

Huber, H., Huber, R., and Stetter, K.O. (2006b). Thermoproteales. In The Prokaryotes, Falkow, S., Rosenberg, E., Schleifer, K.-H., and Stackebrandt, E., eds. (Springer-Verlag, New York), p.959.

Huber, R., Hohn, M.J., Rachel, R., and Stetter, K.O. (2006c). Nanoarchaeota. In The Prokaryotes, Dworkin, M., Falkow, S., Rosenberg, E., Schleifer, K.-H., and Stackebrandt, E., eds. (Springer-Verlag, New York), pp. 274–280.

Huber, J.A., Mark Welch, D., Morrison, H.G., Huse, S.M., Neal, P.R., Butterfield, D.A., and Sogin, M.L. (2007).

Microbial population structures in the deep marine biosphere. Science *318*, 97–100.

Hugenholtz, P., Pitulle, C., Hershberger, K.L., and Pace, N.R. (1998). Novel division level bacterial diversity in a Yellowstone hot spring. J. Bacteriol. *180*, 366–376.

Hutchins, A.M., Holden, J.F., and Adams, M.W.W. (2001). Phosphoenolpyruvate synthetase from the hyperthermophilic Archaeon *Pyrococcus furiosus*. J. Bacteriol. *183*, 709–715.

Jahn, U., Gallenberger, M., Paper, W., Junglas, B., Eisenreich, W., Stetter, K.O., Rachel, R., and Huber, H. (2008). *Nanoarchaeum equitans* and *Ignicoccus hospitalis*: New insights into a unique, intimate association of two Archaea. J. Bacteriol. *190*, 1743–1750.

von Jan, M., Lapidus, A., Glavina Del Rio, T., Copeland, A., Tice, H., Cheng, J.-F., Lucas, S., Chen, F., Nolan, M., Goodwin, L., *et al.* (2010). Complete genome sequence of *Archaeoglobus profundus* type strain (AV18T). Stand. Gen. Sci. *2*, 327–346.

Jannasch, H.W. (1995). Microbial Interactions with hydrothermal fluids. In Seafloor Hydrothermal Systems: Physical, Chemical, Biological, and Geological Interactions, Geophysical Monograph 91, Humphries, S.E., Zierenberg, R.A., Mullineaux, L.S., and Thomson, R.E., eds. (American Geophysical Union, Washington, D.C.), pp. 273–296.

Jenney, J.F.E., and Adams, M.W.W. (2008). Hydrogenases of the model hyperthermophiles. Ann. N.Y. Acad. Sci. *1125*, 252–266.

Kaper, T., van Heusden, H.H., van Loo, B., Vasella, A., van der Oost, J., and de Vos, W.M. (2002). Substrate specificity engineering of β-mannosidase and β-glucosidase from *Pyrococcus* by exchange of unique active site residues. Biochemistry *41*, 4147–4155.

Karakashev, D., Kotay, S.M., Trably, E., and Angelidaki, E. (2009). A strict anaerobic extreme thermophilic hydrogen-producing culture enriched from digested household waste. J. Appl. Microbiol. *106*, 1041–1049.

Karl, D.M. (1995). Ecology of free-living hydrothermal vent microbial communities. In The Microbiology of Deep-Sea Hydrothermal Vents, Karl, D.M., ed. (CRC Press, New York), pp. 35–124.

Kashefi, K., and Lovley, D.R. (2000). Reduction of Fe(III), Mn(IV), and toxic metals at 100 degrees C by *Pyrobaculum islandicum*. Appl. Environ. Microbiol. *66*, 1050–1056.

Kashefi, K., and Lovley, D.R. (2003). Extending the upper temperature limit for life. Science *301*, 934.

Kashefi, K., Tor, J.M., Nevin, K.P., and Lovley, D.R. (2001). Reductive Precipitation of Gold by Dissimilatory Fe(III)-Reducing Bacteria and Archaea. Appl. Environ. Microbiol. *67*, 3275–3279.

Kashefi, K., Holmes, D.E., Reysenbach, A.-L., and Lovley, D.R. (2002a). Use of Fe(III) as an electron acceptor to recover previously uncultured hyperthermophiles: Isolation and characterization of *Geothermobacterium ferrireducens* gen., nov., sp. nov. Appl. Environ. Microbiol. *68*, 1735–1742.

Kashefi, K., Tor, J.M., Holmes, D.E., Gaw Van Praagh, C., Reysenbach, A.-L., and Lovley, D.R. (2002b). *Geoglobus ahangari*, gen. nov., sp. nov., a novel hyperthermophilic archaeon capable of oxidizing organic acids and growing autotrophically on hydrogen with Fe(III) serving as the sole electron acceptor. Int. J. System. Evol. Microbiol. *52*, 719–728.

Kashefi, K., Holmes, D.E., Lovley, D.R., and Tor, J.M. (2004). Potential importance of dissimilatory Fe(III)-reducing microorganisms in hot sedimentary environments. In The Subseafloor Biosphere at Mid-ocean Ridges, Wilcock, W.S.D., DeLong, E.F., Kelley, D.S., Baross, J.A., and Cary, S.C., eds. (American Geophysical Union, Washington, DC), pp. 199–211.

Kashefi, K., Moskowitz, B.M., and Lovley, D.R. (2008a). Characterization of extracellular minerals produced during dissimilatory Fe(III) and U(VI) reduction at 100°C by *Pyrobaculum islandicum*. Geobiology *6*, 147–154.

Kashefi, K., Shelobolina, E.S., Elliott, W.C., and Lovley, D.R. (2008b). Growth of thermophilic and hyperthermophilic Fe (III)-reducing microorganisms on a ferruginous smectite as the sole electron acceptor. Appl. Environ. Microbiol. *74*, 251–258.

Kashima, Y., Mori, K., Fukada, H., and Ishikawa, K. (2005). Analysis of the function of a hyperthermophilic endoglucanase from *Pyrococcus horikoshii* that hydrolyzes crystalline cellulose. Extremophiles *9*, 37–43.

Kawarabayasi, Y., Sawada, M., Horikawa, H., Haikawa, Y., Hino, Y., Amamoto, Y.S., Sekine, M., Baba, S.-I., Kosugi, H., Hosoyama, A., *et al.* (1998). Complete sequence and gene organization of the genome of a hyper-thermophilic archaebacterium, *Pyrococcus horikoshii* OT3. DNA Res. *5*, 55–76.

Kelley, D.S., Baross, J.A., and Delaney, J.R. (2002). Volcanoes, fluids, and life at mid-ocean ridge spreading centers. Annu. Rev. Earth Planet Sci. *30*, 385–491.

Keswani, J., Orkand, S., Premachandran, U., Mandelco, L., Franklin, M.J., and Whitman, W.B. (1996). Phylogeny and taxonomy of mesophilic *Methanococcus* spp. and comparison of rRNA, DNA hybridization, and phenotypic methods. Int. J. Syst. Bacteriol. *46*, 727–735.

Khelifi, N., Grossi, V., Hamdi, M., Dolla, A., Tholozan, J.L., Ollivier, B., and Hirschler-Rea, A. (2010). Anaerobic oxidation of fatty acids and alkenes by the hyperthermophilic sulfate-reducing archaeon *Archaeoglobus fulgidus*. Appl. Environ. Microbiol. *76*, 3057–3060.

Klenk, H.P., Clayton, R.A., Tomb, J.F., White, O., Nelson, K.E., Ketchum, K.A., Dodson, R.J., Gwinn, M., Hickey, E.K., Peterson, J.D., *et al.* (1997). The complete genome sequence of the hyperthennophilic, sulphate-reducing archaeon *Archaeoglobus fulgidus*. Nature *390*, 364–370.

Kletzin, A. (1994). Sulfur oxidation and reduction in Archaea: sulfur oxygenase/-reductase and hydrogenases from the extremely thermophilic and facultatively anaerobic archaeon *Desulfurolobus ambivalens*. Syst. Appl. Microbiol. *16*, 534–543.

Kletzin, A., Urich, T., Müller, F., Bandeiras, T.M., and Gomes, C.M. (2004). Dissimilatory oxidation and reduction of elemental sulfur in thermophilic archaea. J. Bioenerg. Biomembr. *36*, 77–91.

Kotsyurbenko, O.R., Glagolev, M.V., Nozhevnikova, A.N., and Conrad, R. (2001). Competition between homoacetogenic bacteria and methanogenic archaea for hydrogen at low temperature. FEMS Microbiol. Ecol. *38*, 153–159.

Koutsopoulos, S., van der Oost, J., and Norde, W. (2005). Temperature dependant structural and functional features of a hyperthermostable enzyme using elastic neutron scattering. Proteins *61*, 377–384.

Kurr, M., Huber, R., Konig, H., Jannasch, H.W., Fricke, H., Trincone, A., Kristjansson, J.K., and Stetter, K.O. (1991). *Methanopyrus kandleri*, gen. and sp. nov. represents a novel group of hyperthermophilic methanogens, growing at 110°C. Arch. Microbiol. *156*, 239–247.

Lampreia, J., Fauque, G., Speich, N., Dahl, C., Moura, I., Truper, H.G., and Moura, J.J.G. (1991). Spectroscopic studieson APS reductase isolated from the hyperthermophilic sulfate-reducing archaebacterium *Archaeoglobus fulgidus*. Biochem. Biophys. Res. Commun. *181*, 342–374.

Lebedinsky, A.V., Chernyh, N.A., and Bonch-Osmolovskaya, E.A. (2007). Phylogenetic systematics of microorganisms inhabiting thermal environments. Biochemistry 72, 1299–1312.

Lecompte, O., Ripp, R., Puzos-Barbe, V., Duprat, S., Heilig, R., Dietrich, J., Thierry, J.C., and Poch, O. (2001). Genome evolution at the genus level: comparison of three complete genomes of hyperthermophilic archaea. Genome Res. *11*, 981–993.

Lewis, S.M. (1996). Fermentation alcohol. In Industrial Enzymology, Godfrey, T., and West, S., eds. (Stockton Press, New York, NY), pp. 11–48.

Lovley, D.R. (1991). Dissimilatory Fe(III) and Mn(IV) reduction. Microbiol. Rev. *55*, 259–287.

Lovley, D.R., and Phillips, E.J.P. (1986). Organic matter mineralization with reduction of ferric iron in anaerobic sediments. Appl. Environ. Microbiol. *51*, 683–689.

Lovley, D.R., Stolz, J.F., Nord, G.L., and Phillips, E.J.P. (1987). Anaerobic production of magnetite by a dissimilatory iron-reducing microorganism. Nature *330*, 252–254.

Lovley, D.R., Phillips, E.J.P., Gorby, Y.A., and Landa, E.R. (1991). Microbial reduction of uranium. Nature *350*, 413–416.

Lovley, D.R., Kashefi, K., Vargas, M., Tor, J.M., and Blunt-Harris, E.L. (2000). Reduction of humic substances and Fe(III) by hyperthermophilic microorganisms. Chem. Geol. *169*, 289–298.

Lovley, D.R., Holmes, D.E., and Nevin, K.P. (2004). Dissimilatory Fe(III) and Mn(IV) reduction. Adv. Microb. Physiol. *49*, 219–286.

Lundberg, K.S., Shoemaker, D.D., Adams, M.W.W., Short, J.M., Serge, J.A., and Mathur, E.J. (1991). Archaea, hydrothermal, hyperthermophilic, phylogeny, taxonomy. Gene *108*, 1–6.

Lynd, L.R., Laser, M.S., Bransby, D., Dale, B.E., Davison, B., Hamilton, R., Himmel, M., Keller, M., McMillan, J.D., Sheehan, J., *et al.* (2008). How biotech can transform biofuels. Nat. Biotechnol. *26*, 169–172.

Ma, K., Hutchins, A., Sung, S.-J.S., and Adams, M.W.W. (1997). Pyruvate ferredoxin oxidoreductase from the hyperthermophilic archaeon, *Pyrococcus furiosus*, functions as a CoA-dependent pyruvate decarboxylase. Proc. Natl. Acad. Sci. U.S.A. *94*, 9608–9613.

Madigan, M.T., Martinko, J.M., Dunlap, P.V., and Clark, D.P., eds (2009). Brock Biology of Microorganisms, 12th edn. (San Francisco, Pearson Benjamin Cummings).

Maeder, D.L., Weiss, R.B., Dunn, D.M., Cherry, J.L., Gonzalez, J.M., DiRuggiero, J., and Robb, F.T. (1999). Divergence of the hyperthermophilic archaea *Pyrococcus furiosus* and *P. horikoshii* inferred from complete genomic sequences. Genetics *152*, 1299–1305.

Mai, X., and Adams, M.W.W. (1994). Indolepyruvate ferredoxin oxidoreductase from the hyperthermophilic archaeon *Pyrococcus furiosus*. A new enzyme involved in peptide fermentation. J. Biol. Chem. *269*, 16726–16732.

Mattila, P., Korpela, J., Tenkanen, T., and Pitkanen, K. (1991). Fidelity of DNA synthesis by the *Thermococcus litoralis* DNA polymerase-an extremely heat stable enzyme with proofreading activity. Nucleic Acids Res. *19*, 4967–4973.

McKay, D.S., Gibson Jr., E.K., Thomas-Deprta, K.L., Vali, H., Romanek, C.S., Clement, S.J., Chillier, X.D.F., Maechling, C.R., and Zare, R.N. (1996). Search for past life on Mars: possible relic biogenic activity in Martian meteorite ALH84001. Science *273*, 924–930.

Mesbah, N.B., and Wiegel, J. (2008). Life at extreme limits. In Incredible Anaerobes From Physiology to Genomics to Fuels, Wiegel, J., Maier, R.J., and Adams, M.W.W., eds. (Boston, Blackwell Publishing), pp. 44–57.

Miroshnichenko, M.L., Hippe, H.E.S., Kostrikina, N.A., Chernyh, N.A., Jeanthon, C., Nazina, T.N., Belyaev, S.S., and Bonch-Osmolovskaya, E.A. (2001). Isolation and characterization of *Thermococcus sibiricus* sp. nov. from a Western Siberia high-temperature oil reservoir. Extremophiles 5, 85–91.

Möller-Zinkhan, D., and Thauer, R.K. (1990). Anaerobic lactate oxidation to $3CO_2$ by *Archaeoglobus fulgidus* via the carbon monoxide dehydrogenase pathway: Demonstration of the acetyl-CoA carbon–carbon cleavage reaction in cell extracts Arch. Microbiol. *153*, 215–218.

Mori, K., Maruyama, A., Urabe, T., Suzuki, K.-I., and Hanada, S. (2008). *Archaeoglobus infectus* sp. nov., a novel thermophilic, chemolithoheterotrophic archaeon isolated from a deep-sea rock collected at Suiyo Seamount, Izu-Bonin Arc, western Pacific Ocean. Int. J. Syst. Evol. Microbiol. *58*, 810–816.

Mukund, S., and Adams, M.W.W. (1995). Glyceraldehyde-3-phosphate ferredoxin oxidoreductase, a novel Tungsten-containing enzyme with a potential glycolytic role in the hyperthermophilic Archaeon *Pyrococcus furiosus*. J. Biol. Chem. *270*, 8389–8392.

Musfeldt, M., Selig, M., and Schönheit, P. (1999). Acetyl coenzyme A synthetase (ADP forming) from the hyperthermophilic Archaeon *Pyrococcus furiosus*: identification, cloning, separate expression of the encoding genes, *acdAI* and *acdBI*, in *Escherichia coli*, and *in vitro* reconstitution of the active heterotetrameric enzyme from its recombinant subunits. J. Bacteriol. *181*, 5885–5888.

Nakagawa, S., Takai, K., Horikoshi, K., and Sako, Y. (2004). *Aeropyrum camini* sp. nov., a strictly aerobic, hyperthermophilic archaeon from a deep-sea hydrothermal vent chimney. Int. J. Syst. Bacteriol. *54*, 329–335.

Nercessian, O., Bienvenu, N., Moreira, D., Prieur, D., and Jeanthon, C. (2005a). Diversity of functional genes of methanogens, methanotrophs and sulfate reducers in deep-sea hydrothermal environments. Environ. Microbiol. *7*, 118–132.

Nercessian, O., Fouquet, Y., Pierre, C., Prieur, D., and Jeanthon, C. (2005b). Diversity of Bacteria and Archaea associated with a carbonate-rich metalliferous sediment sample from the Rainbow vent field on the Mid-Atlantic Ridge. Environ. Microbiol. 7, 698–714.

Nevin, K.P., and Lovley, D.R. (2000). Potential for nonenzymatic reduction of Fe(III) via electron shuttling in subsurface sediments. Environ. Sci. Technol. *34*, 2472–2478.

Nie, Y.-Q., Liu, H., Du, G.-C., and Chen, J. (2007). Enhancement of acetate production by a novel coupled syntrophic acetogenesis with homoacetogenesis process. Proc. Biochem. *42*, 599–605.

Nishimura, N., and Sako, Y. (2009). Purification and characterization of the oxygen-thermostable hydrogenase from the aerobic hyperthermophilic archaeon *Aeropyrum camini*. J. Biosci. Bioeng. *108*, 299–303.

Nissenbaum, A., Kenyon, D.H., and Oro, J. (1975). On the possible role of organic melanoidin polymers as matrices for prebiotic activity. J. Mol. Evol. *6*, 253–270.

Nozhevnikova, A.N., Zepp, K., Vazquez, F., Zehnder, A.J.B., and Holliger, C. (2003). Evidence for the existence of 10 psychrophilic methanogenic communities in anoxic sediments of deep lakes. Appl. Environ. Microbiol. *69*, 1832–1835.

Orange, F., Westall, F., Disnar, J.-R., Prieur, D., Bienvenu, N., Le Romancer, M., and Défarge, C. (2009). Experimental silicification of the extremophilic Archaea *Pyrococcus abyssi* and *Methanocaldococcus jannaschii*: applications in the search for evidence of life in early Earth and extraterrestrial rocks. Geobiology *7*, 403–418.

Pace, N.R. (1991). Origin of life-facing up to the physical setting. Cell *65*, 531–533.

Pichler, T., and Veizer, J. (1999). Precipitation of Fe(III) oxyhydroxide deposits from shallow-water hydrothermal fluids in Tutum Bay, Ambitle Island, Papua New Guinea. Chem. Geol. *162*, 15–31.

Pihl, T.D., Black, L.K., Schulman, B.A., and Maier, R.J. (1992). Hydrogen-oxidizing electron transport components in the hyperthermophilic archaebacterium *Pyrodictium brockii*. J. Bacteriol. *174*, 137–143.

Pikuta, E.V., Hoover, R.B., and Tang, J. (2007). Microbial extremophiles at the limits of life. Crit. Rev. Microbiol. *33*, 183–209.

Podar, M., Anderson, I., Makarova, K.S., Elkins, J.G., Ivanova, N., Wall, M.A., Lykidis, A., Mavromatis, K., Sun, H., Hudson, M.E., *et al.* (2008). A genomic analysis of the archaeal system *Ignicoccus hospitalis-Nanoarchaeum equitans*. Genome Biol. *9*, R158.

Prieur, D. (1997). Microbiology of deep-sea hydrothermal vents. Trends Biotechnol. *15*, 242–244.

Prokofeva, M.I., V., K.I., Nercessian, O., Tourova, T.P., Kolganova, T.V., Lebedinsky, A.V., Bonch-Osmolovskaya, E.A., Spring, S., and Jeanthon, C. (2005). Cultivated anaerobic acidophilic/acidotolerant thermophiles from terrestrial and deep-sea hydrothermal habitats. Extremophiles *9*, 437–448.

Randau, L., Münch, R., Hohn, M.J., Jahn, D., and Söll, D. (2005). *Nanoarchaeum equitans* creates functional t-RNAs from separate genes for their 50- and 30-halves. Nature *433*, 537–541.

Ravin, N.V., Mardanov, A.V., Beletsky, A.V., Kublanov, I.V., Kolganova, T.V., Lebedinsky, A.V., Chernyh, N.A., Bonch-Osmolovskaya, E.A., and Skryabin, K.G. (2009). Complete Genome Sequence of the Anaerobic, Protein-Degrading Hyperthermophilic Crenarchaeon *Desulfurococcus kamchatkensis*. J. Bacteriol. *191*, 2371–2379.

Reigstad, L.J., Jorgensen, S.L., and Schleper, C. (2010). Diversity and abundance of Korarchaeota in terrestrial hot springs of Iceland and Kamchatka. J. ISME *4*, 346–356.

Reysenbach, A.-L., Ehringer, M., and Hershberger, K. (2000a). Microbial diversity at 83°C in Calcite Spring, Yellowstone National Park: another environmental where the Aquificales and 'Korarchaeota' coexist. Extremophiles *4*, 61–67.

Reysenbach, A.-L., Longnecker, K., and Kirshtein, J. (2000b). Novel bacterial and Archaeal lineages from an *in situ* growth chamber deployed at a mid-Atlantic ridge hydrothermal vent. Appl. Enriron. Microbiol. *66*, 3798–3806.

Reysenbach, A.-L., and Shock, E. (2002). Merging genomes with geochemistry in hydrothermal ecosystems Science *296*, 1077–1082.

Robb, F.T., Maeder, D.L., Brown, J.R., DiRuggiero, J., Stump, M.D., Yeh, R.K., Weiss, R.B., and Dunn, D.M. (2001). Genomic sequence of hyperthermophile, *Pyrococcus furiosus*: Implications for physiology and enzymology. In Methods in Enzymology, Adams, M.W.W., and Kelly, R.M., eds. (London, Academic Press), pp. 134–157.

Ronimus, R.S., Reysenbach, A.L., Musgrave, D.R., and Morgan, H.W. (1997). The phylogenetic position of the Thermococcus isolate AN1 based on 16S rRNA gene sequence analysis: a proposal that AN1 represents a new species, *Thermococcus zilligii* sp. nov. Arch. Microbiol. *168*, 245–248.

Roy, R., Menon, A.L., and Adams, M.W.W. (2001). Aldehyde oxidoreductases from *Pyrococcus furiosus*. Methods Enzymol. *331*, 132–144.

Russell, M.J., Daia, D.E., and Hall, A.J. (1998). The emergence of life from FeS bubbles at alkaline hot springs in an acid ocean. In Thermophiles: The Keys to Molecular Evolution and the Origin of Life? Wiegel, J., and Adams, M.W.W., eds. (Philadelphia, PA, Taylor & Francis Ltd.), pp. 77–126.

Russell, M.J., and Hall, A.J. (2002). Chemiosmotic coupling and transition element clusters in the onset of life and photosynthesis. Geochem. News *113*, 6–12.

Saiki, R.K., Gelfand, D.H., Stoffel, S., Scharf, S.J., Higuchi, R., Horn, G.T., Mullis, K.B., and Erlich, H.A. (1988). Primer-directed enzymatic amplification of DNA with a thermostable DNA polymerase. Science *239*, 487–491.

Sako, Y., Nomura, N., Uchida, A., Ishida, Y., Morii, H., Koga, Y., Hoaki, T., and Maruyama, T. (1996). *Aeropyrum pernix* gen. nov., sp. nov., a novel aerobic hyperthermophilic archaeon growing at temperatures up to 100°C. Int. J. Syst. Bacteriol. *46*, 1070–1077.

Sakuraba, H., and Ohshima, T. (2002). Novel energy metabolism in anaerobic hyperthermophilic archaea: a modified Embden–Meyerhof pathway. J. Biosci. Bioeng. *93*, 441–448.

Sakuraba, H., Goda, S., and Ohshima, T. (2004). Unique sugar metabolism and novel enzymes of hyperthermophilic archaea. Chem. Rec. *3*, 281–287.

Sakuraba, H., Utsumi, E., Schreier, H.J., and Ohshima, T. (2001). Transcriptional regulation of phosphoenolpyruvate synthase by maltose in the hyperthermophilic archaeon, *Pyrococcus furiosus*. J. Biosci. Bioeng. *92*, 108–113.

Scott, D.T., McKnight, D.M., Blunt-Harris, E.L., Kolesar, S.E., and Lovley, D.R. (1998). Quinone moieties act as electron acceptors in the reduction of humic substances by humics-reducing microorganisms. Environ. Sci. Technol. *32*, 2984–2989.

Segerer, A., Neuner, A., Kristjansson, J.K., and Stetter, K.O. (1986). *Acidianus infernus* gen. nov., sp. nov., and *Acidianus brierleyi* comb. nov.: facultatively aerobic, extremely acidophilic thermophilic sulfur-metabolizing archaebacteria. Int. J. Syst. Bacteriol. *36*, 559–564.

Segerer, A.H., Trincone, A., Gahrtz, M., and Stetter, K.O. (1991). *Stygiolobus azoricus* gen. nov., sp. nov. represents a novel genus of anaerobic, extremely thermophilic archaebacteria of the order Sulfolobales. Int. J. Syst. Bacteriol. *41*, 495–501.

Selig, M., and Schonheit, P. (1994). Oxidation of organic compounds to CO_2 with sulfur or thiosulfate as electron acceptor in the anaerobic hypertherrnophilic archaea Thermoproteus tenax and Pyrobaculum islandicum proceeds via the citric acid cycle. Arch. Microbiol. *162*, 286–294.

Shock, E.L. (1996). Hydrothermal systems as environments for the emergence of life. In Evolution of hydrothermal ecosystems on Earth (and Mars?), Bock, G.R., and Goode, J.A., eds. (West Sussex, England, John Wiley & Sons, Ltd.), pp. 40–60.

Siriwongrungson, V., Zeng, R.J., and Angelidaki, I. (2007). Homoacetogenesis as the alternative pathway for H2 sink during thermophilic anaerobic degradation of butyrate under suppressed methanogenesis. Water Res. *41*, 4204–4210.

Slesarev, A.I., Mezhevaya, K.V., Makarova, K.S., Polushin, N.N., Shcherbinina, O.V., Shakhova, V.V., Belova, G.I., Aravind, L., Natale, D.A., Rogozin, I.B., *et al.* (2002). The complete genome of hyperthermophile *Methanopyrus kandleri* AV19 and monophyly of archaeal methanogens. Proc. Natl. Acad. Sci. U.S.A. *99*, 4644–4649.

Slobodkin, A.I., Zavarzina, D.G., Sokolova, T.G., and Bonch-Osmolovskaya, E.A. (1999). Dissimilatory reduction of inorganic electron acceptors by thermophilic anaerobic prokaryotes. Microbiology *68*, 600–622.

Smith, D.R., DoucetteStamm, L.A., Deloughery, C., Lee, H.M., Dubois, J., Aldredge, T., Bashirzadeh, R., Blakely, D., Cook, R., Gilbert, K., *et al.* (1997). Complete genome sequence of *Methanobacterium thermoautotrophicum* ΔH: functional analysis and comparative genomics. J. Bacteriol. *179*, 7135–7155.

Smith, E.T., Blamey, J.M., and Adams, M.W.W. (1994). Pyruvate ferredoxin oxidoreductases of the hyperthermophilic archaeon, *Pyrococcus furiosus*, and the hyperthermophilic bacterium, *Thermotoga maritima*, have different catalytic mechanisms. Biochemistry *33*, 1008–1016.

Sokolova, T.G., Jeanthon, C., Kostrikina, N.A., Chernyh, N.A., Lebedinsky, A.V., Stackebrandt, E., and Bonch-Osmolovskaya, E.A. (2004). The first evidence of anaerobic CO oxidation coupled with H-2 production by a hyperthermophilic archaeon isolated from a deep-sea hydrothermal vent. Extremophiles *8*, 317–323.

Speich, N., and Truper, H.G. (1988). Adenylylsulfate reductase in a dissimilatory sulfate-reducing archaebacterium. J. Gen. Microbiol. *134*, 1419–1425.

Steinsbu, B.O., Thorseth, I.H., Nakagawa, S., Inagaki, F., Lever, M.A., Engelen, B., Ovreas, L., and Pedersen, R.B. (2010). *Archaeoglobus sulfaticallidus* sp. nov., a novel thermophilic and facultatively lithoautotrophic sulfate-reducer isolated from black rust exposed to hot ridge flank crustal fluids. Int. J. Syst. Evol. Microbiol. *60*:2745–2752.

Stetter, K.O. (1988). *Archaeoglobus fulgidus* gen. nov., sp. nov.: a new taxon of extremely thermophilic Archaebacteria. System. Appl. Microbiol. *10*, 172–173.

Stetter, K.O. (1996). Hyperthermophilic procaryotes. FEMS Microbiol. Rev. *18*, 149–158.

Stetter, K.O. (2006a). History of discovery of the first hyperthermophiles. Extremophiles *10*, 357–362.

Stetter, K.O. (2006b). Hyperthermophiles in the history of life. Phil. Trans. Royal Soc. B: Biol. Sci. *361*, 1837–1843.

Stetter, K.O., and Huber, H. (2000). The role of hyperthermophilic prokaryotes in oil fields. Paper presented at: Microbial Biosystems: New Frontiers: Proceedings of the 8th International Symposium on Microbial Ecology (Halifax, Canada, Atlantic Canada Society for Microbial Ecology).

Stetter, K.O., Huber, R., Blochol, E., Kurr, M., Eden, R.D., Fielder, M., Cash, H., and Vance, I. (1993). Hyperthermophilic Archaea are thriving in deep

North-Sea and Alaskan oil-reservoirs. Nature *365*, 743–745.

Stetter, K.O., Koning, H., and Stackebrandt, E. (1983). *Pyrodictium* gen. nov., a new genus of submarine disc-shaped sulphur reducing araebacteria growing optimally at 105 °C. Syst. Appl. Microbiol. *4*, 535–551.

Takagi, M., Nishioka, M., Kakihara, H., Kitabayashi, M., Inoue, H., Kawakami, B., Oka, M., and Imanaka, T. (1997). Characterization of DNA polymerase from *Pyrococcus* sp. strain KOD1 and its application to PCR. Appl. Environ. Microbiol. *63*, 4504–4510.

Takahata, Y., Nishijima, M., Hoaki, T., and Maruyama, T. (2001). *Thermotoga petrophila* sp. nov. and *Thermotoga naphthophila* sp. nov., two hyperthermophilic bacteria from the Kubiki oil reservoir in Niigata, Japan. Int. J. System. Evol. Microbiol. *51*, 1901–1909.

Takai, K., Komatsu, T., Inagaki, F., and Horikoshi, K. (2001). Distribution of archaea in a black smoker chimney structure. Appl. Environ. Microbiol. *67*, 3618–3629.

Takai, K., Gamo, T., Tsunogai, U., Nakayama, N., Hirayama, H., Nealson, K.H., and Horikoshi, K. (2004). Geochemical and microbiological evidence for a hydrogen-based, hyperthermophilic subsurface lithoautotrophic microbial ecosystem (HyperSLiME) beneath an active deep-sea hydrothermal field. Extremophiles *8*, 269–282.

Takai, K., Nakamura, K., Toki, T., Tsunogai, U., Miyazaki, M., Miyazaki, J., Hirayama, H., Nakagawa, S., Nunoura, T., and Horikoshi, K. (2008). Cell proliferation at 122°C and isotopically heavy CH_4 production by a hyperthermophilic methanogen under high-pressure cultivation. Proc. Natl. Acad. Sci. U.S.A. *105*, 10949–10954.

Tivey, M.K., and Delaney, J.R. (1986). Growth of large sulfide structures on the Endeavour segment of the Juan de Fuca Ridge. Earth Plan. Sci. Lett. (EPSL) *77*, 303–317.

Tivey, M.K., Stakes, D.S., Cook, T.L., Hannington, M.D., and Petersen, S. (1999). A model for growth of steep-sided vent structures on the endeavour segment of the Juan de Fuca Ridge: results of a petrologic and geochemical study. Geophys. Res. *104*, 22859–22883.

Tolan, J.S. (1996). Pulp and paper. In Industrial Enzymology, Godfrey, T., and West, S., eds. (New York, N.Y., Stockton Press), pp. 327–338.

Tor, J.M., and Lovley, D.R. (2001). Anaerobic degradation of aromatic compounds coupled to Fe(III) reduction by *Ferroglobus placidus*. Environ. Microbiol. *3*, 281–287.

Tor, J.M., Kashefi, K., and Lovley, D.R. (2001). Acetate oxidation coupled to Fe(III) reduction in hyperthermophilic microorganisms. Appl. Environ. Microbiol. *67*, 1363–1365.

Turner, P., Mamo, G., and Nordberg Karlsson, E. (2007). Potential and utilization of thermophiles and thermostable enzymes in biorefining. Microbial cell Factories 6, 9.

Vargas, M., Kashefi, K., Blunt-Harris, E.L., and Lovley, D.R. (1998). Microbiological evidence for Fe(III) reduction on early Earth. Nature *395*, 65–67.

Ver Eecke, H.C., Kelley, D.S., and Holden, J.F. (2009). Abundances of hyperthermophilic autotrophic Fe(III) oxide reducers and heterotrophs in hydrothermal sulfide chimneys from the northeastern Pacific Ocean. Appl. Environ. Microbiol. 75, 242–245.

Vieille, C., and Zeikus, G.J. (2001). Hyperthermopilic enzymes: Sources, uses, and molecular mechanisms for thermostability. Microbiol. Mol. Biol. Rev. 65, 1–43.

Volkl, P., Huber, R., Drobner, E., Rachel, R., Burggraf, S., Trincone, A., and Stetter, K.O. (1993). *Pyrobaculum aerophilum* sp. nov, a novel nitrate-reducing hyperthermophilic Archaeum. Appl. Environ. Microbiol. *59*, 2918–2926.

Walker, J.C.G. (1980). Atmospheric constraints on the evolution of metabolism. Orig. Life *10*, 93–104.

Walker, J.C.G. (1984). Suboxic diagenesis in banded iron formations. Nature *309*, 340–342.

Walker, J.C.G. (1987). Was the Archaean biosphere upside down? Nature *329*, 710–712.

Waters, E., Hohn, M.J., Ahel, I., Graham, D.E., Adams, M.D., Barnstead, M., Beeson, K.Y., Bibbs, L., Bolanos, R., Martin Keller, M., *et al.* (2003). The genome of *Nanoarchaeum equitans*: insights into early archaeal evolution and derived parasitism. Proc. Natl. Acad. Sci. U.S.A. *100*, 12984–12988.

Whitman, W.B., Boone, D.R., and Koga, Y. (2001). Order Methanococcales. In Bergey's Manual of Systematic Bacteriology, Boone, D.R., Castenholz, R.W., and Garrity, G.M., eds. (New York, Springer), pp. 236–246.

Woese, C.R., and Fox, G.E. (1977). Phylogenetic structure of the prokaryotic domain: the primary kingdoms. Proc. Natl. Acad. Sci. USA. *74*, 5088–5090.

Woese, C.R., Achenbach, L., Rouviere, P., and Mandelco, L. (1991). Archaeal phylogeny: reexamination of the phylogenetic position of *Archaeoglobus fulgidus* in light of certain composition-induced artefacts. System. Appl. Microbiol. *14*, 364–371.

Yagasaki, M., and Hashimoto, S.-I. (2008). Synthesis and application of dipeptides; current status and perspectives. Appl. Microbiol. Biotechnol. *81*, 13–22.

Zhang, P., Lal, S., Miller, C., Reguera, G., Hartzell, P.L., and Kashefi, K. (2011). Isolation and characterization of a hyperthermophilic sulfate- and iron-reducing archaeon from crude oil. Submitted.

Zheng, R., Matsui, E., Shen, Y., Musati, K.V., Feng, Y., Damis, S., Kawarabayashi, Y., Kikuchi, H., Harata, K., and Matsui, I. (2001). The novel function of a short region K253XRXXD259 conserved the exonuclease domain of hyperthermostable DNA polymerase I from *Pyrococcus horikoshii*. Extremophiles 5, 111–117.

Zhou, H., Li, J., Peng, X., Meng, J., Wang, F., and Ai, Y. (2009). Microbial diversity of a sulfide black smoker in main endeavour hydrothermal vent field, Juan de Fuca Ridge. J. Microbiol. *47*, 235–247.

Microbiology of Piezophiles in Deep-sea Environments

10

Chiaki Kato

Abstract

Piezophilic microorganisms, which are defined as 'pressure-loving' microorganisms, are isolated and characterized from high-pressure environments. They grow better at high-hydrostatic pressures than at atmospheric pressures, and only exist at deeper water column environments, particularly in the deep-sea bottoms. Therefore, piezophilic microorganisms are typical deep-sea microorganisms that are well adapted to deep-sea pressure and temperature conditions. These microorganisms have special strategies for surviving in such extreme environments, where gene expression and enzyme activities could be controlled by pressure. Studies on adaptations to high-pressure environments have recently been studied in detail, and the mechanisms involved are being elucidated. In this chapter, the distribution, taxonomy, cultivation and molecular characters of piezophiles are described.

Introduction

It has been suggested that life may have originated in the deep sea some 3.5 to 4 billion years ago, where it was protected from the damaging effects of ultraviolet light. The deep-sea is a particularly high-pressure environment, and hydrostatic pressure could have been a very important stimulus for early forms of life. Scientists have proposed that life might have originated in deep-sea hydrothermal vents (Stetter, 1993) and thus it appears possible that high pressure-adapted mechanisms of gene expression could represent a feature present in early forms of life (Kato and Horikoshi, 1996). It has been reported that the primary chemical reactions involved in the polymerization of organic materials (i.e. amino acids) could have occurred in such high-pressure and high-temperature environments (Imai *et al.*, 1999). Thus, the study of deep-sea microorganisms may not only enhance our understanding of specific adaptations to abyssal and hadal ocean realms, but may also provide valuable insights into the origin and evolution of life on our planet.

In 1949, Zobell and Johnson began investigating the effects of hydrostatic pressure on microbial activities (Zobell and Johnson, 1949). They first used the term 'barophilic' to define organisms whose optimal growth occurred at pressures higher than 0.1 MPa or a requirement for increased pressure for growth. The term 'piezophilic' has been proposed to replace 'barophilic,' as the prefixes 'baro' and 'piezo,' derived from Greek, mean 'weight' and 'pressure,' respectively (Yayanos, 1995). Thus, the word 'piezophilic' is more suitable than 'barophilic' to describe bacteria that grow better at high pressure than at atmospheric pressure. Therefore, I have opted to use the term 'piezophilic' bacteria, meaning high pressure-loving bacteria, in this chapter. The definitions of 'piezophilic' are shown in Fig. 10.1.

Whilst many piezophilic thermophiles have been described, they are not considered here as they are reviewed by Prieur *et al.* (2010). Instead, the distribution and characterization of psychrophilic (cold-loving) piezophilic microorganisms are the main focus of this chapter.

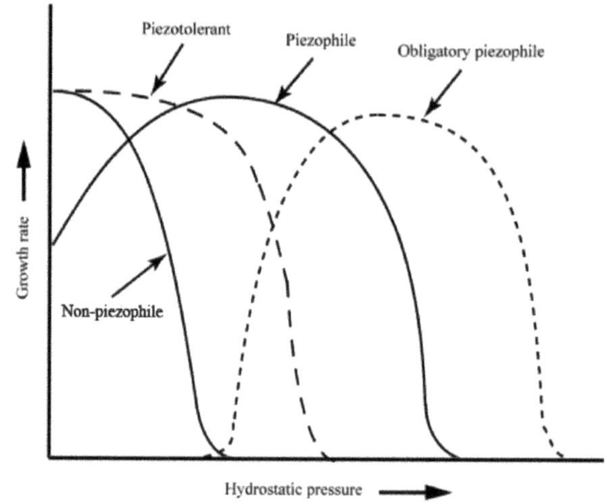

Figure 10.1 Definition of piezophilic growth rate at different hydrostatic pressures.

Distribution and taxonomy of piezophiles

Numerous deep-sea cold-adapted piezophilic microorganisms have been isolated using deep-sea research submersibles and/or several sediment/animal sampling systems. Many of the isolates from the cold sea bottom are also psychrophilic, and we have identified several novel piezophilic species, i.e. *Shewanella violacea, Photobacterium profundum, Moritella japonica, Moritella yayanosii, Psychromonas kaikoi, Psychromonas hadalis* and *Colwellia piezophila*. These five piezophilic genera (*Gammaproteobacteria*) produce significant amounts of unsaturated fatty acids in their cell membrane fractions in order to maintain membrane fluidity in cold and high-pressure environments. Piezophilic microorganisms have been identified in many deep-sea marine environments of the world's oceans. Therefore, these microbes are well distributed on our planet. In this section, the distribution and taxonomy of piezophilic microorganisms and their growth habitats are discussed.

The Mariana Trench, Challenger Deep – the deepest ocean floor

The Mariana Trench, Challenger Deep (11°22'N. 142°25'E), is the deepest ocean floor in the world at a depth of 11,000 m. In 1960, Jacques Piccard and D. Walsh dived to the Challenger Deep aboard the manned submersible *TRIESTE*. They reported seeing several deep-sea animals and fishes there (Piccard and Dietz, 1961). Unfortunately, that dive was not for scientific purposes and no film or photographs were taken, and no samples were collected, so their story could not be confirmed scientifically. In 1996, sediment samples were obtained from this extreme environment by the unmanned submersible KAIKO (Fig. 10.2B; Kyo *et al.*, 1995) in three dives, numbers 21, 22 and 23, at the depth of 11,000 m (Kato *et al.*, 1997b). It seems probable that these were the first sediment samples recovered from the world's deepest point without any microbiological contamination from the water column and surface seawater.

The Mariana Trench is characterized by cold, dark, remoteness and the highest-pressure ocean environment known. Relatively little is known about the natural bacterial communities in the sediment there because of the need to use uncommon instruments to maintain the collected deep-sea organisms at high pressure, and to develop new methods to maintain and cultivate these organisms in the laboratory. Two isolates from the Mariana Trench are *Pseudomonas bathycetes*, and an obligatory piezophilic bacterium strain MT41 (Quigley and Colwell, 1968; Yayanos *et al.*, 1981). Apparently, *P. bathycetes* is not an organism adapted to the deep-sea environment, as its generation time is 33-days under such conditions (2°C, 100 MPa) (Schwarz and Colwell, 1975), and it grows at atmospheric

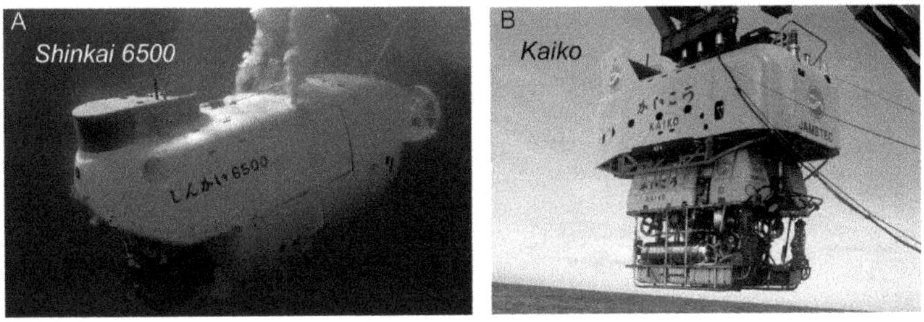

Figure 10.2 Deep-sea research submersibles operated by JAMSTEC. (A) Manned submersible, *Shinkai 6500*, which can dive to a depth of 6500 m. It is the deepest diving manned submersible. (B) The unmanned submersible, *Kaiko*, which means trench in Japanese, can dive to 11,000 m, the depth of the world deepest trench, the Mariana Trench Challenger Deep. *Kaiko* went missing in rough seas during an operation at the Nankai Trough in 2002.

pressure and at temperatures as high as 25°C (Pope *et al.*, 1975). Bacterium strain MT41 was isolated from a deep-sea amphipod (*Hirondellea gigas*) obtained from the Mariana Trench using an insulated trap (Yayanos *et al.*, 1981). It could grow only under pressures greater than 50 MPa, and is closely related to the genus *Colwellia* (Yayanos *et al.*, 1981; DeLong *et al.*, 1997). The obligatory piezophiles, *Moritella yayanosii* DB21MT-5 and *Shewanella benthica* DB21MT-2 were isolated from the sediment of the Mariana Trench (Kato *et al.*, 1998, Nogi and Kato, 1999). In order to understand the piezophily and natural bacterial communities in the sediment of the Mariana Trench, several studies are in progress.

Distribution and diversity of the piezophiles

We have isolated and characterized several piezophilic bacteria from cold deep-sea sediments at depths ranging from 2500 m to 11,000 m using sterilized sediment samplers by means of the submersibles *SHINKAI 6500* and *KAIKO* (Fig. 10.2), systems operated by the Japan Agency for Marine-Earth Science and Technology (JAMSTEC) (Kato, 1999; Kato *et al.*, 1995; 2004).

Most of the isolated strains are not only piezophilic but also psychrophilic, and cannot be cultured at temperatures higher than 20°C. These bacterial strains have been characterized in an effort to understand the interaction between the deep-sea environment and its microbial inhabitants (Yayanos *et al.*, 1979; Kato *et al.*, 1998; Margesin and Nogi, 2004). Thus far, all cold adapted piezophilic bacterial isolates fall into the γ-subgroup of the *Proteobacteria*, according to phylogenetic classifications based on 5S and 16S ribosomal RNA gene sequence information (DeLong *et al.*, 1997; Kato, 1999; Margesin and Nogi, 2004).

Prior to the reports by the JAMSTEC group, two deep-sea piezophilic bacterial species were known: *Shewanella benthica* (Deming *et al.*, 1984; MacDonell and Colwell, 1985) in the genus *Shewanella*, and *Colwellia hadaliensis* (Deming *et al.*, 1988) in the genus *Colwellia*. We have identified several novel piezophilic species within these genera based on the results of chromosomal DNA–DNA hybridization and several other taxonomic properties. Both previously described and novel species of bacteria have been identified among the piezophilic bacterial isolates. Based upon these studies, we have indicated that cultivated psychrophilic and piezophilic deep-sea bacteria could be affiliated with one of five genera within the *Gammaproteobacteria*: *Shewanella*, *Photobacterium*, *Colwellia*, *Moritella*, and *Psychromonas*. Fig. 10.3 shows the phylogenetic relationships between the taxonomically identified piezophilic species and selected other, non-piezophilic, bacteria within the *Gammaproteobacteria*.

The piezophilic *Shewanella* species *S. benthica* and *S. violacea* are also categorized as psychrophilic at atmospheric pressure (Nogi *et al.*, 1998b). *S. gelidimarina* and *S. frigidimarina* isolated from Antarctic ice (Bowman *et al.*, 1997) and *S. hanedai* isolated from the Arctic Ocean (Jensen *et al.*, 1980) are cold-adapted psychrotrophic bacteria

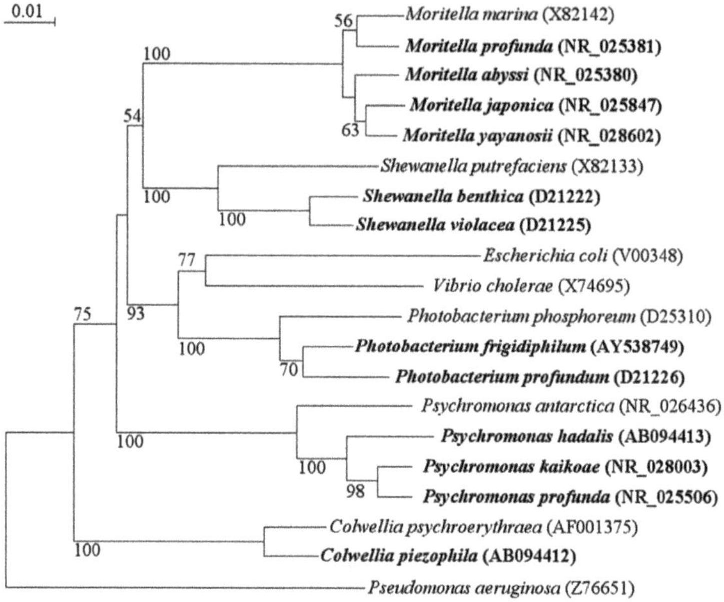

Figure 10.3 Phylogenetic tree showing the relationships between isolated deep-sea piezophilic bacteria (in bold) within the *Gammaproteobacteria* subgroup, determined by comparing 16S rRNA gene sequences using neighbour-joining. Nucleotide sequence accession numbers are in parentheses. The scale represents the average number of nucleotide substitutions per site. Bootstrap values (%) are shown for frequencies above the threshold of 50%.

that grow well at low temperature. A 16S rRNA gene-based phylogenetic tree of these *Shewanella* species within the *Gammaproteobacteria* is shown in Fig. 10.4. On the basis of the properties of *Shewanella* species, our group has proposed that two major branches of the genus *Shewanella* should be recognized taxonomically: *Shewanella* group 1 and group 2 (Fig. 10.4; Kato and Nogi, 2001). The two subgenus branches of *Shewanella* are as follows: *Shewanella* group 1 is characterized as a group of high pressure, cold-adapted species that produce substantial amounts of eicosapentaenoic acid (EPA; a long-chain polyunsaturated fatty acid (PUFA)) and *Shewanella* group 2 is characterized as a group of mostly mesophilic and pressure-sensitive species.

The deep-sea bottom and other cold-temperature environments are probably similar in terms of microbial diversity. Members of *Shewanella* group 1 live in such environments, and most of them show piezophilic or piezotolerant growth properties. In this regard, it is interesting to consider the influence of the ocean circulation, as deep ocean water is derived from polar ice (in the Arctic and/or Antarctic regions) that sinks to the deep-sea bottom (Schmitz, 1995), probably along with microbes. It was reported that *Psychrobacter pacificensis* isolated from seawater of the Japan Trench at a depth of 5000–6000 m was taxonomically similar to the Antarctic isolates *Psychrobacter immobilis, Psychrobacter gracincola,* and *Psychrobacter fridigicola* (Maruyama *et al.*, 2000). The occurrence of *Psychrobacter* in cold seawater deep in the Japan Trench and at the surface of the Antarctic sea suggests that bacterial habitation of the deep-sea and their evolution have been influenced by the global deep ocean circulation linked to the sinking of cooled seawater in polar regions. Thus, it is possible that ocean circulation is one of the major factors influencing microbial diversity on our planet.

Taxonomy of the piezophiles

A listing of cold-adapted piezophilic bacteria isolated from various sources is given in Table 10.1. Cultured deep-sea piezophilic and psychrophilic

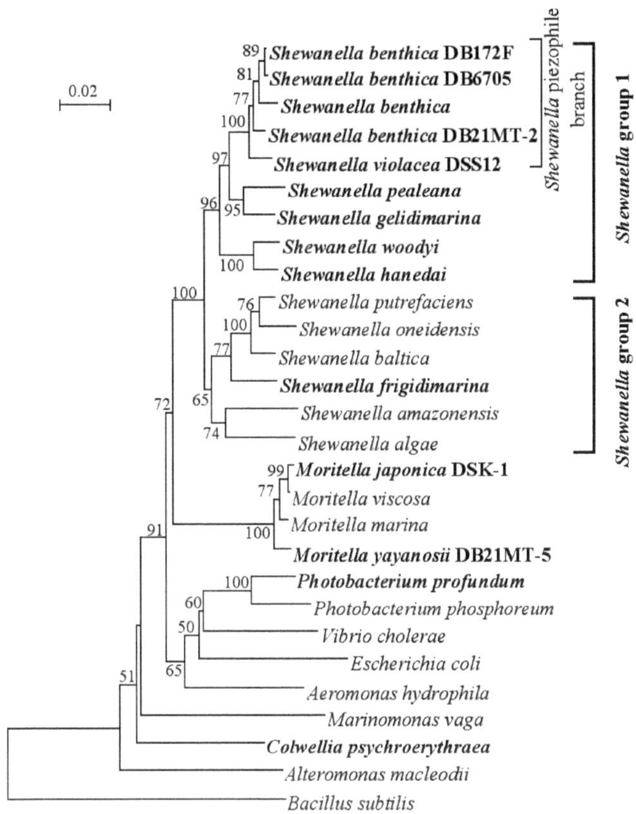

Figure 10.4 Phylogenetic tree showing the relationships of the *Shewanella* species within the *Gammaproteobacteria* subgroup, constructed using 16S rRNA gene sequences with the neighbour-joining method. The scale represents the average number of nucleotide substitutions per site. Bootstrap values (%) were calculated from 1000 trees. Psychrophilic and/or piezophilic bacteria are shown in bold.

bacteria are affiliated with one of five genera within the *Gammaproteobacteria*: *Shewanella, Photobacterium, Colwellia, Moritella,* and *Psychromonas* (Kato *et al.*, 2008a).

The genus *Shewanella*

The type strain of this genus is *Shewanella putrefaciens*, a bacterium formerly known as *Pseudomonas putrefaciens* (Owen *et al.*, 1978; MacDonell and Colwell, 1985). Whilst several novel marine *Shewanella* species have been recently isolated and described, these isolates are not piezophilic species (e.g. *Shewanella piezotolerans*, Wang *et al.*, 2008). *S. benthica* and *S. violacea* are members of the genus *Shewanella* showing piezophilic growth properties (Nogi *et al.*, 1998b). *Shewanella* piezophilic strains, PT-99, DB5501, DB6101, DB6705, and DB6906, DB172F, DB172R, and DB21MT-2 were all identified as members of the same species, *S. benthica* (Nogi *et al.*, 1998b; Kato and Nogi, 2001). *S. violacea* and *S. benthica* both produce eicosapentaenoic acid (EPA). The production of such long-chain polyunsaturated fatty acids (PUFA) is a property shared by many deep-sea bacteria to maintain the cell-membrane fluidity under environmental conditions of extreme cold and high hydrostatic pressure (Fang *et al.*, 2003). Recently, it was reported that EPA in *Shewanella* is not required for bulk bilayer fluidity, but it does play a beneficial role in membrane organization and cell division at low temperature in psychrotrophic *S. livingstonensis* Ac10 (Kawamoto *et al.*, 2009). Therefore, EPA may play an important function in the cell division process during high-pressure growth in piezophilic *Shewanella* (Kawamoto *et al.*, paper in submission).

Table 10.1 List of piezophilic bacteria isolated from various sources

Organisms	Source (depth)	Optimal growth conditions	Investigations	Reference
Genus *Shewanella*				
Shewanella sp. SC2A	E. Pacific Ocean (1957 m)	14 MPa, 20°C	Physiology	Yayanos and Diets (1982)
Shewanella violacea DSS12	Ryukyu Trench (5110 m)	30 MPa, 8°C	Taxonomy	Kato *et al.* (1995a), Nogi *et al.* (1998b), Kato and Nogi (2001)
			Gene expression	Kato *et al.* (1996b, 1997a); Nakasone *et al.* (1998, 2002), Li *et al.* (1998), Yamada *et al.* (2000); Ikegami *et al.* (2000a,b
			Respiratory system	Tamegai *et al.* (1998, 2005)
			Lipid biosynthesis	Fang *et al.* (2003, 2004)
			Genome sequence	Aono *et al.* (2010)
Shewanella benthica WHB46	Weddell Sea (4995 m)	40 MPa, 5°C	Phylogenetics	Liesack *et al.* (1991)
Shewanella benthica F1A	N. Atlantic	41 MPa, 8°C	Phylogenetics	DeLong *et al.* (1997)
Shewanella benthica DB5501	Suruga Bay (2485 m)	50 MPa, 10°C	Taxonomy	Kato *et al.* (1995a), Nogi *et al.* (1998b), Kato and Nogi (2001)
Shewanella benthica DB6101	Ryukyu Trench (5110 m)	50 MPa, 10°C	Taxonomy	Kato *et al.* (1995a), Nogi *et al.* (1998b), Kato and Nogi (2001)
Shewanella benthica DB6705	Japan Trench (6356 m)	50 MPa, 10°C	Taxonomy	Kato *et al.* (1995a), Nogi *et al.* (1998b), Kato and Nogi (2001)
			Gene expression	Kato *et al.* (1995b)
Shewanella benthica DB6906	Japan Trench (6269 m)	50 MPa, 10°C	Taxonomy	Kato *et al.* (1995a), Nogi *et al.* (1998b), Kato and Nogi (2001)
Shewanella benthica PT48	Philippine Trench (6163 m)	62 MPa, 8°C	Phylogenetics	DeLong *et al.* (1997)
Shewanella benthica PT99	Philippine Trench (8600 m)	62 MPa, 8°C	Phylogenetics	DeLong *et al.* (1997)
			Membrane lipids	DeLong and Yayanos (1985, 1986)
Shewanella benthica DB172F	Izu-Bonin Trench (6499 m)	70 MPa, 10°C	Taxonomy	Nogi *et al.* (1998b); Kato *et al.* (1996a)
			Phylogenetics	Kato and Nogi (2001)
			Respiratory	Qureshi *et al.* (1998a,b)
Shewanella benthica DB172R	Izu-Bonin Trench (6499 m)	70 MPa, 10°C	Taxonomy	Nogi *et al.* (1998b); Kato *et al.* (1996a)
			Phylogenetics	Kato and Nogi (2001)
Shewanella benthica DB21MT-2	Mariana Trench (10,898 m)	70 MPa, 10°C	Physiology	Kato *et al.* (1998)
			Taxonomy	Nogi and Kato (1999)
			Lipid biosynthesis	Fang *et al.* (2003)

Genus *Photobacterium*				
Photobacterium profundum DSJ4	Ryukyu Trench (5110 m)	10 MPa, 10°C	Taxonomy	Kato *et al.* (1995a), Nogi *et al.* (1998a)
Photobacterium profundum SS9	Sulu Sea (2551 m)	28 MPa, 15°C	Gene expression	Bartlett *et al.* (1989), Chi and Bartlett (1995), Welch and Bartlett (1996), Welch and Bartlett (1998)
			Membrane lipids	Allen *et al.* (1999), Allen and Bartlett (2000)
			Genome sequence	Vezzi *et al.* (2005)
Photobacterium frigidiphilum SL13	Edison Seamount (1450 m)	10 MPa, 14°C	Taxonomy	Seo *et al.* (2005)
Genus *Colwellia*				
Colwellia piezophila Y223G	Japan Trench (6,278 m)	60 MPa, 10°C	Taxonomy	Nogi *et al.* (2004)
Colwellia hadaliensis	Puerto Rico Trench	93 MPa, 10°C	Physiology	Deming *et al.* (1988)
Colwellia sp. MT41	Mariana Trench	103 MPa, 8°C	Physiology	DeLong *et al.* (1997); Yayanos *et al.* (1981)
			Membrane lipids	DeLong and Yayanos (1986)
Genus *Moritella*				
Moritella profunda	Off W. Africa coast (2815 m)	20–24 MPa, 6°C	Taxonomy	Xu *et al.* (2003a)
Moritella abyssi	Off W. Africa coast (2815 m)	29–30 MPa, 10°C	Taxonomy	Xu *et al.* (2003a)
Moritella PE36	E. Pacific Ocean (3584 m)	41 MPa, 15°C	Phylogenetics	Liesack *et al.* (1991)
Moritella japonica DSK1	Japan Trench (6353 m)	50 MPa, 15°C	Taxonomy	Kato *et al.* (1995a), Nogi *et al.* (1998a)
			Membrane lipids	Fang *et al.* (2003)
			Carbon isotope fractionation	Fang *et al.* (2006)
Moritella yayanosi DB21MT-5	Mariana Trench (10,898 m)	70 MPa, 10°C	Physiology	Kato *et al.* (1998)
			Taxonomy	Nogi and Kato (1999)
			Lipid biosynthesis	Fang *et al.* (2003, 2004)
Genus *Psychromonas*				
Psychromonas profunda	Atlantic Ocean (2770 m)	25 MPa, 10°C	Taxonomy	Xu *et al.* (2003b)
Psychromonas kaikoae	Japan Trench (7434 m)	50 MPa, 10°C	Taxonomy	Nogi *et al.* (2002)
Psychromonas hadalis	Japan Trench (7542 m)	60 MPa, 6°C	Taxonomy	Nogi *et al.* (2007)
Psychromonas strain CNPT3	N. Pacific (5700 m)	52 MPa, 8°C	Physiology	Yayanos *et al.* (1979), Yayanos (1986)
			Membrane lipids	DeLong and Yayanos (1985, 1986)
			Phylogenetics	DeLong *et al.* (1997)

S. violacea DSS12 has been studied extensively, particularly with respect to its molecular mechanisms of adaptation to high pressure (Nakasone *et al.*, 1998, 2002; Kato *et al.*, 2000). This strain is moderately piezophilic, with a fairly constant doubling time at pressures between 0.1 MPa and 70 MPa, whereas the doubling times of most piezophilic *S. benthica* strains change substantially with increasing pressure. As there are few differences in the growth characteristics of strain DSS12 under different pressure conditions, this strain serves as a very convenient model deep-sea piezophilic bacterium for studies on the mechanisms of adaptation to high-pressure environments (e.g. genome analysis; Aono *et al.*, 2010).

The genus *Photobacterium*

The genus *Photobacterium* was one of the earliest known bacterial taxa and was first proposed by Beijerinck in 1889 (Beijerinck, 1889). Phylogenetic analysis based on 16S rRNA gene sequences has shown that the genus *Photobacterium* falls within the *Gammaproteobacteria* and, in particular, is closely related to the genus *Vibrio* (Nogi *et al.*, 1998c). *Photobacterium profundum*, a novel species, was identified through studies of the moderately piezophilic strains DSJ4 and SS9 (Nogi *et al.*, 1998c). *P. profundum* strain SS9 has been extensively studied with regard to the molecular mechanisms of pressure regulation (Bartlett, 1999) and subsequently genome sequencing and expression analysis (Vezzi *et al.*, 2005). *P. frigidiphilum* was reported to be slightly piezophilic: its optimal pressure for growth is 10 MPa (Seo *et al.*, 2005). Thus, *P. profundum* and *P. frigidiphilum* are the only species within the genus *Photobacterium* known to display piezophily and the only two known to produce the long-chain PUFA, EPA. No other known species of *Photobacterium* produces EPA (Nogi *et al.*, 1998c).

The genus *Colwellia*

Species of the genus *Colwellia* are defined as facultative anaerobic and psychrophilic bacteria (Deming *et al.*, 1988). In the genus *Colwellia*, the only deep-sea piezophilic species reported prior to JAMSTEC reports was *C. hadaliensis* BNL-1 (Deming *et al.*, 1988), although no public culture collections maintain this species and/or its 16S rRNA gene sequence information. Bowman *et al.* (1998) reported that *Colwellia* species produce the long-chain PUFA, docosahexaenoic acid (DHA). We have isolated the obligately piezophilic strain Y223G from sediment at the bottom of a deep-sea fissure in the Japan Trench, which was identified as *C. piezophila* (Nogi *et al.*, 2004). This strain did not contain EPA or DHA fatty acids in the membrane, whereas high levels of unsaturated fatty acids (16:1 fatty acids) were produced. This observation suggested that the possession of 'long-chain' PUFA should not be a requirement for classification as a piezophilic bacterium; however, the amount production of unsaturated fatty acids (UFA) could be a common property of piezophiles.

The genus *Moritella*

The type strain of the genus *Moritella,* once known as *Vibrio marinus* (Colwell and Morita, 1964), is one of the most common psychrophilic organisms isolated from marine environments. However, *V. marinus* has now been reclassified as *Moritella marina* gen. nov. comb. nov. (Urakawa *et al.*, 1998). *M. marina* is closely related to the genus *Shewanella,* based on 16S rRNA gene sequence data, and is not a piezophilic bacterium. Strain DSK1, a moderately piezophilic bacterium isolated from the Japan Trench, was identified as *Moritella japonica* (Nogi *et al.*, 1998a). This microbe was the first piezophilic species identified in the genus *Moritella.* Production of the long-chain PUFA, DHA, is a characteristic property of the genus *Moritella.* An extremely piezophilic bacterial strain, DB21MT-5, isolated from a depth of 10,898 m in the Mariana Trench Challenger Deep, was also identified as a *Moritella* species and designated *M. yayanosii* (Nogi and Kato, 1999). *M. yayanosii* strain DB21MT-5 is an obligate peizophile: its optimal pressure for the growth is 80 MPa, it is unable to grow at pressures of less than 50 MPa, but can grow well at pressures as high as 100 MPa (Kato *et al.*, 1998). The fatty acid composition of piezophilic strains changes as a function of pressure and, in general, greater amounts of PUFAs are synthesized at higher growth pressures. Approximately 70% of the membrane lipids in *M. yayanosii* are unsaturated fatty acids, which is a finding consistent with its adaptation to very high pressures (Nogi and Kato, 1999; Fang *et al.*, 2000). Two other species of the genus *Moritella*, *M. abyssi* and *M. profunda*, were isolated from a depth of 2815 m off the West African coast (Xu *et al.*, 2003a). They are moderately piezophilic and display growth profiles similar to *M. japonica*.

The genus *Psychromonas*

The genus *Psychromonas*, which contains psychrophilic bacteria, is closely related to the genera *Shewanella*, *Photobacterium*, and *Moritella* on the basis of 16S rRNA gene sequence data (Fig. 10.3). The type species of the genus *Psychromonas*, *Ps. antarctica*, was isolated as an aerotolerant anaerobic bacterium from a high-salinity pond on the McMurdo ice-shelf in Antarctica (Mountfort *et al.*, 1998). This strain did not display piezophilic properties. *Ps. kaikoae* and *Ps. hadalis*, isolated from sediment collected from the deepest seep environment with chemosynthesis-based animal communities within the Japan Trench (at a depth of 7434 m and 7542 m), respectively, are novel obligatory piezophilic bacteria (Nogi *et al.*, 2002, 2007). The optimal temperature and pressure for growth of *Ps. kaikoae* are 10°C and 50 MPa, respectively, and both the PUFAs, EPA and DHA, are present in the membrane. The optima for *Ps. hadalis* are at 6°C and 60 MPa, but only DHA is produced. *Ps. antarctica* does not contain either EPA or DHA in its membrane layer, consistent with its non-piezophilic status. Another microorganism, strain CNPT-3 was found to belong to an unidentified genus of piezophiles (DeLong *et al.*, 1997). This strain proved to be closely related to *Ps. kaikoae*. *Psychromonas*, the fifth genus within the *Gammaproteobacteria* reported to contain piezophilic species, also contains *Ps. profunda*, a moderately piezophilic bacterium isolated from deep Atlantic sediments at a depth of 2770 m (Xu *et al.*, 2003b). This species is similar to the piezosensitive strain *Ps. marina*, which also produces small amounts of DHA. The only member of the genus *Psychromonas* that produces both EPA and DHA is *Ps. kaikoae*.

PUFA functions in piezophiles

The production of PUFAs varies in different piezophilic psychrophiles: *Shewanella* and *Photobacterium* strains produce EPA (Nogi *et al.*, 1998b,c), *Moritella* strains produce DHA (Nogi *et al.*, 1998a; Nogi and Kato, 1999), and *Ps. kaikoae* produces both EPA and DHA (Nogi *et al.*, 2002), whilst *C. piezophila* does not produce such PUFAs (Nogi *et al.*, 2004). The fatty acid composition of piezophilic strains changes as a function of pressure, and in general greater amounts of PUFAs are synthesized under higher-pressure growth conditions (DeLong and Yayanos, 1985; 1986). Psychrophilic and piezophilic bacteria were believed to produce one of the long-chain PUFAs, either EPA or DHA, and particularly *Shewanella* sp. might require EPA for growth at low-temperature and high-pressure conditions (Kawamoto *et al.*, 2009, 2011). But PUFAs do not appear to be obligatory. For example, Allen *et al.* (1999) reported that monounsaturated fatty acids, but not PUFAs, are required for the growth of the piezophilic bacterium *P. profundum* SS9, based on the analysis of pressure-sensitive mutants. In their mutant experiment, 18:1 fatty acids proved to be necessary for growth under low-temperature and/or high-pressure conditions. In the case of *C. piezophila*, this species had no 18:1 fatty acids but contained a large amount of 16:1 fatty acids in its cell membrane layer. All piezophilic and psychrophilic bacteria analysed so far have 16:1 fatty acids; thus, this fatty acid would appear to be one of the important components required for high-pressure growth. The fatty acid composition of piezophilic strains are distinct, being genus-dependent, and high amounts of unsaturated fatty acids (49~71%) are commonly present in their membrane layer.

Methods for the isolation of piezophiles

Piezophiles living at the deep-ocean floor have adapted to a high-pressure environment and are able to grow better under high hydrostatic pressure conditions than at atmospheric pressure. In the laboratory, piezophiles can be isolated and cultivated under high-pressure conditions, and therefore special high-pressure equipment, such as pressure vessels, hydrostatic pumps, etc., are necessary. When sampling for viable piezophiles from deep-sea sediments or seawater to obtain living piezophiles, a pressure-retaining sampler is requires to maintain the environmental pressure and temperature, because some piezophiles are sensitive to drastic changes in these parameters. JAMSTEC has been developing a 'deep-sea baro-piezophile and thermophile isolation and cultivation system,' referred to as the 'DEEP-BATH' system, for the handling and study of piezophiles (Kyo *et al.*, 1991). In this section, the traditional methods for the handling of piezophilic microorganisms are described, along with novel systems.

Procedures for isolation and cultivation

In 1957, Zobell and Morita were first considering how to handle piezophilic microorganisms living under the deep-sea floor. They developed a titanium pressure vessel that could produce pressure of up to 100 MPa for the handling and cultivation of such microbes (Zobell and Morita, 1957; see Fig. 10.5A). They attempted several times to isolate piezophiles from deep-sea samples, but they could isolate only piezotolerant deep-sea microbes, which showed better growth under atmospheric pressure condition but also grew at high pressure. One reason for their inability to isolate piezophiles was that such extremophiles can be sensitive to drastic changes in pressure and temperature, and it may be very difficult to maintain the microbes at atmospheric pressure. Nevertheless, Zobell and Morita made a great contribution to high-pressure microbiology, and many subsequent researchers began to study biological physiology under pressurised conditions (for a review, see Marquis, 1976).

In 1979, Yayanos and co-workers used pressure-retaining traps to successfully isolate piezophilic microorganisms from amphipods recovered from a depth of 5782 m in the Philippine Trench (Yayanos *et al.*, 1979). This was the first report on the isolation of a piezophilic microbe. They modified their pressure vessels for easier handling, using a pin system to retain the pressure instead of the original screw system (Fig. 10.5B). This modification solved the problem of pressure leakage due to wearing and metal fatigue of the screw, and the new system required little power for operation. Using such pressure vessels, the author's JAMSTEC group has isolated numerous piezophiles during the past 20 years (Kato *et al.*, 2000, 2004, 2008a).

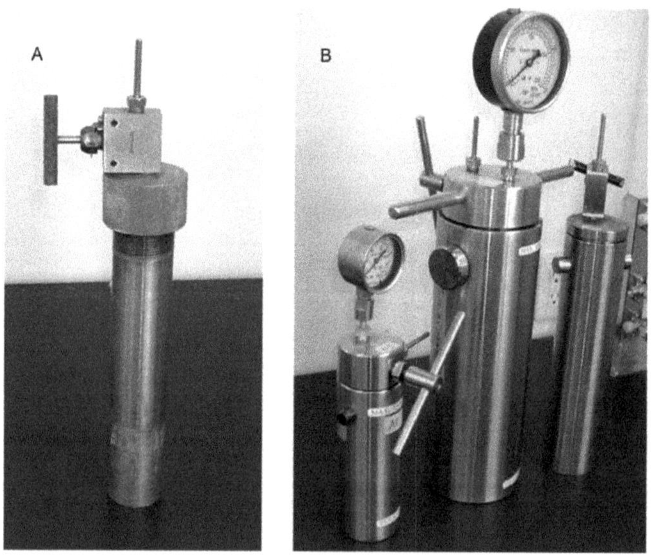

Figure 10.5 Pressure vessels used in the JAMSTEC project. (A) Zobell-Morita-type pressure vessel, kindly provided with permission from Professor Richard Y. Morita. (B) Yayanos-type pressure vessels produced by Shin-corporation Inc. (Kyoto, Japan). Yayanos modified the pressure-retaining system using a pin for easier handling.

The procedures used for sampling deep-sea sediment are described below.

Sampling of deep-sea sediments

For sampling from the deep-sea bottom, several submersible systems are available to JAMSTEC: the manned submersible *Shinkai 6500* (maximum depth 6500 m, accommodating one scientist and two pilots during operation), and the unmanned submersible *Kaiko* (maximum depth 11,000 m, containing launcher and vehicle systems) (Fig. 10.2). The sterilized sediment (SS) sampler (Fig. 10.6A) and/or core sampler is controlled by submersibles' remote manipulators to obtain the desired samples (Fig. 10.6B). Once collected, the samples are carried to the mother ship laboratories, where they are placed under environmental pressure and temperature conditions as soon as possible.

Mixed cultivation of microorganisms at environmental pressures and temperatures

Deep-sea sediment samples obtained using the submersibles are mixed with culture medium (e.g. Marine broth medium 2216, Difco Co.), and high-pressure cultivation initiated using pressure vessels maintained under environmental pressure and temperature conditions. After several days of cultivation, the mixed cultivations are transferred to fresh media to continue culturing under different conditions, if necessary. This mixed cultivation procedure can eliminate any atmospheric adapted microorganisms derived from pelagic water, and consequently only high pressure adapted microorganisms are enriched. Agar plate isolation is suitable to isolate moderate piezophiles and piezotolerant microbes, and dilution-to-extinction is good for thermophilic piezophiles.

Many obligate piezophiles have been isolated using the so-called pressure bag method. In this approach, a mixed cultivation is diluted and combined with low melting-point agar (0.7% agar) medium in sterilized plastic bags (10 cm in diameter, approximately 20 ml of agar medium; see Fig. 10.7A) at lower than 20°C. Then the bags are sealed and transferred to ice-cold water for solidification of the medium. After solidification, the bags are placed into the pressure vessels, pressurized to environmental pressure, and kept at environmental temperature for several days. After pressure cultivation, isolated microbial colonies can be identified in the bag, as shown in Fig. 10.7B. Each colony is placed into a separate disposable syringe under a stereoscopic microscope, and cultivation under both high pressure and atmospheric pressure conditions are initiated. If the isolates grow better under elevated pressure conditions than at atmospheric pressure, they are identified as piezophiles.

Characterization of isolates

To determine the optimal growth conditions of isolates, multiple (20–30) pressure vessels

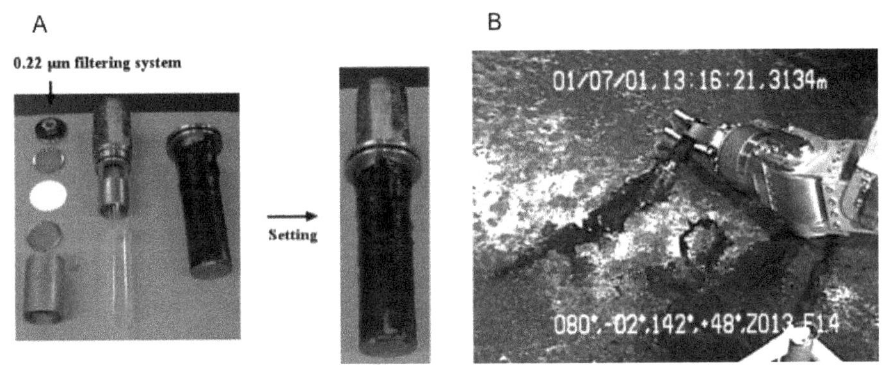

Figure 10.6 (A) Sterilized sediment (SS) sampler. After assembly (setting), the SS sampler is sterilized and taken to the deep-sea bottom. The 0.22 m filtering system protects against surface water contamination. (B) Sediment sampling *in situ* from the deep-sea floor using a SS sampler with the *Shinkai 6500* manipulator. This sampling was performed at a bacterial mat site in the northeast Sea of Japan at a depth of 3134 m (1 July 2001; dive 6K#624; observer, C. Kato).

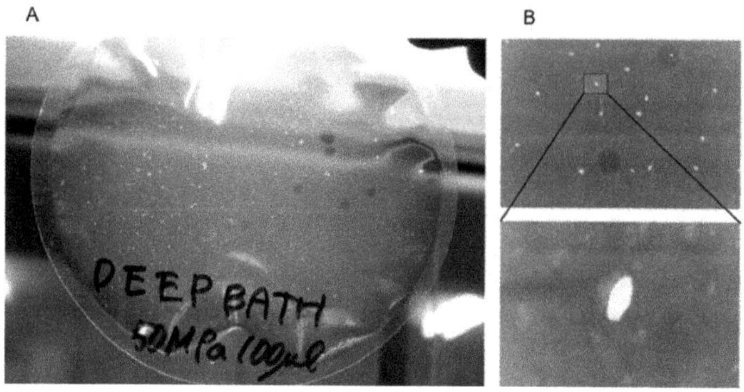

Figure 10.7 Sterilized pressure bag methods. (A) Plastic bag after high-pressure cultivation. Several microbial colonies that formed at a pressure of 50 MPa are visible. (B) Close-up of piezophile colonies in the bag. A single colony can be suctioned into a sterilized disposable syringe for subsequent studies.

are used to avoid the effects of repeated pressurization and decompression on the cells. If the pressure of a vessel is released, the same vessel is not re-pressurized to ensure that the results are reproducible. In the case of cultivation under aerobic conditions, oxygenated fluorinert (FC-72, Sumitomo-3M Co., Japan) is added to the cultures (25% of total volume; Kato *et al.*, 1995a). To study the physiological and molecular properties of the isolates, 200-ml scale cultivations are performed using a large pressure vessel.

Using the above methods, we have performed several studies on the taxonomy of piezophiles, gene expression controlled by pressure conditions, and genomic analysis of piezophilic microorganisms (Nakasone *et al.*, 1998, 2002; Kato *et al.*, 2000, 2004, 2008a–c). However, the 200-ml culture scale is sometimes insufficient to yield the desired amounts of protein, DNA, lipids, etc. because the maximum cell density obtained in piezophilic cultures is not very high (ca. 10^8 cells/ml). Therefore, we developed DEEPBATH, a large high-pressure (HP) cultivation system.

The DEEPBATH system

The DEEPBATH system consists of four separate devices: (1) a pressure-retaining sampling device; (2) a pressurized dilution device; (3) an isolation device; and (4) a cultivation device (Kyo *et al.*, 1991). The system is controlled by central regulation systems, and the pressure and temperature ranges of the devices are, respectively, from 0.1 to 65 MPa and 0 to 150°C. The scale of the cultivation devices (2 sets) is 1.5 l each, allowing cultures of up to 3 l to be used. The construction of the system and the sample stream are shown in Fig. 10.8. The principle of the isolation of piezophiles is the dilution-to-extinction procedure, although in some cases microbes may be attached so firmly to the sediment that it is difficult to obtain a pure culture using this approach. Thus we have combined the traditional procedure explained earlier and the DEEPBATH system protocol for isolation studies.

High pressure cultivations using the DEEPBATH system

The DEEPBATH system is an excellent tool for studying piezophilic cultivations, not only for scale-up experiments, but also continuous cultivation. The light scattering system of DEEPBATH is able to detect a low level of growth (ca. 10^7 cells/ml) in the culture under high-pressure conditions without the need for decompression. Three HP cultivation experiments using the DEEPBATH system are described below.

Changes in the microbial community of Japan Trench sediment during cultivation without decompression

A deep-sea sediment sample (about 5 ml) was obtained from the Mannequin Valley in the Japan Trench (6292 m depth) using the pressure-retaining sampling device of the *Shinkai 6500*

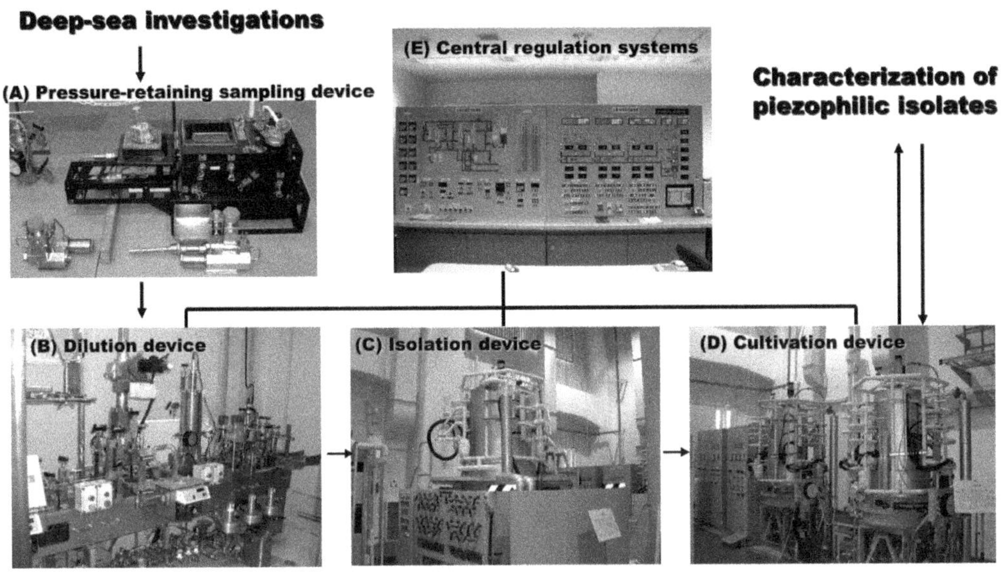

Figure 10.8 The DEEPBATH system. The four devices that make up the system (A–D) and the sample stream are shown. The monitoring and control console (E) controls the system.

submersible (dive #373, June 1997, observer C. Kato) under conditions of 65 MPa and 2°C. The sample was diluted almost 100-fold with sterilized seawater with the dilution device, at the same pressure and temperature, and 10 ml of this dilution inoculated into the cultivation device (containing 1.5 l of Marine broth 2216 medium) with no change in pressure and temperature. Subcultivation was repeated five times at 65 MPa and 10°C without decompression, and each time cultivation was continued for approximately 4 days until the early stationary phase had been reached, as confirmed with a laser beam microbe density measurement system (Kyo *et al.*, 1991). Each cultivated bacterial mixture was centrifuged, and DNA was purified from the mixture to study microbial diversity. The changes in the microbial community are shown in Table 10.2 (Yanagibayashi *et al.*, 1999). Initially, 15 bacterial recombinants with different 16S rRNA gene sequences were identified in the samples. However, only two groups of the bacterial genera *Shewanella* and *Moritella* were identified after cultivation at 65 MPa. Interestingly, the genus *Moritella* became abundant during consecutive cultivation without decompression, indicating that it may be better adapted than piezophilic *Shewanella* species to growth under limiting oxygen conditions at high pressure.

High pressure cultivation of deep-sea psychrophilic piezophiles

Most of isolated deep-sea strains may not only be piezophilic or piezotolerant, but also psychrophiles unable to be grown at temperatures above 20°C. Typical growth profiles of such bacteria as a function of several pressures and temperatures are shown in Fig. 10.9A. A piezophilic bacterium, *Shewanella benthica* DB6705, was not able to grow at atmospheric pressure (0.1 MPa) at temperatures above 10°C. However, this strain grew better at higher than lower temperatures when incubated at high pressure (>50 MPa). Growth of the moderately piezophilic bacterium, *Moritella japonica* DSK1, was also better at a higher temperature (15°C) compared with a lower one (10°C) at pressures above 50 MPa. These results illustrate the generalization that deep-sea bacteria show their strongest piezophilic responses near their upper temperature limits for growth (typically ~15°C) (Kato *et al.*, 2001). The effects

Table 10.2 Number of 16S rRNA gene recombinants of different groups identified in the first to fifth subcultures of Japan Trench samples consecutively cultivated at 65 MPa. The genera *Shewanella* and *Moritella* are typical piezophilic groups found in the deep-sea environment

Bacterial group	First	Second	Third	Fourth	Fifth	Total recombinants
Genus *Shewanella*	63%	33%	10%	0%	0%	9
Genus *Moritella*	37%	67%	90%	100%	100%	37
Total recombinants	8	9	10	10	9	46

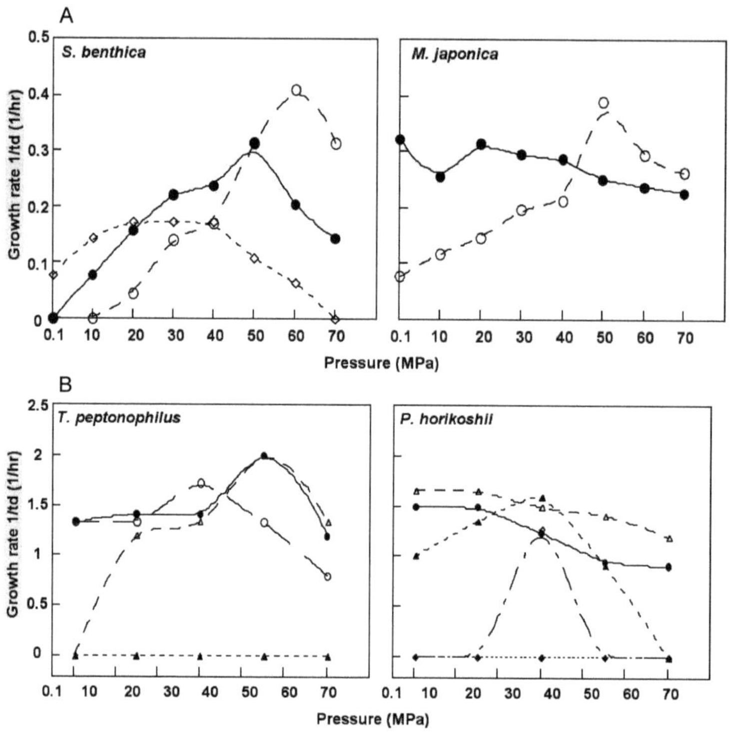

Figure 10.9 Growth profiles of psychrophilic piezophiles under different pressure and temperature conditions, determined using the DEEPBATH system. The growth rate is shown as $1/t_d$, where t_d represents the doubling time (in hours). (A) Psychro-piezophilic bacterium *Shewanella benthica* strain DB6705 and psychro-piezotolerant bacterium *Moritella japonica* strain DSK1. Growth at 15°C (open circles), 10°C (filled circles) and 4°C (diamonds) is shown. (B) Thermo-piezophilic archaea *Thermococcus peptonophilus* and *Pyrococcus horikoshii*. Growth at 85°C (open circles), 90°C (filled circles), 95°C (open triangles), 100°C (filled triangles), 103°C (open diamonds) and 105°C (filled diamonds).

of pressure and temperature on cell growth of isolated deep-sea piezophilic strains are similar in that they all become more piezophilic at higher temperatures (Yayanos, 1986). These studies indicate that all piezophilic isolates are obligately so above the temperature that permits growth at atmospheric pressure, implying that the upper temperature limit for growth can be extended by high pressure. Likewise, piezophilic bacteria reproduce more rapidly at a lower temperature (such as 2°C) when the pressure is less than that found at their capture depth. It also appears to be true as a general rule that the pressure of maximal rate of reproduction at 2°C may reflect the true habitat depth of an isolate (Yayanos *et al.*, 1982). The reproduction rate of piezophiles at close to the pressure of their capture depth increased with increasing temperature over an interval of 6–10°C. Why do piezophilic cells grow best at temperatures that might never be encountered in the cold deep-sea? There are three possible explanations. One is that this behaviour is simply a consequence of adaptation to high pressure and low temperature. Secondly, the response may have been inherited by the bacteria from ancestors who lived in a warmer environment (Deming *et al.*, 1984). A third explanation is that bacteria of the cold deep-sea do periodically encounter warmer temperatures.

High pressure cultivation of deep-sea hyperthermophilic microorganisms

The DEEPBATH system can be used at temperatures of up to 150°C, allowing high-pressure cultivation studies of hyperthermophilic microorganisms, which can grow at around 100°C. Erauso *et al.* (1993) reported the cultivation of the hyperthermophilic archaeon *Pyrococcus abyssi*, isolated from a deep-sea hydrothermal vent, under elevated pressure conditions. They indicated that there were problems associated with the effects of repeated decompression and repressurization processes, which affected the quality of the growth results. DEEPBATH is currently the only system available that permits sampling without any change in pressure and temperature. Using this system, we have performed high-pressure and high-temperature (HP-HT) cultivation studies of two hyperthermophilic archaea isolated from submarine hydrothermal vents, *Thermococcus peptonophilus* (isolated from the Izu-Bonin Trough at a depth of 1400 m; Gonzalez *et al.*, 1995) and *Pyrococcus horikoshii* (isolated from the Okinawa Trough at a depth of 1300 m; Gonzalez *et al.*, 1998). The optimal growth temperatures of *T. peptonophilus* and *P. horikoshii* are 85°C and 95°C, respectively. The results of HP-HT growth studies are shown in Fig. 10.9B (Canganella *et al.*, 1997; Kato, 1997b). It was interesting that the temperature for growth of both hyperthermophilic archaea could be increased under higher-pressure conditions, and their growth profiles appeared piezophilic. In particular, the growth profile of *P. horikoshii* at 103°C was the same as that of an obligatory piezophilic microorganism. Such piezophilic growth profiles in deep-sea hyperthermophiles are common, as indicated by Deming and Baross (1993), who suggested that these microbes might originate from deep subsurface environments. The DEEPBATH system should also be a useful tool in the field of subsurface microbiology.

Other equipment for high-pressure studies

To study the morphology and molecular biology of piezophilic microorganisms under high-pressure conditions, several pieces of equipment were developed. A high-pressure microscopic system (Fig. 10.10A) is a powerful tool to observe morphological changes and swimming habits under pressurized conditions (Miwa *et al.*, 2002). To analyse protein–protein interactions and their activities, a high-pressure fluorescent spectrophotometer (Fig. 10.10B), high-pressure spectrophotometer (Fig. 10.10C) and high-pressure electrophoresis apparatus (Fig. 10.10D) were constructed. Using these systems, several features of piezophilic proteins have been elucidated (Abe and Horikoshi, 1998; Kawano *et al.*, 2004). Akasaka (2006) developed high-pressure nuclear magnetic resonance systems (Fig. 10.10E) to analyse protein dynamics and structural changes. As a result of the discoveries made using the new techniques listed above, the field of high-pressure bioscience and biotechnology has become one of the most important and promising fields of bioscience.

High pressure and piezophiles

The molecular mechanisms of regulation in microorganisms isolated from the deep-sea have been investigated to determine how they adapt to high hydrostatic pressures. As a simple model to examine these mechanisms, the *Escherichia coli* system has been studied to establish the relationship between gene expression and pressure. For example, filament formation in *E. coli* has been observed when it is grown at increased hydrostatic pressure (see later in this chapter; Zobel and Cobet, 1963; Tamura *et al.*, 1992). Despite

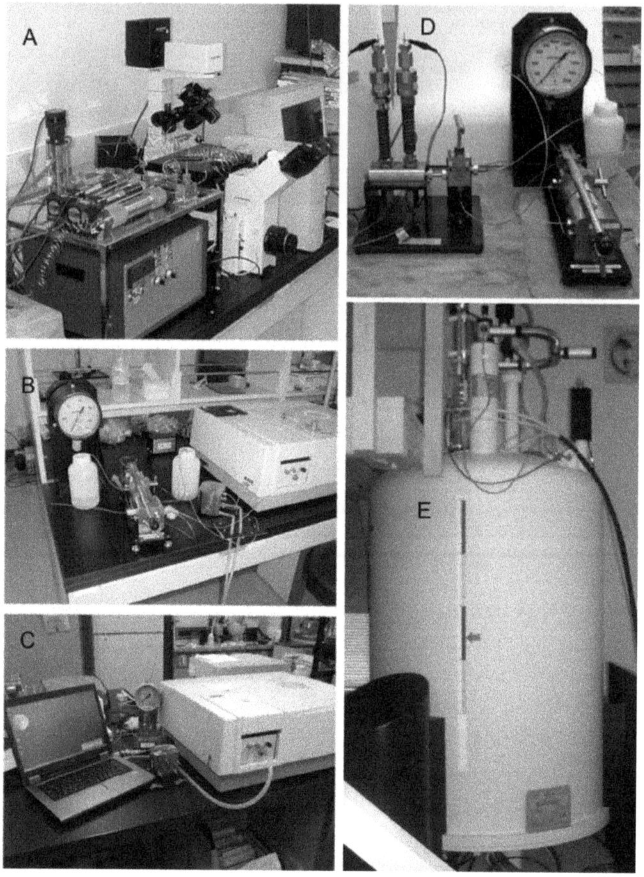

Figure 10.10 Equipment modified to work at elevated pressures. (A) high pressure (HP) microscopic system (HPMS); (B) high-pressure florescent spectrophotometer (HPFSP); (C) high-pressure spectrophotometer (HPSP); (D) high-pressure electrophoresis apparatus (HPEA); and (E) high-pressure NMR systems (HP-NMR).

Table 10.3 Effect of high pressure on gene expression in *Escherichia coli*

Gene (plasmid)	Gene regulation by elevated pressure	References
lac (pUC13)	+++ (30 MPa)	Kato *et al.* (1994)
		Sato *et al.* (1995)
tac (pKK223–3)	+++ (50 MPa)	Kato *et al.* (1994)
		Sato *et al.* (1995)
tet, antitet, amp (pBR322)	N	Kato *et al.* (1994)
cat (pACYC184)	N	Kato *et al.* (1994)
*omp*F, *omp*C	– (30 MPa)	Nakashima *et al.* (1995)
envZ, *omp*R	N	Nakashima *et al.* (1995)
*mal*B operon	– (30 MPa)	Sato *et al.* (1996)
asd	N	Kato *et al.* (1997c)
*lac*I	N	Kato (1997a)

+, high pressure-inducible gene expression; N, pressure-independent gene expression; –, high pressure-repressible gene expression.

the fact that *E. coli* is a piezo-sensitive species, several gene regulation mechanisms controlled by high-pressure conditions have been identified (Table 10.3). These observations suggest that the phenomenon of pressure regulation could occur in any microorganism (Abe and Kato, 1999).

Deep-sea piezophilic species are adapted to, and grow well under, high-pressure conditions. One such species, the psychrophilic, moderately piezophilic bacterium *S. violacea* DSS12 grows optimally at 30 MPa and 8°C, but also grows well at pressures ranging from atmospheric (0.1 MPa) to 70 MPa (Kato *et al.*, 1995a, 2008c; Nogi *et al.*, 1998b). We have examined this strain to elucidate the molecular basis for gene regulation at different pressures, as it is useful as a model prokaryote for comparing the various features of bacterial physiology under pressure conditions. Proteins from such deep-sea adapted piezophiles could be, in general, active under high pressures (Kato *et al.*, 2008b), whilst atmospheric pressure-adapted proteins can be inactive under such conditions. In the case of respiratory proteins, the cell divisional protein FtsZ, the RNA polymerase subunit, dihydrofolate reductase (DHFR), and isopropylmalate dehydrogenase (IPMDH), piezophilic proteins were uniquely adapted to (and some were much more stable and active under) high-pressure environments. The pressure-regulated mechanisms of gene expression in the strain DSS12 and the features of piezophilic proteins are described below.

Pressure-regulated gene expression systems

A pressure-regulated operon from *S. violacea* DSS12 whose promoter was activated by growth under high pressure was cloned and characterized. This operon, which has five transcription initiation sites, is controlled at the transcriptional level by elevated pressure (Nakasone *et al.*, 1998, 1999). Moreover, transcriptional analysis showed that expression of the genes in the pressure-regulated operon is positively controlled at the transcriptional level by elevated pressure and further suggested that most transcripts from the operon at atmospheric pressure coincided with expression. The consensus sequence for the RNA polymerase sigma factor, σ^{54}, was found upstream from this operon, and the *S. violacea* σ^{54} was shown to bind to this region (Nakasone *et al.*, 1999). The σ^{54}-containing RNA polymerase has been shown to be responsible for the transcription of several genes, for example, nitrogen metabolic genes like the *glnA* operon (Merric and Edwards, 1995). Glutamine synthetase is one of the enzymes involved in nitrogen metabolism. Expression of the *glnA* operon is also controlled by elevated pressure conditions at the transcriptional level in *S. violacea*, particularly by factor σ^{54} (Ikegami *et al.*, 2000b). These results suggest that σ^{54} might play an important role in pressure-regulated transcription in piezophilic bacteria, although the expression of σ^{54} itself is not pressure regulated (Ikegami *et al.*, 2000a; Nakasone *et al.*, 2002).

Our approach towards understanding the molecular basis for gene expression under defined conditions is via detailed characterization of the components of the transcriptional machinery and the accessory factors involved. In piezophilic *S. violacea*, this regulation is mediated by one of the sigma factors, σ^{54}, and a two-component regulatory system composed of the bacterial signal-transducing protein NtrB and the bacterial enhancer-binding protein NtrC. In other bacteria transcription from the σ^{54}-dependent promoter, such as in the case of *glnAp2*, is regulated by enhancer-binding protein NtrC (Ninfa *et al.*, 1987). The NtrC-binding sites are essential for the regulation of transcription by the σ^{54}-containing RNA polymerase. Two NtrC-binding consensus sites were also found in the promoter region of *glnA* in *S. violacea*. From elecrophoretic mobility shift assay studies, we confirmed that the *S. violacea* NtrC protein specifically recognizes the element containing the NtrC consensus sequence on the *S. violacea glnA* operon (Ikegami *et al.*, 2000b,c).

To reconstitute the two-component regulatory system and characterize the autophosphorylation of NtrB and transphosphorylation of phosphorylated NtrB to NtrC *in vitro*, we purified both of the proteins from the *S. violacea ntrB* and *ntrC* genes using recombinant DNA techniques. The results of the phosphorylation experiments at several temperatures indicated that autophosphorylation of NtrB occurred only at low temperatures (0–10°C), whereas no activity was detected at 37°C. Autophosphorylation activity of NtrB at such low temperatures was not detected in *E. coli*. Furthermore, we have detected transcriptional activity at low temperatures in *S. violacea*. Therefore, this piezophilic bacterium adapts to the psychrosphere (low-temperature environment) and may have evolved a low-temperature-adapted system in the deep-sea environment. The phosphorylated relay between NtrB and NtrC of the two-component system in *S. violacea* could be detected particularly at low-temperature and high-pressure conditions (Nakasone *et al.*, 2002).

In transcription of the σ^{54}-dependent promoter, such as the pressure-regulated and the *glnA* operons, σ^{54}-containing RNA polymerase holoenzyme activates transcription at the promoter assisted by the activity of NtrC, which in turn is controlled by NtrB. As mentioned above, these *trans*-acting factors (σ^{54}, NtrC, or NtrB) might play an important role in pressure-regulated transcription at the σ^{54}-dependent promoter in this piezophilic bacterium. The σ^{54} in *S. violacea* is expressed at a relatively constant level under both atmospheric and high-pressure conditions, suggesting that the level of functional σ^{54} molecules is possibly regulated by the availability of NtrC. We examined the expression of the NtrC protein in *S. violacea* by Western blot analysis under different pressure conditions. The results indicated that the levels of this factor expressed at high pressure were greater than at atmospheric pressure (Ikegami *et al.*, 2000c).

Consideration of the results of pressure regulation by σ^{54} led us to propose a possible model for the mechanism of regulated expression of the pressure-regulated and the *glnA* operons in the deep-sea piezophilic bacterium *S. violacea*, as shown in Fig. 10.11. Because *S. violacea* σ^{54} is expressed at consistent levels at both atmospheric and high pressure, it is suggested that the intracellular levels of σ^{54}-containing RNA polymerase holoenzyme under both conditions are constant. This observation also strongly suggests that the transcriptional activity at this σ^{54}-dependent promoter is proportional to the amount of NtrC factor and that this factor regulates gene expression under pressure conditions. This model suggests that NtrB might function as a pressure sensor, and that it would be autophosphorylated under high-pressure conditions. Subsequently, the phosphorylated relay affects the NtrC protein followed by phosphorylated NtrC activation of the σ^{54}-dependent promoters. This model explaining gene expression at high pressure should be confirmed using molecular-genetic approaches

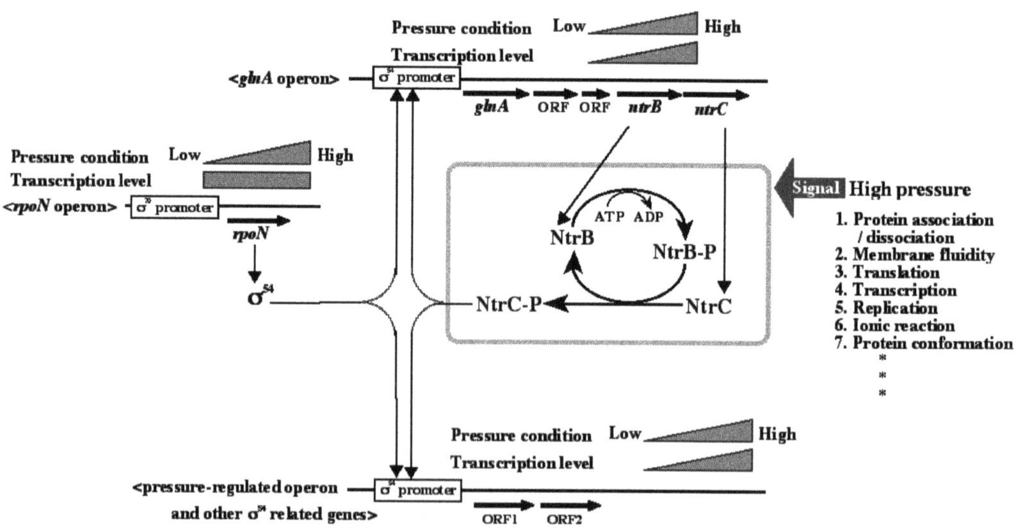

Figure 10.11 Proposed model for the transcriptional mechanisms of pressure-regulated gene expression in piezophilic *S. violacea* strain DSS12.

analysing several pressure-and low-temperature-sensitive mutants in these transcription systems. In the case of *P. profundum* SS9, the ToxR-S system functioned as a pressure sensor (Bartlett, 1999). Both NtrC and ToxR-S proteins were located in the bacterial cell membrane, suggesting that it is the location where a change in pressure is first sensed. Following a pressure change, those membrane proteins could be activated and/or inactivated. This mechanism might be a common one for pressure sensing in piezophilic bacteria.

Respiratory systems

As evidence of piezo-adaptation, a pressure-regulated operon was observed in *S. violacea*, as described above. Downstream from this operon, an open reading frame homologous to the *cydD* gene of *E. coli* was found, and the significance of the gene in bacterial growth under high pressure has been suggested (Kato *et al.*, 1996b). The gene product of *cydD* in *E. coli* is thought to be required for the assembly of respiratory components (Poole *et al.*, 1989, 1993, 1994). Further, the expression of a respiratory system was regulated by hydrostatic pressure in *S. violacea* (Tamegai *et al.*, 1998; Yamada *et al.*, 2000; Nakasone *et al.*, 2001) and in another piezophilic bacterium, *Shewanella* sp. strain DB-172F (Qureshi *et al.*, 1998a,b). These were the first reports concerning respiratory systems in a deep-sea bacterium, and the first evidence that expression of genes for respiratory components are regulated by physical parameters, such as hydrostatic pressure. Generally, bacteria have branched respiratory chains. Specifically, *Shewanella* strains have many respiratory components for adaptation to environmental changes (Heidelberg *et al.*, 2003). These observations and the results of previous studies suggest that pressure regulation for the expression of respiratory systems in *S. violacea* plays an important role in bacterial adaptation to high pressure.

Cytochrome *bd* is one of the members of the quinol oxidases, distinct from the haem-copper oxidase super family. In *E. coli*, two types of quinol oxidases, cytochrome *bo* and cytochrome *bd* exist and both share roles in respiration. Cytochrome *bo* is expressed in log phase and cytochrome *bd* is expressed in stationary phase (Kita *et al.*, 1984a,b). Cytochrome *bd* shows higher affinity for O_2 as compared with cytochrome *bo* and the former acts as a terminal oxidase under low oxygen concentrations (Kita *et al.*, 1984b). For the biosynthesis of cytochrome *bd*, structural genes (encoded by the *cydAB* operon) and genes for assembly of mature enzyme (encoded by the *cydDC* operon) are required (Georgiou *et al.*, 1987; Poole *et al.*, 1989). Expression of *cydAB* in *E. coli* is regulated by ArcA and Fnr, common O_2-regulated transcriptional regulators (Cotter *et al.*, 1997; Govantes *et al.*, 2000), whilst *cydDC* expression is regulated by NarL, involved in the two-component regulatory system for nitrate respiration (Tamegai *et al.*, 2008), as well as Fnr (Cook *et al.*, 1997). However, in *S. violacea*, no cytochrome *bd* has been detected spectrophotometrically under atmospheric pressure even during stationary phase. Surprisingly, cytochrome *bd* has been detected only at growth at high hydrostatic pressures (Tamegai *et al.*, 1998). Thus, transcriptional regulation of cytochrome *bd* in *S. violacea* may be different from other organisms, and this fact may be important for bacterial adaptation to high pressure. Furthermore, cytochrome *bd*-encoding *cydAB* genes were identified in *S. violacea*. Transcriptional analysis was carried out for *cydAB* and *cydCD* operons, and it was observed that transcription of the *cydDC* operon was strongly regulated by hydrostatic pressure (Tamegai *et al.*, 2005).

There are two types of soluble cytochrome *c* in *S. violacea*. One of them, cytochrome c_A, is constitutively expressed regardless of growth pressure. In comparison, the expression of cytochrome c_B is repressed under high hydrostatic pressure (Yamada *et al.*, 2000).

Considering the above observations, a model was proposed for the pressure-regulated respiratory systems involving cytochrome *bd* and *c* (Fig. 10.12). At atmospheric pressures, the cytochrome *c* oxidase system could be main pathway, whilst at high pressures the cytochrome *bd* oxidase pathway could dominate. It appears as though pressure regulation of the expression of the respiratory systems in *S. violacea* plays an important role in bacterial adaptation to high hydrostatic pressure.

The cell division protein FtsZ

Some rod-shaped bacteria, including *E. coli*, exhibit cell filamentation without septum

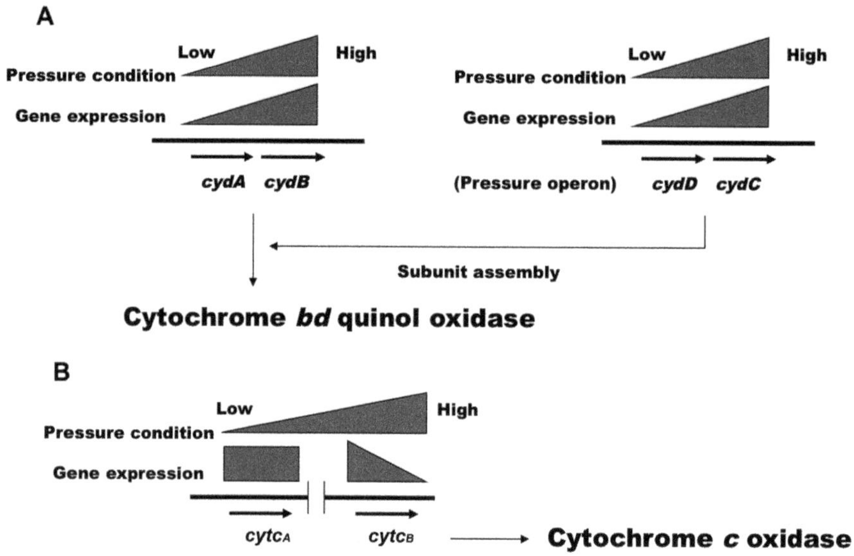

Figure 10.12 Respiration model of *S. violacea* under different pressure conditions. (A) Cytochrome *bd* quinol oxidase system. (B) Cytochrome c oxidase system.

formation under high hydrostatic pressures, indicating a pressure effect on the cell division process (Marquis, 1976). The cell divisional protein, FtsZ, is one of the house keeping proteins in bacterial cells, and FtsZ ring formation is essential for cell growth. We examined the effects of high pressure on FtsZ-ring formation in *E. coli* cells by indirect immunofluorescence microscopy. Elevated pressure completely repressed *E. coli* colony formation at 40 MPa and the cells exhibited obviously filamentous shapes. In the elongated cells, normal cell division processes appeared to be inhibited, as no FtsZ-rings were observed by indirect immunofluorescent staining. In addition, we also observed that hydrostatic pressure dissociated the *E. coli* FtsZ (ecFtsZ) polymers *in vitro* (Ishii *et al.*, 2004). Cell division of lactic acid bacteria, *Lactococcus lactis* and *Lactobacillus sanfranciscensis* was also studied as another example. Immunofluorescence microscopy demonstrated an inhibitory effect of high pressure on the formation of the FtsZ ring in these terrestrial microorganisms, as seen in *E. coli* (Molina-Hoppner *et al.*, 2003). These results suggest that the high pressure directly affects cell survival and morphology through the dissociation of cytoskeletal frameworks.

To analyse transcription upstream from the *ftsZ* gene in *S. violacea* DSS12, Northern blot and primer extension analyses were performed. The results indicated that gene expression was not pressure dependent. Western blot analysis also showed that the *S. violacea* FtsZ protein (svFtsZ) was equally expressed at several pressures, ranging from atmospheric (0.1 MPa) to high (70 MPa) (Ishii *et al.*, 2002). Using immunofluorescence microscopy, the svFtsZ ring was observed in the centre of cells at pressure conditions of 0.1 to 50 MPa (Fig. 10.13). These results imply that the svFtsZ protein function is not affected by elevated pressure in this piezophilic bacterium.

Those observations indicated that polymerization of ecFtsZ and svFtsZ occurred under high-pressure conditions, while the ecFtsZ and svFtsZ ring formations were, respectively, inhibited and not inhibited under HP conditions. The C-terminus of ecFtsZ and svFtsZ, the region essential for polymerization activity, was not conserved (Ishii *et al.*, 2002). Such differences might be the cause of the differences in cell division observed between the two species under high-pressure conditions. Therefore, we expect that characterization of the biochemical features and polymerization activities of FtsZ and other terrestrial bacteria *in vitro* will help us to understand

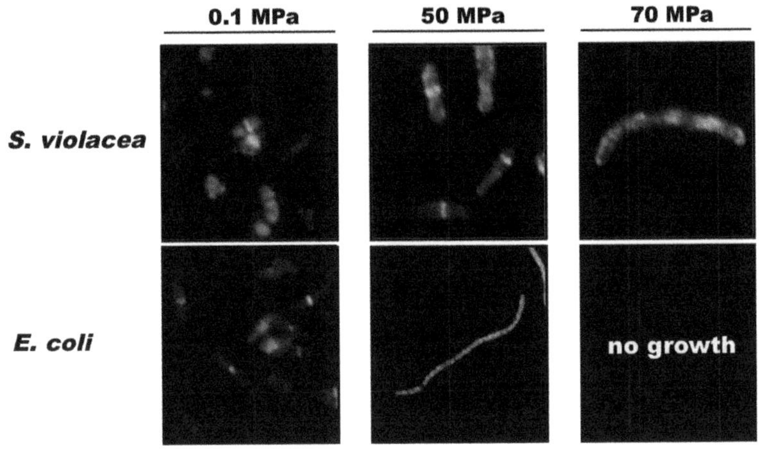

Figure 10.13 Observation of FtsZ ring formation by immunofluorescence microscopy in *S. violacea* (upper panel), and comparison with *E. coli* (lower panel), grown at different pressure conditions. The FtsZ ring is the brighter, linear region seen along the short axis of the cell.

the effect of pressure on the stages of cell division *in vivo*.

Stability of RNA polymerase under high-pressure conditions

RNA polymerase is one of the most important oligomeric enzymes in transcription. It is composed of the α, β, β′, and one of several σ subunits, σ^{70}, and governs the selectivity of the promoter gene sequence (Yura and Ishihama, 1979). The subunits of *E. coli* RNA polymerase (ecRNAP) have been shown to be dissociated by high pressure (Erijman and Clegg, 1995). This led us to investigate the stability of the RNA polymerase of the piezophile *S. violacea* (svRNAP) under high-pressure conditions. To examine the subunit dissociation of RNA polymerases under high pressures, a high-pressure electrophoresis apparatus (HPEA) was developed by a modification of the method previously reported by Erijman and Clegg (1995) (Fig. 10.10D; Kawano *et al.*, 2004). In this system, the cathode and anode are attached separately above and below the sample chamber. Pressures of up to 200 MPa can be applied within 1 minute with silicon oil KF-96-1.5CS (Shin-Etsu Chemical, Tokyo, Japan), using a hand pump. First, native polyacrylamide gel electrophoresis (PAGE) is performed under pressure in capillary glass tubes (75 mm length, 0.85 mm inner diameter) using the high-pressure electrophoresis apparatus (HPEA) described earlier. Following electrophoresis, the gel is decompressed, and electrophoresis carried out in a second dimension (10% SDS-PAGE) at atmospheric pressure.

Using the HPEA system, the relative stability of the *S. violacea* RNA polymerase (svRNAP) and *E. coli* RNA polymerase (ecRNAP) under high pressures was analysed. The results showed that a pressure of 140 MPa caused dissociation of ecRNAP but not of svRNAP (Fig. 10.14A and B). On the other hand, the core enzyme of svRNAP, which lacked the σ^{70}, was dissociated at 140 MPa (Fig. 10.14C; Kawano *et al.*, 2004). These results suggest that the sigma factor is required for stabilization of svRNAP under high-pressure conditions. The sigma subunit is known to change the quaternary structure of ecRNAP (Wu *et al.*, 1976; Greiner *et al.*, 1996). It is probable that the *S. violacea* σ^{70} stabilizes the core enzyme through alteration of the quaternary structure of RNA polymerase, resulting in piezotolerance. In this context, the predicted β-sheet domain of σ^{70} subunit in svRNAP, which is not observed in the *E. coli* and *S. oneidensis* σ^{70}, may have a role in stabilization of RNA polymerase at high pressure (Kawano *et al.*, 2004). Further experimentation is required to determine the significance of the β-sheet domain by comparing the structure with

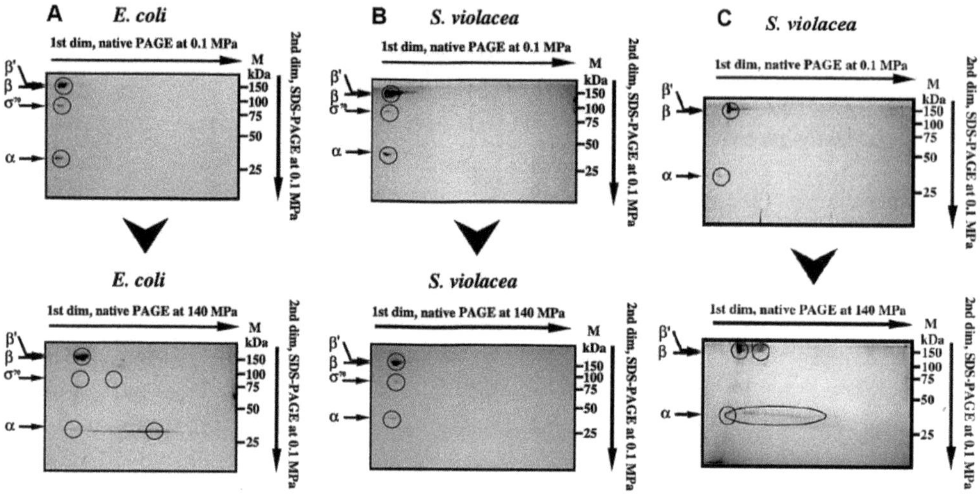

Figure 10.14 Effects of high hydrostatic pressure on subunit association in RNA polymerase in *E. coli* (A), *S. violacea* (B) and *S. violacea* without sigma factor (C). Native PAGE [first dimension] was performed at 0.1 MPa (upper panels) or 140 MPa (lower panels), followed by SDS-PAGE [second dimension] at 0.1 MPa. Proteins were visualized by silver staining. Each subunit of RNA polymerase is shown by an open circle. M, molecular weight size markers; first dimension, 1-D separation; second dimension, 2-D separation.

that of other mesophilic *Shewanella* strains, and by analysis of the effects of mutations within the domain on the piezotolerance of RNA polymerase in terms of transcriptional activity and subunit association. For investigations of the molecular adaptation of proteins to high hydrostatic pressure, the HPEA is a powerful tool when used in combination with techniques based on molecular biology and bioinformatics.

Enzymes from piezophiles and their activities under high pressures

Enzyme activities of deep-sea piezophilic bacteria are not only pressure tolerant, but also activated under higher-pressure conditions. Two experimental examples are described below.

Dihydrofolate reductase (DHFR)

A novel dihydrofolate reductase (svDHFR) has been purified from *S. violacea* DSS12. In contrast with *E. coli* DHFR (ecDHFR), the enzyme activity of svDHFR increased with increasing hydrostatic pressure up to 100 MPa, suggesting that the enzyme kinetics and structural fluctuation of svDHFR are adapted to a high-pressure environment. Fig. 10.15A shows the relative activities of the two DHFR enzymes as a function of hydrostatic pressure. The activity of svDHFR increases by up to 30% with pressure (at 100 MPa), and then gradually decreases, although it maintains a higher activity at 250 MPa than at atmospheric pressure. This behaviour is significantly different from that of ecDHFR, which exhibits a steady, linear decrease in activity, even under low pressure (Ohmae *et al.*, 2004). Therefore, it can be expected that svDHFR has specific enzyme kinetics and structural fluctuation that allow it to adapt itself to deep-sea conditions. The activation volume of enzyme reaction, $\Delta V^{\#}$, for svDHFR was negative for pressures below 100 MPa, meaning that the activated state has a smaller volume than the reactant in the catalytic reaction coordinate. At pressures above 100 MPa, this was reversed, as shown by a positive $\Delta V^{\#}$ value (also for ecDHFR). These results predict that svDHFR, being highly flexible at pressures below 100 MPa, changes its conformation to be more rigid (like ecDHFR) at higher pressures, although the protein may be denatured at pressures above a few hundred MPa.

To determine whether a negative $\Delta V^{\#}$ of DHFR is a common property in piezophiles, *P. profundum* DHFR (ppDHFR), *M. yayanosii* DHFR

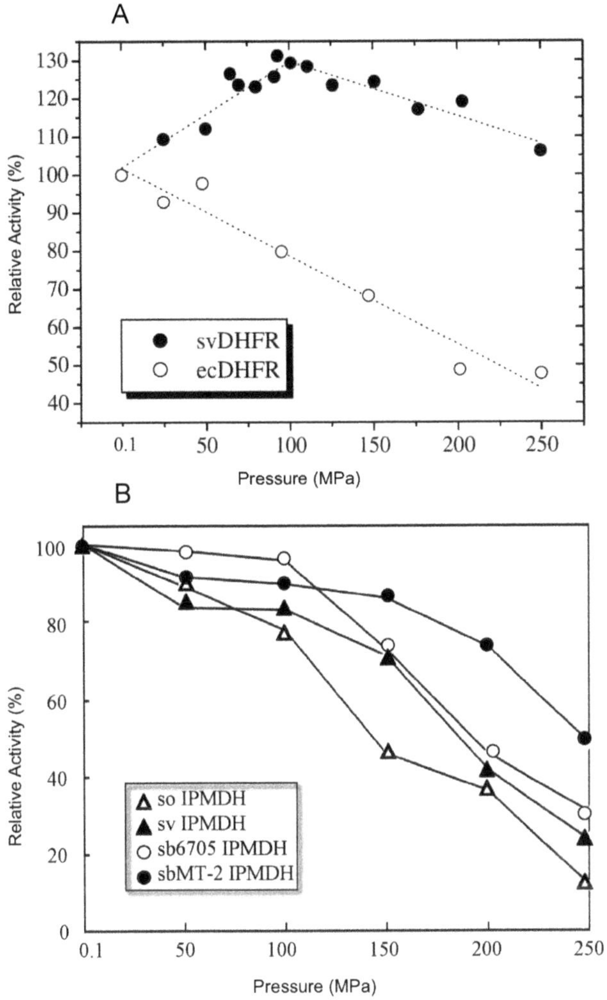

Figure 10.15 (A) Pressure dependency of the enzymatic activities of svDHFR (filled circles) and ecDHFR (open circles). Dashed lines represent the trend lines for the data. (B) Pressure-dependent activity of isopropylmalate dehydrogenase (IPMDH) from various bacterial sources. Relative activities were measured in pressurized cells 2 min after the application of pressure. Symbols indicate the origin of IPMDH: open triangles, *S. oneidensis* (so); filled triangles, *S. violacea* DSS12 (sv); open circles, *S. benthica* DB6705 (sb6705); filled circles, *S. benthica* DB21MT-2 (sbMT-2).

(myDHFR) and *M. japonica* DHFR (mjDHFR) were analysed. The results are shown in Table 10.4 (Murakami *et al.*, 2010). The enzyme activity of ppDHFR and mjDHFR (moderate piezophilic bacteria) as well as ecDHFR decreased as pressure increased, while svDHFR and myDHFR (piezophilic bacteria) showed a significant tolerance to pressure. These results suggest that DHFRs from deep-sea bacteria possess specific enzymatic properties adapted to life under high pressure. *M. profunda* DHFR (mpDHFR) was analysed by high-pressure NMR (Hata *et al.*, 2004). The data suggested that maximum stability of mpDHFR was around 5°C at 0.1 MPa. Despite the relatively low overall stability, the structure of the core part of the protein remained intact up to ~200 MPa, giving some rigidity to the folded conformation. However, from the crystal structure of mpDHFR, it was difficult to identify any obvious structural features that would suggest pressure adaptation

Table 10.4 Activation volumes of ecDHFR and deep-sea DHFR enzymatic reactions[a]

DHFR[b]	$\Delta V^{\#}$ (ml/mol)[c]
ecDHFR	7.5±0.2 (0.1–250 MPa)
svDHFR	−2.7±1.1 (0.1–75 MPa); 8.9±0.6 (75–250 MPa)
ppDHFR	13.8±0.1 (0.1–250 MPa)
myDHFR	1.7±0.6 (0.1–75 MPa); 16.5±0.6 (75–250 MPa)
mjDHFR	38.7±0.3 (0.1–250 MPa)

[a]Reactions carried out at 25°C and pH 7.0 in 20 mM Tris-HCl (pH 7.0), 0.1 mM EDTA, 0.1 mM dithiothreitol, 250 mM NADPH and 250 mM DHF.

[b]DHFR, dihydrofolate reductase; ec, *E. coli*; sv, *S. violacea*; pp, *P. profundum*; my, *M. yayanosii*; mj, *M. japonica*.

[c]The values in parentheses indicate the pressure range used for calculations.

(Hay *et al.*, 2009). Determining the volume changes in each process and the rate-limiting process, which is in progress, should provide more detailed understanding of the adaptation mechanism of the enzyme reaction.

Isopropylmalate dehydrogenase (IPMDH)

The isopropylmalate dehydrogenase (IPMDH) enzymes from various *Shewanella* species adapted to different pressure environments have been extensively studied: the 0.1 MPa-adapted *S. oneidensis* MR-1 (Vankateswaran *et al.*, 1999), the piezophilic species *S. violacea* DSS12, *S. benthica* DB6705 and *S. benthica* DB21MT-2, isolated from depths of 5110, 6356, and 10,989m, respectively (Kato *et al.*, 1995a; 1998; Nogi *et al.*, 1998b; Nogi and Kato, 1999). All IPMDH genes were cloned and sequenced as well as recombinantly expressed to yield purified protein for further biochemical studies. The greatest variation in IPMDH amino acid sequences was found between *S. oneidensis* and the two deep-sea *Shewanella* species, which also possessed more hydrophobic amino acid residues in their proteins (De Poorter *et al.*, 2004).

The activity of IPMDH was also investigated as a function of temperature and pressure. Activity measurements performed between 10 and 72°C showed a positive correlation between the optimal temperature for enzyme activity and the optimal pressure for the growth of piezophilic *Shewanella* species. In contrast, *S. oneidensis* IPMDH had a very broad temperature optimum. Also, no direct correlation was observed between temperature-induced denaturation and pressure adaptation (De Poorter *et al.*, 2004).

The application of hydrostatic pressure reduced the activity of enzymes from all species (Fig. 10.15B). Pressure-dependent stability was higher for IPMDH enzymes from organisms isolated at greater depths. Although *S. benthica* DB6705 IPMDH has a higher relative activity than strain DB21MT-2 for pressures up to 100 MPa, it is rapidly deactivated at higher pressures. DB21MT-2 IPMDH might not be very stable at low pressures because it is naturally adapted to very high pressure (Kasahara *et al.*, 2009). These results suggest that the activity of enzymes from piezophilic *Shewanella* species is driven by pressure acclimatization.

S. violacea genome analysis

To understand pressure adaptation in deep-sea bacteria, we performed genome analysis of the piezophilic bacterium *S. violacea* DSS12 as a model deep-sea bacterium (Aono *et al.*, 2010). The genome consists of a single circular chromosome of 4962,103 bp, with an average G+C content of 44.7%. Table 10.5 summarizes the general features of the genome. A total of 4346 coding sequences (CDS), with an average size of 970 bp, were predicted. These coding regions cover 81.6% of the genome. Among the predicted CDS, 2219 (51.1%) were assigned putative functions, 973 (22.4%) match predicted CDS of unknown function from other organisms, and 761 (17.5%)

Table 10.5 General features of the *S. violacea* genome

Feature	Value
Genome size	4,962,103 bp
G+C content	44.7%
Number of predicted CDS[a]	4,346
Average size of CDS	970 bp
Percentage of protein-coding region	81.6%
Number of rRNA operons	14
Number of tRNAs	126
Number of CDS similar to proteins of known function	2,219
Number of CDS similar to proteins of unknown function[b]	393
Number of conserved hypothetical proteins[c]	973
Number of hypothetical proteins (no database match)[d]	761

[a]CDS, coding sequences.

[b]Proteins of unknown function that have a significant similarity to a known protein for which there is currently no known function.

[c]Conserved hypothetical proteins that have sequence similarity to another CDS, but for which there is currently no evidence that a protein is expressed.

[d]Hypothetical proteins that have no significant similarity to other proteins in any database.

Data from Aono *et al.* (2010).

had no database match. The genome includes 14 copies of rRNA operons and 126 tRNA genes representing all 20 amino acids. The number of rRNA operons is the largest among all completely sequenced *Shewanella* genomes (http://www.ncbi.nlm.nih.gov/genomes/MICROBES/microbial_taxtree.html) and comparable to the number present in the deep-sea bacterium *Photobacterium profundum* (15 copies, Vezzi *et al.*, 2005). The number of ribosomal operons might correspond to extreme environmental conditions, and It is probable that the ability of *S. violacea* to grow under high pressure is due to the transcriptional efficiency of numerous ribosomal proteins under such stress conditions. Analyses of the annotation suggested that almost 50% of the CDS are hypothetical or unknown functional proteins. This result indicates that many genes from deep-sea microbes could be novel and might also possess previously undescribed functions. Therefore, genomic analysis of marine extremophiles may lead to the discovery of new functional genes.

Following *S. violacea* genome analysis, we are planning to initiate a postgenome study called the 'PIEZOME project,' which will focus on the proteome, transcriptome, and metabolome of piezophiles affected by high-pressure environments. We are looking forward to furthering our understanding of the molecular mechanisms involved in high-pressure life on our planet.

Conclusion

Piezophiles are isolated from deep-sea high-pressure environments. From the cold deep-sea, psychrophilic and piezophilic bacteria belong to one of five genera, *Shewanella, Photobacterium, Moritella, Colwellia,* and *Psychromonas,* all within the gammaproteobacterial branch. Components of the membrane lipids of these piezophiles consist of unsaturated fatty acids (UFAs), but PUFAs, like EPA and DHA, are commonly involved. To study high-pressure microbiology in more detail, we have constructed several kinds of high-pressure equipment and have developed novel protocols. Using this equipment, mechanisms of pressure adaptation have been elucidated at the molecular level.

Future trends

The history of high-pressure microbiology began in the second half of the last century. The first isolation of piezophiles was recorded in 1979 (Yayanos *et al.*, 1979), so the field is around 40 years old. Therefore, this field is still at an early stage, with many undiscovered investigations surely awaiting discovery. For example, recently Takai *et al.* (2009) reported the isolation of novel piezophilic and thermophilic bacteria from a deep-sea hydrothermal vent, including *Piezobacter thermophila* 108, an isolate from the alphaproteobacteria. As described in this chapter, all previously described psychrophilic and piezophilic bacteria belonged to the gammaproteobacteria. Hence, the study of Takai *et al.* (2009) suggests that piezophilic bacterial diversity could be more extensive in the deep-sea environment than previously noted. Many novel piezophiles will very probably be discovered by the application of isolation procedures that mimic the different environmental conditions of pressure, temperature, nutrients, etc. found in the deep sea. An obligate piezophilic hyperthermophile, *Pyrococcus* sp. CH1 was also recently isolated from the deepest hydrothermal vent (Zeng *et al.*, 2009). Considering these new discoveries, more investigations focusing on piezophilic life and its molecular basis will be necessary in the future.

Web resources

Japan Agency for Marine-Earth Science and Technology (http://www.jamstec.go.jp/e/index.html)

Genome information of *S. violacea* (http://gib.genes.nig.ac.jp/single/index.php?spid=Svio_DSS12)

Acknowledgements

I am very grateful to Professor Koki Horikoshi for continued support of my extremophile studies. I also thank my colleagues, Drs Maria Smorawinska, Lina Li, Takako Sato, Juan Gonzalez, Yuichi Nogi, Kaoru Nakasone, Hideyuki Tamegai, Mohammad Hassan Qureshi, Francesco Canganella, Jiasong Fang, Eiji Ohmae, Akihiko Ikegami, Akihiro Ishii and Hiroaki Kawano, for excellent collaboration. Finally, I appreciate the crews of the research ship, members of the submersible operation division and technicians of the DEEPBATH system at JAMSTEC for their efforts in collecting deep-sea samples and conducting high-pressure cultivation experiments.

References

Abe, F., and Horikoshi, K. (1998). Analysis of intracellular pH in the yeast *Saccharomyces cerevisiae* under elevated hydrostatic pressure: a study in baro- (piezo-) physiology. Extremophiles 2, 223–228.

Abe, F., and Kato, C. (1999). Barophysiology (Piezophysiology). In Extremophiles in Deep-Sea Environments, Horikoshi, K., and Tsujii, K., eds. (Tokyo, Japan, Springer-Verlag), pp. 227–248.

Akasaka, K. (2006). Probing conformational fluctuation of proteins by pressure perturbation. Chem. Rev. *106*, 1814–1835.

Allen, E.E., and Bartlett, D.H. (2000). FabF is required for piezoregulation of *cis*-vaccenic acid levels and piezophilic growth of the deep-sea bacterium *Photobacterium profundum* strain SS9. J. Bacteriol. *182*, 264–1271.

Allen, E.E., Facciotti, D., and Bartlett, D.H. (1999). Monounsaturated but not polyunsaturated fatty acids are required for growth of the deep-sea bacterium *Photobacterium profundum* SS9 at high pressure and low temperature. Appl. Environ. Microbiol. *65*, 1710–1720.

Aono, E., Baba, T., Ara, T., Nishi, T., Nakamichi, T., Inamoto, E., Toyonaga, H., Hasegawa, M., Takai, Y., Okumura, Y., *et al.* (2010). Complete genome sequence and comparative analysis of *Shewanella violacea*, a psychrophilic and piezophilic bacterium from deep sea floor sediments. Mol. Bio. Syst. *6*, 1216–1226.

Bartlett, D.H. (1999). Microbial adaptations to the psychrosphere/piezosphere. J. Mol. Microbiol. Biotechnol. *1*, 93–100.

Bartlett, D.H., Wright, M., Yayanos, A.A., and Silverman, M. (1989). Isolation of a gene regulated by hydrostatic pressure in a deep sea bacterium. Nature *342*, 572–574.

Beijerinck, M.W. (1889). Le *Photobacterium luminosum*, Bactérie luminosum de la Mer Nord. (in French) Arch. Néerl. Sci. 23, 401–427.

Bowmam, J.P., McCammon, S.A., Nichols, D.S., Skerratt, J.H., Rea, S.M., Nichols, P.D., and McMeekin, T.A. (1997). *Shewanella gelidimarina* sp. nov. and *Shewanella frigidimarina* sp. nov., novel Antarctic species with the ability to produce eicosapentaenoic acid (20:5ω3) and grow anaerobically by dissimilatory Fe(III) reduction. Int. J. Syst. Bacteriol. *47*, 1040–1047.

Bowman, J.P., Gosink, J.J., McCammon, S.A., Lewis, T.E., Nichols, D.S., Nichols, P.D., Skerratt, J.H., Staley, J.T., and McMeekin, T.A. (1998). *Colwellia demingiae* sp. nov., *Colwellia hornerae* sp. nov., *Colwellia rossensis* sp. nov. and *Colwellia psychrotropica* sp. nov.: psychrophilic Antarctic species with the ability to synthesize docosahexaenoic acid (22:6ω3). Int. J. Syst. Bacteriol. *48*, 1171–1180.

Canganella, F., Gonzalez, J.M., Yanagibayashi, M., Kato, C., and Horikoshi, K. (1997). Pressure and temperature effects on growth and viability of the hyperthermophilic archaeon *Thermococcus peptonophilus*. Arch. Microbiol. *168*, 1–7.

Chi, E., and Bartlett, D.H. (1995). An *rpoE*-like locus controls outer membrane protein synthesis and growth at cold temperatures and high pressures in the deep-sea bacterium *Photobacterium* sp. strain SS9. Mol. Microbiol. *17*, 713–726.

Colwell, R.R., and Morita, R.Y. (1964). Reisolation and emendation of description of *Vibrio marinus* (Russell) Ford. J. Bacteriol. *88*, 831–837.

Cook, G.M., Membrillo-Hernandez, J., and Poole, R.K. (1997). Transcriptional regulation of the *cydDC* operon, encoding a heterodimeric ABC transporter required for assembly of cytochromes *c* and *bd* in *Escherichia coli* K-12: Regulation by oxygen and alternative electron acceptors. J. Bacteriol. *179*, 6525–6530.

Cotter, P.A., Melville, S.B., Albrecht, J.A., and Gunsalus, R.P. (1997). Aerobic regulation of cytochrome *d* oxidase (*cydAB*) operon expression in *Escherichia coli*: roles of Fnr and ArcA in repression and activation. Mol. Microbiol. *25*, 605–615.

DeLong, E.F., and Yayanos, A.A. (1985). Adaptation of the membrane lipids of a deep-sea bacterium to changes in hydrostatic pressure. Science *228*, 1101–1103.

DeLong, E.F., and Yayanos, A.A. (1986). Biochemical function and ecological significance of novel bacterial lipids in deep-sea prokaryotes. Appl. Environ. Microbiol. *51*, 730–737.

DeLong, E.F., Franks, D.G., and Yayanos, A.A. (1997). Evolutionary relationship of cultivated psychrophilic and barophilic deep-sea bacteria. Appl. Environ. Microbiol. *63*, 2105–2108.

Deming, J.W., and Baross, J.A. (1993). Deep-sea smokers: Windows to a subsurface biosphere? Geochim. Cosmochim. Acta. *57*, 3219–3230.

Deming, J.W., Hada, H., Colwell, R.R., Luehrsen, K.R., and Fox, G.E. (1984). The nucleotide sequence of 5S rRNA from two strains of deep-sea barophilic bacteria. J. Gen. Microbiol. *130*, 1911–1920.

Deming, J.W., Somers, L.K., Straube, W.L., Swartz, D.G., and Macdonell, M.T. (1988). Isolation of an obligately barophilic bacterium and description of a new genus, *Colwellia* gen. nov. System. Appl. Microbiol. *10*, 152–160.

De Poorter, L.M.I., Suzaki, Y., Sato, T., Tamegai, H., and Kato, C. (2004). Effects of pressure on the structure and activity of isopropylmalate dehydrogenases from deep-sea *Shewanella* species. Mar. Biotechnol. *6*, s190–194.

Erauso, G., Reysenbach, A-L., Godfroy, A., Meunier, J.R., Crump, B., Partensky, F., Baross, J.A., Marteinsson, V., Barbier, G., Pace, N.R., and Prieur, D. (1993). *Pyrococcus abyssi* sp. nov., a new hyperthermophilic archaeon isolated from a deep-sea hydrothermal vent. Arch. Microbiol. *160*, 338–349.

Erijman, L., and Clegg, R.M. (1995). Heterogeneity of *E. coli* RNA polymerase revealed by high pressure. J. Mol. Biol. 253, 259–265.

Fang, J., Barcelona, M.J., Nogi, Y., and Kato, C. (2000). Biochemocal function and geochemical significance of novel phospholipids of the extremely barophilic bacteria from the Mariana Trench at 11,000 meters. Deep-sea Res. *147*, 1173–1182.

Fang, J., Chan, O., Kato, C., Sato, T., Peeples, T., and Niggemeyer, K. (2003). Phospholipid FA of piezophilic bacteria from the deep sea. Lipids *38*, 885–887.

Fang, J., Kato, C., Sato, T., Chan, O., and McKay, D.S. (2004). Polyunsaturated fatty acids in piezophilic bacteria: biosynthesis or dietary uptake? Comp. Biochem. Physiol. B *137*, 455–461.

Fang, J., Uhle, M., Billmark, K., Bartlett, D.H., and Kato, C. (2006). Fractionation of carbon isotopes in biosynthesis of fatty acids by a piezophilic bacterium *Moritella japonica* DSK1. Geochim. Cosmochim. Acta. *70*, 1753–1760.

Georgiou, C.D., Fang, H., and Gennis, R.B. (1987). Identification of the *cydC* locus required for expression of the functional form of the cytochrome *d* terminal oxidase complex in *Escherichia coli*. J. Bacteriol. *169*, 2107–2112.

Gonzalez, J.M., Kato, C., and Horikoshi, K. (1995). *Thermococcus peptonophilus* sp. nov., a fast-growing, extremely thermophilic archaebacterium isolated from deep-sea hydrothermal vents. Arch. Microbiol. *164*, 159–164.

Gonzalez, J.M., Masuchi, Y., Robb, F.T., Ammerman, J.W., Maeder, D.L., Yanagibayashi, M., Tamaoka, J., and Kato, C. (1998). *Pyrococcus horikoshii* sp. nov., a hyperthermophilic archaeon isolated from a hydrothermal vent at the Okinawa Trough. Extremophiles 2, 123–130.

Govantes, F., Albrecht, J.A., and Gunsalus, R.P. (2000). Oxygen regulation of the *Escherichia coli* cytochrome *d* oxidase (*cydAB*) operon: roles of multiple promoters and the Fnr-1 and Fnr-2 binding sites. Mol. Microbiol. *37*, 1456–1469.

Greiner, D.P., Hughes, K.A., Gunasekera, A.H., and Meares, C.F. (1996). Binding of the σ^{70} protein to the core subunits of *Escherichia coli* RNA polymerase, studied by iron-EDTA protein footprinting. Proc. Natl. Acad. Sci. U.S.A. *93*, 71–75.

Hata, K., Kono, R., Fujisawa, M., Kitahara, R., Kamatari, Y.O., Akasaka, K., and Xu, Y. (2004). High pressure NMR study of dihydrofolate reductase from a deep-sea bacterium *Moritella profunda*. Cell. Mol. Biol. *50*, 311–316.

Hay, S., Evans, R.M., Levy, C., Loveridge, E.J., Wang, X., Leys, D., Allemann, R.K., and Scrutton, N.S. (2009). Are the catalytic properties of enzymes from piezophilic organisms pressure adapted? Chembiochem *10*, 2348–2357.

Heidelberg, J.F., Paulsen, I.T., Nelson, K.E., Gaidos, E.J., Nelson, W.C., Read, T.D., Eisen, J.A., Seshadri, R., Ward, N., Methe, B., *et al.* (2003). Genome sequence of the dissimilatory metal ion-reducing bacterium *Shewanella oneidensis*. Nat. Biotechnol. *20*, 1118–1123.

Ikegami, A., Nakasone, K., Fujita, M., Fujii, S., Kato, C., Usami, R., and Horikoshi, K. (2000a). Cloning and characterization of the gene encoding RNA polymerase sigma factor σ^{54} of deep-sea piezophilic

Shewanella violacea. Biochim. Biophys. Acta. *1491*, 315–320.

Ikegami, A., Nakasone, K., Kato, C., Nakamura, Y., Yoshikawa, I., Usami, R., and Horikoshi, K. (2000b). Glutamine synthetase gene expression at elevated hydrostatic pressure in a deep-sea piezophilic *Shewanella violacea*. FEMS Microbiol. Lett. *192*, 91–95.

Ikegami, A., Nakasone, K., Kato, C., Usami, R., and Horikoshi, K. (2000c). Structural analysis of *ntrBC* genes of deep-sea piezophilic *Shewanella violacea*. Biosci. Biotech. Biochem. *64*, 915–918.

Imai, E., Honda, H., Hatori, K., Brack, A., and Matsuno, K. (1999). Elongation of oligopeptides in a simulated submarine hydrothermal system. Science *283*, 831–833.

Ishii, A., Nakasone, K., Sato, T., Wachi, M., Sugai, M., Nagai, K., and Kato, C. (2002). Isolation and characterization of the *dcw* cluster from the piezophilic deep-sea bacterium *Shewanella violacea*. J. Biochem. *132*, 183–188.

Ishii, A., Sato, T., Wachi, M., Nagai, K., and Kato, C. (2004). Effects of high hydrostatic pressure on bacterial cytoskeleton FtsZ polymers *in vivo* and *in vitro*. Microbiology *150*, 1965–1972.

Jensen, M.J., Tebo, B.M., Baumann, P., Mandel, M., and Nealson, K.H. (1980). Characterization of *Alteromonas hanedai* (sp. nov.), a non-fermentative luminous species of marine origin. Curr. Microbiol. 3, 311–315.

Kasahara, R., Sato, T., Tamegai, H., and Kato, C. (2009). Piezo-adapted 3-isopropylmalate dehydrogenase of the obligate piezophile *Shewanella benthica* DB21MT-2 isolated from the 11,000-m depth of the Mariana Trench. Biosci. Biotechnol. Biochem. 73, 2541–2543.

Kato, C. (1997a). Gene regulations of microorganisms in high hydrostatic pressure environment (in Japanese). In Recent Progress of Medical Microbiology, Nakano, M., ed. (Saikon Press, Tokyo, Japan) pp. 209–217.

Kato, C. (1997b). Genomic analyses of hyperthermophilic archaea (in Japanese). J. Japan Oil Chem. Soc. *46*, 517–523.

Kato, C. (1999). Barophiles (Piezophiles). In Extremophiles in Deep-Sea Environments, Horikoshi, K., and Tsujii, K., eds. (Springer-Verlag, Tokyo, Japan), pp. 91–111.

Kato, C., and Horikoshi, K. (1996). Gene expression under high pressure. In Progress in Biotechnology 13, High Pressure Bioscience and Biotechnology, Hayashi, R., and Balny, C., eds. (Elsevier Science BV, Amsterdam, Netherlands), pp. 59–66.

Kato, C., and Nogi, Y. (2001). Correlation between phylogenetic structure and function: examples from deep-sea *Shewanella*. FEMS Microbiol. Ecol. 35, 223–230.

Kato, C., Sato, T., Smorawinska, M., and Horikoshi, K. (1994). High pressure conditions stimulate expression of chloramphenicol acetyltransferase regulated by the *lac* promoter in *Escherichia coli*. FEMS Microbiol. Lett. *122*, 91–96.

Kato, C., Sato, T., and Horikoshi, K. (1995a). Isolation and properties of barophilic and barotolerant bacteria from deep-sea mud samples. Biodiv. Conserv. *4*, 1–9.

Kato, C., Smorawinska, M., Sato, T., and Horikoshi, K. (1995b). Cloning and expression in *Escherichia coli* of a pressure-regulated promoter region from a barophilic bacterium, strain DB6705. J. Mar. Biotechnol. *2*, 125–129.

Kato, C., Masui, N., and Horikoshi, K. (1996a). Properties of obligately barophilic bacteria isolated from a sample of deep-sea sediment from the Izu-Bonin trench. J. Mar. Biotechnol. *4*, 96–99.

Kato, C., Tamegai, H., Ikegami, A., Usami, R., and Horikoshi, K. (1996b). Open reading frame 3 of the barotolerant bacterium strain DSS12 is complementary with *cydD* in *Escherichia coli*: *cydD* functions are required for cell stability at high pressure. J. Biochem. *120*, 301–305.

Kato, C., Ikegami, A., Smorawinska, M., Usami, R., and Horikoshi, K. (1997a). Structure of genes in a pressure-regulated operon and adjacent regions from a barotolerant bacterium strain DSS12. J. Mar. Biotechnol. *5*, 210–218.

Kato, C., Li, L., Tamaoka, J., and Horikoshi, K. (1997b). Molecular analyses of the sediment of the 11,000m deep Mariana Trench. Extremophiles *1*, 117–123.

Kato, C., Smorawinska, M., and Horikoshi, K. (1997c). Comparison of the gene expression of aspartate β-D-semialdehyde dehydrogenase at elevated hydrostatic pressure in deep-sea bacteria. J. Biochem. *121*, 717–723.

Kato, C., Li, L., Nakamura, Y., Nogi, Y., Tamaoka, J., and Horikoshi, K. (1998). Extremely barophilic bacteria isolated from the Mariana Trench, Challenger Deep, at a depth of 11,000 meters. Appl. Environ. Microbiol. *64*, 1510–1513.

Kato, C., Nakasone, K., Qureshi, M.H., and Horikoshi, K. (2000). How do deep-sea microorganisms respond to changes in environmental pressure? In Cell and Molecular Response to Stress, Vol. 1. Environmental Stressors and Gene Responses, K.B. Storey, and J.M. Storey, eds. (Amsterdam, Netherland: Elsevier Science BV), pp. 277–291.

Kato, C., Li, L., Nogi, Y., Nakasone, K., and Bartlett, D.H. (2001). 10. Marine microbiology: Deep sea adaptations. In Biological Systems under Extreme Conditions, Structure and Function, Biological and Medical Physics Series, Y. Taniguchi, H.E. Stanley and H. Ludwig, eds. (Heidelberg, Germany: Springer-Verlag), pp. 205–220.

Kato, C., Sato, T., Nogi, Y., and Nakasone, K. (2004). Piezophiles: High pressure-adapted marine bacteria. Mar. Biotechnol. *6*, s195–201.

Kato, C., Nogi, Y., and Arakawa, S. (2008a). Isolation, cultivation, and diversity of deep-sea piezophiles. In High Pressure Microbiology, Michiels, C., Bartlett, D.H., and Aertsen, A., eds. (ASM Press, Washington DC, USA), pp. 203–217.

Kato, C., Sato, T., Abe, F., Ohmae, E., Tamegai, H., Nakasone, K., Siddiqui, K.S., and Thomas, T. (2008b). Protein adaptation to high pressure environments.

In Protein Adaptation in Extremophiles, Molecular Anatomy and Physiology of Proteins Series, Thomas, T., and Siddiqui, K.S., eds. (Nova Science Publisher, New York, USA), pp. 167–191.

Kato, C., Sato, T., Nakasone, K., and Tamegai, H. (2008c). Chapter 17. Molecular Biology of the Model Piezophile, *Shewanella violacea* DSS12. In High pressure Microbiology, Michiels, C., Bartlett, D.H., and Aertsen, A., eds. (ASM Press, Washington DC, USA), pp. 305–317.

Kawamoto, J., Kurihara, T., Yamamoto, K., Nagayasu, M., Tani, Y., Mihara, H., Hosokawa, M., Baba, T., Sato, S.B., and Esaki, N. (2009). Eicosapentaenoic acid plays a beneficial role in membrane organization and cell division of a cold-adapted bacterium, *Shewanella livingstonensis* Ac10. J. Bacteriol. *191*, 632–640.

Kawamoto, J., Saso, T., Nakasone, K., Kato, C., Mihara, H., Esaki, N., and Kurihara, T. (2011). Favorable effects of eicosapentaenoic acid on the late-step of the cell division in a piezophilic bacterium, *Shewanella violacea* DSS12, at high-hydrostatic pressure. Environ. Microbiol. doi: 10.1111/j.1462-2920.2011.02487.x

Kawano, H., Nakasone, K., Matsumoto, M., Usami, R., Kato, C., and Abe, F. (2004). Differential pressure resistance in the activity of RNA polymerase isolated from *Shewanella violacea* and *Escherichia coli*. Extremophiles *8*, 367–375.

Kita, K., Konishi, K., and Anraku, Y. (1984a). Terminal oxidases of *Escherichia coli* aerobic respiratory chain I. purification and properties of cytochrome b_{562}- *o* complex from cells in the early exponential phase of aerobic growth. J. Biol. Chem. *259*, 3368–3374.

Kita, K., Konishi, K., and Anraku, Y. (1984b). Terminal oxidases of *Escherichia coli* aerobic respiratory chain II. purification and properties of cytochrome b_{558}- *d* complex from cells grown with limited oxygen and evidence of branched electron-carrying systems. J. Biol. Chem. *259*, 3375–3381.

Kyo, M., Miyazaki, E., Tsukioka, S., Ochi, H., Amitani, Y., Tsuchiya, T., Aoki, T., and Takagawa, S. (1995). The sea trial of 'KAIKO', the full ocean depth research ROV. Oceans '95 *3*, 1991–1996.

Kyo, M., Tuji, T., Usui, H., and Itoh, T. (1991). Collection, isolation and cultivation system for deep-sea microbes study: concept and design. Oceans '91 *1*, 419–423.

Li, L., Kato, C., Nogi, Y., and Horikoshi, K. (1998). Distribution of the pressure-regulated operons in deep-sea bacteria. FEMS Microbiol. Lett. *159*, 159–166.

Liesack, W.H., Weyland, W., and Stackebrandt, E. (1991). Potential risks of gene amplification by PCR as determined by 16S rDNA analysis of a mixed-culture of strict barophilic bacteria. Microb. Ecol. *21*, 191–198.

MacDonell, M.T., and Colwell, R.R. (1985). Phylogeny of the Vibrionaceae, and recommendation for two new genera, *Listonella* and *Shewanella*. Syst. Appl. Microbiol. *6*, 171–182.

Margesin, R., and Nogi, Y. (2004). Psychropiezophilic microorganisms. Cell. Mol. Biol. *50*, 429–436.

Marquis, R.E. (1976). High pressure microbial physiology. Adv. Microb. Physiol. *14*, 159–241.

Maruyama, A., Honda, D., Yamamoto, H., Kitamura, K., and Higashihara, T. (2000). Phylogenetic analysis of psychrophilic bacteria isolated from the Japan Trench, including a description of the deep-sea species *Psychrobacter pacificensis* sp. nov. Int. J. Syst. Evol. Microbiol. *50*, 835–846.

Merrick, M.J., and Edwards, R.A. (1995). Nitrogen control in bacteria. Microbiol. Rev. *59*, 604–622.

Miwa, T., Sato, T., Kato, C., Aizawa, M., and Horikoshi, K. (2002). Restoration of *Escherichia coli* from high hydrostatic pressure –A study of the FtsZ-ring formation using confocal laser microscopy. In Progress in Biotechnology 19, Trends in High Pressure Bioscience and Biotechnology, Hayashi, R., ed. (Elsevier Science BV, Amsterdam, The Netherlands), pp. 227–231.

Molina-Hoppner, A., Sato, T., Kato, C., Ganzle, M.G., and Vogel, R.F. (2003). Effects of pressure on cell morphology and division of lactic acid bacteria. Extremophiles *7*, 511–516.

Mountfort, D.O., Rainey, F.A., Burghardt, J., Kasper, F., and Stackebrant, E. (1998). *Psychromonas antarcticus* gen. nov., sp. nov., A new aerotolerant anaerobic, halophilic psychrophile isolated from pond sediment of the McMurdo ice shelf, Antarctica. Arch. Microbiol. *169*, 231–238.

Murakami, C., Ohmae, E., Tate, S., Gekko, K., Nakasone, K., and Kato, C. (2010). Cloning and characterization of dihydrofolate reductases from deep-sea bacteria. J. Biochem. *147*, 591–599.

Nakashima, K., Horikoshi, K., and Mizuno, T. (1995). Effect of hydrostatic pressure on the synthesis of outer membrane proteins in *Escherichia coli*. Biosci. Biotechnol. Biochem. *59*, 130–132.

Nakasone, K., Ikegami, A., Kato, C., Usami, R., and Horikoshi, K. (1998). Mechanisms of gene expression controlled by pressure in deep-sea microorganisms. Extremophiles *2*, 149–154.

Nakasone, K., Ikegami, A., Kato, C., Usami, R., and Horikoshi, K. (1999). Analysis of *cis*-elements upstream of the pressure-regulated operon in the deep-sea barophilic bacterium *Shewanella violacea* strain DSS12. FEMS Microbiol. Lett. *176*, 351–356.

Nakasone, K., Yamada, M., Qureshi, M.H., Kato, C., and Horikoshi, K. (2001). Piezoresponse of the *cyo*-operon coding for quinol oxidase subunits in a deep-sea piezophilic bacterium, *Shewanella violacea*. Biosci. Biotechnol. Biochem. *65*, 690–693.

Nakasone, K., Ikegami, A., Kawano, H., Usami, R., Kato, C., and Horikoshi, K. (2002). Transcriptional regulation under pressure conditions by the RNA polymerase σ^{54} factor with a two component regulatory system in *Shewanella violacea*. Extremophiles *6*, 89–95.

Ninfa, A.J., Reitzer, L.J., and Magasanik, B. (1987). Initiation of transcription at the bacterial *glnAp2* promoter by purified *E. coli* components is facilitated by enhancers. Cell *50*, 1039–1046.

Nogi, Y., and Kato, C. (1999). Taxonomic studies of extremely barophilic bacteria isolated from the Mariana Trench, and *Moritella yayanosii* sp. nov., a new barophilic bacterial species. Extremophiles *3*, 71–77.

Nogi, Y., Kato, C., and Horikoshi, K. (1998a). *Moritella japonica* sp. nov., a novel barophilic bacterium isolated from a Japan Trench sediment. J. Gen. Appl. Microbiol. *44*, 289–295.

Nogi, Y., Kato, C., and Horikoshi, K. (1998b). Taxonomic studies of deep-sea barophilic *Shewanella* species, and *Shewanella violacea* sp. nov., a new barophilic bacterial species. Arch. Microbiol. *170*, 331–338.

Nogi, Y., Masui, N., and Kato, C. (1998c). *Photobacterium profundum* sp. nov., a new, moderately barophilic bacterial species isolated from a deep-sea sediment. Extremophiles *2*, 1–7.

Nogi, Y., Kato, C., and Horikoshi, K. (2002). *Psychromonas kaikoae* sp. nov., a novel piezophilic bacterium from the deepest cold-seep sediments in the Japan Trench. Int. J. Syst. Evol. Microbiol. *52*, 1527–1532.

Nogi, Y., Hosoya, S., Kato, C., and Horikoshi, K. (2004). *Colwellia piezophila* sp. nov., isolation of novel piezophilic bacteria from the deep-sea fissure sediments of the Japan Trench. Int. J. Syst. Evol. Microbiol. *54*, 1627–1631.

Nogi, Y., Hosoya, S., Kato, C., and Horikoshi, K. (2007). *Psychromonas hadalis* sp. nov., a novel piezophilic bacterium isolated from the bottom of the Japan Trench. Int. J. Syst. Evol. Microbiol. *57*, 1360–1364.

Ohmae, E., Kubota, K., Nakasone, K., Kato, C., and Gekko, K. (2004). Pressure-dependent activity of dihydrofolate reductase from a deep-sea bacterium *Shewanella violacea* strain DSS12. Chem. Lett. *33*, 798–799.

Owen, R., Legros, R.M., and Lapage, S.P. (1978). Base composition, size and sequence similarities of genome deoxyribonucleic acids from clinical isolates of *Pseudomonas putrefaciens*. J. Gen. Microbiol. *104*, 127–138.

Piccard, J., and Dietz, R.S. (1961). Seven Miles Down. (G.P. Putnum and Sons, New York, USA).

Poole, R.K., Williams, H.D., Downie, A., and Gibson, F. (1989). Mutations affecting the cytochrome *d*-containing oxidase complex of *Escherichia coli* K12: identification and mapping of a fourth locus, *cydD*. J. Gen. Microbiol. *135*, 1865–1874.

Poole, R.K., Hatch, L., Cleeter, M.W.J., Gibson, F., Cox, G.B., and Wu, G. (1993). Cytochrome *bd* biosynthesis in *Escherichia coli*: the sequences of the *cydC* and *cydD* genes suggest that they encode the components of an ABC membrane transporter. Mol. Microbiol. *10*, 421–430.

Poole, R.K., Gibson, F., and Wu, G. (1994). The *cydD* gene product, component of a heterodimeric ABC transporter, is required for assembly of periplasmic cytochrome *c* and of cytochrome *bd* in *Escherichia coli*. FEMS Microbiol. Lett. *117*, 217–224.

Pope, D.H., Smith, W.P., Swartz, R.W., and Landau, J.V. (1975). Role of bacterial ribosomes in barotolerance. J. Bacteriol. *121*, 664–669.

Prieur, D., Jebbar, M., Bartlett, D., Kato, C., and Oger, Ph. (2010). Piezophilic prokaryotes, In Comparative High Pressure Biology, Sebert, P., ed. (Science Publishers, New Hampshire, USA), pp. 281–318.

Quigley, M.M., and Colwell, R.R. (1968). Proposal of a new species *Pseudomonas bathycetes*. Int. J. Syst. Bacteriol. *18*, 241–252.

Qureshi, M.H., Kato, C., and Horikoshi, K. (1998a). Purification of a *ccb* type quinol oxidase specifically induced in a deep-sea barophilic bacterium, *Shewanella* sp. strain DB-172F. Extremophiles 2, 93–99.

Qureshi, M.H., Kato, C., and Horikoshi, K. (1998b). Purification of two pressure-regulated *c*-type cytochromes from a deep-sea bacterium, *Shewanella* sp. strain DB-172F. FEMS Microbiol. Lett. *161*, 301–309.

Sato, T., Kato, C., and Horikoshi, K. (1995). Effect of high pressure on gene expression by *lac* and *tac* promoters in *Escherichia coli*. J. Mar. Biotechnol. 3, 89–92.

Sato, T., Nakamura, Y., Nakashima, K.K., Kato, C., and Horikoshi, K. (1996). High pressure represses expression of the *malB* operon in *Escherichia coli*. FEMS Microbiol. Lett. *135*, 111–116.

Schmitz, W.J., Jr. (1995). On the interbasin-scale thermohaline circulation. Rev. Geophys. *33*, 151–173.

Schwarz, J.R., and Colwell, R.R. (1975). Abstracts, 75th Annual meeting of the American Society for Microbiology. American Society for Microbiology, Washington DC, *p*.162.

Seo, H.J., Bae, S.S., Lee, J.H., and Kim, S.J. (2005). *Photobacterium frigidiphilum* sp. nov., a psychrophilic, lipolytic bacterium isolated from deep-sea sediments of Edison Seamount. Int. J. Syst. Evol. Microbiol. *55*, 1661–1666.

Stetter, K.O. (1993). Life at the upper temperature border. In Frontiers of Life, Van Tran Than, J., Van Tran Than, K., Mounolou, J.C., Schneider, J., and McKay, C., eds. (Frontières, Gif-sur-Yvette), pp. 195–219.

Takai, K., Miyazaki, M., Hirayama, H., Nakagawa, S., Querellou, J., and Godfroy, A. (2009). Isolation and physiological characterization of two novel, piezophilic, thermophilic chemolithoautotrophs from a deep-sea hydrothermal vent chimney. Environ. Microbiol. *11*, 1983–1997.

Tamegai, H., Kato, C., and Horikoshi, K. (1998). Pressure-regulated respiratory system in barotolerant bacterium, *Shewanella* sp. strain DSS12. J. Biochem. Mol. Biol. Biophys. *1*, 213–220.

Tamegai, H., Kawano, H., Ishii, A., Chikuma, S., Nakasone, K., and Kato, C. (2005). Pressure-regulated biosynthesis of cytochrome *bd* in piezo- and psychrophilic deep-sea bacterium *Shewanella violacea* DSS12. Extremophiles *9*, 247–253.

Tamegai, H., Chikuma, S., Ishii, M., Nakasone, K., and Kato, C. (2008). The *narQP* genes for a two-component regulatory system from the deep-sea bacterium *Shewanella violacea* DSS12. DNA Seq. *19*, 308–312.

Tamura, K., Shimizu, T., and Kourai, H. (1992). Effects of ethanol on the growth and elongation of *Escherichia coli* under high pressures up to 40 MPa. FEMS Microbiol. Lett. 99, 321–324.

Urakawa, H., Kita-Tsukamoto, K., Steven, S.E., Ohwada, K., and Colwell, R.R. (1998). A proposal to transfer *Vibrio marinus* (Russell 1891) to a new genus *Moritella* gen. nov. as *Moritella marina* comb. nov. FEMS Microbiol. Lett. *165*, 373–378.

Venkateswaran, K., Moser, D.P., Dollhopf, M.E., Lies, D.P., Saffarini, D.A., MacGregor, B.J., Ringelberg, D.B., White, D.C., Nishijima, M., Sano, H., *et al.* (1999). Polyphasic taxonomy of the genus *Shewanella* and description of *Shewanella oneidensis* sp. nov. Int. J. Syst. Bacteriol. *49*, 705–724.

Vezzi, A., Campanaro, S., D'Angelo, M., Simonato, F., Vitulo, N., Lauro, F.M., Cestaro, A., Malacrida, G., Simionati, B., Cannata, N., *et al.* (2005). Life at depth: *Photobacterium profundum* genome sequence and expression analysis. Science *307*, 1459–1461.

Wang, F., Wang, J., Jian, H., Zhang, B., Li, S., Wang, F., Zeng, X., Gao, L., Bartlett, D.H., Yu, J., *et al.* (2008) Environmental adaptation: genomic analysis of the piezotolerant and psychrotolerant deep-sea iron reducing bacterium *Shewanella piezotolerans* WP3. PLoS ONE 3, e1937.

Welch, T.J., and Bartlett, D.H. (1996). Isolation and characterization of the structural gene for OmpL, a pressure-regulated porin-like protein from the deep-sea bacterium *Photobacterium* species strain SS9. J. Bacteriol. *178*, 5027–5031.

Welch, T.J., and Bartlett, D.H. (1998). Identification of a regulatory protein required for pressure-responsive gene expression in the deep- sea bacterium *Photobacterium* species strain SS9. Mol. Microbiol. *27*, 977–985.

Wu, F.Y., Yarbrough, L.R., and Wu, C.W. (1976). Conformational transition of *Escherichia coli* RNA polymerase induced by the interaction of sigma subunit with core enzyme. Biochemistry *15*, 3254–3258.

Xu, Y., Nogi, Y., Kato, C., Liang, Z., Rüger, H-J., Kegel, D.D., and Glansdorff, N. (2003a). *Moritella profunda* sp. nov. and *Moritella abyssi* sp. nov., two psychropiezophilic organisms isolated from deep Atlantic sediments. Int. J. Syst. Evol. Microbiol. *53*, 533–538.

Xu, Y., Nogi, Y., Kato, C., Liang, Z., Rüger, H-J., Kegel, D.D., and Glansdorff, N. (2003b). *Psychromonas profunda* sp. nov., a psychropiezophilic bacterium from deep Atlantic sediments. Int. J. Syst. Evol. Microbiol. *53*, 527–532.

Yamada, M., Nakasone, K., Tamegai, H., Kato, C., Usami, R., and Horikoshi, K. (2000). Pressure-regulation of soluble cytochromes *c* in a deep-sea piezophilic bacterium, *Shewanella violacea*. J. Bacteriol. *182*, 2945–2952.

Yanagibayashi, M., Nogi, Y., Li, L., and Kato, C. (1999). Changes in the microbial community in Japan Trench sediment from a depth of 6292 m during cultivation without decompression. FEMS Microbiol. Lett. *170*, 271–279.

Yayanos, A.A. (1986). Evolutional and ecological implications of the properties of deep-sea barophilic bacteria. Proc. Natl. Acad. Sci. U.S.A. *83*, 9542–9546.

Yayanos, A.A. (1995). Microbiology to 10,500 meters in the deep sea. Annu. Rev. Microbiol. *49*, 777–805.

Yayanos, A.A., and Dietz, A.S. (1982). Death of a hadal deep-sea bacterium after decompression. Science *220*, 497–498.

Yayanos, A.A., Dietz, A.S., and Boxtel, R.V. (1979). Isolation of a deep-sea barophilic bacterium and some of its growth characteristics. Science *205*, 808–810.

Yayanos, A.A., Dietz, A.S., and Boxtel, R.V. (1981). Obligately barophilic bacterium from the Mariana Trench. Proc. Natl. Acad. Sci. U.S.A. *78*, 5212–5215.

Yayanos, A.A., Dietz, A.S., and Boxtel, R.V. (1982). Dependence of reproduction rate on pressure as a hallmark of deep-sea bacteria. Appl. Environ. Microbiol. *44*, 1356–1361.

Yura, T., and Ishihama, A. (1979). Genetics of bacterial RNA polymerases. Annu. Rev. Genet. *13*, 59–97.

Zeng, X., Birrien, J-L., Fouquet, Y., Cherkashov, G., Jebbar, M., Querellou, J., Oger, Philippe, Cambon-Bonavia, M-A., Xiao, X., and Prieur, D. (2009). *Pyrococcus* CH1, an obligate piezophilic hyperthermophile: extending the upper pressure-temperature limits for life. ISME J. 3, 873–876.

Zobel, C.E., and Cobet, A.B. (1963). Filament formation by *Escherichia coli* at increased hydrostatic pressures. J. Bacteriol. *87*, 710–719.

Zobell, C.E., and Johnson, F.H. (1949). The influence of hydrostatic pressure on the growth and viability of terrestrial and marine bacteria. J. Bacteriol. *57*, 179–189.

Zobell, C.E., and Morita, R.Y. (1957). Barophilic bacteria in some deep-sea sediments. J. Bacteriol. *73*, 563–568.

Physiological Adaptations and Biotechnological Applications of Acidophiles

11

Mark Dopson

Abstract

Acidophilic microorganisms are capable of growth at low pH and are defined as having an optimum below pH 5, with some extreme acidophiles capable of growth at pH 0. Acidophiles have an important role in ecology by catalysing the generation of acidic, metal-containing solutions that can inhibit plant and animal growth, and in biotechnology via their propensity to solubilize metals from sulfide minerals. This review summarizes the most important aspects of acidophile physiology, including growth substrates (e.g. inorganic and organic carbon) and energy sources; temperature optima (from cold environments to high temperature thermal pools); and adaptations to metals in solution and low growth pH. The exploitation of acidophiles in biotechnology (e.g. in biomining) and potential future trends in research are also discussed. Owing to a lack of general genetic systems in acidophiles, much of the latest research has been generated by systems biology approaches and these data have been focused upon.

Introduction

Acidophilic microorganisms grow optimally between pH 0.5 and 5 while extreme acidophiles grow with a pH optimum < 3. Though acidophiles are most widely distributed in the *Bacterial* and *Archaeal* domains they are also present in the *Eukarya*, with acidophilic yeasts having been recently identified (Gadanho and Sampaio, 2006; Russo *et al.*, 2008). Acidophilic microorganisms thrive over a wide temperature range, from a single identified psychrotolerant species that can grow at 5°C (no psychrophilic acidophiles have been characterized), through mesophiles, to thermophiles that can grow at 80°C.

Acidophiles are of interest for a number of reasons. They may represent some of the earliest microorganisms on Earth, as it has been hypothesized that acidic conditions may have existed during the Archaean and Proterozoic periods (circa 3800–500 million years ago). For the same reasons, acidophiles are also relevant to the search for life on other planets, termed 'Astrobiology'. Another area of interest is in bioenergetics, as acidophiles have a large pH gradient across the cytoplasmic membrane that has to be maintained while producing ATP (see the 'pH homeostasis' section). Acidophilic microorganisms can impact the environment via generation of acid and metal laden solutions from mining environments and acid soils (Johnson, 2005), and by affecting global geochemical cycles (e.g. iron and sulfur) by mediating oxidation and reduction reactions that can affect the biological availability of elements (Druschel *et al.*, 2004) (see the 'Growth substrates and their sources' section). Acidophiles have also been exploited in biotechnology, as exemplified by their role in the catalysis of sulfide mineral dissolution for recovery of metals, and also the production of enzymes for other biotechnological processes (see the 'Biotechnological applications' section).

This chapter will present an overview of acidophile ecology, metabolism, adaptations to their environment, and their biotechnological applications. This is a broad area for which several more specific reviews have been published, including on acidophile ecology (Hallberg and Johnson, 2001; Rawlings and Johnson, 2007), metabolism (Rohwerder and Sand, 2007; Ghosh and Dam, 2009; Johnson and Hallberg, 2009; Quatrini *et al.*, 2009), metal resistance (Dopson *et al.*, 2003), pH homeostasis (Slonczewski *et al.*, 2009), and biotechnology (Rawlings and Johnson, 2007). Owing to a lack of universal acidophile genetic systems, much of the latest investigations have utilized systems biology approaches and therefore, this review will focus on the most recent data generated with 'omics' techniques.

Ecology and metabolism

Acidophiles are found across all three domains of life (Table 11.1 and Fig. 11.1) although significantly more information is available for the acidophilic bacteria and archaea as compared with the eukarya. The following sections will focus on the two prokaryotic domains and acidophiles have been identified from bacterial and archaeal phyla including *Acidianus, Acidimicrobium, Acidiphilium, Acidithiobacillus, Alicyclobacillus, Ferroplasma, Leptospirillum, Metallosphaera, Picrophilus, Sulfobacillus, Sulfolobus,* and *Thermoplasma*. However, it is worth mentioning that eukaryotic 18S rRNA genes with homology to acidophilic fungi and protists genes have been identified from the extremely acidic (pH <0.9) Richmond mine site at Iron Mountain (CA, USA) (Baker *et al.*, 2009). The analysis was consistent with previous data from this site suggesting that the total acidophile diversity was low, presumably due to the extremely acidic pH. A second environment containing acidophilic eukarya is the naturally acidic Rio Agrio-Lake Caviahue volcanic area (northwestern Patagonia) where two novel yeast species have been identified (Russo *et al.*, 2008). Very little information is available concerning eukaryotic adaptations to acidic pH. The acidophilic yeast *Clamydomonas* that has a pH optimum of 2 maintains a cytoplasmic pH close to neutral (as do acidophilic bacteria and archaea; see the 'pH homeostasis' section) with an increased energy demand associated with growth at low pH (Messerli *et al.*, 2005).

Natural and anthropomorphic environments

Areas characterized by low pH encompass both natural and manmade environments. Examples of naturally low pH environments include locations where sulfide minerals are exposed to air and water, acid soils, and marine hydrothermal vents. Anthropomorphic acidic environments include biomining assemblies, coal mine drainage, and some industrial process waters such as from the metallurgy industry. The pH in these environments ranges from moderately acidic, as in some coal run-off with relatively low sulfur content (that generates acidity by acidophile mediated oxidation to sulfuric acid), to extreme acid mine drainage (AMD) with negative pH values. One of the most extreme acidic environments is the Iron Mountain AMD site where pH values as low as −3.6 have been measured (Nordstrom and Alpers, 1999), though no microorganisms have been found capable of growth at that pH. Another very low pH environment is the Frasassi cave system (Italy) with pH values between 0 and 1 due to oxidation of H_2S (Macalady *et al.*, 2007).

Growth substrates and their sources

Various acidophile species are able to grow aerobically or anaerobically by oxidation of inorganic energy sources such as ferrous iron (Fe^{2+}) and inorganic sulfur compounds (ISCs) as well as organic carbon such as yeast extract. Both chemoautotrophic and chemoheterotrophic acidophiles exist, which are able to fix CO_2 and/or assimilate organic carbon for cell growth, respectively.

Iron oxidation, reduction, and energy conservation

Many acidophiles use ferrous iron oxidation to generate energy because Fe^{2+} ions are produced during sulfide mineral dissolution (see the 'Biomining' section), although a too high intracellular concentration can be toxic. Therefore, the intracellular Fe^{2+} concentration needs to be regulated (see the 'Metal and metalloid homeostasis and resistance' section). The increased solubility of Fe^{2+} over Fe^{3+} at moderately acidic pH values and lower rate of abiotic Fe^{2+} oxidation by oxygen

Table 11.1 Selected acidophilic microorganisms, focusing on those whose genome has been sequenced and/or discussed in this review. The table has been adapted and updated from Johnson (1998), Hallberg and Johnson (2001), and Siezen and Wilson (2009)

Organism	Energy source	Metabolism	Temperature range	Oxygen utilization	Genome sequence
***Bacteria* (Gram-negative)**					
Acidiphilium acidophilum	ISC and organic C	A, M, H	Mesophile	Aerobic	NA
Acidiphilium cryptum	Organic C	H	Mesophile	Facultative anaerobe	http://genome.ornl.gov/
Acidiphilium multivorum	Organic C	H	Mesophile	Aerobe	Incomplete
Acidithiobacillus caldus	ISC	A	Thermophile	Aerobe	Valdes *et al.* (2009)
Acidithiobacillus ferrivorans	Fe^{2+} and ISC	A	Psychrotolerant	Facultative anaerobe	Liljeqvist *et al.* (2011)
Acidithiobacillus ferrooxidans	Fe^{2+} and ISC	A	Mesophile	Facultative anaerobe	Unpublished
Acidithiobacillus thiooxidans	ISC	A	Mesophile	Aerobe	Unpublished
Leptospirillum ferriphilum	Fe^{2+}	A	Mesophile	Aerobe	NA
Leptospirillum ferrooxidans	Fe^{2+}	A	Mesophile	Aerobe	NA
'*Leptospirillum rubarum*'	Fe^{2+}	A	Mesophile	Aerobe	Tyson *et al.* (2004)
Methylacidiphilium infernorum	CH_4	A	Thermophile	Aerobe	Hou *et al.* (2008)
Thiomonas cuprina	ISC and organic C	A, M, H	Mesophile	Aerobe	NA
***Bacteria* (Gram-positive)**					
Acidimicrobium ferrooxidans	Fe^{2+}	A, M, H	Thermophile	Aerobe	Clum *et al.* (2009)
Sulfobacillus acidophilus	Fe^{2+}, ISC, organic C	A, M, H	Thermophile	Aerobe	Incomplete
Su. thermosulfidooxidans	Fe^{2+}, ISC	A, M, H	Thermophile	Aerobe	Incomplete
Acidobacterium capsulatum	Organic C	H	Mesophile	Aerobe	Incomplete

Table 11.1 continued

Organism	Energy source	Metabolism	Temperature range	Oxygen utilization	Genome sequence
Archaea					
Acidianus brierleyi	Fe^{2+} and ISC	A,M,H	Thermophile	Facultative anaerobe	Incomplete
'*Ferroplasma acidarmanus*'	Fe^{2+} and organic C	M,H	Mesophile	Facultative anaerobe	http://genome.ornl.gov/
Ferroplasma acidiphilum	Fe^{2+} and organic C	A,M,H	Mesophile	Facultative anaerobe	N.A.
Ferroplasma Type II	Fe^{2+} and organic C	H	Mesophile	Unknown	Tyson *et al.* (2004)
Metallosphaera sedula	Fe^{2+} and ISC	A,M,H	Thermophile	Aerobe	Auernik *et al.* (2008)
Picrophilus oshimae	Organic C	H	Thermophile	Aerobe	N.A.
Picrophilus torridus	Organic C	H	Thermophile	Aerobe	Fütterer *et al.* (2004)
Sulfolobus acidocaldarius	Organic C	H	Thermophile	Aerobe	Chen *et al.* (2005)
Sulfolobus metallicus	Fe^{2+} and ISC	A	Thermophile	Aerobe	N.A.
Sulfolobus solfataricus	S^0 and organic C	M, H	Thermophile	Aerobe	She *et al.* (2001)
Sulfolobus tokodaii	Fe^{2+} and organic C	M, H	Thermophile	Aerobe	Kawarabayasi *et al.* (2001)
Thermoplasma acidophilum	Organic C	H	Thermophile	Facultative anaerobe	Ruepp *et al.* (2000)
Thermoplasma volcanium	Organic C	H	Thermophile	Facultative anaerobe	Kawashima *et al.* (2000)
Eukarya					
Chlamydomonas sp.	Organic C, light	H	Mesophile	Aerobe	N.A.

A, autotroph; M, mixotroph; H, heterotroph; NA, not available.

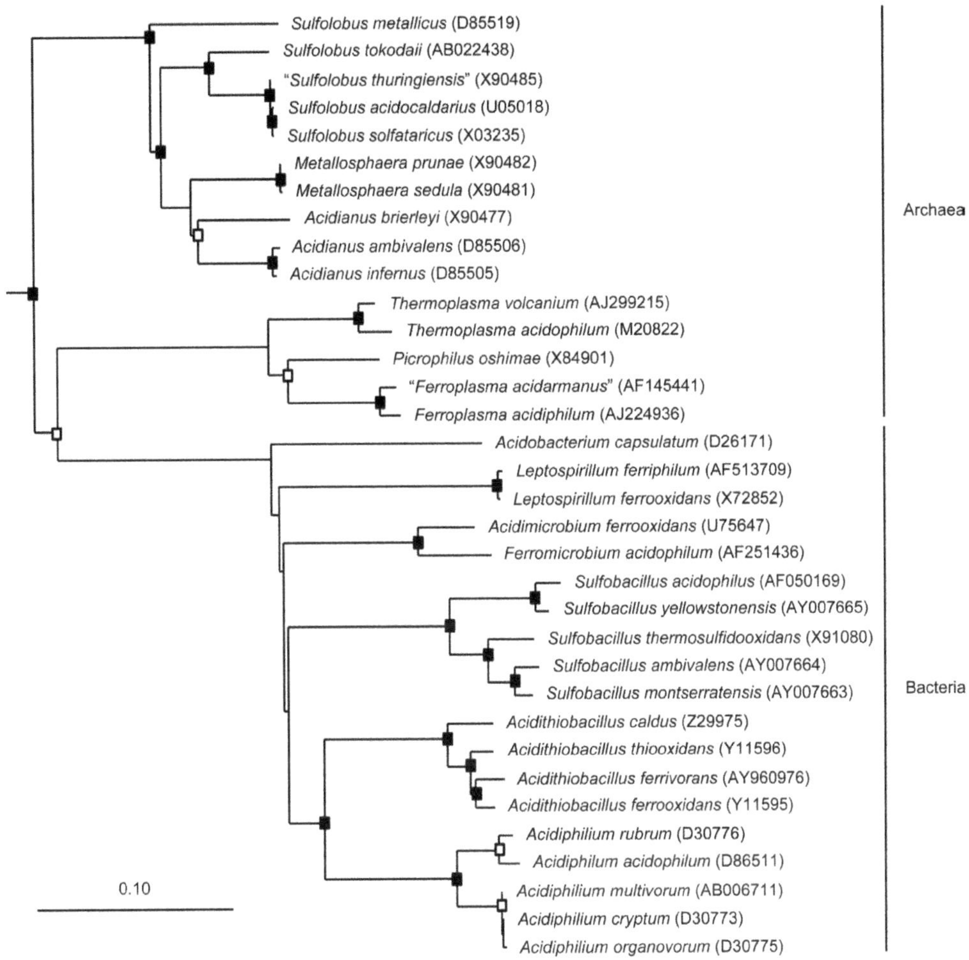

Figure 11.1 16S rRNA gene phylogenetic tree of selected acidophilic archaea and bacteria. Three trees were constructed using neighbour joining (presented), maximum likelihood, and parsimony methods using the ARB phylogenetic program (Ludwig *et al.*, 2004). The tree was rooted with korarchaeote pBA5 (accession number AF176347; not shown) and nodes where all three trees (■) and two trees (□) agreed are marked. The scale bar represents 10% sequence divergence.

make it easily accessible. However, oxidation of Fe^{2+} at acidic pH offers a lower energy gain from the Fe^{2+}/O_2 couple compared with that obtained at pH 7.0.

The best characterized Fe^{2+} oxidation system is that of *Acidithiobacillus ferrooxidans* (Fig. 11.2A) due to its role as a model acidophilic microorganism (Ferguson and Ingledew, 2008). The paradigm for *A. ferrooxidans* Fe^{2+} oxidation has evolved based on studies that utilize bioinformatic analysis of the genome (Valdes *et al.*, 2008; Quatrini *et al.*, 2009), transcriptomics (Quatrini *et al.*, 2006, 2009;), proteomics (Ramirez *et al.*, 2004; Chi *et al.*, 2007), and biochemistry (Castelle *et al.*, 2008). The *A. ferrooxidans* genes responsible for Fe^{2+} oxidation and 'downhill' electron transfer for ATP production are encoded in the *petI* and *rus* transcriptional units. The *petI* unit comprises the genes for the bc_1 complex, a short chain dehydrogenase for which the function has not been confirmed, and cytochrome c_4 that may accept the electrons from Fe^{2+} oxidation. The *rus* transcriptional unit consists of a conserved hypothetical gene encoding the Cup protein, two

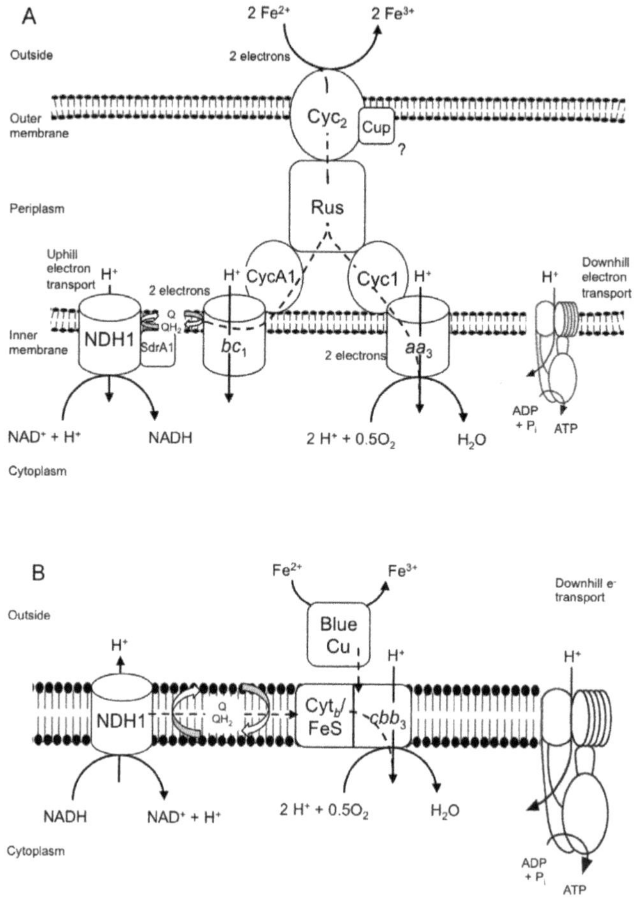

Figure 11.2 Models of Fe^{2+} oxidation and energy conservation in (A) *Acidithiobacillus ferrooxidans*, with both up- and downhill electron transport shown (adapted from Chi *et al.*, 2007, and Quatrini *et al.*, 2009); and (B) '*F. acidarmanus*' Fer1 (adapted from Dopson *et al.* (2005). Electron flow is indicated by dashed lines. The structure of the model components and electron flow is highly speculative in '*F. acidarmanus*' Fer1, for which much less data is available to support the model.

c-type cytochromes, an aa_3 cytochrome oxidase, and rusticyanin (Quatrini *et al.*, 2009). The role of the *petI* and *rus* encoded proteins in Fe^{2+} oxidation is suggested by the transcriptomic studies mentioned above, as well as Bruscella *et al.* (2007) and Yarzabal *et al.* (2004). The products of the *rus* and *petI* gene clusters form a complex for Fe^{2+} oxidation that spans the outer membrane, periplasm, and inner membrane to pass electrons from Fe^{2+} to oxygen (Castelle *et al.*, 2008). The role of the conserved hypothetical protein (Cup) may be to deliver copper to the aa_3 cytochrome oxidase or rusticyanin (Rus), or it may be involved in electron transfer (Quatrini *et al.*, 2009). In addition, several of the suggested Fe^{2+} oxidation proteins have been identified using a high throughput proteomic analysis of periplasmic proteins, including *c*-type cytochromes (Cyc1, Cyc2, CycA-1, and Cyc A-2), Rus, and the Iro iron oxidase (Chi *et al.*, 2007). The newly described species, *Acidithiobacillus ferrivorans* does not contain a typical *rusA* gene encoding rusticyanin and instead has a *rusB* homologue (Hallberg *et al.*, 2009).

An alternative pathway for electron transfer also occurs when *A. ferrooxidans* is grown on Fe^{2+} alone (termed 'reverse' or 'up-hill' electron transfer), as the Fe^{2+}/O_2 redox couple is too weak to drive reduction of $NAD(P)^+$ to NAD(P)H during CO_2 fixation by the Calvin cycle. Up-hill electron transport is driven by the proton motive force generated by electron transfer from CycA1 through the bc_1 complex, the quinone pool, and to the NADH dehydrogenase complex that is potentially facilitated by the *sdrA1* gene product (Quatrini *et al.*, 2009). Experimental evidence for up-hill electron transfer is presented by Elbehti *et al.* (2000), although the results have been questioned (Ferguson and Ingledew, 2008). The mechanism by which the flow of electrons between downhill and up-hill electron transport is regulated is unknown but is suggested to be thermodynamic (Ferguson and Ingledew, 2008).

The Fe^{2+} oxidation system of '*Ferroplasma acidarmanus*' Fer1 has been reconstructed from its genome sequence (Tyson *et al.*, 2004) and subsequently updated after proteomic investigation of proteins induced by the presence of Fe^{2+} (Dopson *et al.*, 2005). The putative model (Fig. 11.2B) contains a blue copper-haem sulfocyanin that may act in a similar manner to *Sulfolobus acidocaldarius* SoxE and rusticyanin from *A. ferrooxidans* by passing electrons from Fe^{2+} to a cbb_3 terminal oxidase (Dopson *et al.*, 2005). Tyson *et al.* (2004) provide genomic evidence for both downhill and up-hill electron transport in the obligatory Fe^{2+} oxidizing *Leptospirillum* group II ('*Leptospirillum rubarum*') and a Fe^{2+} oxidation system with the first electron acceptor being a red cytochrome that passes the electron to a cytochrome *b*/FeS and onto a *bd* type cytochrome oxidase. The *Leptospirillum* group II and III genome sequences also contain *c*-type cytochromes, genes encoding a putative bc_1 complex, and a cytochrome cbb_3 complex (Goltsman *et al.*, 2009). A different method for Fe^{2+} oxidation is present in the archaeon *Sulfolobus metallicus*, where genes up-regulated during Fe^{2+} oxidation were identified using subtractive hybridization, indicating that the *foxABC* locus is probably involved in electron transport. Homologous genes in the genomes of other *Sulfolobus* spp. were only identified in *Sulfolobus tokodaii*, the only other species from the *Sulfolobus* genus capable of oxidizing Fe^{2+} (Bathe and Norris, 2007). In *Metallosphaera sedula* the *fox* and *soxNL-cbsABA* terminal/quinol oxidase gene clusters were up-regulated during growth on Fe^{2+}, though it is not known how electrons from Fe^{2+} oxidation are linked to energy conservation (Auernik and Kelly, 2008). Finally, *A. ferrivorans* can grow utilizing the S^0/Fe^{3+} redox couple under aerobic conditions (Kupka *et al.*, 2009), and the *rus* operon of the halotolerant acidophile '*Thiobacillus prosperus*' does not contain a *cyc1* homologue (coding for a previously characterized key cytochrome *c*) suggesting it has a modified Fe^{2+} oxidizing pathway (Nicolle *et al.*, 2009).

A. ferrooxidans ferric iron (Fe^{3+}) reduction was identified in 1976 (Brock and Gustafson, 1976) and has since been identified in other acidophiles, including *Acidiphilium* spp. (Bridge and Johnson, 1998; Kusel *et al.*, 1999), *Acidocella facilis* (Johnson and McGinness, 1991; Coupland and Johnson, 2008), and *Ferroplasma* spp. (Dopson *et al.*, 2006). *A. ferrooxidans* Fe^{3+} reduction may be mediated by tetrathionate hydrolase, though a direct coupling has not been demonstrated (Sugio *et al.*, 2009). In contrast to the type strain, *A. ferrooxidans* strain JCM 7811 is also able to grow via the H_2/Fe^{3+} couple linked to a *c*-type cytochrome (Ohmura *et al.*, 2002). Bioinformatic analysis of the *A. ferrooxidans* major anaerobic transcriptional regulator (FNR) suggests it regulates genes associated with energy metabolism, nitrogen fixation, and carbohydrate metabolism (Osorio *et al.*, 2009). Proteomic analysis of '*F. acidarmanus*' Fer1 anaerobic growth supported by organic carbon oxidation linked to Fe^{3+} reduction identified proteins related to the reductive tricarboxylic acid pathway and a putative flavoprotein in electron transport (Dopson *et al.*, 2006).

Inorganic sulfur compound (ISC) oxidation, reduction and energy conservation

Typical ISCs oxidized by acidophiles include sulfide (S^{2-}), elemental sulfur (S^0), tetrathionate ($S_4O_6^{2-}$), thiosulfate ($S_2O_3^{2-}$), and sulfite (SO_3^{2-}), all of which may be present in sulfide mineral dissolution environments. The pathway of ISC oxidation has been biochemically identified in microorganisms including *Acidithiobacillus* spp., *Acidiphilium acidophilum*, *Sulfobacillus* spp., and *Acidianus ambivalens* (reviewed by Rohwerder and Sand (2007)). ISC oxidation to sulfuric acid by Gram-negative microorganisms occurs through a concerted action of periplasmic and cytoplasmic enzymes. More energy is generated from the oxidation of one mole of many ISCs than one mole of Fe^{2+} as, for example, oxidation of S^{2-} (oxidation state −2) to SO_4^{2-} (oxidation state +6) generates eight electrons compared with a single electron generated from Fe^{2+} oxidation. From an experimental perspective, the multiple species of ISCs, the many biological oxidation pathways, and the existence of abiotic oxidation renders elucidation of the ISC oxidation systems quite complicated.

As suggested for Fe^{2+} oxidation, the recently proposed species *A. ferrivorans* (made up of strains previously attributed to *A. ferrooxidans*) may further add to the complexity of the array of *A. ferrooxidans* ISC oxidation genes and pathways (Fig. 11.3). Genes for *A. ferrooxidans* ISC oxidation include the sulfide/quinone oxidoreductase (*sqr*) that oxidizes sulfide to sulfur and feeds electrons into the quinone pool (Wakai *et al.*, 2004). Oxidation of thiosulfate to tetrathionate is proposed to be mediated by two homologues of the *doxDA* genes coding for a thiosulfate/quinone oxidoreductase that also feeds electrons into the quinone pool. The two *doxDA* sets are located in different parts of the genome and their individual roles have not been experimentally elucidated (Valdes *et al.*, 2008). A thiosulfate regenerating tetrathionate hydrolase (*tetH*) has also been identified (Kanao *et al.*, 2007). Other genes involved in ISC oxidation include five open reading frames encoding a thiosulfate sulfur transferase (rhodanese), although its physiological role has not been resolved (Gardner and Rawlings, 2000). Also, relevant enzyme activities have been identified biochemically, such as sulfur dioxygenase that oxidizes S^0 to SO_3^{2-} (Rohwerder and Sand, 2003) and sulfite oxidase (Vestal and Lundgren, 1971). Energy conservation from ISC oxidation is possible via three alternative oxidases encoded either by the *pet-II* operon (*hip-petA2B2C2-sdrA2-cycA2*), a *bd* ubiquinol oxidase (*cydAB*), or a cytochrome bo_3 ubiquinol oxidase (*cyoABCD*) (Brasseur *et al.*, 2004; Bruscella *et al.*, 2007). Of

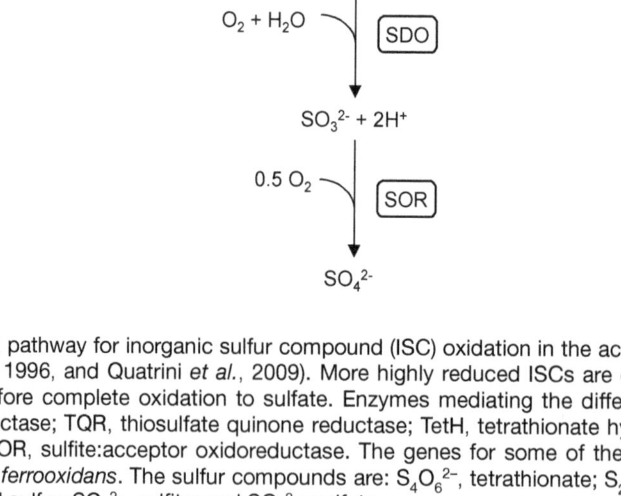

Figure 11.3 General pathway for inorganic sulfur compound (ISC) oxidation in the acidithiobacilli (adapted from Hallberg *et al.*, 1996, and Quatrini *et al.*, 2009). More highly reduced ISCs are oxidized to S^0 as the key intermediate before complete oxidation to sulfate. Enzymes mediating the different steps are: SQR, sulfide quinone reductase; TQR, thiosulfate quinone reductase; TetH, tetrathionate hydrolase; SDO, sulfur dioxygenase; and SOR, sulfite:acceptor oxidoreductase. The genes for some of these proteins have not been identified in *A. ferrooxidans*. The sulfur compounds are: $S_4O_6^{2-}$, tetrathionate; $S_2O_3^{2-}$, thiosulfate; S^{2-}, sulfide; S^0, elemental sulfur; SO_3^{2-}, sulfite; and SO_4^{2-}, sulfate.

the genes coding for proteins involved in ISC oxidation, a high throughput proteomic analysis identified tetrathionate hydrolase, HiPIP (high-potential iron protein), and cytochrome c_4 (Chi *et al.*, 2007).

The ISC oxidation pathway has been biochemically identified in *Acidithiobacillus caldus* and was found to be similar to that of *A. ferrooxidans* (Hallberg *et al.*, 1996). Genes encoding proteins in the pathway have been identified in the *A. caldus* genome, including two separate clusters of incomplete SOX system genes for S^0 oxidation (*soxYZB-hyp-resB-soxAX-resC* and *soxYZA-hyp-SoxB*); sulfide/quinone oxidoreductase (*sqr*); sulfur oxygenase-reductase gene (*sor*); tetrathionate hydrolase (*tetH*) and a thiosulfate quinone oxidoreductase (*doxD*) (Valdes *et al.*, 2009). The *A. caldus* gene cluster containing tetrathionate hydrolase (IS*ac1-rsrRS-tetH-doxD*) was characterized and shown to be up-regulated in the presence of tetrathionate, thiosulfate, and pyrite (Rzhepishevska *et al.*, 2007). ISC oxidation is linked to energy conservation by a bo_3-type quinol oxidase (*cyoBACD*), a *bd*-type ubiquinol oxidase (*cydAB*), and a putative aa_3-type quinol oxidase (*quoxBACD*) (Valdes *et al.*, 2009). In addition, inhibitor studies suggest that *A. caldus* also contains a cytochrome *c* oxidase (cytochrome aa_3 or cytochrome cbb_3) and a ubiquinol oxidase (Dopson *et al.*, 2002).

Other acidophile ISC utilization components that have been characterized include a sulfur oxygenase-reductase whose gene (*sor*) is up-regulated when *S. metallicus* is grown on S^0, concomitant with down-regulation of most of the genes involved in Fe^{2+} oxidation (Bathe and Norris, 2007). *M. sedula* genes encoding the SoxABCD quinol oxidase and SoxM oxidase supercomplex were up-regulated during growth on S^0 and yeast extract, respectively, suggesting that regulation of the respiratory complexes is substrate-dependent (Kappler *et al.*, 2005). Other identified *M. sedula* genes encode the thiosulfate quinol oxidase (DoxDA), two sulfite:acceptor oxidoreductase-like proteins, sulfide:quinone oxidoreductase proteins, and a membrane-associated TetH, though the *sor* gene was not present (Auernik *et al.*, 2008). All of the sequenced *Sulfolobus* spp. genomes contain ISC metabolic genes for sulfite reductase, 5′-adenylylsulfate sulfotransferase, and sulfate adenylyltransferase, but only *S. tokodaii* seems to contain a sulfur oxygenase that oxidizes S^0 (Chen *et al.*, 2005).

There are only a few reported cases of acidophiles capable of growth via ISC reduction, including *A. ferrooxidans* strain JCM 7811 that can grow via the H_2/S^0 couple (Ohmura *et al.*, 2002) and a single acidophilic sulfate reducer (isolate CL4) that grows at pH ≥3.0, utilizing glycerol as an electron donor (Johnson *et al.*, 2009).

Growth on sulfide minerals

Some acidophiles are able to catalyse the oxidation of sulfide minerals to produce Fe^{2+} and ISCs (see the 'Biomining' section) and certain genes/proteins are up-regulated during this mode of growth. Rusticyanin is expressed when *A. ferrooxidans* is grown in the presence of iron-containing sulfide minerals but not zinc sulfide (Ramirez *et al.*, 2004), and a rhodanese-like protein was expressed during growth on sulfide minerals and ISCs but absent when grown on Fe^{2+} (Ramirez *et al.*, 2002). The *S. metallicus* Fe^{2+} oxidation mediating *fox* genes (expressed during Fe^{2+} oxidation) were also highly expressed on pyrite, while *foxGH* were more highly expressed on pyrite than Fe^{2+} (Bathe and Norris, 2007). In contrast to the above genes, the *M. sedula* SoxABCD quinol oxidase is down-regulated when grown on pyrite (Kappler *et al.*, 2005).

Inorganic and organic carbon

Many acidophiles are capable of fixing CO_2 for autotrophic growth (Table 11.1). Described pathways for CO_2 fixation include the Calvin–Benson–Bassham cycle, the acetyl-CoA pathway, the reductive tricarboxylic acid cycle, or the 3-hydroxypropionate cycle. As mentioned above, acquisition of organic carbon from CO_2 explains why the obligate autotroph *A. ferrooxidans* requires up-hill electron transport during growth on Fe^{2+} alone.

The most common CO_2 fixation pathway in the acidophiles is the Calvin-Benson-Bassham cycle that takes place in either the cytoplasm or within inclusion bodies called carboxysomes. The formation of carboxysomes has been observed in *A. ferrooxidans* and *Acidithiobacillus thiooxidans* under CO_2 limiting conditions (Shively *et al.*, 1998). The first step in the Calvin-Benson-Bassham cycle is mediated by the key enzyme in the cycle, ribulose bisphosphate carboxylase (RubisCO), that catalyses the reaction of CO_2 and ribulose bisphosphate to form two molecules of phosphoglyceric acid (one of which contains the carbon from CO_2). The cycle continues to reform ribulose bisphosphate and the stoichiometry is such that for each six CO_2 molecules, one fructose-6-phosphate is produced and incorporated into the cell. *A. ferrooxidans, A. thiooxidans,* and *A. caldus* all contain genes for CO_2 fixation via the Calvin-Benson-Bassham cycle (Valdes *et al.*, 2007a). The presence of RubisCO has been biochemically confirmed in *A. ferrooxidans* (Gale and Beck, 1967) and *A. thiooxidans* (Suzuki and Werkman, 1958). *A. ferrooxidans* contains two copies of the form I and one copy of the form II RubisCO genes, whereas *A. thiooxidans* contains only a single copy of each form (Levican *et al.*, 2008). Microarray analysis of *A. ferrooxidans* gene clusters coding for CO_2 fixation and carboxysome construction were up-regulated when cells were grown on sulfur in comparison to growth on Fe^{2+} (Appia-Ayme *et al.*, 2006). Form I of RubisCO dominates during growth under autotrophic conditions, whereas form II is expressed during all conditions and may be used to maintain redox balance (Appia-Ayme *et al.*, 2006). The Gram-positive bacteria *Sulfobacillus thermosulfidooxidans* and *Sulfobacillus acidophilus* grow autotrophically with different requirements for CO_2, and *Su. acidophilus* has been shown to contain 8 putative Calvin-Benson-Bassham cycle genes in a single cluster (Caldwell *et al.*, 2007).

Reconstruction of metagenomic sequencing of *Leptospirillum* group II ('*L. rubarum*') identified the presence of RubisCO (Tyson *et al.*, 2004), though these species have not been biochemically demonstrated to fix CO_2 by this mechanism. In contrast, *Leptospirillum ferriphilum* DSM 17947 contains the four key enzymes necessary for utilization of the reductive TCA cycle for CO_2 fixation (Levican *et al.*, 2008), in which the cycle runs in reverse to fix two molecules of CO_2, producing acetyl CoA. The majority of the genes coding for the reductive TCA cycle are found in two genomic clusters, with the remaining genes present on other loci (Levican *et al.*, 2008).

Some archaea use the 3-hydroxypropionate cycle for CO_2 fixation. The key enzymes of this cycle, acetyl-CoA carboxylase and propionyl-CoA carboxylase, are present in *Acidianus brieryleyi* (Ishii *et al.*, 1996) and *S. metallicus* (Burton *et al.*, 1999). These activities are carried out by a single enzyme in *M. sedula* as part of a modified pathway (Hugler *et al.*, 2007). The case of CO_2 fixation in *Ferroplasma* spp. has been the subject of conflicting reports, with the original publication by Golyshina *et al.* (2000) reporting that fixation of CO_2 occurred. Subsequent studies have not found any evidence for CO_2 fixation (Dopson *et al.*, 2004, 2005), though the absence of data to support a hypothesis does not constitute a proof that it does not occur. Also, key enzymes in the archaeal 3-hydroxypropionate cycle, the Ljungdahl–Wood pathway, and the reverse TCA cycle were absent from *Ferroplasma* spp. as judged by comparative genomics and metabolic reconstruction (Cárdenas *et al.*, 2009). However, bioinformatic analysis of *Ferroplasma* genomes has suggested a putative chimerical pathway combining the Ljungdahl–Wood pathway and the 'serine cycle' for CO_2 fixation, though it has not been biochemically verified (Cárdenas *et al.*, 2009).

The use of organic carbon for carbon and energy sources is widespread, with many acidophiles capable of growth under heterotrophic or mixotrophic conditions (Table 11.1). Pathways for organic carbon utilization have been reconstructed for several acidophiles including *A. ferrooxidans* (Valdes *et al.*, 2008), *Sulfolobus* spp. (Kawarabayasi *et al.*, 2001; She *et al.*, 2001; Chen *et al.*, 2005), *Ferroplasma* spp. (Tyson *et al.*, 2004), and *Picrophilus* spp. (Fütterer *et al.*, 2004; Angelov and Liebl, 2006). Analysis of *Ferroplasma* spp. revealed that they could grow in continuous culture in the presence of yeast extract alone, and a complementing proteomic analysis of heterotrophic versus mixotrophic growth (yeast extract vs. yeast extract plus Fe^{2+}, respectively) identified up-regulated proteins associated with central metabolic pathways (Dopson *et al.*, 2005). This same study also identified differentially regulated electron transport components in response to heterotrophic and mixotrophic growth (Dopson *et al.*, 2005).

Methane and hydrogen

Only a single acidophilic methane oxidizer has been described, *Methylacidiphilum infernorum* V4 that was isolated from Hell's Gate geothermal area, New Zealand (Dunfield *et al.*, 2007). The genome sequence of this aerobic bacterium suggests an autotrophic central metabolism utilizing methane, and includes genes encoding a methane monooxygenase homologous to those found in other methanotrophs (Hou *et al.*, 2008).

Several acidophiles are able to utilize hydrogen as an energy source, including *A. ferrooxidans* during aerobic (Drobner *et al.*, 1990) and anaerobic growth (Ohmura *et al.*, 2002). The *A. ferrooxidans* genome contains four sets of hydrogenases, including genes encoding a Group I hydrogenase (and the related genes required for its production) that underlies the ability to utilize hydrogen as an energy source (Valdes *et al.*, 2008). The arrangement of these hydrogenase genes is the same as that found in *Ac. ambivalens* (Valdes *et al.*, 2008).

Temperature effects on growth

All microorganisms grow within a certain temperature range. Acidophile growth is split into several classifications: a psychrotolerant acidophile that grows at 5°C but with a temperature optimum >15°C, mesophiles that have temperature optima between 15°C and 40°C, and thermophiles that have an optimum temperature above 40°C. Differences in growth temperature affect cell physiology, as exemplified by protein structure and codon usage alterations.

A. ferrivorans is the only characterized species of psychrotolerant acidophile (Hallberg *et al.*, 2010) and has already been discussed with respect to its aerobic and anaerobic growth (see the 'Growth substrates and their sources' section). The type strain, *A. ferrivorans*^{T}, is capable of growth between 4°C and 37°C, with an optimum around 28°C to 30°C (Hallberg *et al.*, 2010). *A. ferrivorans* is similar to *A. ferrooxidans* in that it is capable of autotrophic growth on both Fe^{2+} and ISCs (Hallberg *et al.*, 2010), though *A. ferrivorans* SS3 appears to preferentially oxidize Fe^{2+} (Kupka *et al.*, 2007; Kupka *et al.*, 2009). The ability of *A. ferrivorans* SS3 to catalyse mineral dissolution at low temperatures has also been investigated and another strain (T7) was found to dominate a low-temperature mixed biomining culture (Dopson *et al.*, 2007). Other reports of Fe^{2+} and ISC oxidation activity at acidic pH from low-temperature environments in Canada and elsewhere were probably due to psychrotolerant or psychrophilic microorganisms, but the strains have not been isolated and characterized (Leduc *et al.*, 1993, 2002; Langdahl and Ingvorsen, 1997).

Acidophilic thermophiles have specific adaptations to life at high temperature, such as their generally smaller genome size and higher coding density dictated by the often higher nucleic acid processing error rate in thermophiles (Angelov and Liebl, 2006). How some thermoacidophiles maintain their DNA structure is poorly understood, as the genomes of *Thermoplasma acidophilum* and *Picrophilus torridus* do not contain homologues of DNA repair proteins that are essential in other species (Angelov and Liebl, 2006). Membrane transport also appears to be adapted in thermoacidophiles, as *Thermoplasma* spp. and *Sulfolobus* spp. have a much higher number of membrane potential driven secondary transporters, compared with the higher preponderance of ATP driven transporters present in neutrophilic thermophiles (Albers *et al.*, 2001). General protein structure adaptations to high temperature include: (i) larger numbers of hydrogen bonds; (ii) denser packing of amino acids to avoid exposure to the solvent; (iii) smaller external surface loops; and (iv) polar content in the exposed areas. However, thermoacidic protein adaptations are difficult to discern due to the low number of solved crystal structures of proteins continuously exposed to the acidic environment, such as in the periplasm or membranes. Hopefully this lack of knowledge with respect to acidophiles may be addressed by protein structure modelling utilizing the genomes from the moderately thermophilic *A. caldus* (Valdes *et al.*, 2009), the mesophiles *A. ferrooxidans* (Valdes *et al.*, 2007b) and *A. thiooxidans* (Valdes *et al.*, 2007b), and the psychrotolerant *A. ferrivorans* SS3 (Liljeqvist *et al.*, 2011).

Biofilms and mixed populations

The majority of the studies described in the 'Ecology and metabolism' section concern pure, homogeneous cultures grown under aseptic conditions without the formation of chemical gradients. However, this is not how microorganisms usually thrive in nature, where mostly mixed cultures of interacting populations exist, such as the acidophilic communities used for biomining (Rawlings and Johnson, 2007). These interactions include: (i) predation from acidophilic protozoa; (ii) competition for growth substrates; (iii) mutualism between autotrophic and heterotrophic acidophiles whereby heterotrophs metabolize 'toxic' organic carbon excreted by autotrophs; and (iv) synergism, where mixed populations of acidophiles are capable of extracting metals more efficiently than a single species (reviewed in Johnson (1998)). The existence of chemical and physical gradients (e.g. growth substrates, temperature, and oxygen) often results in acidophiles co-inhabiting the same niches in a manner allowing for communal interaction, such as in a biofilm. The ecological importance of biofilms in typical acidic environments may be due to increased resistance to toxic metals and low pH (see below).

Quorum sensing

Biofilms are complex, dynamic, three-dimensional communities inhabited by microorganisms with distinct roles (reviewed in Nadell *et al.* (2009)). The formation of biofilms by some microorganisms is mediated by autoinducer signal molecules released to the medium, that when they reach a certain concentration trigger biofilm formation, termed 'quorum sensing' (reviewed in Waters and Bassler (2005)). A functional AI-1 type quorum sensing system (*afeIR*) that is mainly involved in intra-species communication is present in *A. ferrooxidans* (Farah *et al.*, 2005; Rivas *et al.*, 2005). Biofilm formation is regulated by AfeIR, encoding a synthase that produces several differing side chain length acyl-homoserine lactones and a transcriptional regulator, respectively (Farah *et al.*, 2005). A similar quorum sensing system has also been identified in *A. thiooxidans* but not in *L. ferrooxidans* (Ruiz *et al.*, 2008). No acyl-homoserine lactones were identified in spent media of '*F. acidarmanus*' Fer1 cultures (Baker-Austin *et al.*, 2010).

Initial stages of biofilm formation

Cell attachment is the initial step in biofilm formation and acidophile attachment to minerals is a crucial step in biomining (see the 'Biomining' section). Attachment is mediated by extracellular polymeric substances (EPS) (Sand and Gehrke, 2006) that form a zone in which the Fe^{3+} concentration is elevated, thus increasing the leaching rate (Kinzler *et al.*, 2003). In the case of pyrite, attachment is facilitated via an electrochemical interaction of Fe^{3+} ions in the EPS with the negatively charged surface of the mineral. However, if *A. ferrooxidans* is grown on S^0 the hydrophobicity of the cell surface increases, rendering the microorganism unable to attach to pyrite (Gehrke *et al.*, 1998). An *A. ferrooxidans* gene cluster mediating EPS formation from the modified sugars UDP-glucose, UDP-galactose, GDP-mannose, and dTDP-rhamnose (formed via the Leloir pathway) has been identified and shown to be up-regulated during growth on pyrite compared with S^0 (Barreto *et al.*, 2005). Genes with homology to some of the *A. ferrooxidans* EPS forming enzymes have also been identified in *M. sedula* (Auernik *et al.*, 2008). In one study, the majority of the acidophile species and strains tested more rapidly attached to pyrite than glass (Afzal Ghauri *et al.*, 2007).

Mature biofilms

Interactions between acidophilic microorganisms in a mature AMD biofilm from the Iron Mountain site were studied using microscopy (Wilmes *et al.*, 2008), metagenomics (Tyson *et al.*, 2004; Dick *et al.*, 2009; Goltsman *et al.*, 2009), and metaproteomics (Goltsman *et al.*, 2009; Ram *et al.*, 2005). During initial pellicle formation the biofilm is dominated by *Leptospirillum* group II ('*L. rubarum*') that, as the biofilm matures, contains an increasing population of archaea such as *Ferroplasma* spp. and some eukaryotes (Wilmes *et al.*, 2008). The *Ferroplasma* spp. comprised up to 85% of the total microbial population (Edwards *et al.*, 2000b). In the mature biofilm, *Leptospirillum* group II ('*L. rubarum*') is packed in a layer at the base of the biofilm, with the archaea positioned above (Wilmes *et al.*, 2008). Metagenomic sequence analysis revealed that it was possible to assign the sequences to distinct species, despite the fact that the microorganisms co-existed in a similar environment (Dick *et al.*, 2009). The metagenome revealed the potential within the *Leptospirillum* group II ('*L. rubarum*') and '*F. acidarmanus*' Fer1 for utilization of growth substrates and that *Leptospirillum* spp. were responsible for both carbon and nitrogen fixation (Tyson *et al.*, 2004), although a putative CO_2 fixation pathway for *Ferroplasma* spp. has now been proposed (see the 'Growth substrates and their sources' section). Metaproteomic analysis of the same site revealed that nitrogen fixation and polymer synthesis for biofilm production were divided among the community members (Ram *et al.*, 2005). In a separate study, *Leptospirillum* groups II ('*L. rubarum*') and III ('*L. ferrodiazotrophum*') from the AMD biofilm were also analysed, demonstrating differences in the presence and expression of genes encoding for chemosensing, metabolic potential, and signal transduction (Goltsman *et al.*, 2009).

Another characterized acidophile biofilm community is that inhabiting the Frasassi cave system in Italy, which is different to the Iron Mountain mine in being based on H_2S oxidation rather than AMD (Macalady *et al.*, 2007). The Frasassi biofilms also contain a mixed population, including autotrophic acidithiobacilli, heterotrophs such as *Ferroplasma* spp., and mixotrophic *Sulfobacillus*-like bacteria (Macalady *et al.*, 2007). Similarly, the microbial constituents of a macroscopic filament from the naturally acidic Rio Tinto river in Spain include *A. ferrooxidans*, *Leptospirillum ferrooxidans*, and *Acidiphilium* spp., matching the species present in the water column (Garcia-Moyano *et al.*, 2007). However, in addition to the typical microorganisms found in acidic environments, other species were also evident as sequences that aligned with non-acidophilic microorganisms such as *Aeromonas* and *Acinetobacter* (Garcia-Moyano *et al.*, 2007).

Adaptations to life at low pH

Oxidation of sulfide minerals results in release of metals into solution and production of ISCs that are further oxidized by acidophiles to sulfuric acid. These reactions create a low pH environment containing elevated levels of biologically available metals. Consequently, acidophiles are some of the most acid and metal tolerant microorganisms in existence and their survival requires homeostatic mechanisms to limit influx of these metals and protons into the cell.

Metal and metalloid homeostasis and resistance

Metals have a number of cellular functions, including: (i) to stabilize protein structure, as has been postulated for *Ferroplasma acidiphilum* Y^T that uses iron as a 'rivet' to stabilize the three-dimensional structures of metalloproteins (Golyshina *et al.*, 2006; Ferrer *et al.*, 2007); (ii) in redox proteins that shuttle electrons in the electron transfer chain, such as the blue copper-haem protein sulfocyanin in *Ferroplasma* spp. (Dopson *et al.*, 2005); and (iii) acting as co-factors in enzyme-catalysed reactions, such as in succinate dehydrogenase, that bears iron sulfur centres and haem in many acidophiles (Tyson *et al.*, 2004). Metal resistance mechanisms in acidophilic microorganisms can be divided into five categories: (i) enzymatic conversion of the metal redox state; (ii) production of less sensitive cellular targets; (iii) intra- or extracellular sequestration; (iv) exclusion by permeability barriers; and (v) efflux of the metal. Acidophiles can be over 1000-fold more resistant to metals than microorganisms that exist at neutral pH despite often using the same resistance mechanisms. For example, *A. ferrooxidans* is resistant to 1071 mM Zn^{2+} (Kondratyeva *et al.*, 1995) compared with 1 mM in *Escherichia coli* (Nies, 1999). However, the mechanism(s) behind this increased resistance is(are) yet to be elucidated. The best studied acidophile metal homeostasis and resistance systems are those concerning metals such as arsenic, iron, copper, and zinc typically found in sulfide minerals (reviewed in Dopson *et al.* (2003)). With the exception of arsenic respiring microorganisms such as *Chrysiogenes arsenatis* (Macy *et al.*, 1996; Macy *et al.*, 2000) and *Thiomonas* sp. (Duquesne *et al.*, 2008), arsenic is the only one of these metals that does not have a biological function.

Arsenic

The biologically important redox states of inorganic arsenic are arsenate (H_3AsO_4; As(V)) and arsenite ($As(OH)_3$; As(III)). These compounds enter *E. coli* via, respectively, the phosphate uptake system (Rosenberg *et al.*, 1977) and the GlpF aquaglyceroporin (Meng *et al.*, 2004),though their uptake method in acidophiles is unknown. The toxicity of As(V) is due to it acting as a phosphate analogue while As(III) binds to cysteine residues in enzymes (Paez-Espino *et al.*, 2009). As(III) and As(V) toxicity to acidophiles has been previously summarized in Dopson *et al.* (2003).

Many acidophiles have an arsenic resistance system containing the *arsRBC* genes which encode a negative repressor (ArsR), a membrane potential driven efflux pump (ArsB), and an As(V) reductase (ArsC) that converts As(V) to As(III) before it is effluxed by the ArsB protein (reviewed in Bhattacharjee and Rosen (2007)). In addition, the *ars* operon may also contain a gene encoding an ATPase (*arsA*) that increases the ArsB As(III) efflux efficiency (Kuroda *et al.*, 1997), and ArsD that has two functions, as a metallochaperone that transfers As(III) to the ArsA ATPase (Lin *et al.*, 2006) and as a weak repressor of the *ars* operon (Wu and Rosen, 1993). The *ars* operon (either *arsRBC* or with the additional *arsAD*) has been cloned and characterized from *A. ferrooxidans* (Butcher *et al.*, 2000; Butcher and Rawlings, 2002), *A. caldus* (Tuffin *et al.*, 2004; Kotze *et al.*, 2006), *Acidiphilium multivorum* (Suzuki *et al.*, 1997, 1998), and *L. ferriphilum* (Tuffin *et al.*, 2006). The gene encoding the efflux protein (*arsB*) has also been identified by Southern hybridization in *A. thiooxidans, Ap. acidophilum, Thiomonas cuprina,* and *Ac. facilis* (Dopson *et al.*, 2001). Deletion of the *A. caldus arsB* gene resulted in a three-fold decrease in As(III) resistance as compared with the wild type strain (Van Zyl *et al.*, 2008). As(V) reduction to As(III) and subsequent efflux have been biochemically demonstrated in *A. caldus* (Dopson *et al.*, 2001).

In contrast, the archaea '*F. acidarmanus*' Fer1 (Gihring *et al.*, 2003) and *T. acidophilum* (Ruepp *et al.*, 2000) only contain complete genes with homology to *arsRB* that confers resistance to As(III) but not As(V). Despite lacking an *arsC* and the inability to reduce As(V) to As(III), '*F. acidarmanus*' Fer1 is resistant to 133 mM As(V) by an unknown mechanism (Baker-Austin *et al.*, 2007). Previously, *S. metallicus* BC has been demonstrated to enzymatically oxidize As(III) (Sehlin and Lindström, 1992) and *in silico* studies suggested that *Sulfolobus* spp. do not contain a typical *ars* operon (Dopson *et al.*, 2003). However, recent genomic data suggests that *Sulfolobus* spp. (NCBI database) and *M. sedula* (Auernik *et al.*, 2008) do indeed contain genes with similarity to *arsB*, although the encoded protein bears similarity to citrate transporters and the true activity has not been elucidated *in vivo*.

Iron

Fe^{2+} is solubilized during dissolution of many sulfide minerals (see the 'Biomining' section) where its concentration can reach 160 g l^{-1} (Osorio *et al.*, 2008). Fe^{2+} toxicity is mediated via free radical-induced oxidative damage (Touati, 2000). As already discussed, some acidophiles oxidize Fe^{2+} to Fe^{3+} to gain energy (see the 'Growth substrates and their sources' section). The low growth pH for acidophiles results in more Fe^{3+} remaining in solution (Fe^{3+} precipitates above pH 1.6), necessitating that they be resistant to this form of iron. However, some acidophiles still require systems for iron acquisition as they can also grow at near neutral pH. *Acidithiobacillus* spp. have genes for iron uptake that are regulated by Fur in *A. ferrooxidans* (Quatrini *et al.*, 2007), though the microorganisms lack the ability to produce known siderophores (Osorio *et al.*, 2008). In contrast, the leptospirilli lack the Feo and MntH Fe^{2+} uptake systems found in the acidithiobacilli and may instead take up Fe^{2+} by an Ftr-like permease (Osorio *et al.*, 2008). The absence of the Feo and MntH systems may restrict Fe^{2+} uptake such that the intracellular concentration does not become toxic (Osorio *et al.*, 2008).

Tolerance to Fe^{3+} has been suggested to explain why *Leptospirillum* spp. outcompete *A. ferrooxidans* (Rawlings *et al.*, 1999), and *Su. acidophilus* outcompetes *Acidimicrobium ferrooxidans* (Clark and Norris, 1996). Many acidophiles are tolerant to high levels of Fe^{2+} and Fe^{3+}, although Fe^{3+} inhibits Fe^{2+} oxidation in *A. ferrooxidans* (Boon *et al.*, 1999) and *L. ferrooxidans* (van Scherpenzel *et al.*, 1988). One tolerance mechanism used by acidophiles to combat high levels of Fe^{2+} is the production of several oxidative stress proteins (presumably due to free radical formation) in '*F. acidarmanus*' Fer1 (Dopson *et al.*, 2005). Iron efflux has also been suggested to play a role in alleviating toxicity associated with elevated intracellular concentrations. This hypothesis is supported by the presence of homologues of the iron efflux proteins FieF and MamB (Grunberg *et al.*, 2001; Grass *et al.*, 2005) in the acidithiobacilli and leptospirilli (Osorio *et al.*, 2008).

Copper

Copper is also solubilized during dissolution of some sulfide minerals, particularly from chalcopyrite ($CuFeS_2$) and chalcocite (Cu_2S), and is required in trace amounts by many microorganisms (e.g. in cytochrome-*c* oxidase). Since the cytoplasm is reducing, Cu^{2+} is reduced to Cu^{+}, which is toxic via promotion of hydroxyl radical production and exhaustion of sulfhydryls (Magnani and Solioz, 2007). Copper homeostasis mechanisms include sequestration, efflux by P-type cation transporters, and employment of metallochaperones (reviewed in Magnani and Solioz, 2007). Also, copper is inhibitory to Fe^{2+} oxidation in the acidophile *Su. thermosulfidooxidans* (Vartanyan *et al.*, 1990).

Almost all tested acidophiles are more resistant to copper than *E. coli* (Dopson *et al.*, 2003) including *A. ferrooxidans*, which can tolerate 800 mM Cu^{2+} (Das *et al.*, 1997). Investigations of acidophilic copper resistance mechanisms revealed that *A. ferrooxidans* had decreased Cu^{2+} uptake after adaptation to high concentrations (Boyer *et al.*, 1998), accompanied by induction of cell surface components (Das *et al.*, 1998). At least ten genes related to copper homeostasis are present in the *A. ferrooxidans* genome, including putative copper P-Type ATPases (Navarro *et al.*, 2009), while a CPX P-ATPase for Cu^{2+} and Zn^{2+} efflux has been identified in the *M. sedula* genome (Auernik *et al.*, 2008). Copper resistance can also involve production of polyphosphate that sequesters heavy metals before transport out of the cell (Kornberg *et al.*, 1999). A potential *A. ferrooxidans* copper efflux mechanism (Boyer *et al.*, 1998) was supported by increased polyphosphate production and exopolyphosphatase activity in the presence of Cu^{2+} (Alvarez and Jerez, 2004). In conjunction with the earlier data, a transcriptional study suggests copper resistance is partly or wholly due to novel copper chaperones and a poly-P resistance system (Navarro *et al.*, 2009). Evidence for the polyphosphate copper resistance mechanism is also found in the archaeon *S. metallicus*, in which intracellular polyphosphate levels decrease in the presence of Cu^{2+} (Remonsellez *et al.*, 2006). In contrast, the *Sulfolobus solfataricus* P2 predicted *cop* cluster, which encodes a metallochaperone (CopM), a P-type copper transporter (CopA), and a transcriptional regulator (CopT), is up-regulated in the presence of growth-inhibiting concentrations of Cu^{2+} (Ettema *et al.*, 2006). Homologues of this cluster also exist in the *M. sedula* genome (Auernik *et al.*, 2008). A *copYZB* operon encoding a copper transporter was up-regulated in '*F. acidarmanus*' Fer1 in response to the presence of Cu^{2+}, along with proteins associated with DNA repair, heat shock, and protein chaperones (Baker-Austin *et al.*, 2005). This suggests Cu^{2+} resistance in '*F. acidarmanus*' Fer1 is mediated by a combination of efflux to reduce the intra-cellular concentration and stress proteins to ameliorate the damage (Baker-Austin *et al.*, 2005).

Zinc

Zinc is present in some enzymes and its toxicity is attributed to it forming complexes with a range of cellular components. For example, Zn^{2+} competitively inhibits Fe^{2+} oxidation in *Su. thermosulfidooxidans* (Vartanyan *et al.*, 1990). Zn^{2+} toxicity to *A. ferrooxidans* is dependent on the substrate with greater resistance when growing on Fe^{2+} compared with thiosulfate (Trevors *et al.*, 1985). Prokaryotic Zn^{2+} homeostasis mechanisms include import, export, and sequestration by chaperones (reviewed in Rensing and Mitra, 2007).

Zinc uptake and efflux mechanisms can be divided into primary and secondary systems that utilize chemical energy and energy stored in electrochemical gradients, respectively (Blencowe and Morby, 2003). Bioinformatic analysis of a range of acidophiles identified proteins (of varied levels of similarity) assigned to a number of Zn^{2+} homeostatic mechanisms (Ödling and Dopson, unpublished). These proteins include the complete or incomplete Znu primary import system, the MgtA broad spectrum metal ion importer, and the MntH secondary Zn^{2+} import mechanism in *Acidithiobacillus* spp., *Leptospirillum* spp., '*F. acidarmanus*' Fer1, and *T. acidophilum*. Proteins of the ZntA primary efflux system are also present in *Acidithiobacillus* spp., *Leptospirillum* spp., '*F. acidarmanus*' Fer1, and *T. acidophilum*, whereas the Czc secondary efflux system was only identified in the Gram-negative acidophiles *A. ferrooxidans*, *A. caldus*, and *Leptospirillum* Group II ('*L. rubarum*') (Ödling and Dopson, unpublished). The only evaluated species from the above list that contained a protein with a significant hit to the SmtA metallochaperone is *A. caldus*. In other acidophiles, plasmids from *Acidocella* GS19h (Ghosh *et al.*, 1997, 2000) and '*Acidiphilium symbioticum*' (Mahapatra *et al.*, 2002) have been shown to confer Zn^{2+} resistance.

pH homeostasis

All microorganisms posses mechanisms to maintain their intracellular pH, termed pH homeostasis. This is especially important in extreme acidophiles where the external pH may be 2 or lower (Nordstrom *et al.*, 2000). Acidophiles maintain a near neutral cytoplasm that is typically 4 to 5 pH units above the external pH. For instance, *S. solfataricus* has an internal pH of approximately 6.5 with an external growth pH of 2 to 4 (Moll and Schafer, 1988), and *A. ferrooxidans* maintains a cytoplasmic pH approximately 6.5 to 7 at external pH values of 1 to 7 (Cox *et al.*, 1979). The large pH gradient across the cytoplasmic membrane can drive ATP generation and transport across the membrane. The mechanisms employed for pH homeostasis include cell buffering, cell membranes highly resistant to the influx of protons, proton transport, and metabolically connected proton utilization (reviewed in Slonczewski *et al.*, 2009).

Cell buffering

The cytoplasm of all microorganisms has a buffering capacity derived from small molecules such as amino acids (individually or as part of a protein) and inorganic phosphate that can be used to maintain pH homeostasis. It has been postulated that the majority of acidophile buffering is carried out by amino acid side chains (Zychlinsky and Matin, 1983). The buffering capacity of microorganisms is not dependent on the pH of the external milieu and appears to be a common mechanism for acidophiles, neutrophiles, and alkaliphiles (Zychlinsky and Matin, 1983; Krulwich *et al.*, 1985; Rius and Loren, 1998). Cytoplasmic buffering contributes to the total cell pH homeostasis, but without additional active mechanisms (detailed below) the cytoplasm quickly becomes acidified and the cell dies. Examples of measured buffering capacities include 97 and 85 nmol H^+ per mg protein for an acidophilic strain PW2 and *E. coli*, respectively (Goulbourne *et al.*, 1986). In the absence of cytoplasmic buffering, the amount of protons added in this experiment would have lowered the cytoplasm to pH 2.

Cell membranes highly resistant to the influx of protons

Energy production (e.g. ATP) requires proton transport across the cell membrane, whilst simultaneously maintaining a near neutral cytoplasm by balancing proton influx with consumption (Matin *et al.*, 1982; Dopson *et al.*, 2002). Therefore, the cytoplasmic membrane has to be resistant to uncontrolled influx, and indeed, acidophile membranes are resistant to the flow of protons down the concentration gradient from outside to inside the cell (Konings *et al.*, 2002). Archaeal membranes are particularly resistant to proton translocation, as tight packing and lipid tetraether linkages renders them less permeable (Elferink *et al.*, 1994). However, in the absence of an uncoupler, protons will not flow down their concentration gradient (acidic pH outside, compared with pH 7 inside) without an inflow of permeant anions or outflow of cations (Ferguson and Ingledew, 2008). In addition to a poorly permeable membrane, acidophiles resist the influx of protons by an internal positive membrane potential (neutrophiles have an internal negative membrane potential) (Matin *et al.*, 1982). The formation of this potential has been attributed to potassium influx, an idea supported by the requirement for potassium during respiration-associated proton translocation in *Sulfolobus* spp. (Schäfer, 1996) and membrane potential generation in *A. thiooxidans* (Suzuki,

1999). A third mechanism whereby acidophile membranes may be resistant to low pH is adaptation of membrane channels such that they hinder proton influx. One example is the up-regulation of an *A. ferrooxidans* outer membrane protein (Omp40) upon a drop in growth pH from 3.5 to 1.5 (Amaro *et al.*, 1991). The *A. ferrooxidans* Omp40 has several adaptations (compared to its *E. coli* homologue) suggested to allow growth at low pH, including an external amino acid loop that may control ion entrance into the porin channel and a positive isoelectric point (the *E. coli* homologue has a negative isoelectric point) to hinder proton passage (Guiliani and Jerez, 2000).

Proton transport

Proton transport out of the cell is achieved via both primary and secondary proton pumps involved in energy production and transport. The ATP-producing activity of primary proton pumps involved in electron transport has been demonstrated in *A. ferrooxidans* (Ferguson and Ingledew, 2008), '*F. acidarmanus*' Fer1 (Dopson *et al.*, 2005), and *Sulfolobus* spp. (Schäfer *et al.*, 1999), as well as the identification of primary pump encoding genes in the genome of *L. ferriphilum* (Tyson *et al.*, 2004).The proton translocating oxidases in *A. ferrooxidans* and '*F. acidarmanus*' Fer1 appear to be atypical as they lack residues in the proton translocating k channel that may be the result of adaptations to energy conservation at low pH (Ferguson and Ingledew, 2008). The generation of a proton motive force has been linked to pH homeostasis in *Bacillus acidocaldarius* and *T. acidophilum* (Michels and Bakker, 1985).

Other primary proton pumps include a putative P-type ATPase identified in *A. ferrooxidans* (Valdes *et al.*, 2008). Secondary proton transporters, such as symporters using the membrane potential for solute transport, have been identified in *A. ferrooxidans* (Ferguson and Ingledew, 2008) and for lactose transport in *B. acidocaldarius* (Krulwich *et al.*, 1978). The possible importance of potassium in the membrane potential may be reflected in the large number of secondary transporters identified in sequenced acidophile genomes (She *et al.*, 2001; Tyson *et al.*, 2004; Valdes *et al.*, 2008).

Metabolically connected proton utilization

Metabolism can generate compounds such as small organic acids that are particularly toxic to acidophiles, because they can pass through the membrane in the neutral, protonated form whereupon they dissociate to release a proton inside the cytoplasm. This results in depletion of the cell's buffering capacity (Alexander *et al.*, 1987; Kishimoto *et al.*, 1990). The most toxic of the tested organic acids to *A. ferrooxidans* was chloroacetic acid (Alexander *et al.*, 1987). Fermentation products are potentially toxic to low-pH adapted microorganisms, which might explain why no fermentative acidophiles have been identified and why the most extreme acidophiles are heterotrophs capable of oxidizing organic acids (Dopson *et al.*, 2004; Ciaramella *et al.*, 2005; Angelov and Liebl, 2006).

Biotechnological applications

Acidophiles have been used in the biotechnological application of 'biomining' for hundreds, if not thousands, of years. For example, metals such as copper were extracted from sulfide minerals exposed to the surface, and the copper recovered by precipitation on iron (see the 'Biomining' section). Since then other applications have been developed, including bioremediation of contaminated environments and industrial process waters (see the 'Biomediation' section), and utilization of enzymes able to withstand harsh conditions in some industrial processes (see the 'Extremozymes' section).

Biomining

The generic term 'biomining' encompasses both 'bioleaching' and 'biooxidation'. Bioleaching is where the desired metal(s) is part of the sulfide mineral and is released into solution [e.g. copper from chalcopyrite ($CuFeS_2$)]. Biooxidation is where the metal of interest is trapped as microscopic particles within the mineral structure but is not itself part of the latter (e.g. gold release from arsenopyrite (FeAsS)). At low pH, the sulfide minerals pyrite (FeS_2; eqn 11.1) and chalcopyrite (eqn 11.2) are abiotically oxidized by Fe^{3+}, generating Fe^{2+} and an ISC product. The role of acidophilic microorganisms is to catalyse mineral breakdown (Singer and Stumm, 1970) by enzymatic oxidation of Fe^{2+} to regenerate Fe^{3+} (eqn 11.3) and enzymatically remove the ISC to produce the acidic (H_2SO_4) environment in which the microorganisms exist (eqns 11.4 and 11.5 for pyrite and chalcopyrite, respectively).

$$FeS_2 + 6Fe^{3+} + 3H_2O \rightarrow 7Fe^{2+} + S_2O_3^{2-} + 6H^+ \quad (11.1)$$

$$CuFeS_2 + 4Fe^{3+} \rightarrow Cu^{2+} + 5Fe^{2+} + 2S^0 \quad (11.2)$$

$$4Fe^{2+} + O_2 + 4\,H^+ \rightarrow 4Fe^{3+} + 2H_2O \quad (11.3)$$

$$S_2O_3^{2-} + 2O_2 + H_2O \rightarrow 2SO_4^{2-} + 2H^+ \quad (11.4)$$

$$2S^0 + 3O_2 + 2H_2O \rightarrow 2SO_4^{2-} + 2H^+ \quad (11.5)$$

The initial ISC product formed is either thiosulfate, $S_2O_3^{2-}$ (e.g. from FeS_2; eqn 11.1) or S^0 (e.g. from $CuFeS_2$; eqn 11.2) and is dependent on the mineralogy of the sulfide minerals (Schippers and Sand, 1999). One role of ISC-oxidizing acidophiles in biomining is to remove the S^0 that, along with other products, can inhibit access of Fe^{3+} to the mineral surface in what is termed 'passivation' (Sasaki *et al.*, 1995; Fowler and Crundwell, 1998; Edwards *et al.*, 2000a). The passivating S^0 layer can be removed by sulfur-oxidizing microorganisms such as *A. caldus* (Dopson and Lindström, 1999; Edwards *et al.*, 2000a) and *A. ferrooxidans* (Fowler *et al.*, 1999), thereby allowing bioleaching to proceed (Dopson and Lindström, 1999). The expression of genes and production of proteins in response to the presence of Fe^{2+} and ISCs during mineral breakdown has been detailed in the 'Growth substrates and their sources' section.

Metal dissolution is mediated by three methods, termed 'contact', 'non-contact', and 'cooperative' leaching (Fig. 11.4) (Rawlings, 2002). The contact mechanism is mediated by attached, Fe^{2+}oxidizing microorganisms (as discussed in the 'Biofilms and mixed populations' section), where the distance between the outer membrane of the acidophile and the mineral is 10–100 nm, and is filled with EPS (Rodriguez-Leiva and Tributsch, 1988). The Fe^{3+} concentration inside the EPS increases and this allows a higher leaching rate. In the case of acid soluble minerals, the metal sulfide bonds can also be broken during contact bioleaching by protons. The non-contact method is mediated in the same way as the contact method, with Fe^{3+} oxidizing the mineral, but without cellular attachment to the mineral surface and the concomitant localized increase in Fe^{3+} within the EPS. The final method is cooperative leaching, whereby planktonic cells oxidize ISCs and mineral fragments, yielding Fe^{3+} and protons for non-contact leaching.

Industrial biomining operations

Industrial biomining operations are widespread around the world, with some of the most intensive use of the technique in Australia, Chile, and South Africa. These operations can be classified into the broad definitions of *in situ*, heap (dump), and tank biomining.

In situ biomining is unique in that the ore is not removed from the ground. Instead, cracks and fissures are created in the ore body by blasting and then acidic water and acidophiles are sprayed over the ore body. The acid water trickles through the ore and the mineral is oxidized, releasing metals into solution that can be collected at the base of the ore. This technique requires specific geological conditions to be feasible and is not widespread.

The second biomining method is by crushing and then stacking the mined ore in heaps or dumps. Once again acidic water is sprayed over the surface, percolates downwards, and is collected by an impermeable membrane at the base of the heap. To help increase the leaching rate, air can be pumped into the heap from the bottom. Heap bioleaching utilizes relatively cheap equipment and has low running costs, and therefore is generally used for large-scale treatment of lower grade ores. However, the major drawback of such operations is the long reaction time that can run into years. The first commercial biohydrometallurgical operation used this method in 1980 for copper recovery in Chile (Bustos *et al.*, 1993). Since that first operation, many industrial copper recovery plants have been commissioned

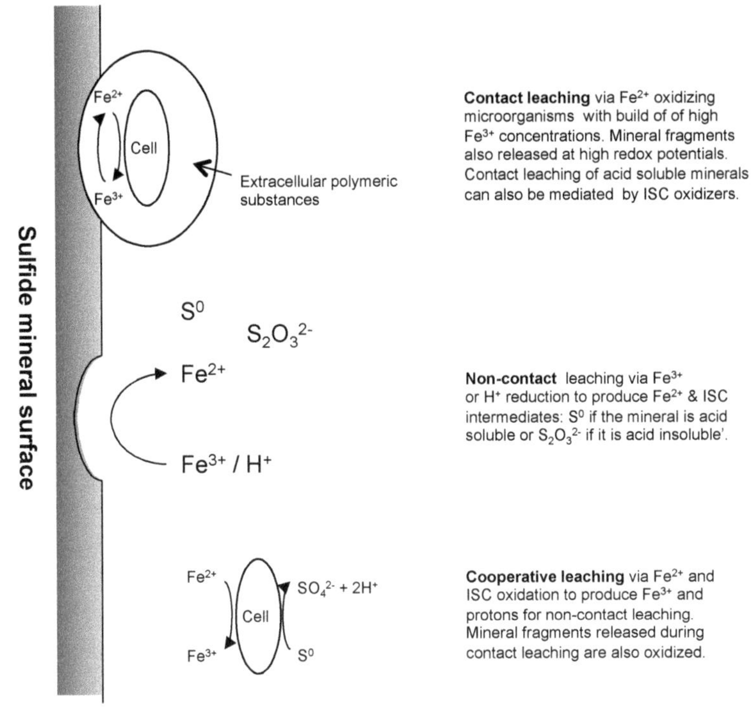

Figure 11.4 Contact, non-contact, and cooperative mechanisms for biomining by Fe^{2+} and ISC oxidizing microorganisms (adapted and re-drawn from Rawlings, 2002).

and presently >20% of the copper produced worldwide is obtained by hydrometallurgical techniques (Watling, 2006).

The third biomining method is tank bioleaching, which typically consists of several continuously operating reactors in series. Owing to the high capital costs associated with constructing and maintaining these tanks, this method is generally used to treat high grade ores and concentrates. The first commercial operation was started in 1986 to treat refractory gold containing arsenopyrite in South Africa (Brierley and Briggs, 2002). Conditions in the tanks are optimized for metal dissolution by maintaining the correct pH and temperature while injecting air. The tanks are stirred to ensure the mineral slurry stays in suspension and no local gradients occur. The maintenance of optimum conditions allied with finely ground minerals providing a large surface area increases the oxidation rate such that a typical bioleaching run takes days rather than the months/years required for heap bioleaching.

Microorganisms present in biomining

The first acidophile recognized as playing a role in biomining was *A. ferrooxidans* (Colmer *et al.*, 1950). The advent of molecular phylogenetic techniques (primarily based on the 16S rRNA gene) has since identified many other acidophiles in biomining areas. However, the difficulty of removing and lysing microorganisms attached to the surface of sulfide minerals that are encased in a thick layer of EPS (see the 'Biofilms and mixed populations' section) has made reliable identification of the entire microbial populations challenging.

A wide variety of bacteria and archaea are found in heap leaching operations due to the gradients of temperature (exothermic reactions can cause a bioheap to reach 70°C in the centre), oxygen, and nutrients. A yearlong study found that a copper containing test heap was dominated by microorganisms similar to *A. ferrooxidans*, *L. ferriphilum*, and *F. acidiphilum*, with lesser amounts of microbes more distantly related to

Sulfurisphaera spp. and *Sulfobacillus* spp. (Demergasso *et al.*, 2005). These populations were not fixed over the year, but varied over time depending on the season and degree of mineral oxidation (Demergasso *et al.*, 2005). A second studied example of a heap bioleaching operation is in Talvivaara, Finland (Puhakka *et al.*, 2007), where the heap is subjected to extremes of temperature from –30°C at the surface in winter to 60°C in the centre of the heap during summer. Column leaching studies using an inoculum from the Talvivaara heap at 7°C identified 16S rRNA gene sequences related to the psychrotolerant *A. ferrivorans*, whilst columns maintained at 21°C contained sequences related to *A. ferrivorans*, an uncultured bacterium clone DX30, *L. ferrooxidans*, and *Sulfobacillus thermotolerans* (Dopson *et al.*, 2007).

Lower population diversity tends to develop during tank bioleaching compared with bioheaps because the stirring creates homogeneous conditions that typically select for a small group of acidophiles. A study of moderately thermophilic pyrite, chalcopyrite, and arsenopyrite tank bioleaching identified 16S rRNA gene sequences that aligned with *A. caldus* C-SH12, *Su. thermosulfidooxidans* AT-1, '*Su. montserratensis*' L15, and an uncultured thermal soil bacterium (Dopson and Lindström, 2004). These same species were present independent of the mineral substrates tested (Dopson and Lindström, 2004). A study of moderately thermophilic continuous tank bioleaching of a polymetallic ore identified *L. ferriphilum*, *A. caldus*, *Sulfobacillus* sp., and *F. acidiphilum* (Okibe *et al.*, 2003). The relative abundance of these species was not uniform through the three bioleaching tank series, with approximately equal amounts of *L. ferriphilum*, *A. caldus*, *Sulfobacillus* sp. in the first tank where the metal and organic carbon concentrations were low. By the time the mineral reached the third tank, the soluble metal and organic carbon content had increased and the culture was almost completely dominated by *F. acidiphilum* (Okibe *et al.*, 2003). Finally, a study of a thermophilic tank leaching environment treating $CuFeS_2$ at 78°C identified *Sulfobales*-like microorganisms, including potentially new species most closely related to *Sulfolobus shibitae*, *Sulfurisphaera ohwakuensis*, *Stygiolobus azoricus*, *Metallosphaera* sp., and *Acidianus infernus* (Mikkelsen *et al.*, 2006).

Bioremediation

Sulfide minerals exposed to air and water may be oxidized to produce metal-rich, acidic solutions, a process termed acid mine drainage (AMD) or acid rock drainage (ARD). Some industrial waters can also be acidic, especially those generated by the metal processing industry. Some of the most acidic environments known have already been mentioned in this chapter, and include AMD from the Iron Mountain site where negative pH values have been recorded (Nordstrom *et al.*, 2000). The environmental effects of this pollution include lowering the pH of rivers and lakes, contamination of groundwater, and precipitation of metals that can smother plant and animal life (reviewed in Johnson and Hallberg, 2003). These polluting waters can be treated chemically, typically by addition of lime to neutralize the pH and precipitate the metals. Alternatively, bioremediation strategies for AMD and ARD using acidophiles can be employed. These approaches are primarily based upon oxidation of Fe^{2+} to Fe^{3+} by species such as *A. ferrooxidans* in an aerobic process, with the subsequent precipitation of the metals as hydroxides when the pH is raised (often by lime addition). A commercial plant using this strategy has been operated in Japan (Yabuuchi and Imanaga, 1976). One drawback of precipitation of AMD/ARD streams with lime is that the mixed metal sludge produced has poor recycling potential, as it is difficult to separate the metals. This drawback could be alleviated by the use of biological sulfate reduction to sulfide, which generates alkalinity to raise the pH of the treated water (Koschorreck, 2008). Using this technique, the pH can be controlled to selectively precipitate the metals as metal sulfides rather than metal hydroxides (Bijmans *et al.*, 2009b). This process has been demonstrated with acid tolerant microorganisms down to pH 5 (Bijmans *et al.*, 2009a). Significantly, an acidophilic sulfate-reducing bacterial species has recently been discovered (Johnson *et al.*, 2009), opening the way for this type of application at even lower pH values.

Other bioremediation applications of acidophiles include exploitation of the *A. ferrooxidans mer* resistance system encoding (amongst other proteins) a mercury reducing flavoprotein that converts Hg^{2+} to Hg^0, which then volatizes out of the cell (reviewed in Dopson *et al.*, 2003). This system has been considered for bioremediation of mercury contaminated acidic soils (Takeuchi *et al.*, 2001). Waste streams contaminated with a range of aliphatic compounds can also be remediated with heterotrophic acidophiles (Renia and Christopher, 2000). Finally, it has been proposed that the acidophilic methane oxidizer *Me. infernorum* V4 (see the 'Growth substrates and their sources' section) be utilized as a filter to remove the greenhouse gas methane before it enters the atmosphere (Dunfield *et al.*, 2007).

Extremozymes

Extremozymes are proteins obtained from extremophiles that are stable under extreme conditions and may be used in industrial processes, for example, at low or high temperatures and/or acidic pH (reviewed in Eichler (2001)). Biotechnological applications from acidophiles include: high temperature trehalose production (used to stabilize vaccines) employing *Sulfolobus* enzymes; an *S. acidocaldarius* proteinase; and *S. solfataricus* alcohol dehydrogenases used to produce optically active alcohols that are a stepping stone in the production of other industrially important compounds (Cowan, 1996).

Other biotechnological applications

It has been suggested that *A. ferrooxidans* and *A. thiooxidans* be utilized for pyrite removal from coal to lower sulfur dioxide formation (Hoffman *et al.*, 1981; Roberto *et al.*, 1991). *A. thiooxidans* has also been used to remove hydrogen sulfide from natural gas in a continuous bioreactor before combustion (Lizama and Sankey, 1983). Finally, *Acidiphilium cryptum* has been used in a microbial fuel cell for electricity production using Fe^{2+} for electron transfer to the electrode (Boon and Heijnen, 1998).

Conclusions

The understanding of acidophile microbiology has accelerated recently with the application of molecular phylogenetic techniques and systems biology approaches though traditional microbiological techniques still yield new data. The first biotechnological application of acidophiles was biomining, which has developed from poorly understood dump leaching of ores to stirred tank reactors treating mineral concentrates. The search for acidophiles with novel phenotypes still identifies new species, such as recently described acidophilic yeasts (Russo *et al.*, 2008), psychrotolerant species (Hallberg *et al.*, 2010), and acidophilic sulfate reducers (Johnson *et al.*, 2009), thereby suggesting that novel biotechnological processes may still be discovered.

One of the most heavily studied areas in acidophile microbiology is the Fe^{2+} oxidation system in *A. ferrooxidans* (reviewed in Ferguson and Ingledew, 2008), though the complication of chemical and biological reactions by ISCs has hindered progress in elucidating their metabolism. Many of the studies have been performed with pure cultures, so there is much to investigate in how these species interact in mixed populations, a deficiency that is just beginning to be addressed by biofilm studies (Wilmes *et al.*, 2008).

Future trends

The study of acidophilic bacteria has been hindered by the lack of practical DNA manipulation techniques. There have been a few specific cases of genetic transfer in acidophiles, such as in *A. ferrooxidans* (Kusano *et al.*, 1992; Liu *et al.*, 2000) and *A. caldus* (Kotze *et al.*, 2006) and null mutants have been constructed in these two microorganisms (Liu *et al.*, 2000; Van Zyl *et al.*, 2008). Other acidophiles in which foreign genetic elements have been expressed include *Ap. cryptum* and *Ap. acidophilum* (Quentmeier and Friedrich, 1994). Unfortunately, several of these systems have proved to be strain specific. The construction of efficient mobile genetic elements and gene knockout mutants for these species will greatly facilitate understanding of these biotechnologically important microorganisms. This experimental limitation has been partially circumvented by the use of systems biology approaches that are bound to increase in significance as more sequenced genomes become available. It will also be of importance to not only sequence the type strains but also environmental isolates specifically adapted to specific conditions, such as high levels of toxic metals.

One of the potentially profitable areas within biohydrometallurgy is the bioleaching of chalcopyrite, as much of the world's copper reserves are in this mineral. The bottleneck associated with the challenges of passivation is being investigated and a potential solution lies within controlling the oxidation–reduction potential. A second biotechnological application is the further development and industrial application of low pH sulfate reduction for metal recovery, which will allow increased re-use of the metals.

Web resources

BioMine Wiki: a Wiki-based resource for the application of acidophiles to biohydrometallurgy (http://wiki.biomine.skelleftea.se/wiki/index.php/Main_Page).

Extremophiles journal: a Springer journal focusing on all extremophiles including articles on acidophiles (http://www.springerlink.com/content/100494/).

Genamics acidophile genomes: Links to available acidophile genomes (http://genamics.com/cgi-bin/genamics/genomes/genomesearch.cgi?field=Phenotype&query=Acidophile).

Nasa Astrobiology Institute: Nasa's homepage for the possibility of life on other planets (http://astrobiology.nasa.gov/nai/).

Science Education Resource Centre: An education site containing acidophile information for teachers (http://serc.carleton.edu/microbelife/extreme/acidic/).

SGM micro-encyclopaedia: a general resource for microbiology related facts on acidophiles (http://www.sgm.ac.uk/micro_encyc/05beneficial.cfm).

Acknowledgements

Members of my research group are acknowledged for stimulating discussions and for proof reading this chapter. Research in the group is funded by Vetenskapsrådet (The Swedish Research Council) and Kempe Foundation.

References

Afzal Ghauri, M., Okibe, N., and Johnson, D.B. (2007). Attachment of acidophilic bacteria to solid surface: the significance of species and strain variations. Hydrometallurgy *85*, 72–80.

Albers, S.V., Van deVossenberg, J.L., Driessen, A.J., and Konings, W.N. (2001). Bioenergetics and solute uptake under extreme conditions. Extremophiles *5*, 285–294.

Alexander, B., Leach, S., and Ingledew, W.J. (1987). The relationship between chemiosmotic parameters and sensitivity to anions and organic acids in the acidophile *Thiobacillus ferrooxidans*. J. Gen. Microbiol. *133*, 1171–1179.

Alvarez, S., and Jerez, C.A. (2004). Copper ions stimulate polyphosphate degradation and phosphate efflux in *Acidithiobacillus ferrooxidans*. Appl. Environ. Microbiol. *70*, 5177–5182.

Amaro, A.M., Chamorro, D., Seeger, M., Arredondo, R., Peirano, I., and Jerez, C.A. (1991). Effect of external pH perturbations on *in vivo* protein synthesis by the acidophilic bacterium *Thiobacillus ferrooxidans*. J. Bacteriol. *173*, 910–915.

Angelov, A., and Liebl, W. (2006). Insights into extreme thermoacidophily based on genome analysis of *Picrophilus torridus* and other thermoacidophilic archaea. J. Biotechnol. *126*, 3–10.

Appia-Ayme, C., Quatrini, R., Denis, Y., Denizot, F., Silver, S., Roberto, F., Veloso, F., Valdes, J., Cardenas, J.P., Esparza, M., *et al.* (2006). Microarray and bioinformatic analyses suggest models for carbon metabolism in the autotroph *Acidithiobacillus ferrooxidans*. Hydrometallurgy *83*, 273–280.

Auernik, K.S., and Kelly, R.M. (2008). Identification of components of electron transport chains in the extremely thermoacidophilic crenarchaeon *Metallosphaera sedula* through iron and sulfur compound oxidation transcriptomes. Appl. Environ. Microbiol. *74*, 7723–7732.

Auernik, K.S., Maezato, Y., Blum, P.H., and Kelly, R.M. (2008). The genome sequence of the metal-mobilizing, extremely thermoacidophilic archaeon *Metallosphaera sedula* provides insights into bioleaching-associated metabolism. Appl. Environ. Microbiol. *74*, 682–692.

Baker-Austin, C., Dopson, M., Wexler, M., Sawers, G., and Bond, P.L. (2005). Molecular insight into extreme copper resistance in the extremophilic archaeon '*Ferroplasma acidarmanus*' Fer1. Microbiology *151*, 2637–2646.

Baker-Austin, C., Dopson, M., Wexler, M., Sawers, R.G., Stemmler, A., Rosen, B.P., and Bond, P.L. (2007). Extreme arsenic resistance by the acidophilic archaeon '*Ferroplasma acidarmanus*' Fer1. Extremophiles *11*, 425–434.

Baker-Austin, C., Potrykus, J., Wexter, M., Bond, P.L., and Dopson, M. (2010). Biofilm development in the extremely acidophilic archeon '*Ferroplasma acidermanus*' Fer1. Extremophiles *14*, 485–491.

Baker, B.J., Tyson, G.W., Goosherst, L., and Banfield, J.F. (2009). Insights into the diversity of eukaryotes in acid mine drainage biofilm communities. Appl. Environ. Microbiol. *75*, 2192–2199.

Barreto, M., Jedlicki, E., and Holmes, D.S. (2005). Identification of a gene cluster for the formation of extracellular polysaccharide precursors in the chemolithoautotroph *Acidithiobacillus ferrooxidans*. Appl. Environ. Microbiol. *71*, 2902–2909.

Bathe, S., and Norris, P.R. (2007). Ferrous iron- and sulfur-induced genes in *Sulfolobus metallicus*. Appl. Environ. Microbiol. *73*, 2491–2497.

Bhattacharjee, H., and Rosen, B.P. (2007). Arsenic metabolism in prokaryotic and eukaryotic microbes.

In Molecular Microbiology of Heavy Metals, Nies, D.H., and Silver, S., eds. (Springer-Verlag, Berlin), pp. 371–406.

Bijmans, M.F.M., Dopson, M., Peeters, T.W.T., Lens, P.N.L., and Buisman, C.J.N. (2009a). Sulfate reduction at pH 5 in a high-rate membrane bioreactor: Reactor performance and microbial community analyses. J. Microbiol. Biotechnol. *19*, 698–708.

Bijmans, M.F.M., van Helvoort, P.J., Dar, S.A., Dopson, M., Lens, P.N.L., and Buisman, C.J.N. (2009b). Selective recovery of nickel over iron from a nickel-iron solution using microbial sulfate reduction in a gas-lift bioreactor. Water Res. *43*, 853–861.

Blencowe, D.K., and Morby, A.P. (2003). Zn(II) metabolism in prokaryotes. FEMS Microbiol. Rev. *27*, 291–311.

Boon, M., and Heijnen, J.J. (1998). Chemical oxidation kinetics of pyrite in bioleaching processes. Hydrometallurgy *48*, 27–41.

Boon, M., Meeder, T.A., Thione, C., Ras, C., and Heijnen, J.J. (1999). The ferrous iron oxidation kinetics of *Thiobacillus ferrooxidans* in continuous culture. Appl. Microbiol. Biotechnol. *51*, 820–826.

Boyer, A., Magnin, J.P., and Ozil, P. (1998). Copper ion removal by *Thiobacillus ferrooxidans* biomass. Biotechnol. Lett. *20*, 187–190.

Brasseur, G., Levican, G., Bonnefoy, V., Holmes, D., Jedlicki, E., and Lemesle-Meunier, D. (2004). Apparent redundancy of electron transfer pathways via bc_1 complexes and terminal oxidases in the extremophilic chemolithoautotrophic *Acidithiobacillus ferrooxidans*. Biochim. Biophys. Acta. *1656*, 114–126.

Bridge, T.A.M., and Johnson, D.B. (1998). Reduction of soluble iron and reductive dissolution of ferric iron-containing minerals by moderately thermophilic iron-oxidizing bacteria. Appl. Environ. Microbiol. *64*, 2181–2186.

Brierley, C.A., and Briggs, A.P. (2002). Selection and sizing of biooxidation equipment and circuits. In Mineral Processing Plant Design, Practice and Control, Mular, A.L., Halbe, D.N., and Barret, D.J., eds. (Littleton, Colo., Society of Mining Engineers), pp. 1540–1568.

Brock, T.D., and Gustafson, J. (1976). Ferric iron reduction by sulfur- and iron-oxidizing bacteria. Appl. Environ. Microbiol. *32*, 567–571.

Bruscella, P., Appia-Ayme, C., Levican, G., Ratouchniak, J., Jedlicki, E., Holmes, D.S., and Bonnefoy, V. (2007). Differential expression of two bc_1 complexes in the strict acidophilic chemolithoautotrophic bacterium *Acidithiobacillus ferrooxidans* suggests a model for their respective roles in iron or sulfur oxidation. Microbiology *153*, 102–110.

Burton, N.P., Williams, T.D., and Norris, P.R. (1999). Carboxylase genes of *Sulfolobus metallicus*. Arch. Microbiol. *172*, 349–353.

Bustos, S., Castro, S., and Montealegre, S. (1993). The Sociedad Mineral Pudahuel bacterial thin-layer leaching process at Lo Aguirre. FEMS Microbiol. Rev. *11*, 231–236.

Butcher, B.G., and Rawlings, D.E. (2002). The divergent chromosomal *ars* operon of *Acidithiobacillus ferrooxidans* is regulated by an atypical ArsR protein. Microbiology *148*, 3983–3992.

Butcher, B.G., Deane, S.M., and Rawlings, D.E. (2000). The chromosomal arsenic resistance genes of *Thiobacillus ferrooxidans* have an unusual arrangement and confer increased arsenic and antimony resistance to *Escherichia coli*. Appl. Environ. Microbiol. *66*, 1826–1833.

Caldwell, P.E., MacLean, M.R., and Norris, P.R. (2007). Ribulose bisphosphate carboxylase activity and a Calvin cycle gene cluster in *Sulfobacillus* species. Microbiology *153*, 2231–2240.

Cárdenas, J.P., Martínez, V., Covarrubias, P., Holmes, D.S., and Quatrini, R. (2009). Predicted CO/CO_2 fixation in *Ferroplasma* spp. via a novel chimaeric pathway. Adv. Mat. Res. *71–73*, 219–222.

Castelle, C., Guiral, M., Malarte, G., Ledgham, F., Leroy, G., Brugna, M., and Giudici-Orticoni, M.T. (2008). A new iron-oxidizing/O_2-reducing supercomplex spanning both inner and outer membranes, isolated from the extreme acidophile *Acidithiobacillus ferrooxidans*. J. Biol. Chem. *283*, 25803–25811.

Chen, L., Brugger, K., Skovgaard, M., Redder, P., She, Q., Torarinsson, E., Greve, B., Awayez, M., Zibat, A., Klenk, H.P., *et al.* (2005). The genome of *Sulfolobus acidocaldarius*, a model organism of the Crenarchaeota. J. Bacteriol. *187*, 4992–4999.

Chi, A., Valenzuela, L., Beard, S., Mackey, A.J., Shabanowitz, J., Hunt, D.F., and Jerez, C.A. (2007). Periplasmic proteins of the extremophile *Acidithiobacillus ferrooxidans*: a high throughput proteomics analysis. Mol. Cell Proteomics *6*, 2239–2251.

Ciaramella, M., Napoli, A., and Rossi, M. (2005). Another extreme genome: how to live at pH 0. Trends Microbiol. *132*, 49–51.

Clark, D.A., and Norris, P.R. (1996). *Acidimicrobium ferrooxidans* gen. nov., sp. nov.: mixed-culture ferrous iron oxidation with *Sulfobacillus* species. Microbiology *142*, 785–790.

Clum, A., Nolan, M., Lang, E., Glavina Del Rio, T., Tice, H., Copeland, A., Cheng, J.-F., Lucas, S., Chen, F., Bruce, D., *et al.* (2009). Complete genome sequence of *Acidimicrobium ferrooxidans* type strain (ICP^T). Stand. Genomic Sci. *1*, 1.

Colmer, A.R., Temple, K.L., and Hinkle, M.E. (1950). An iron-oxidizing bacterium from the acid drainage of some bituminous coal mines. J. Bacteriol. *59*, 317–328.

Coupland, K., and Johnson, D.B. (2008). Evidence that the potential for dissimilatory ferric iron reduction is widespread among acidophilic heterotrophic bacteria. FEMS Microbiol. Lett. *279*, 30–35.

Cowan, D. (1996). Industrial enzyme technology. Trends Biotechnol. *14*, 177–178.

Cox, J.C., Nicholls, D.G., and Ingledew, W.J. (1979). Transmembrane electrical potential and transmembrane pH gradient in the acidophile *Thiobacillus ferrooxidans*. Biochem. J. *178*, 195–200.

Das, A., Modak, J.M., and Natarajan, K.A. (1997). Studies on multi-metal ion tolerance of *Thiobacillus ferrooxidans*. Miner Eng. *10*, 743–749.

Das, A., Modak, J.M., and Natarajan, K.A. (1998). Surface chemical studies of *Thiobacillus ferrooxidans* with reference to copper tolerance. Anton Van Leeuwen *73*, 215–222.

Demergasso, C.S., Galleguillos, P.A.P., Escudero, L.V.G., Zepeda, V.J.A., Castillo, D., and Casamayor, E.O. (2005). Molecular characterization of microbial populations in a low-grade copper ore bioleaching test heap. Hydrometallurgy *80*, 241–253.

Dick, G.J., Andersson, A.F., Baker, B.J., Simmons, S.L., Thomas, B.C., Yelton, A.P., and Banfield, J.F. (2009). Community-wide analysis of microbial genome sequence signatures. Genome Biol. *10*, R85.

Dopson, M., and Lindström, E.B. (1999). Potential role of *Thiobacillus caldus* in arsenopyrite bioleaching. Appl. Environ. Microbiol. *65*, 36–40.

Dopson, M., and Lindström, E.B. (2004). Analysis of community composition during moderately thermophilic bioleaching of pyrite, arsenical pyrite and chalcopyrite. Microb. Ecol. *48*, 19–28.

Dopson, M., Lindström, E.B., and Hallberg, K.B. (2001). Chromosomally encoded arsenical resistance of the moderately thermophilic acidophile *Acidithiobacillus caldus*. Extremophiles *5*, 247–255.

Dopson, M., Lindstrom, E.B., and Hallberg, K.B. (2002). ATP generation during reduced inorganic sulfur compound oxidation by *Acidithiobacillus caldus* is exclusively due to electron transport phosphorylation. Extremophiles *6*, 123–129.

Dopson, M., Baker-Austin, C., Koppineedi, P.R., and Bond, P.L. (2003). Growth in sulfidic mineral environments: metal resistance mechanisms in acidophilic micro-organisms. Microbiology *149*, 1959–1970.

Dopson, M., Baker-Austin, C., Hind, A., Bowman, J.P., and Bond, P.L. (2004). Characterization of *Ferroplasma* isolates and *Ferroplasma acidarmanus* sp. nov., extreme acidophiles from acid mine drainage and industrial bioleaching environments. Appl. Environ. Microbiol. *70*, 2079–2088.

Dopson, M., Baker-Austin, C., and Bond, P.L. (2005). Analysis of differential protein expression during growth states of *Ferroplasma* strains and insights into electron transport for iron oxidation. Microbiology *151*, 4127–4137.

Dopson, M., Baker-Austin, C., and Bond, P. (2006). Towards determining details of anaerobic growth coupled to ferric iron reduction by the acidophilic archaeon '*Ferroplasma acidarmanus*' Fer1. Extremophiles, 159–168.

Dopson, M., Halinen, A.K., Rahunen, N., Ozkaya, B., Sahinkaya, E., Kaksonen, A.H., Lindstrom, E.B., and Puhakka, J.A. (2007). Mineral and iron oxidation at low temperatures by pure and mixed cultures of acidophilic microorganisms. Biotechnol. Bioengin. *97*, 1205–1215.

Drobner, E., Huber, H., and Stetter, K.O. (1990). *Thiobacillus ferrooxidans*, a facultative hydrogen oxidizer. Appl. Environ. Microbiol. *56*, 2922–2923.

Druschel, G.K., Baker, B.J., Gihring, T.M., and Banfield, J.F. (2004). Acid mine drainage biogeochemistry at Iron Mountain, California. Geochem. Trans. *5*, 13–32.

Dunfield, P.F., Yuryev, A., Senin, P., Smirnova, A.V., Stott, M.B., Hou, S., Ly, B., Saw, J.H., Zhou, Z., Ren, Y., *et al.* (2007). Methane oxidation by an extremely acidophilic bacterium of the phylum Verrucomicrobia. Nature *450*, 879–882.

Duquesne, K., Lieutaud, A., Ratouchniak, J., Muller, D., Lett, M.C., and Bonnefoy, V. (2008). Arsenite oxidation by a chemoautotrophic moderately acidophilic *Thiomonas* sp.: from the strain isolation to the gene study. Environ. Microbiol. *10*, 228–237.

Edwards, K.J., Bond, P.L., and Banfield, J.F. (2000a). Characteristics of attachment and growth of *Thiobacillus caldus* on sulphide minerals: a chemotactic response to sulphur minerals? Environ. Microbiol. *2*, 324–332.

Edwards, K.J., Bond, P.L., Gihring, T.M., and Banfield, J.F. (2000b). An archaeal iron-oxidizing extreme acidophile important in acid mine drainage. Science *287*, 1796–1799.

Eichler, J. (2001). Biotechnological uses of archaeal extremozymes. Biotechnol. Adv. *19*, 261–278.

Elbehti, A., Brasseur, G., and Lemesle-Meunier, D. (2000). First evidence for existence of an uphill electron transfer through the bc_1 and NADH-Q oxidoreductase complexes of the acidophilic obligate chemolithotrophic ferrous ion-oxidizing bacterium *Thiobacillus ferrooxidans*. J. Bacteriol. *182*, 3602–3606.

Elferink, M.G.L., de Wit, J.G., Driessen, A.J.M., and Konings, W.N. (1994). Stability and proton-permeability of liposomes composed of archaeal tetraether lipids. Biochimica et Biophysica Acta (BBA) – Biomembranes *1193*, 247–254.

Ettema, T.J., Brinkman, A.B., Lamers, P.P., Kornet, N.G., de Vos, W.M., and van der Oost, J. (2006). Molecular characterization of a conserved archaeal copper resistance (*cop*) gene cluster and its copper-responsive regulator in *Sulfolobus solfataricus* P2. Microbiology *152*, 1969–1979.

Farah, C., Vera, M., Morin, D., Haras, D., Jerez, C.A., and Guiliani, N. (2005). Evidence for a functional quorum-sensing type AI-1 system in the extremophilic bacterium *Acidithiobacillus ferrooxidans*. Appl. Environ. Microbiol. *71*, 7033–7040.

Ferguson, S.J., and Ingledew, W.J. (2008). Energetic problems faced by micro-organisms growing or surviving on parsimonious energy sources and at acidic pH: I. *Acidithiobacillus ferrooxidans* as a paradigm. Biochim. Biophys. Acta (BBA) – Bioenergetics *1777*, 1471–1479.

Ferrer, M., Golyshina, O.V., Beloqui, A., Golyshin, P.N., and Timmis, K.N. (2007). The cellular machinery of *Ferroplasma acidiphilum* is iron-protein-dominated. Nature *445*, 91–94.

Fowler, T.A., and Crundwell, F.K. (1998). Leaching of zinc sulfide by *Thiobacillus ferrooxidans*: experiments with a controlled redox potential indicate no direct bacterial mechanism. Appl. Environ. Microbiol. *64*, 3570–3575.

Fowler, T.A., Holmes, P.R., and Crundwell, F.K. (1999). Mechanism of pyrite dissolution in the presence of

Thiobacillus ferrooxidans. Appl. Environ. Microbiol. *65*, 2987–2993.

Fütterer, O., Angelov, A., Liesegang, H., Gottschalk, G., Schleper, C., Schepers, B., Dock, C., Antranikian, G., and Liebl, W. (2004). Genome sequence of *Picrophilus torridus* and its implications for life around pH 0. Proc. Natl. Acad. Sci. USA *101*, 9091–9096.

Gadanho, M., and Sampaio, J.P. (2006). Microeukaryotic diversity in the extreme environments of the Iberian Pyrite Belt: a comparison between universal and fungi-specific primer sets, temperature gradient gel electrophoresis and cloning. FEMS Microbiol. Ecol. *57*, 139–148.

Gale, N.L., and Beck, J.V. (1967). Evidence for the Calvin cycle and hexose monophosphate pathway in *Thiobacillus ferrooxidans*. J. Bacteriol. *94*, 1052–1059.

Garcia-Moyano, A., Gonzalez-Toril, E., Aguilera, A., and Amils, R. (2007). Prokaryotic community composition and ecology of floating macroscopic filaments from an extreme acidic environment, Rio Tinto (SW, Spain). Syst. Appl. Microbiol. *30*, 601–614.

Gardner, M.N., and Rawlings, D.E. (2000). Production of rhodanese by bacteria present in bio-oxidation plants used to recover gold from arsenopyrite concentrates. J. Appl. Microbiol. *89*, 185–190.

Gehrke, T., Telegdi, J., Thierry, D., and Sand, W. (1998). Importance of extracellular polymeric substances from *Thiobacillus ferrooxidans* for bioleaching. Appl. Environ. Microbiol. *64*, 2743–2747.

Ghosh, W., and Dam, B. (2009). Biochemistry and molecular biology of lithotrophic sulfur oxidation by taxonomically and ecologically diverse bacteria and archaea. FEMS Microbiol. Ecol. 33, 999–1043.

Ghosh, S., Mahapatra, N.R., and Banerjee, P.C. (1997). Metal resistance in *Acidocella* strains and plasmid-mediated transfer of this characteristic to *Acidiphilium multivorum* and *Escherichia coli*. Appl. Environ. Microbiol. *63*, 4523–4527.

Ghosh, S., Mahapatra, N.R., Ramamurthy, T., and Banerjee, P.C. (2000). Plasmid curing from an acidophilic bacterium of the genus *Acidocella*. FEMS Microbiol. Lett. *183*, 271–274.

Gihring, T.M., Bond, P.L., Peters, S.C., and Banfield, J.F. (2003). Arsenic resistance in the archaeon '*Ferroplasma acidarmanus*': New insights into the structure and evolution of the *ars* genes. Extremophiles 7, 123–130.

Goltsman, D.S., Denef, V.J., Singer, S.W., VerBerkmoes, N.C., Lefsrud, M., Mueller, R.S., Dick, G.J., Sun, C.L., Wheeler, K.E., Zemla, A., *et al.* (2009). Community genomic and proteomic analyses of chemoautotrophic iron-oxidizing '*Leptospirillum rubarum*' (Group II) and '*Leptospirillum ferrodiazotrophum*' (Group III) bacteria in acid mine drainage biofilms. Appl. Environ. Microbiol. 75, 4599–4615.

Golyshina, O.V., Pivovarova, T.A., Karavaiko, G.I., Kondrat'eva, T.F., Moore, E.R.B., Abraham, W.R., Lunsdorf, H., Timmis, K.N., Yakimov, M.M., and Golyshin, P.N. (2000). *Ferroplasma acidiphilum* gen. nov., sp. nov., an acidophilic, autotrophic, ferrous-iron-oxidizing, cell-wall-lacking, mesophilic member of the *Ferroplasmaceae* fam. nov., comprising a distinct lineage of the Archaea. Int. J. Syst. Evol. Microbiol. *50*, 997–1006.

Golyshina, O.V., Golyshin, P.N., Timmis, K.N., and Ferrer, M. (2006). The 'pH optimum anomaly' of intracellular enzymes of *Ferroplasma acidiphilum*. Environ. Microbiol. *8*, 416–425.

Goulbourne, E.J., Matin, M., and Zychlinsky, E. (1986). Mechanism of delta pH maintenance in active and inactive cells of an obligately acidophilic bacterium. J. Bacteriol. *166*, 59–65.

Grass, G., Otto, M., Fricke, B., Haney, C.J., Rensing, C., Nies, D.H., and Munkelt, D. (2005). FieF (YiiP) from *Escherichia coli* mediates decreased cellular accumulation of iron and relieves iron stress. Arch. Microbiol. *183*, 9–18.

Grunberg, K., Wawer, C., Tebo, B.M., and Schuler, D. (2001). A large gene cluster encoding several magnetosome proteins is conserved in different species of magnetotactic bacteria. Appl. Environ. Microbiol. *67*, 4573–4582.

Guiliani, N., and Jerez, C.A. (2000). Molecular cloning, sequencing, and expression of *omp-40*, the gene coding for the major outer membrane protein from the acidophilic bacterium *Thiobacillus ferrooxidans*. Appl. Environ. Microbiol. *66*, 2318–2324.

Hallberg, K.B., and Johnson, D.B. (2001). Biodiversity of acidophilic prokaryotes. Adv. Appl. Microbiol. *49*, 37–84.

Hallberg, K.B., Dopson, M., and Lindstrom, E.B. (1996). Reduced sulfur compound oxidation by *Thiobacillus caldus*. J. Bacteriol. *178*, 6–11.

Hallberg, K.B., Amouric, A., Brochier-Armanet, C., Bonnefoy, V., and Johnson, D.B. (2009). Physiological and phylogenetic heterogeneity among iron-oxidizing *Acidithiobacillus* spp., and characteristics of the novel species *Acidithiobacillus ferrivorans*. Adv. Mat. Res. *71–73*, 167–170.

Hallberg, K.B., González-Toril, E., and Johnson, D.B. (2010). *Acidithiobacillus ferrivorans*, sp. nov., facultatively anaerobic, psychrotolerant iron- and sulfur-oxidizing acidophiles isolated from metal mine-impacted environments. Extremophiles *14*, 9–19..

Hoffman, H.R., Faust, B.C., Panda, F.A., Koo, H.H., and Tsuchiya, H.M. (1981). Kinetics of the removal of iron pyrite from coal by microbial catalysis. Appl. Environ. Microbiol. *42*, 259–271.

Hou, S.B., Makarova, K.S., Saw, J.H.W., Senin, P., Ly, B.V., Zhou, Z.M., Ren, Y., Wang, J.M., Galperin, M.Y., Omelchenko, M.V., *et al.* (2008). Complete genome sequence of the extremely acidophilic methanotroph isolate V4, *Methylacidiphilum infernorum*, a representative of the bacterial phylum Verrucomicrobia. Biology Direct 3, 26.

Hugler, M., Huber, H., Molyneaux, S.J., Vetriani, C., and Sievert, S.M. (2007). Autotrophic CO_2 fixation via the reductive tricarboxylic acid cycle in different lineages within the phylum *Aquificae*: evidence for two ways of citrate cleavage. Environ. Microbiol. *9*, 81–92.

Ishii, M., Miyake, T., Satoh, T., Sugiyama, H., Oshima, Y., Kodama, T., and Igarashi, Y. (1996). Autotrophic carbon dioxide fixation in *Acidianus brierleyi*. Arch. Microbiol. *166*, 368–371.

Johnson, D.B. (1998). Biodiversity and ecology of acidophilic microorganisms. FEMS Microbiol. Ecol. *27*, 307–317.

Johnson, D.B. (2005). Biohydrometallurgy and the environment: Intimate and important interplay. Paper presented at: 16th International Biohydrometallurgy Symposium (IBS2005) (Cape Town, South Africa).

Johnson, D.B., and Hallberg, K.B. (2003). The microbiology of acidic mine waters. Res. Microbiol. *154*, 466–473.

Johnson, D.B., and Hallberg, K.B. (2009). Carbon, iron and sulfur metabolism in acidophilic micro-organisms. Adv. Microb. Physiol. *54*, 201–255.

Johnson, D.B., and McGinness, S. (1991). Ferric iron reduction by acidophilic heterotrophic bacteria. Appl. Environ. Microbiol. *57*, 207–211.

Johnson, D.B., Jameson, E., Rowe, O.F., Wakeman, K., and Hallberg, K.B. (2009). Sulfidogenesis at low pH by acidophilic bacteria and its potential for the selective recovery of transition metals from mine waters. Adv. Mat. Res. *71–73*, 693–696.

Kanao, T., Kamimura, K., and Sugio, T. (2007). Identification of a gene encoding a tetrathionate hydrolase in *Acidithiobacillus ferrooxidans*. J. Biotechnol. *132*, 16–22.

Kappler, U., Sly, L.I., and McEwan, A.G. (2005). Respiratory gene clusters of *Metallosphaera sedula* – differential expression and transcriptional organization. Microbiology *151*, 35–43.

Kawarabayasi, Y., Hino, Y., Horikawa, H., Jin-no, K., Takahashi, M., Sekine, M., Baba, S., Ankai, A., Kosugi, H., Hosoyama, A., *et al.* (2001). Complete genome sequence of an aerobic thermoacidophilic crenarchaeon, *Sulfolobus tokodaii* strain7. DNA Res. *8*, 123–140.

Kawashima, T., Amano, N., Koike, H., Makino, S., Higuchi, S., Kawashima-Ohya, Y., Watanabe, K., Yamazaki, M., Kanehori, K., Kawamoto, T., *et al.* (2000). Archaeal adaptation to higher temperatures revealed by genomic sequence of *Thermoplasma volcanium*. Proc. Nat. Acad. Sci. U.S.A. *97*, 14257–14262.

Kinzler, K., Gehrke, T., Telegdi, J., and Sand, W. (2003). Bioleaching – a result of interfacial processes caused by extracellular polymeric substances (EPS). Hydrometallurgy *71*, 83–88.

Kishimoto, N., Inagaki, K., Sugio, T., and Tano, T. (1990). Growth-inhibition of *Acidiphilium* species by organic-acids contained in yeast extract. J. Ferment. Bioengin. *70*, 7–10.

Kondratyeva, T.F., Muntyan, L.N., and Karavaiko, G.I. (1995). Zinc-resistant and arsenic-resistant strains of *Thiobacillus ferrooxidans* have increased copy numbers of chromosomal resistance genes. Microbiology *141*, 1157–1162.

Konings, W.N., Albers, S.V., Koning, S., and Driessen, A.J.M. (2002). The cell membrane plays a crucial role in survival of bacteria and archaea in extreme environments. Antonie Van Leeuwenhoek Int. J. Gen. Mol. Microbiol. *81*, 61–72.

Kornberg, A., Rao, N.N., and Ault-Riche, D. (1999). Inorganic polyphosphate: A molecule of many functions. Ann. Rev. Biochem. *68*, 89–125.

Koschorreck, M. (2008). Microbial sulphate reduction at a low pH. FEMS Microbiol. Ecol. *64*, 329–342.

Kotze, A.A., Tuffin, I.M., Deane, S.M., and Rawlings, D.E. (2006). Cloning and characterization of the chromosomal arsenic resistance genes from *Acidithiobacillus caldus* and enhanced arsenic resistance on conjugal transfer of *ars* genes located on transposon Tn*AtcArs*. Microbiology *152*, 3551–3560.

Krulwich, T.A., Davidaon, L.F., Flip, J., S. J., Ziuckenan, R.S., and Guffanti, A.A. (1978). The proton motive force and β-galactoside transport in *Bacillus acidocaldarius*. J. Biol. Chem. *253*, 4599–4603.

Krulwich, T.A., Agus, R., Schneier, M., and Guffanti, A.A. (1985). Buffering capacity of bacilli that grow at different pH ranges. J. Bacteriol. *162*, 768–772.

Kupka, D., Rzhepishevska, O.I., Dopson, M., Lindstrom, E.B., Karnachuk, O.V., and Tuovinen, O.H. (2007). Bacterial oxidation of ferrous iron at low temperatures. Biotechnol. Bioengin. *97*, 1470–1478.

Kupka, D., Liljeqvist, M., Nurmi, P., Puhakka, J.A., Tuovinen, O.H., and Dopson, M. (2009). Oxidation of elemental sulfur, tetrathionate, and ferrous iron by the psychrotolerant *Acidithiobacillus* strain SS3. Res. Microbiol. *160*, 767–774.

Kuroda, M., Dey, S., Sanders, O.I., and Rosen, B.P. (1997). Alternate energy coupling of ArsB, the membrane subunit of the Ars anion-translocating ATPase. J. Biol. Chem. *272*, 326–331.

Kusano, T., Sugawara, K., Inoue, C., Takeshima, T., Numata, M., and Shiratori, T. (1992). Electrotransformation of *Thiobacillus ferrooxidans* with plasmids containing a mer determinant. J. Bacteriol. *174*, 6617–6623.

Kusel, K., Dorsch, T., Acker, G., and Stackebrandt, E. (1999). Microbial reduction of Fe(III) in acidic sediments: Isolation of *Acidiphilium cryptum* JF-5 capable of coupling the reduction of Fe(III) to the oxidation of glucose. Appl. Environ. Microbiol. *65*, 3633–3640.

Langdahl, B.R., and Ingvorsen, K. (1997). Temperature characteristics of bacterial iron solubilisation and ^{14}C assimilation in naturally exposed sulfide ore material at Citronen Fjord, North Greenland (83°N). FEMS Microbiol. Ecol. *23*, 275–283.

Leduc, L.G., Trevors, J.T., and Ferroni, G.D. (1993). Thermal characterization of different isolates of *Thiobacillus ferrooxidans*. FEMS Microbiol. Lett. *108*, 189–194.

Leduc, D., Leduc, L.G., and Ferroni, G.D. (2002). Quantification of bacterial populations indigenous to acidic drainage streams. Water Air Soil Pollut. *135*, 1–21.

Levican, G., Ugalde, J.A., Ehrenfeld, N., Maass, A., and Parada, P. (2008). Comparative genomic analysis of carbon and nitrogen assimilation mechanisms in three indigenous bioleaching bacteria: predictions and validations. BMC Genomics 9, 581.

Liljeqvist, M., Valdes, J., Holmes, D.S., and Dopson, M. (2011). Draft genome of the psychotolerant acidophile *Acidithiobacillus ferrivorans* SS3. J. Bacteriol. In press.

Lin, Y.-F., Walmsley, A.R., and Rosen, B.P. (2006). An arsenic metallochaperone for an arsenic detoxification pump. Proc. Natl. Acad. Sci. U.S.A. *103*, 15617–15622.

Liu, Z.Y., Guiliani, N., Appia-Ayme, C., Borne, F., Ratouchniak, J., and Bonnefoy, V. (2000). Construction and characterization of a recA mutant of *Thiobacillus ferrooxidans* by marker exchange mutagenesis. J. Bacteriol. *182*, 2269–2276.

Lizama, H.M., and Sankey, B.M. (1983). Conversion of hydrogen sulphide by acidophilic bacteria. Appl. Microbiol. Biotechnol. *40*, 438–441.

Ludwig, W., Strunk, O., Westram, R., Richter, L., Meier, H., Yadhukumar, A., Buchner, A., Lai, T., Steppi, S., Jacob, G., *et al.* (2004). ARB: A software environment for sequence data. Nucleic Acids Res. *32*, 1363–1371.

Macalady, J.L., Jones, D.S., and Lyon, E.H. (2007). Extremely acidic, pendulous cave wall biofilms from the Frasassi cave system, Italy. Environ. Microbiol. *9*, 1402–1414.

Macy, J.M., Nunan, K., Hagen, K.D., Dixon, D.R., Harbour, P.J., Cahill, M., and Sly, L.I. (1996). *Chrysiogenes arsenatis* gen. nov., sp. nov., a new arsenate-respiring bacterium isolated from gold mine wastewater. Int. J. Syst. Bacteriol. *46*, 1153–1157.

Macy, J.M., Santini, J.M., Pauling, B.V., O'Neill, A.H., and Sly, L.I. (2000). Two new arsenate/sulfate-reducing bacteria: mechanisms of arsenate reduction. Arch. Microbiol. *173*, 49–57.

Magnani, D., and Solioz, M. (2007). How bacteria handle copper. In Molecular Microbiology of Heavy Metals, Nies, D.H., and Silver, S., eds. (Springer-Verlag, Berlin), pp. 259–286.

Mahapatra, N.R., Ghosh, S., Deb, C., and Banerjee, P.C. (2002). Resistance to cadmium and zinc in *Acidiphilium symbioticum* KM2 is plasmid mediated. Curr. Microbiol. *45*, 180–186.

Matin, A., Wilson, B., Zychlinsky, E., and Matin, M. (1982). Proton motive force and the physiological basis of delta pH maintenance in *Thiobacillus acidophilus*. J. Bacteriol. *150*, 582–591.

Meng, Y.-L., Liu, Z., and Rosen, B.P. (2004). As(III) and Sb(III) uptake by GlpF and efflux by ArsB in *Escherichia coli*. J. Biol. Chem. *279*, 18334–18341.

Messerli, M.A., Amaral-Zettler, L.A., Zettler, E., Jung, S.K., Smith, P.J., and Sogin, M.L. (2005). Life at acidic pH imposes an increased energetic cost for a eukaryotic acidophile. J. Exp. Biol. *208*, 2569–2579.

Michels, M., and Bakker, E.P. (1985). Generation of a large, protonophore-sensitive proton motive force and pH difference in the acidophilic bacteria *Thermoplasma acidophilum* and *Bacillus acidocaldarius*. J. Bacteriol. *161*, 231–237.

Mikkelsen, D., Kappler, U., McEwan, A.G., and Sly, L.I. (2006). Archaeal diversity in two thermophilic chalcopyrite bioleaching reactors. Environ. Microbiol. *8*, 2050–2055.

Moll, R., and Schafer, G. (1988). Chemiosmotic-H^+ cycling across the plasma-membrane of the thermoacidophilic archaebacterium *Sulfolobus acidocaldarius*. FEBS Lett. *232*, 359–363.

Nadell, C.D., Xavier, J.B., and Foster, K.R. (2009). The sociobiology of biofilms. FEMS Microbiol. Rev. *33*, 206–224.

Navarro, C.A., Orellana, L.H., Mauriaca, C., and Jerez, C.A. (2009). Transcriptional and functional studies of *Acidithiobacillus ferrooxidans* genes related to survival in the presence of copper. Appl. Environ. Microbiol. *75*, 6102–6109.

Nicolle, J.L.C., Simmons, S., Bathe, S., and Norris, P.R. (2009). Ferrous iron oxidation and rusticyanin in halotolerant, acidophilic '*Thiobacillus prosperus*'. Microbiology *155*, 1302–1309.

Nies, D.H. (1999). Microbial heavy-metal resistance. Appl. Microbiol. Biotechnol. *51*, 730–750.

Nordstrom, D.K., and Alpers, C.N. (1999). Negative pH, efflorescent mineralogy, and consequences for environmental restoration at the iron mountain superfund site, California. Proc. Natl. Acad. Sci. U.S.A. *96*, 3455–3462.

Nordstrom, D.K., Alpers, C.N., Ptacek, C.J., and Blowes, D.W. (2000). Negative pH and extremely acidic mine waters from Iron Mountain, California. Env. Sci. Technol. *34*, 254–258.

Ohmura, N., Sasaki, K., Matsumoto, N., and Saiki, H. (2002). Anaerobic respiration using Fe^{3+}, S^0, and H_2 in the chemolithoautotrophic bacterium *Acidithiobacillus ferrooxidans*. J. Bacteriol. *184*, 2081–2087.

Okibe, N., Gericke, M., Hallberg, K.B., and Johnson, D.B. (2003). Enumeration and characterization of acidophilic microorganisms isolated from a pilot plant stirred-tank bioleaching operation. Appl. Environ. Microbiol. *69*, 1936–1943.

Osorio, H., Martinez, V., Nieto, P.A., Holmes, D.S., and Quatrini, R. (2008). Microbial iron management mechanisms in extremely acidic environments: comparative genomics evidence for diversity and versatility. BMC Microbiol. *8*, 203.

Osorio, H., Cardenas, J.P., Valdes, J., and Holmes, D.S. (2009). Prediction of FNR regulated genes and metabolic pathways potentially involved in anaerobic growth of *Acidithiobacillus ferrooxidans*. Adv. Mat. Res. *71–73*, 195–198.

Paez-Espino, D., Tamames, J., de Lorenzo, V., and Canovas, D. (2009). Microbial responses to environmental arsenic. Biometals *22*, 117–130.

Puhakka, J.A., Kaksonen, A.H., and Riekkola-Vanhanen, M. (2007). Heap leaching of black schist. In Biomining, Rawlings, D.E., and Johnson, D.B., eds. (Springer-Verlag, Berlin), pp. 139–151.

Quatrini, R., Appia-Ayme, C., Denis, Y., Ratouchniak, J., Veloso, F., Valdes, J., Lefimil, C., Silver, S., Roberto, F., Orellana, O., *et al.* (2006). Insights into the iron and sulfur energetic metabolism of *Acidithiobacillus ferrooxidans* by microarray transcriptome profiling. Hydrometallurgy *83*, 263–272.

Quatrini, R., Lefimil, C., Veloso, F.A., Pedroso, I., Holmes, D.S., and Jedlicki, E. (2007). Bioinformatic prediction and experimental verification of Fur-regulated genes in the extreme acidophile *Acidithiobacillus ferrooxidans.* Nucleic Acids Res. *35,* 2153–2166.

Quatrini, R., Appia-Ayme, C., Denis, Y., Jedlicki, E., Holmes, D., and Bonnefoy, V. (2009). Extending the models for iron and sulfur oxidation in the extreme acidophile *Acidithiobacillus ferrooxidans.* BMC Genomics *10,* 394.

Quentmeier, A., and Friedrich, C.G. (1994). Transfer and expression of degradative and antibiotic resistance plasmids in acidophilic bacteria. Appl. Environ. Microbiol. *60,* 973–978.

Ram, R.J., VerBerkmoes, N.C., Thelen, M.P., Tyson, G.W., Baker, B.J., Blake, R.C., Shah, M., Hettich, R.L., and Banfield, J.F. (2005). Community proteomics of a natural microbial biofilm. Science *308,* 1915–1920.

Ramirez, P., Toledo, H., Guiliani, N., and Jerez, C.A. (2002). An exported rhodanese-like protein is induced during growth of *Acidithiobacillus ferrooxidans* in metal sulfides and different sulfur compounds. Appl. Environ. Microbiol. *68,* 1837–1845.

Ramirez, P., Guiliani, N., Valenzuela, L., Beard, S., and Jerez, C.A. (2004). Differential protein expression during growth of *Acidithiobacillus ferrooxidans* on ferrous iron, sulfur compounds, or metal sulfides. Appl. Environ. Microbiol. *70,* 4491–4498.

Rawlings, D.E. (2002). Heavy metal mining using microbes. Annu. Rev. Microbiol. *56,* 65–91.

Rawlings, D.E., and Johnson, D.B. (2007). The microbiology of biomining: development and optimization of mineral-oxidizing microbial consortia. Microbiology *153,* 315–324.

Rawlings, D.E., Tributsch, H., and Hansford, G.S. (1999). Reasons why '*Leptospirillum*'-like species rather than *Thiobacillus ferrooxidans* are the dominant iron-oxidizing bacteria in many commercial processes for the biooxidation of pyrite and related ores. Microbiology *145,* 5–13.

Remonsellez, F., Orell, A., and Jerez, C.A. (2006). Copper tolerance of the thermoacidophilic archaeon *Sulfolobus metallicus*: possible role of polyphosphate metabolism. Microbiology *152,* 59–66.

Renia, T.G., and Christopher, J.K. (2000). Utilisation of aliphatic compounds by acidophilic heterotrophic bacteria. The potential for bioremediation of acidic wastewaters contaminated with toxic organic compounds and heavy metals. FEMS Microbiol. Lett. *192,* 185–190.

Rensing, C., and Mitra, B. (2007). Zinc, cadmium, and lead resistance and homeostasis. In Molecular Microbiology of Heavy Metals, Nies, D.H., and Silver, S., eds. (Springer-Verlag, Berlin), pp. 321–341.

Rius, N., and Loren, J.G. (1998). Buffering capacity and membrane H^+ conductance of neutrophilic and alkalophilic Gram-positive bacteria. Appl. Environ. Microbiol. *64,* 1344–1349.

Rivas, M., Seeger, M., Holmes, D.S., and Jedlicki, E. (2005). A lux-like quorum sensing system in the extreme acidophile *Acidithiobacillus ferrooxidans.* Biol. Res. *38,* 283–297.

Roberto, F.F., Glenn, A.W., Bulmer, D., Bruhn, D.F., and Ward, T.E. (1991). Genetic manipulation of acidophilic bacteria which are potentially applicable in coal beneficiation. Paper presented at: American Chemical Society (ACS) Fuels Division meeting (Washington, DC, USA).

Rodriguez-Leiva, M., and Tributsch, H. (1988). Morphology of bacterial leaching patterns by *Thiobacillus ferrooxidans* on synthetic pyrite. Arch. Microbiol. *149,* 401–405.

Rohwerder, T., and Sand, W. (2003). The sulfane sulfur of persulfides is the actual substrate of the sulfur-oxidizing enzymes from *Acidithiobacillus* and *Acidiphilium* spp. Microbiology *149,* 1699–1709.

Rohwerder, T., and Sand, W. (2007). Oxidation of inorganic sulfur compounds in acidophilic prokaryotes. Eng.Life Sci. *7,* 301–309.

Rosenberg, H., Gerdes, R.G., and Chegwidden, K. (1977). Two systems for uptake of phosphate in *Escherichia coli.* J. Bacteriol. *131,* 505–511.

Ruepp, A., Graml, W., Santos-Martinez, M.L., Koretke, K.K., Volker, C., Mewes, H.W., Frishman, D., Stocker, S., Lupas, A.N., and Baumeister, W. (2000). The genome sequence of the thermoacidophilic scavenger *Thermoplasma acidophilum.* Nature *407,* 508–513.

Ruiz, L.M., Valenzuela, S., Castro, M., Gonzalez, A., Frezza, M., Soulere, L., Rohwerder, T., Queneau, Y., Doutheau, A., Sand, W., *et al.* (2008). AHL communication is a widespread phenomenon in biomining bacteria and seems to be involved in mineral-adhesion efficiency. Hydrometallurgy *94,* 133–137.

Russo, G., Libkind, D., Sampaio, J.P., and van Broock, M.R. (2008). Yeast diversity in the acidic Rio Agrio-Lake Caviahue volcanic environment (Patagonia, Argentina). FEMS Microbiol. Ecol. *65,* 415–424.

Rzhepishevska, O.I., Valdés, J., Marcinkeviciene, L., Algora Gallardo, C., Meskys, R., Bonnefoy, V., Holmes, D.S., and Dopson, M. (2007). Regulation of a novel *Acidithiobacillus caldus* gene cluster involved in reduced inorganic sulfur compound metabolism. Appl. Environ. Microbiol. *73,* 7367–7372.

Sand, W., and Gehrke, T. (2006). Extracellular polymeric substances mediate bioleaching/biocorrosion via interfacial processes involving iron(III) ions and acidophilic bacteria. Res. Microbiol. *157,* 49–56.

Sasaki, K., Tsunekawa, M., Ohtsuka, T., and Konno, H. (1995). Confirmation of a sulfur-rich layer on pyrite after oxidative dissolution by Fe(III) ions around pH 2. Geochim. Cosmochim. Acta *59,* 3155–3158.

Schäfer, G. (1996). Bioenergetics of the archaebacterium *Sulfolobus.* Biochim. Biophys. Acta. *1277,* 163–200.

Schäfer, G., Engelhard, M., and Muller, V. (1999). Bioenergetics of the Archaea. Microbiol. Mol. Biol. Rev. *63,* 570–620.

van Scherpenzel, D.A., Boon, M., Ras, C., Hansford, G.S., and Heijnen, J.J. (1988). Kinetics of ferrous iron oxidation by *Leptospirillium* bacteria in continuous culture. Biotechnol. Prog. *14,* 425–433.

Schippers, A., and Sand, W. (1999). Bacterial leaching of metal sulfides proceeds by two indirect mechanisms via thiosulfate or via polysulfides and sulfur. Appl. Environ. Microbiol. *65,* 319–321.

Sehlin, H.M., and Lindström, E.B. (1992). Oxidation and reduction of arsenic by *Sulfolobus acidocaldarius* strain BC. FEMS Microbiol. Lett. 93, 87–92.

She, Q., Singh, R.K., Confalonieri, F., Zivanovic, Y., Allard, G., Awayez, M.J., Chan-Weiher, C.C.Y., Clausen, I.G., Curtis, B.A., De Moors, A., *et al.* (2001). The complete genome of the crenarchaeon *Sulfolobus solfataricus* P2. Proc. Nat. Acad. Sci. USA 98, 7835–7840.

Shively, J.M., van Keulen, G., and Meijer, W.G. (1998). Something from almost nothing: carbon dioxide fixation in chemoautotrophs. Annu. Rev. Microbiol. *52*, 191–230.

Siezen, R.J., and Wilson, G. (2009). Bioleaching genomics. Microbial Biotechnol. *2*, 297–303.

Singer, P.C., and Stumm, W. (1970). Acid mine drainage: The rate-determining step. Science *167*, 1121–1123.

Slonczewski, J.L., Fujisawa, M., Dopson, M., and Krulwich, T.A. (2009). Cytoplasmic pH measurement and homeostasis in bacteria and archaea. Adv. Microb. Physiol. *55*, 1–79.

Sugio, T., Taha, T.M., and Takeuchi, F. (2009). Ferrous iron production mediated by tetrathionate hydrolase in tetrathionate-, sulfur-, and iron-grown *Acidithiobacillus ferrooxidans* ATCC 23270 cells. Biosci. Biotechnol. Biochem. *73*, 1381–1386.

Suzuki, I. (1999). Oxidation of inorganic sulfur compounds: Chemical and enzymatic reactions. Can. J. Microbiol. *45*, 97–105.

Suzuki, I., and Werkman, C.H. (1958). Chemoautotrophic carbon dioxide fixation by extracts of *Thiobacillus thiooxidans*. Arch. Biochem. Biophys. *76*, 103–111.

Suzuki, K., Wakao, N., Sakurai, Y., Kimura, T., Sakka, K., and Ohmiya, K. (1997). Transformation of *Escherichia coli* with a large plasmid of *Acidiphilium multivorum* AIU 301 encoding arsenic resistance. Appl. Environ. Microbiol. *63*, 2089–2091.

Suzuki, K., Wakao, N., Kimura, T., Sakka, K., and Ohmiya, K. (1998). Expression and regulation of the arsenic resistance operon of *Acidiphilium multivorum* AIU 301 plasmid pKW301 in *Escherichia coli*. Appl. Environ. Microbiol. *64*, 411–418.

Takeuchi, F., Iwahori, K., Kamimura, K., Negishi, A., Maeda, T., and Sugio, T. (2001). Volatilization of mercury under acidic conditions from mercury-polluted soil by a mercury-resistant *Acidithiobacillus ferrooxidans* SUG 2-2. Biosci. Biotechnol. Biochem. *65*, 1981–1986.

Touati, D. (2000). Iron and oxidative stress in bacteria. Arch. Biochem. Biophys. *373*, 1–6.

Trevors, J.T., Oddie, K.M., and Belliveau, B.H. (1985). Metal resistance in bacteria. FEMS Microbial. Rev. *32*, 39–54.

Tuffin, I.M., Hector, S.B., Deane, S.M., and Rawlings, D.E. (2006). Resistance determinants of a highly arsenic-resistant strain of *Leptospirillum ferriphilum* isolated from a commercial biooxidation tank. Appl. Environ. Microbiol. *72*, 2247–2253.

Tuffin, M., de Groot, P., Deane, S.M., and Rawlings, D.E. (2004). Multiple sets of arsenic resistance genes are present within highly arsenic-resistant industrial strains of the biomining bacterium, *Acidithiobacillus caldus*. Paper presented at: third International Conference of Comparative Physiology and Biochemistry (KwaZulu Natal, South Africa).

Tyson, G.W., Chapman, J., Hugenholtz, P., Allen, E.E., Ram, R.J., Richardson, P.M., Solovyev, V.V., Rubin, E.M., Rokhsar, D.S., and Banfield, J.F. (2004). Community structure and metabolism through reconstruction of microbial genomes from the environment. Nature *428*, 37–43.

Valdes, J., Pedroso, I., Quatrini, R., Hallberg, K.B., Valenzuela, P.D.T., and Holmes, D.S. (2007a). Insights into the metabolism and ecophysiology of three acidithiobacilli by comparative genome analysis. Paper presented at: 17th International Biohydrometallurgy Symposium (Frankfurt am Main, Germany).

Valdes, J., Pedroso, I., Quatrini, R., and Holmes, D.S. (2007b). Comparative genome analysis of *Acidithiobacillus ferrooxidans, A. thiooxidans* and *A. caldus*: Insights into their metabolism and ecophysiology. Paper presented at: 17th International Biohydrometallurgy Symposium (Frankfurt am Main, Germany).

Valdes, J., Pedroso, I., Quatrini, R., Dodson, R.J., Tettelin, H., Blake, R., Eisen, J.A., and Holmes, D.S. (2008). *Acidithiobacillus ferrooxidans* metabolism: from genome sequence to industrial applications. BMC Genomics *9*, 597.

Valdes, J., Quatrini, R., Hallberg, K., Dopson, M., Valenzuela, P.D., and Holmes, D.S. (2009). Draft genome sequence of the extremely acidophilic bacterium *Acidithiobacillus caldus* ATCC 51756 reveals metabolic versatility in the genus *Acidithiobacillus*. J. Bacteriol. *191*, 5877–5878.

Van Zyl, L.J., Van Munster, J.M., and Rawlings, D.E. (2008). Construction of *arsB* and *tetH* mutants of the sulfur-oxidizing bacterium *Acidithiobacillus caldus* by exchange marker. Appl. Environ. Microbiol. *74*, 5686–5694.

Vartanyan, N.S., Karavaiko, G.I., Pivovarova, T.A., and Dorofeev, A.G. (1990). Resistance of *Sulfobacillus thermosulfidooxidans* subspecies *asporogenes* to Cu^{2+}, Zn^{2+} and Ni^{2+} ions. Microbiology *59*, 399–404.

Vestal, J.R., and Lundgren, D.G. (1971). The sulfite oxidase of *Thiobacillus ferrooxidans* (*Ferroobacillus ferrooxidans*). Can. J. Biochem. *49*, 1125–1130.

Wakai, S., Kikumoto, M., Kanao, T., and Kamimura, K. (2004). Involvement of sulfide:quinone oxidoreductase in sulfur oxidation of an acidophilic iron-oxidizing bacterium, *Acidithiobacillus ferrooxidans* NASF-1. Biosci. Biotechnol. Biochem. *68*, 2519–2528.

Waters, C.M., and Bassler, B.L. (2005). Quorum sensing: cell-to-cell communication in bacteria. Annu. Rev. Cell. Dev. Biol. *21*, 319–346.

Watling, H.R. (2006). The bioleaching of sulphide minerals with emphasis on copper sulphides – A review. Hydrometallurgy *84*, 81–108.

Wilmes, P., Remis, J.P., Hwang, M., Auer, M., Thelen, M.P., and Banfield, J.F. (2008). Natural acidophilic biofilm communities reflect distinct organismal and functional organization. ISME J. 3, 266–270.

Wu, J.H., and Rosen, B.P. (1993). The *arsD* gene encodes a second trans-acting regulatory protein of the plasmid-encoded arsenical operon. Mol. Microbiol. *8*, 615–623.

Yabuuchi, E., and Imanaga, Y. (1976). Oxidation of ferrous ions in mine drainage by iron-oxidizing bacteria. World Min. Met. Technol. *2*, 943–956.

Yarzabal, A., Appia-Ayme, C., Ratouchniak, J., and Bonnefoy, V. (2004). Regulation of the expression of the *Acidithiobacillus ferrooxidans rus* operon encoding two cytochromes c_1 a cytochrome oxidase and rusticyanin. Microbiology *150*, 2113–2123.

Zychlinsky, E., and Matin, A. (1983). Cytoplasmic pH homeostasis in an acidophilic bacterium, *Thiobacillus acidophilus*. J. Bacteriol. *156*, 1352–1355.

Index

A

B

O

P

R

S

T

V

W

X

Y

Current Books of Interest

Bacterial Spores: Current Research and Applications	2012
Small DNA Tumour Viruses	2012
Extremophiles: Microbiology and Biotechnology	2012
Bacillus: Cellular and Molecular Biology	2012
Microbial Biofilms: Current Research and Applications	2012
Bacterial Glycomics: Current Research, Technology and Applications	2012
Non-coding RNAs and Epigenetic Regulation of Gene Expression	2012
Brucella: Molecular Microbiology and Genomics	2012
Molecular Virology and Control of Flaviviruses	2012
Bacterial Pathogenesis: Molecular and Cellular Mechanisms	2012
Bunyaviridae: Molecular and Cellular Biology	2011
Emerging Trends in Antibacterial Discovery: Answering the Call to Arms	2011
Epigenetics: A Reference Manual	2011
Metagenomics: Current Innovations and Future Trends	2011
Nitrogen Cycling in Bacteria: Molecular Analysis	2011
The Biology of Paramyxoviruses	2011
Helicobacter pylori	2011
Microbial Bioremediation of Non-metals: Current Research	2011
Lactic Acid Bacteria and Bifidobacteria: Current Progress in Advanced Research	2011
Viruses and Interferon: Current Research	2011
Essentials of Veterinary Parasitology	2011
Hepatitis C: Antiviral Drug Discovery and Development	2011
Streptomyces: Molecular Biology and Biotechnology	2011
Alphaherpesviruses: Molecular Virology	2011
Recent Advances in Plant Virology	2011
Vaccine Design: Innovative Approaches and Novel Strategies	2011
Salmonella: From Genome to Function	2011
PCR Troubleshooting and Optimization: The Essential Guide	2011
Insect Virology	2010
Environmental Microbiology: Current Technology and Water Applications	2011
Sensory Mechanisms in Bacteria: Molecular Aspects of Signal Recognition	2010
Bifidobacteria: Genomics and Molecular Aspects	2010
Molecular Phylogeny of Microorganisms	2010
Nanotechnology in Water Treatment Applications	2010
Iron Uptake and Homeostasis in Microorganisms	2010
Caliciviruses: Molecular and Cellular Virology	2010
Epstein-Barr Virus: Latency and Transformation	2010
Anaerobic Parasitic Protozoa: Genomics and Molecular Biology	2010
Lentiviruses and Macrophages: Molecular and Cellular Interactions	2010
Microbial Population Genetics	2010
Borrelia: Molecular and Cellular Biology	2010

www.ingramcontent.com/pod-product-compliance
Lightning Source LLC
LaVergne TN
LVHW071336210926
848486LV00003B/159
* 9 7 8 1 9 0 4 4 5 5 9 8 1 *